STATISTICS & PROBABILITY & THEIR APPLICATIONS

Patrick Brockett

University of Texas at Austin

Arnold Levine

Tulane University

 SAUNDERS COLLEGE PUBLISHING

Philadelphia   New York   Chicago
San Francisco   Montreal   Toronto
London   Sydney   Tokyo   Mexico City
Rio de Janeiro   Madrid

Address orders to:
383 Madison Avenue
New York, NY 10017

Address editorial correspondence to.
West Washington Square
Philadelphia, PA 19105

Text Typeface: 10/12 Times Roman
Compositor: Progressive Typographers
Acquisitions Editor: Leslie Hawke
Developmental Editor: Lloyd Black
Project Editor: Robin C. Bonner
Copyeditor: Elaine L. Honig
Art Director: Carol C. Bleistine
Art/Design Assistant: Virginia A. Bollard
Text Design: Lisa Delgado
Cover Design: Lawrence R. Didona
Text Artwork: Philatek
Managing Editor & Production Manager: Tim Frelick
Assistant Production Manager: Maureen Iannuzzi

Cover credit: Abstract. copyright © G. Gove/The Image Bank

Library of Congress Cataloging in Publication Data

Brockett, Patrick.
    Statistics and probability and their applications.

    Includes index.
    1. Mathematical statistics.   2. Probabilities.
I. Levine, Arnold.   II. Title.
QA276.B69   1984   519.5   84-5384
ISBN-0-03-053406-2

STATISTICS AND PROBABILITY AND THEIR APPLICATIONS   ISBN 0-03-053406-2

567   016   98765432

CBS COLLEGE PUBLISHING
Saunders College Publishing
Holt, Rinehart and Winston
The Dryden Press

Preface

A noncalculus statistics text of the kind that is useful to both students and practitioners in a wide variety of fields faces a number of stringent demands. The traditional material (random variables, estimation, hypothesis testing, regression, correlation, and so on) must be covered together with the appropriate notation so that a student who wishes to continue along either the mathematical road or any other route to learning statistics is in no way penalized. Moreover, the material must be written to take into account the considerable differences between readers with respect to their facility with basic algebra.

In addition, two other demands are increasingly being made by those departments that desire their students to obtain a basic foundation in statistics. First, they insist that their students *understand* the material. The applications of statistics studied by undergraduates in their special fields have become sufficiently profound that a "cookbook" course no longer suffices as background. The other demand concerns the *coverage* of the material. Generally, textbooks emphasize the analysis of data. The design of the data collection process, the checking, editing, and reduction of data generally merit little attention. By not being exposed to these methodologies, the student obtains a distorted view of statistical activities. As a result, there is an increasing demand that nonsuperficial discussions of sampling techniques and sampling instruments (such as questionnaires) be included in a first course.

We have attempted to satisfy the conditions noted above. A careful attempt has been made to explain notation and conventions clearly; we have included discussions of "data collection" techniques, including questionnaire design, survey and sampling design, experimental design, and so on, to provide a balance of coverage not available in texts of this level. The area of "data collection" is a large one; we have attempted to give the student insight into those techniques used both in scientific experiments and in surveys.

In addition, we have taken pains to include a great deal of material of importance to practitioners. We present examples and problems taken from actual situations in agronomy, business, biology, economics and finance, education, engineering, epidemiology, law, medical research, social science, physics, and psychology. The inclusion of numerous applications should give the reader a strong sense of "applied statistics" and through this sense a better understanding of statistical theory.

We have also sought to increase this understanding by using the concept of the "mathematical model" to present efficiently the basic theory. This concept facilitates both the discussion of probability and the models underlying the statistical analyses.

The text has been designed to give the instructor considerable latitude in the organization of course material. Some independent pathways are plotted in the course outlines that follow. One can begin, for example, with probability, the preference of many instructors, rather than with the "presentation of data." The instructor, the one most knowledgeable about the strengths and weaknesses of the students, may find none of these particular suggestions applicable. There is sufficient material to derive many other course sequences.

ORGANIZATION OF COURSES

The following are suggestions for course outlines. However, the instructor may have a time schedule or a need for emphasis totally different from that described below. For example, one of the authors had to present a two-week course for quality-control engineers at an aircraft plant in Los Angeles; this course had to be integrated with other two-week courses in physics and engineering given by different instructors. In this case, none of the outlines below were directly applicable, yet the material selected for this short statistics course was gleaned from the material in this text.

I. One-Year Course (2 semesters)

    A. Emphasis on statistics, with light coverage of probability.
        1. First semester: Chapters 1 to 9. Omit sections 3.8, 4.8 to 4.11, 4.14 to 4.16. Assign the easiest problems in Chapters 3, 4, and 5. If 4.15 is omitted, then sections 7.7 and 8.6 should be omitted also.
        2. Second semester: Chapters 10 to 15. Note that Chapters 8, 10, and 11 have sections that discuss data collection in scientific experiments whereas Chapters 13 and 14 deal mostly with survey sampling. The instructor can select those topics that are of interest to the type of student taking the course since these topics are discussed independently of each other.

    B. Emphasis on probability.
        1. First semester: Chapters 3 to 7. Chapter 2 may be included if desired. Explain all models including those in the problem sets of Chapters 3 and 4.

      2. Second semester: Chapters 7 to 13. Sections 13.3 to 13.6 may be omitted if the instructor wishes to have time to include sections of interest from Chapters 14 and 15.

  C. A course intermediate between A and B.
      1. First semester: Chapters 3 to 8 omitting those same sections listed in (A) but including the more difficult problems from the problem sets in Chapters 3 and 4.
      2. Second semester: Chapters 9 to 15, omitting 13.3 to 13.6.

II. One-Semester Course

  A. Emphasis on statistics. Light coverage of Chapters 3 and 4 followed by 5 to 9 and 11. Only the rules of probability, conditional probability, independence, and expectation should be covered in Chapters 3 and 4 with the assignment of the easiest problems. Omit 7.4 to 7.7, 8.2; either 8.4 or 8.5, or both; 8.6; 9.5; and Chapter 11, except 11.1 and 11.2.

  B. Emphasis on probability. A one-semester course can be given from Chapters 3 to 5 alone if one goes deeply into the problem sets and all topics.

ACKNOWLEDGMENTS

We express thanks to our editors, Leslie Hawke and Lloyd Black, for their useful ideas and substantive help. A number of colleagues have made helpful criticisms and suggestions. We name some of them below as a sign of our gratitude but emphasize that they are not responsible for any weaknesses that may remain in the text.

Mr. Edward Dowd, Mr. Pascal Diethelm, and Dr. Dev Ray of the World Health Organization have contributed significantly to our store of knowledge of survey techniques and statistical computation. Attorneys Walter Christy, Robert Glass, John Reed, and Robert Mitchell have provided us with invaluable insights into statistical applications to the Law.

Special gratitude is reserved for Professor Julia Norton of the California State University at Hayward and Professor William Notz of Purdue University who provided excellent suggestions and generous words of encouragement. Professor James Kepner of the University of Florida at Gainsville, through his trenchant reviews, greatly aided us in getting things right. We were fortunate to have other conscientious reviewers who provided incisive and constructive reviews. We take this opportunity to thank them:

William D. Bandes (San Diego Mesa College)
William E. Geeslin (University of New Hampshire at Durham)

Ken Goldstein (Miami-Dade Community College)
Michael L. Levitan (Villanova University)
Robert L. Page (University of Maine at Augusta)
Gerald E. Rubin (Marshall University)
Mark F. Schilling (University of Southern California at Los Angeles)
Donna Skane (Catonsville Community College)
Stephen A. Vardeman (Iowa State University)
Wilbur Waggoner (Central Michigan University)
I. Weiss (University of Colorado)

Our ardent thanks to Hester Paternostro, Meredith Mickel, Susan Lam, and Karen L. Bowers for their expert help joyously given in preparing the manuscript.

Contents Overview

Contents

CHAPTER 4 **Random Variables, Probability Distributions, and Probability Models 107**

CHAPTER 5   The Normal Distribution 157

CHAPTER 6   Estimation; Confidence Intervals 189

CHAPTER 7   **Hypothesis Testing About the Mean and Variance of a Population 217**

CHAPTER 8   **Comparing Two Populations 263**

CHAPTER 9 **Prediction and Association: Regression and Correlation 303**

CHAPTER 10 **Evaluation of Categorical Data; Contingency Tables 341**

CHAPTER 11 Analysis of Variance 369

CHAPTER 12 Nonparametric Statistics 399

APPENDIX A   **Statistical Tables 527**

APPENDIX B   **Derivation of the Binomial Distribution 547**

APPENDIX C   **Confidence Belts for Proportions 549**

Answers 551

Index 573

CHAPTER · 1

The Nature and Purposes of Statistics

"Let us sit on this log at the roadside, says I, and forget the inhumanity and the ribaldry of the poets. It is in the glorious columns of ascertained facts and legalized measures that beauty is to be found. In this very log that we sit upon, Mrs. Sampson, says I, is statistics more wonderful than any poem."

O. Henry
The Handbook of Hymen

Statistical information plays an important part in a wide range of fields, many of which intimately affect our lives. In this chapter we study two main areas in statistics: descriptive and inferential statistics. We also explore two concepts basic to all our subsequent discussions: the "population" and the "sample."

1 ■ 1   INTRODUCTION

The news media have made us aware of the importance of statistics in our everyday lives. "Cost of living" indices are printed regularly, unemployment rates are frequently announced, accident statistics are published before holiday weekends in the hopes of reducing carnage on the roads, crime statistics are periodically reported to an anxious public, and so forth.

Although this information is important, it forms merely the tiniest tip of an immense "iceberg." A great deal of information directly connected to our personal welfare is collected, of which the general public usually has only a vague knowledge. For example, the World Health Organization supports a program for the worldwide collection and analysis of information on adverse reactions to commercially marketed drugs. The organization's purpose is to prevent tragedies such as the one in the late 1960's, when numerous babies were born deformed as a result of their mothers taking the drug thalidomide during pregnancy.

As another example, considerable information on weather conditions over our entire planet is collected hourly by satellites. Statistical analysis of these data now permits more accurate weather predictions. The influence of statistics has become so great in this field that it has even entered into the language of television weather reporters who no longer simply announce, "Well, it looks like rain," but add something like "with a 40% probability."

The preceding paragraphs do not begin to tell the story of how the field of statistics affects every aspect of our lives, from choosing a suitable car to family planning, or how it enters almost every field of study. Although we could give you a list of fields in which statistics finds invaluable application — business administration, economics, medicine, psychology, public health, and so on — such a list would be sterile, not conveying the challenges and excitement found in collecting and interpreting the information. It would not explain why you, the reader, should spend many hours in mastering new concepts and a new language, the language of statisticians. To assert, for example, that statistics has applications to health does not present the unusual experiences and problems one encounters in this area, such as those of one of the authors who spent several months in the northern mountains of Iran collecting statistics on major health problems found in distant, almost inaccessible villages. To say "statistics finds applications in science" tells us nothing about the marvels it has uncovered, for instance, the discovery in small-particle physics of remarkable elements in nature that possess negative mass. Through examples and anecdotes we hope to communicate the concepts and substantive body of statistics as well as the excitement and emotion generated from making sense of seemingly disconnected information — the excitement of discovery.

As our story unfolds, we show you the opportunities that exist for interesting, rewarding careers in statistics and, for those who do not intend to become statisticians, how essential the techniques of statistics are to your field of interest, whether it is the natural sciences, social sciences, engineering, business, the law, or many other areas. Further, we intend to show that a knowledge of statistics can provide insights that are useful in everyday living.

1 ■ 2   WHAT STATISTICS IS

Despite the extensive use of statistics, there are many widespread misconceptions about it. Often we hear comments such as "one can prove almost anything by using statistics," and "figures don't lie, but liars figure." The former statement is untrue. In fact, statisticians are usually very careful to state the limitations of their data to ensure that no conclusions are unjustifiably made. The second statement, unfortunately, is true (witness some of the advertising claims made by manufacturers). An auxiliary purpose of this text is to give the reader the knowledge to recognize those claims of advertisers or others that are actually warranted by the data they present and those claims that are merely a "liar's figures." Perhaps one reason certain people can get away with presenting misleading statistical information is that many individuals tend to view users of mathematics and statistics with awe, believing that numerical methods of analysis somehow magically lend credence to their results. Once the concepts are understood, though, we see that the magic disappears, leaving us with a reliable experimental tool. In fact, the study of statistics should prove useful to everyone who is a consumer of statistics. In today's society this includes practically everyone.

Let us now focus on what statistics actually is. *Statistics* is a field of study with two basic objectives: (1) to describe (or summarize) data and (2) to provide a valid method for making generalizations or inferences from a sample about an entire population. In this text we explore both objectives.

1 ■ 3   EXAMPLES OF DESCRIPTIVE STATISTICS

Perhaps the best way to explain the role of descriptive statistics in the real world is by giving a few examples. Often there is so much information in the data that not even the most powerful mind can intuitively grasp it all. Some method must be found for bringing order to the chaos of raw data. Suppose we have a 10-problem test in a statistics class of 150 students. After grading the exams, the professor is faced with 1500 numbers, namely, a score on each of 10 problems for each of 150 students. Before deciding who did well and who did poorly, the instructor must put the data in some order. The data need to be summarized further before he or she can understand the information.

Summary measures such as total test scores, the average (arithmetic mean) score of the class on the exam, the proportion of students in the class with scores above 80%, and so on are necessary for easy understanding of the 1500 pieces of raw data. *These descriptive statistics give a rough idea of the data at a glance.* By computing averages and percentages, or by using graphical techniques, one can often make the data more easily comprehensible. For example, more than 200 million people are questioned when data are accumulated for the U.S. Census Bureau. It is obvious that summarizing the information is the only way to digest the results. In the next chapter we learn how to compute such summary measures.

As another illustration of how useful descriptive statistics can be, a friend of one of the authors, to earn some extra income while studying statistics in graduate school, took a job selling shoes at a nationally known shoe store. One manager after another had been fired because this particular shoe store kept losing money. Thus before long this friend was given the apparently dubious promotion to manager.

Consider the problem faced by the manager of a large shoe store — he must choose the number, sizes, styles, and so on of shoes to order for the next week. Rather than using his intuition as to how many shoes he thought would sell, as the previous managers had done, this friend used his statistical training. By reviewing sales records for the previous weeks, and the same period for the previous year, he obtained a list of every shoe sold, with the color, style, and size noted. The collection of data, however, was just too immense to work with, so he had to find relatively fewer numbers that would summarize the pertinent information. He calculated the *average* number of shoes sold of a particular style, the *percentage* of this style that were blue and the *percentage* of the shoes sold in this style that were size 9. With these three numbers alone, he described the previous sales per week of blue shoes that were size 9 in this style. Using statistics, he was then able to make a more educated prediction of the next week's sales and to order accordingly. (Incidentally, his store went on to become the most profitable store in the chain, so the company's executives came to investigate how he determined his orders. After explaining his method, he was promoted to assistant comptroller of the company, and so quit graduate school for better things.)

In short, *the purpose of descriptive statistics is to represent the features of a mass of data, portraying these features graphically or through summary numerical measures, or both.* Raw data as such are often incomprehensible, but descriptive statistical techniques allow us to grasp the pertinent characteristics at a glance.

1 ∎ 4 INFERENTIAL STATISTICS; SAMPLES, POPULATIONS

Statistical Inference

Almost every field of endeavor has use for inferential statistical methods. *The term "infer" means to deduce or draw as a conclusion.* Accordingly, in the subject of inferential statistics, we start with a *sample* or subcollection from some *population* of interest. Based upon the characteristics of the members of the sample, we then try to *infer*, or draw conclusions about the characteristics of the entire population. Opinion polls and political surveys are perfect examples of this process. Using the answers given by those who are sampled, the poll designer *infers* what the general population's answers might be. After all, the sample is presumably representative of the entire population.

In most instances it is impossible or prohibitively expensive to examine every single population member, and inference must be made from a sample. For example, automobile manufacturers must meet certain crash resistance standards — a crash at 5 miles per hour (mi/hr) can do no more than $200 damage. Clearly, since the testing is destructive, a manufacturer cannot test every member of the car population. Thus a number of cars, a sample, are selected to be tested, and inferential statistics are used to evaluate the cost of crashes for the entire population. Similarly, the Environmental Protection Agency's (EPA) gas mileage estimates that are listed on all new cars are based upon a sample of cars.

Samples and Populations

Since the majority of this text concerns inferential statistical methods, it is important that the terms "sample" and "population" be understood.

> A *population* is a collection of objects that are the underlying subject of the statistical study.

It is the collection about which we want information and about which inference is to be made. Note that a population does not necessarily refer to a group of people but is only the subject of investigation. For example, when investigating health care costs at a hospital, a population could consist of all the medical bills sent out by the hospital.

Usually we are interested in certain measured values (blood pressure, gas mileage, and so on) concerning the population members. The objects in the population that are actually measured are called *units:* These might be people, cars, businesses, or whatever is of interest in the study.

A *sample* is a subcollection of units from the population. Usually we cannot measure the characteristics of every unit in the population, so we select what we hope is a representative sample from the population, measure those units in the sample, and infer or deduce what the characteristics of the entire population must be.

Numerical characteristics of a population (e.g., average income, percentage of minority employees, and so on) are called *parameters of the population.* The corresponding numerical characteristics calculated from a sample are called *sample statistics.* Our procedure in inferential statistics is to calculate the numerical value of a sample statistic and then try to infer or deduce the value of the corresponding population parameter.

Consider the example of the Nielsen ratings of television programs during the 8 PM Saturday time slot. The units in this investigation are the families with television sets. The interest here is to determine those stations that are being watched by the largest numbers of people. A collection of families is selected to report their viewing habits for the 8 PM time slot and to record the station number. This set of units is the sample. Based upon this sample an inference is made about the viewing habits of *all* units (in this case, families) in the viewing area.

For a further example, consider the testing of a new vaccine. A set of 100 people is selected and each person receives a vaccination. Each person is then given a specific exposure to the disease. A unit in this experiment is a person. Based upon the protection provided by the vaccine to this sample, inferences can be made about the effectiveness of this vaccine for protecting the total population.

1 ▪ 5   STEPS IN REACHING THE OBJECTIVE

We have described two basic objectives in statistics. One objective is to describe a population in a comprehensible way. This is the problem faced by the U.S. Census Bureau every 10 years after its complete census of the population is taken. The other is to draw statistical inferences about a population from a sample. In either case, data must first be collected before the objective can be reached. How to implement the collection of data, how much of it to collect, and from which parts of the population to collect it are matters that are important to the validity of the results, whether descriptive or inferential. The study of which parts of the population to sample and how much to sample is called *sampling design* or *design of experiments.* The study of how to implement the collection of data (whether or not to use interviewers, how to train interviewers, what kinds of questionnaires, if any, to use) is called, in some quarters, *survey methodology,* but it does not have a universally accepted name. The procedure for checking, editing, and tabulating data is called *data reduction.* The methodology for reaching statistical inferences is *statistical analysis.* In short, the steps in achieving statistical objectives are: (1) the design of the experiment or sampling procedure; (2) the design and implementation of a methodology that permits the collection of useful data; (3) data reduction; and, in the case of reaching statistical inferences, (4) the choice and implementation of appropriate statistical analyses.

In this text we shall emphasize step (4) but discuss the other steps in some detail as well.

Summary

In this chapter we have discussed the reasons for studying statistics and the two major functions of statistics: descriptive and inferential. Descriptive statistics such as graphs, tables, charts, and summary measures (like averages and percentages) help display large amounts of data in a compact, understandable form. We shall consider descriptive statistics in Chapter 2. "Inferential statistics" takes the information that is gathered in a sample and utilizes it to make statements concerning the population from which the sample was selected.

Statistical methods were seen to have application in many of the diverse areas of research that touch our everyday lives. Some references at the end of the chapter give further information on the topics discussed in this chapter, providing real examples of applications of statistical techniques to a variety of fields.

Key Terms

descriptive function of statistics

inferential function of statistics

population

sample

experimental unit

inference

sample statistics

population parameter

References

Campbell, S. K.: *Flaws and Fallacies in Statistical Thinking.* Englewood Cliffs, NJ, Prentice-Hall, 1974. A book outlining the possible abuses of statistics.

Careers in Statistics. American Statistical Association and the Institute of Mathematical Statistics, 1974. Discusses the nature and roles of statistics and the job opportunities available to people who know statistics.

Haber, A., et al. (eds.): *Readings in Statistics.* Reading, MA, Addison-Wesley, 1970. Interesting applications of statistics.

Huck, S. W., W. H. Cormier, and W. G. Bounds, Jr.: *Reading Statistics and Research.* New York, Harper & Row, 1974.

Huff, D.: *How to Lie with Statistics.* New York, Norton, 1954. A cute, easily read booklet for the layperson that discusses what to look out for when reading statistics.

Larsen, R. J., and D. F. Stroup: *Statistics in the Real World, A Book of Examples.* New York, Macmillan, 1976. A short workbook consisting of examples of applications of statistics of the type we present in this book. The examples come from published articles in many fields.

Mansfield, E. (ed.): *Elementary Statistics for Business and Economics: Selected Readings.* New York, Norton, 1970.

Peters, W. (ed.): *Readings in Applied Statistics.* Englewood Cliffs, NJ, Prentice-Hall, 1969.

Tanur, J. M., et al. (eds.): *Statistics, A Guide to the Unknown.* San Francisco, Holden-Day, 1972. Short articles from many fields where statistics has proven useful.

Runyon, R. P.: *Winning with Statistics.* Reading, MA, Addison-Wesley, 1977.

Exercises*

Sections 1.1 – 1.3

1. Each member of the class should select an article from within one of the references cited previously and report how statistics was used to answer a question arising in his or her field of study. Haber (1970), Larsen and Stroup (1976), Mansfield (1970), Peters (1969), and Tanur et al. (1972) contain many short articles for this purpose.

Presentation of Data:

2. "One hundred percent of those polled prefer PREP toothpaste to any other toothpaste on the market." This statement may be misleading. Why?

3. The Equal Employment Opportunity Act was passed by Congress in 1964. This act outlaws discrimination by race, sex, or national origin in job hiring. An employer claims that his company is not discriminating against women because "in 1966, the number of women in the company's work force increased by 100%."
   a. Could this statement be misleading?
   b. Discuss what additional information might be needed in order to determine if the statement is or is not misleading.
   c. A company was accused of discriminating against women because less than one half its work force were females. Can such an accusation be justified solely upon this fact?

4. Table 2.1 presents the blood pressure readings of 100 patients.
   a. Describe the variation of the numbers on the list by counting how many different numbers there are.
   b. Describe this variation by finding how many numbers are less than 90 and how many are greater than 90.
   c. Describe this variation by counting how many numbers are in each of the following intervals: less than 80; between 80 and 90, 90 and 100, 100 and 110; and greater than 110.
   d. For each description (a), (b), and (c), why may the description not adequately describe the variation in the data?

5. The following example further illustrates the need to summarize the data before trying to understand them.

   In *Explorations in Role Analysis,* the authors discuss a questionnaire designed to determine what motivates people who make decisions. A group of 106 subjects were given 37 different situations and asked to evaluate a particular action regarding the situations. The permitted answers were "absolutely must," "preferably should," "may or may not," "preferably should not," and "absolutely must not" take the indicated action. The data obtained from the questionnaire then consist of $106 \times 37 = 3922$ different answers. No one can compare all 3922 answers simultaneously, and so it becomes difficult to draw conclusions from the data as they stand. Give a list of some summary measures that might be used in making these results more understandable. (Gross, N., W. S. Mason, and A. W. McEachen: *Explorations in Role Analysis.* New York, John Wiley & Sons, 1958, p. 297.)

6. Each member of the class should go through newspapers or news magazines and find an

---

* The exercises that follow are meant to be thought-provoking as well as instructive. The reader at this stage of study may have difficulty thinking through some of the problems. Do not be discouraged. The remainder of the the text is designed to clarify the ideas that may be troublesome at this point.

example of the use of descriptive statistics to summarize data.

Section 1.4

7. Statistical inference is an inference made from a sample about the entire population. Which of the following are inferences that are statistical inferences?
   a. It is noted from the 1980 census that the percentage of those in the population that belong to minority R is 15%, whereas in the 1970 census this percentage is 5%. The conclusion or inference is that minority R has increased relative to the general population in the period 1970 to 1980.
   b. A sample of 200 is selected from the list of 100,000 names of all registered voters in a given town. It is found that 20 of the 200 are of French origin. We infer that 10% of the 100,000 voters are of French origin.
   c. Twenty-five percent of all patients in 1980 in Hospital X who were in bed more than 1 week suffered bedsores. On the basis of this information, the hospital administrators concluded that preventive treatment for bedsores was not being properly administrated. They fired the responsible supervisor and revamped the staff.

8. A population can be
   a. a class of students
   b. a set of hospital patient records
   c. the total output of a factory over a given period

   Give additional examples of populations. For each of the previous examples, and for your examples, state some *numerical* characteristics of the population that would be interesting to know. For instance, with respect to example (a), the grade-point average of all the students in a class might be interesting.

   For each example you have given, describe how you might sample from the population to obtain an *estimate* of the numerical characteristics of interest.

9. Sometimes the presentation of data is mis-leading because the population from which the data have been taken has not been clearly described. For example, consider this quote from the *New York Times,* 8 April 1979, concerning the safety of color television sets: "In a survey of Long Island homes, covering the sets of 37 manufacturers, at least one color receiver of each brand was found to be giving off radiation. . . ."
   a. Does the population consist of
      (i) Long Island homes
      (ii) color television sets
      (iii) color television sets of 37 manufacturers in all Long Island homes?
   b. If the answer to (a) is (iii), can the number of such sets in the sample be determined from the *New York Times* article?
   c. How many color television sets of each manufacturer were in the sample?
   d. Does the article imply that all brands are defective? If so, is the information convincing as previously presented?

10. Each student should select a statistical problem that arises in his or her area of interest and state how statistical methodology might be utilized to help answer the question under study.

11. It is sometimes difficult or even impossible to define the population from which an observation was taken. Consider the case of *People* vs. *Collins* in which a black man with a beard and a white woman with blond hair in a ponytail were accused of robbing a bank in Los Angeles. The getaway car was partly yellow. The prosecution established (to some extent) that the description given of the bank robbers fit the Collins couple. It was the contention of prosecution that such a couple was so "rare" that the Collins couple must have been the one that robbed the bank. We shall subsequently learn in detail how the prosecution defined rare but for the moment let us suppose rare means there is only one such couple in the population. But which population, that of Los Angeles? A perusal of the mar-

riage statistics in Los Angeles informs us that there are thousands of interracial couples in which the male is black and the female is white with blond hair. It is doubtful that Los Angeles is the population the prosecution had in mind. Is the population "all potential or actual bank robbers"? Perhaps, but it would be impossible for anyone to be able to determine who belongs to this population. What population do you think the prosecution had in mind, if any?

Collection of Data: The following problems introduce the need for a methodology for collecting data.

Sections 1.4 and 1.5

12. The city of Cali, Colombia, lies in a valley surrounded by mountains of the Andes range. The city has two hospitals, but the major facility is the overcrowded University Hospital. The hospital administration knows from general experience that a large number of children between 1 and 6 years old who are admitted to the hospital suffer from either pneumonia, diarrheal diseases, or malnutrition. We are concerned here only with the patient population of children between 1 and 6 years of age.

    In order to determine the personnel needed, and to plan an expansion of facilities for this hospital, the administrators must determine the number of patients in each illness category. For example, if 90% of admissions were pneumonia cases, the administrators might feel justified in expanding the pulmonary ward, buying additional X-ray machines, hiring technicians to operate them, and so on. The simplest way to determine the number of patients in each illness category is to examine the medical records and count the number in each illness category found in these records. With this goal in mind, the administrators wish to look

at the records over the last 5 years. There are 27,000 such records. It is estimated that examining every record would require 6 months for a staff of three, or $3 \times 6 = 18$ person-months. There is only 1 month to do the task and only the money to hire one person, or only one person-month is available. Hence only $27,000/18 = 1500$ records can be examined. How shall these 1500 records be selected? Assume the records are filed chronologically and keep in mind that pneumonia and diarrheal diseases are seasonal diseases; the former occurs mainly in the winter, the latter in the summer.

    a. Should the first 1500 records be examined?
    b. Should $1500/5 = 300$ records be looked at in each of the 5 years? If so, should the first 300 records of each year be examined?
    c. If 300 records are to be examined for each year, what would be the advantage over the procedure in (b) in selecting $300/12 = 25$ from each month? Should the first 25 in each month be examined?
    d. Suppose there are 10 times as many reported illnesses in June and July as in December and January. How can selecting 25 records in each month lead to a distortion of the final results?

13. Often data are collected (such as in problem 12) that are not generated from scientific experiments. Give an example of economic data that are collected, summarized, and reported nationwide and that are pertinent to the budget of many families.

14. The collection of data can be influenced by the personnel collecting the data. "Tax collectors were sent to interview farmers as to the potential size of the crop in the following year. The results indicated that a famine was to be expected." Using this quote, explain how the collection of information can be influenced by those who do the collecting (although it may not be their fault).

CHAPTER · 2

Describing Samples and Populations

A picture is worth a thousand words.

Old Chinese proverb

How do we organize and describe a large collection of information, whether it is from a population or a sample? We study two ways of reducing data: through pictorial methods—tables and graphs—and through the use of numerical summaries of the information. In the course of the discussion we shall define *statistics,* which are numerical summaries of samples, and *parameters,* which are numerical summaries of populations.

2 ■ 1 THE VARIABLE NATURE OF DATA

It is a fact that repeated experiments and experiences yield varying results. For example, if we drop a tennis ball several times, even if we drop it from the same height each time, the height to which the ball bounces will vary from trial to trial. It is this variable nature of data, how to describe it, how to draw conclusions from or make predictions in the face of it, that motivates the study of statistics. To ascertain what is "normal" or "high" systolic blood pressure, a physician may take readings on 100 patients and obtain the data listed in Table 2.1. By just looking at the scores in Table 2.1, one cannot clearly determine what the "normal range" of blood pressure should be. The data in Table 2.1, then, must be organized and summarized in some way before we can begin to determine the distribution or variability of the systolic blood pressure readings.

Although there is variation in the data, a recognizable pattern exists in many cases. The techniques for recognizing and characterizing these patterns can generally be divided into two types: graphical and numerical characterizations of data.

Table 2 ■ 1 Systolic Blood Pressure Scores for 100 Male Patients, Age 21–59

100	140	137	148	90
160	162	107	158	118
130	130	120	108	152
120	130	136	114	128
158	110	130	124	120
105	110	174	106	94
110	147	142	129	134
110	119	110	150	174
115	122	108	125	115
134	120	128	118	115
120	131	149	96	126
168	141	120	106	117
138	120	99	117	98
120	110	110	121	115
122	138	119	130	123
130	124	132	107	141
110	140	125	121	129
120	122	131	132	165
120	126	117	151	138
115	165	112	126	140

2 ▪ 2   GRAPHICAL CHARACTERIZATIONS OF DATA

Frequency Distribution; Relative Frequency

The data on systolic blood pressure (Table 2.1) are difficult to analyze because they are completely unorganized. To arrange these scores into a more understandable collection, we divide the range of blood pressure scores into a reasonable number of intervals and count the number of scores that fall into each interval. The number of scores in each interval is called the *frequency* of the scores in the interval.

For illustrative purposes, we pick the score intervals 90 to 99, 100 to 109, 110 to 119, and so forth, as shown in Table 2.2. We count the number of values in Table 2.1 that fall in each of these intervals and present the tally in Table 2.2.

From Table 2.2 we can see many things at a glance that were difficult to ascertain from Table 2.1. For example, almost half the men (49%) have systolic blood pressure from 110 to 129, whereas only 12% of the men have readings of at least 150.

Table 2.2 permits us to see a pattern that we could not otherwise detect. This type of table is called a *frequency distribution.*

> Frequency distribution: A description of the data that shows the frequency with which the data values fall into selected intervals.

Table 2 ▪ 2 Frequency Distribution of Systolic Blood Pressure (BP) for 100 Males, Age 29–59			
Apparent BP Interval	*Tally or No. of Patients*	*Frequency (f)*	*Relative Frequency = Frequency/Total No. of Patients = (f/N)*
90–99	卌	5	5/100
100–109	卌 |||	8	8/100
110–119	卌 卌 卌 卌 ||	22	22/100
120–129	卌 卌 卌 卌 卌 ||	27	27/100
130–139	卌 卌 卌 ||	17	17/100
140–149	卌 ||||	9	9/100
150–159	卌	5	5/100
160–169	卌	5	5/100
170–179	||	2	2/100
Total N = 100		100	1

A frequency distribution is constructed by counting the number of scores in each interval and listing them as shown in Table 2.2. The *relative frequency* of an interval is calculated by dividing the number of scores in the interval by the total number of scores in the entire sample.

The numbers in the left-hand column of Table 2.2 represent the dividing lines between the intervals in the data and are called the *apparent interval limits.* Thus 90 and 99 are the apparent interval limits of the first interval.

For data such as the blood pressure scores the apparent interval limits are determined by how much rounding off took place before we obtained the numbers. Blood pressure scores here were rounded off to the nearest integer. Thus the first interval actually contains all blood pressure scores from 89.5 up to but not including 99.5, since these scores when rounded off fall between the apparent interval limits 90 and 99. The numbers 89.5 to 99.5 are called the *real interval limits.*

The choice of the number and size of the intervals to use in a frequency distribution table is not entirely arbitrary. In choosing these intervals, one can use the following criteria as guidelines:

Guidelines for Constructing a Frequency Distribution
1. **Every score must fall into just one interval. There can be no borderline cases.**
2. **The widths of the intervals are usually made equal, thus permitting a simple interpretation of the table. Unequal intervals should be used only when special reasons demand such divisions.**
3. **If too many intervals are chosen, there may be only several (or no) values in some of the intervals; if too few intervals are used, the resulting table oversimplifies the data presentation. Usually, between 5 and 20 intervals are sufficient. As more scores become available, more intervals are possible.**

These rules suffice for the presentation in this text.*

Drawing a Histogram

A phrase applicable here is "one picture is worth a thousand words." We often wish to translate the information in the frequency distribution so that it can be expressed visually. One type of presentation that is frequently used is the histogram (also called a bar graph).

To construct a *histogram,* we modify the frequency distribution of Table 2.2 slightly in order to use the real interval limits. Since the systolic blood pressure was rounded to an integer, any reading larger than or equal to 89.5 but smaller than 90.5 was recorded as 90, any reading larger than or equal to 98.5 but

---

* Additional rules may be found in Lutz (1949) and Tukey (1977), listed in the reference section.

smaller than 99.5 was recorded as 99, and so on. Modifying the intervals of Table 2.2 to account for this "rounding off," we obtain the real intervals listed in Table 2.3.

Notice that we have formed our new intervals by using real interval limits so they "touch" and have added an adjacent interval at both ends of the range. These two new intervals are of the same length as the other modified intervals and have zero frequency. They are put there to show us where the table begins and ends.

To draw the histogram that corresponds to the data in Table 2.3, we draw a vertical axis of *relative frequencies* between 0 and 1 and a horizontal axis of systolic blood pressure scores. Above each real interval we make a horizontal line, the height of which corresponds to the relative frequency observed for that interval. These lines are connected to form bars as shown in Figure 2.1. For example, the relative frequency of the interval 89.5 to 99.5 is 0.05, so we draw a horizontal line segment of height 0.05 above the interval 89.5 to 99.5. The intervals over which the bars are drawn are called *class intervals*.

We note at once from Figure 2.1 that systolic blood pressures above a value of 170 are probably too high, whereas a normal systolic blood pressure value is about 120. The picture permits us to see at once the relative proportion of patients in each blood pressure interval without having to compare numbers pairwise. The eye can synthesize all the information in one glance.

The Density Histogram

A type of histogram that we will employ frequently is one using a density scale. The height of this histogram over each class interval is the percentage of cases in

Apparent Interval Limits	Real Interval Limits	Midpoint	Frequency	Relative Frequency
		Table 2 ▪ 3 Frequency Distribution of Systolic Blood Pressure with Intervals Modified for Construction of Histogram		
	79.5–89.5	84.5	0	0
90–99	89.5–99.5	94.5	5	5/100 = 0.05
100–109	99.5–109.5	104.5	8	8/100 = 0.08
110–119	109.5–119.5	114.5	22	22/100 = 0.22
120–129	119.5–129.5	124.5	27	27/100 = 0.27
130–139	129.5–139.5	134.5	17	17/100 = 0.17
140–149	139.5–149.5	144.5	9	9/100 = 0.09
150–159	149.5–159.5	154.5	5	5/100 = 0.05
160–169	159.5–169.5	164.5	5	5/100 = 0.05
170–179	169.5–179.5	174.5	2	2/100 = 0.02
	179.5–189.5	184.5	0	0

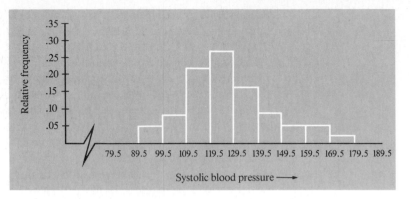

Figure 2 ▪ 1
Relative frequency
histogram of systolic blood
pressure from 100 males,
age 29 to 59.

the class interval divided by the length of the interval. When the class intervals are of equal size, there is no difference in the shape of a histogram using a density scale and that of the relative frequency histogram (although the height of the blocks will differ between the two histograms). When the class intervals are not of equal size, there may be more cases in a larger interval simply because the interval is larger. The histogram employing a density scale compensates for differences in interval sizes and so gives a true picture of the actual distribution of the sample along the intervals. To illustrate this, consider again the data of Table 2.2, which show the distribution of systolic blood pressure among 100 patients. Combine intervals 90 to 99 and 100 to 109. The relative frequency for the new interval 90 to 109 is 13/100 or 13%. Combine the intervals 140 to 149, 150 to 159, 160 to 169, and 170 to 179. The relative frequency for this new interval is (9 + 5 + 5 + 2)/100, which equals 21/100 or 21%. Figure 2.2a displays the relative frequency histogram for the case in which the class intervals are unequal. It appears that a large number of patients have very high values of systolic blood pressure. We know from the data that this is not so. This relative frequency histogram distorts the picture. However, note that the histogram in Figure 2.2b uses a *density* scale and presents a more accurate picture of the actual situation.

One technique used by unscrupulous people in "lying with statistics" is to draw a frequency or relative frequency histogram by using a set of unequal intervals in order to distort the presentation to their own advantage. Watch out for this!

The Population Histogram

We have shown how to draw a histogram of any finite collection of values. The collection of values could be a sample from a population or an entire popula-

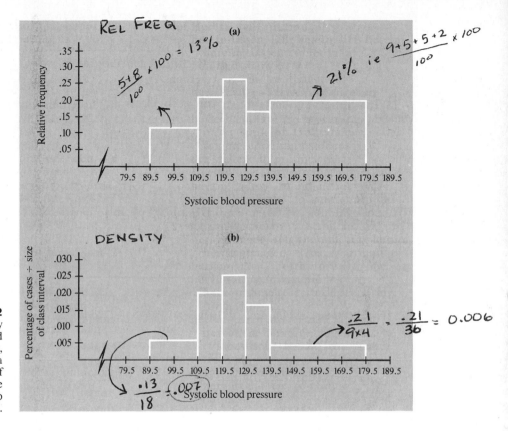

Figure 2 ■ 2
(a) Relative frequency histogram of systolic blood pressure from 100 males, age 29 to 59. (b) Histogram (using density scale) of systolic blood pressure from 100 males, age 29 to 59.

tion.* A density histogram that is a pictorial display of an entire population is called a *population histogram,* a *population density,* or a *population distribution.* We shall always graphically display populations by using histograms with density scales in order to avoid distortions created by unequal class intervals. One important feature of these histograms is that the *area* of each bar above a class interval represents the proportion of the population in that interval. This can be understood from the formula

$$\text{Area of bar} = \text{height of bar} \times \text{size of class interval}$$

$$= \frac{\text{proportion of population in class interval}}{\text{size of class interval}} \times \frac{\text{size of class}}{\text{interval}}$$

---

* In Chapter 1 we defined a population as a collection of objects that are the underlying subject of the statistical study. These "objects" could be numbers. For example, the blood pressure values of all men between 45 and 55 years of age have constituted the population of interest in many studies of hypertension in males. In this chapter and in most of this text the objects will be numbers.

Consequently, the areas of the bars add up to the total proportion, which is 1 or 100%.

> In a density histogram, the area of the bar over the class interval is equal to the proportion of the population in that interval. Consequently, the sum of the areas of all the bars is 1.

Finite Populations

Figure 2.3 is a density histogram of the population consisting of the numbers 0.0, 0.1, 0.2, 0.3, 0.4, 0.5, 0.6, 0.7, 0.8, 0.9. The relative frequency is the same for each number, namely, 1/10. Since the relative frequency is "uniform," the population histogram is called a *uniform density histogram.* We say these 10 values are *equally likely.*

Infinite Populations That Are Continuous

Consider a population consisting of the 100 numbers 0.00, 0.01, 0.02, up to 0.99 that are equally likely. Figure 2.4a presents the population density of this population.

Suppose the numbers 0.000, 0.001, 0.002, up to 0.999 are equally likely. The density histogram of this population is shown in Figure 2.4b. The bars are so close in Figure 2.4b that the histogram appears to be continuous. This observation leads us to the concept of the density histogram for *infinite continuous populations,* that is, for populations that take on all the values in a given interval. For example, a population consisting of all the values from 0 to 1 is an

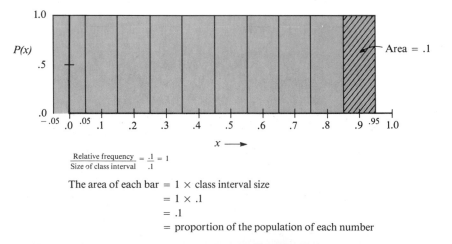

$$\frac{\text{Relative frequency}}{\text{Size of class interval}} = \frac{.1}{.1} = 1$$

The area of each bar $= 1 \times$ class interval size
$$= 1 \times .1$$
$$= .1$$
$$= \text{proportion of the population of each number}$$

Figure 2 ▪ 3
Density histogram: uniform distribution over values .0, .1, .2, .3, .4, .5, .6, .7, .8, .9.

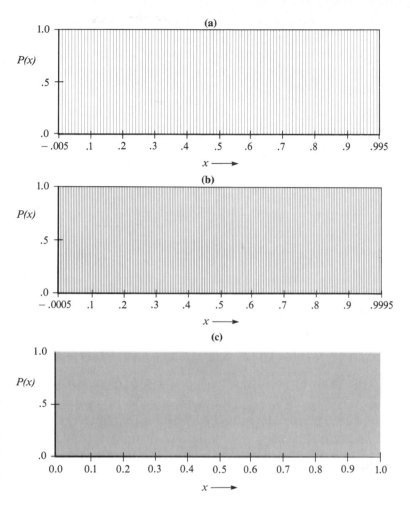

Figure 2 ■ 4
(a) Density histogram: uniform distribution over values .00, .01, .02, up to .99. (b) Histogram: uniform distribution over values .000, .001, .002, up to .999. (c) Histogram: uniform distribution over the interval 0 to 1.

infinite continuous population. If the points all have the same relative frequency (if they are equally likely), then the preceding discussion implies that the density histogram of the population consisting of *all* numbers from 0 to 1 is the *uniform density distribution,* as shown in Figure 2.4c. This distribution is often called simply the *uniform distribution.*

The uniform density is called a *continuous density distribution* because it is the density histogram of a population that takes on all values in a given interval.

> Continuous distribution: A density histogram of a population that takes on all values in a given interval is called a *continuous density distribution.* It is often called simply a *continuous distribution.*

As with all density histograms, the total area under the continuous density is equal to 1. The area under the curve over any interval is the proportion of the population that has values in that interval. For example, in Figure 2.4c the area over the interval 0 to 0.5 is 0.5. This means that one half the population takes on values between 0 and 0.5.

2 ■ 3   NUMERICAL CHARACTERIZATIONS OF DATA

Although graphical methods of representing data yield a quick overall impression of the distribution under study, it is not always possible to use a graph. For example, when trying to describe a distribution to someone over the telephone, the use of graphical techniques is impossible. In order to make quantitative statements, we need to determine how to utilize numbers to describe the data. Such numerical descriptions should enable the listener to picture the histogram mentally.

Observing the histogram in Figure 2.1, we see that it is "centered" somewhere around the point 124.5 and spread out fairly evenly around this point. We now discuss several methods of measuring both the "center" and the "spread" of a distribution numerically.

2 ■ 4   MEASURES OF CENTRAL TENDENCY

Measures of Central Tendency in Samples

One of the most important properties of a histogram is the location of its center. This central value is often taken as a "representative" member of the sample. Numerical measures of this type are called *measures of central tendency* of the sample, since they are used to determine a center for the histogram.

There are, of course, many different ways of defining what a representative numerical value for a sample might be. For example, the following are four different measures of central tendency:

1. A representative numerical value for a sample is the number that occurs most often in the sample. For example, in the sample of numbers 1, 2, 3, 3, 4, 7 we might say the number 3 is representative of the sample since it appears most often.
2. A representative numerical value for a sample is the number that divides the sample in half so that half the numbers in the sample are less than the number and half are larger than this number. It is the center of the distribution or "middle value." For example, in the sample of numbers 1, 2, 3, 5, 7, 15, 27 the number 5 might be considered representative since there are just as many numbers above 5 as there are below 5 (namely, three of them).

3. A representative number for a sample is the number that is halfway between the largest and smallest values in the sample. For example, in the sample of numbers 1, 2, 5, 8, 27, 36 we might say $(1 + 36)/2 = 18.5$ is representative.

4. A representative number for a sample is the arithmetic average of all the members in the sample. For example, in the sample 4, 2, 5, 7, 8, 10 the number $(4 + 2 + 5 + 7 + 8 + 10)/6 = 6$ might be called representative.

Handwritten margin notes:
1 ,2,3,3,4,5,9,11,12
median – 4 - middle #
(n+1)%-2
mode – 3 occurs most.
mean = 5.56 = ave of the #s

Each of these measures of central tendency has a name and is useful for different purposes.

1. The sample mode

> The measure obtained by using definition (1) is called the *sample mode;* it is the number that occurs most often in the sample.

If you put all the numbers in the sample into a box, shake them up, and select a number from the box, the mode will be picked out with the highest probability. Thus if you are betting on the outcome of the selection and a perfect guess is the only time you are paid off, then your best bet is to select the mode. You will be paid off most often.

There can be more than one mode. For example, in the set of numbers 1, 5, 6, 6, 7, 8, 9, 9, 10 there are two numbers, 6 and 9, that occur most often, and this set is called bimodal.

Example 2 ▪ 1   Consider the systolic blood pressure readings of the 100 men given in Table 2.1. The mode of these numbers is 120 since this number occurs 10 times in the table and no other number occurs this often.

2. The sample median.
The second definition of a representative number for a sample leads to the definition of the *sample median.*

> The sample median is the number that is at the middle point of the sample when the sample values are arranged in order of increasing value.

Example 2 ▪ 2   To find the median of the sample, 3, 15, 7, 27, 1, 5, 2, we first order the sample values by size, obtaining 1, 2, 3, 5, 7, 15, 27. There are seven values. The median is the number located at the $(7 + 1)/2 =$ fourth position when the sample values are ordered by size. The median is equal to 5. As you can see, there are an equal number of values on either side of 5.

Consider the following sample, 15, 2, 1, 3, 27, 3, 30, 9, that when the sample values are ordered by size, is 1, 2, 3, 3, 9, 15, 27, 30. There are eight values. The median is the number located at position $(8 + 1)/2 = 4\frac{1}{2}$; that is, it is halfway between the fourth and fifth positions. Since 3 is at position 4 and 9 is at position 5, the median is equal to $(3 + 9)/2 = 6$. Note that there are an equal number of values on either side of 6.

> In general, to locate the median when there are n sample values, order the n values by size; the median is at position $(n + 1)/2$.

Example 2 ■ 3 A psychologist wished to determine whether reward or punishment is a more effective method of inducing learning. The psychologist selected 11 mice and ran them through two sets of mazes. The first maze was run three times and a food reward was presented after the mouse successfully navigated the maze. The second maze was also run three times, but for this maze the mouse was given a small electrical shock every time it made a wrong turn. After a 1-day rest, each mouse was timed going through each maze and the difference between the time to run the first maze and the second maze was recorded. The results of one of these experiments were in seconds: -1.1, 2.0, 1.3, -0.9, -1.9, 0.8, 1.2, -1.2, 4, 0.2, -0.1. The median is the sixth $[(11 + 1)/2]$ observation in order; that is, the median is 0.2.

Example 2 ■ 4 To calculate the median of the 100 systolic blood pressures listed in Table 2.1, we first arrange these numbers in order and then, because there is an even number of observations, average the 50th and 51st observations. This yields the median $[(122 + 123)/2 = 122.5]$.

3. *The sample midrange.* If we utilize the third definition of a representative numerical value, we are led to the measure of central tendency, known as the *sample midrange.* It is obtained by taking the largest and smallest numbers in the sample and averaging them. For example, in the systolic blood pressure of Table 2.1 the largest number is 174 and the smallest number is 90, so the midrange is $(174 + 90)/2 = 132$. The main benefit of using the midrange as a measure of central tendency is that it is very quick and easy to calculate when time and money are significant factors.

4. *The sample mean.* The fourth definition of representative is most commonly used. It leads to the measure known as the arithmetic mean or simply the *mean* of the sample. We denote the mean of a sample $x_1, x_2, \ldots, x_n$ by $\bar{x}$. Thus

> **The mean of a sample of numbers $x_1, \ldots, x_n$ is**
>
> $$\bar{x} = \frac{x_1 + x_2 + \cdots + x_n}{n}$$
>
> **The number $\bar{x}$ is called the *sample mean* or sample average.**

The sample mean $\bar{x}$ is the measure of central tendency that will be used most often throughout this text. It has the advantage of utilizing information about the magnitude of *all* the numbers in the sample. The median only used information about the order of the numbers, whereas the other measures lost information by using only a few numbers in the sample. Since it employs all the information in the sample, the sample mean is sometimes considered as the best measure of central tendency for many practical situations.

Example 2 ▪ 5   Consider again the psychologist's experiment outlined in example 2.3. The psychologist wanted to determine whether reward or punishment was more successful as a learning tool. The difference between the "rewarded" trials and the "punished" trials were $-1.1, 2.0, 1.3, -0.9, -1.9, 0.8, 1.2, -1.2, 4.0, 0.2,$ and -0.1 sec. The average or *mean* of the differences $\bar{x}$ is calculated as follows:

$$\bar{x} = \frac{-1.1 + 2.0 + 1.3 - 0.9 - 1.9 + 0.8 + 1.2 - 1.2 + 4.0 + 0.2 - 0.1}{11}$$

$$= \frac{4.3}{11} = 0.39 \text{ sec}$$

Thus the rewarded group averages 0.39 sec faster than the punished group. Methods of ascertaining whether this is actually a significant difference or merely due to chance will be considered in the chapter on hypothesis testing.

Example 2 ▪ 6   When calculating the mean of the 100 systolic blood pressures in Table 2.1, we obtain

$$\bar{x} = \frac{100 + 160 + 130 + \cdots + 140}{100} = \frac{12637}{100} = 126.37$$

Since we shall frequently be using the mean $\bar{x}$ and other sums in statistics, it is advantageous to use shorthand notation for the addition of the numbers $x_1, x_2, \ldots, x_n$ in the sample. This is accomplished by using summation notation. The sum $x_1 + x_2 + \cdots + x_n$ is represented by the shorthand symbol $\sum_{i=1}^{n} x_i$

handwritten notes in margin:
n — how many #'s to sum up
$\sum x_i$
$1=i$
$i=1$ ie begin with the first #
$i=4$ ie begin with the fourth #.

and is read "sum up the numbers x_i, starting with the lower number $i = 1$ and continuing until $i = n$." For example, if $x_1 = 1$, $x_2 = 5$, $x_3 = 7$, $x_4 = 12$, $x_5 = 2$, $x_6 = 8$, then $\sum_{i=1}^{3} x_i = 1 + 5 + 7$, $\sum_{i=2}^{4} x_i = 5 + 7 + 12$, $\sum_{i=1}^{6} x_i = 1 + 5 + 7 + 12 + 2 + 8$.

Σ: The symbol Σ (read "sigma") is the Greek letter corresponding to S in our alphabet and stands for "sum." Thus

$$\sum_{i=1}^{3} x_i$$

is read "sum up all the numbers from x_1 to x_3," which is the sum $x_1 + x_2 + x_3$. The symbol

$$\sum_{i=1}^{n} x_i$$

is read "sum up all the numbers from x_1 to x_n," which is the sum

$$x_1 + x_2 + \cdots + x_n$$

where the three dots $\cdots$ indicate the intermediate terms in the sum.

Further practice using the summation symbolism may be found in the exercises.

One way to interpret the mean of a sample is as the center of gravity of the histogram; that is, if the histogram were put on a teeter-totter beam, the fulcrum should be put at the mean to ensure that the beam will balance (see Fig. 2.5a).

It should be noted that as a "balance point" for the histogram, the mean is sensitive to extreme values such as those exhibited by the histograms of incomes in Figure 2.5b. Here 5% of the sample had unusually large incomes that were perhaps not entirely characteristic of this group, and hence the mean for the group seems higher than what might be considered as representative.

The general relationship among the mean, median, and mode for various shapes of sample histograms is expressed by Figure 2.6. The most common general shapes that occur in practice, and their names, are also shown. A histogram is symmetric if the side to the right of the median is the mirror image of the side to the left of the median (Fig. 2.6b); otherwise it is called skewed. If a long skinny "tail" extends to the right, the histogram is called skewed to the right (Fig. 2.6a), and if the tail goes to the left, it is called skewed to the left (Fig.

Figure 2 ▪ 5
The mean as a "balancing point" or the "center of gravity" of the histogram. (In part a the ticks indicate the apparent interval limit.)

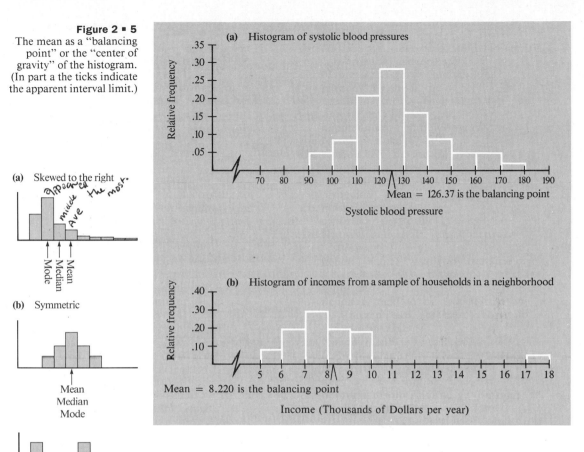

(a)   Histogram of systolic blood pressures

Mean = 126.37 is the balancing point

Systolic blood pressure

(b)   Histogram of incomes from a sample of households in a neighborhood

Mean = 8.220 is the balancing point

Income (Thousands of Dollars per year)

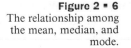

(a)   Skewed to the right

appeared
much the most.
Ave

Mode
Median
Mean

(b)   Symmetric

Mean
Median
Mode

Mode
Mean
Median
Mode

(c)   Skewed to the left

Mode
Median
Mean

Figure 2 ▪ 6
The relationship among the mean, median, and mode.

2.6c). Note that for skewed histograms the mode is at the hump and the median and then the mean trail down the long (skewed) tail.

Choosing Between the Mode, Median, and Mean

The 1964 Civil Rights Act prohibits discrimination by race, sex, or national origin in hiring and promotions. If a person is not hired or not promoted, illegally according to the act, the person is entitled to monetary compensation. The issue as to how much money the person is entitled has been fought in a number of cases.* Let us consider an example taken from one of these cases in which a number of people (called the plaintiffs) were found to have been discriminated against in hiring for clerical positions by the defendant company. To make matters simple, let us say they applied but were not hired in 1975. The decision that they were discriminated against was handed down in 1981. The

* *EEOC* v. *Federal Reserve Bank* 30 FEP Cases 1137–1169.

attorney for the plaintiffs claimed that each plaintiff should receive the job's salary for the entire 7-year period of the litigation. The defendant objected on the grounds that the turnover of those jobs was such that not one of the hundreds of people who had held those types of jobs had ever lasted in the job for as long as 7 years. Because of time limitations, it was not possible to look at the history of all employees. A sample of 319 employee records was taken and the following information pertinent to the job turnover rate was obtained:

Number of months worked before termination	.5	1	2	3	4	5	8	10	11	14	18	20	24	30	35	40	60
Number of employees	15	21	27	33	42	9	17	32	18	12	22	26	11	9	6	8	11

The plaintiffs' attorney pointed out that the mean $(\bar{x})$ of the number of months worked is 12.08 months. The yearly salary was approximately $9500.00, and there were 20 plaintiffs; hence, the company owed $(12.08/12) \times 9500 \times 20 = \$191,266.67$. The defendant's attorney pointed out that the median is 3 months, so that the company owed $(3/12) \times 9500 \times 20 = \$47,500$, a considerably smaller sum. The defendants also noted that using the mean of 12.08 months was "ridiculous because the great majority (67%) of the sample had been terminated before 12 months." You can see that the fight over which descriptive statistic to use was not just academic but involved substantial sums of money.

Why is the mean so different from the median in this example? We see that we have a few employees who have worked a long time. There are 11 extreme values of 60 months, 8 of 40 months. The mean is sensitive to extreme values, the median is not. We can illustrate this fact by considering the two sets of numbers:

1, 3, 3, 7, 8
1, 3, 3, 8, 500

The median of both sets is 3, whereas the mean of the first set is 4.4 and that of the second set is 103.

As with the median, the mode is not sensitive to extreme values. In the example just given, the mode is 3 in both sets of numbers. In those situations in which we do not want extreme values to influence our description of central tendency, it is preferable to use the median or the mode rather than the mean. The mode is to be preferred to the median if the location of most of the sample is important. For example, a politician traveling along a mile-long parade route lined with crowds of different density might not stop to give a speech at the median point of the distribution of the crowd (which could conceivably be an empty lot) but would certainly stop to speak at the mode (the place where most people are standing).

Despite its weakness with respect to extreme values, the mean has a number of advantages over the median and mode that will be discussed extensively in subsequent chapters.

Grouped Data

In many situations that arise in applications the data are given to us already grouped into intervals, such as the data in the frequency distribution of Table 2.2. How does one calculate a mean when the individual sample numbers are not given?

To calculate the sample mean from interval grouped data, let m_i denote the midpoint of the ith interval and f_i denote the frequency of observations that fall in the ith interval. If we assume that all f_i of observations in the ith interval are evenly spread throughout the interval, then the midpoint m_i is simultaneously a mean and median for the interval, and hence best represents the numbers in the ith interval. Thus the total contribution to the calculation of the sample mean from the ith interval is approximately $f_i m_i$. Summing over each of the intervals yields the following formula for calculating the sample mean from grouped data:

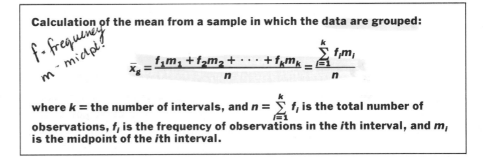

Calculation of the mean from a sample in which the data are grouped:

$$\bar{x}_g = \frac{f_1 m_1 + f_2 m_2 + \cdots + f_k m_k}{n} = \frac{\sum_{i=1}^{k} f_i m_i}{n}$$

where k = the number of intervals, and $n = \sum_{i=1}^{k} f_i$ is the total number of observations, f_i is the frequency of observations in the ith interval, and m_i is the midpoint of the ith interval.

Example 2 ▪ 7   Calculate $\bar{x}_g$ from the frequency distribution of blood pressures given in Table 2.3. Here we have $k = 11$ intervals with midpoints 84.5, 94.5, 104.5, 114.5, 124.5, 134.5, 144.5, 154.5, 164.5, 174.5, and 184.5 and corresponding frequencies 0, 5, 8, 22, 27, 17, 9, 5, 5, 2, and 0. Thus we calculate

$$\bar{x}_g = \frac{0 \cdot 84.5 + 5 \cdot 94.5 + 8 \cdot 104.5 + \cdots + 2 \cdot 174.5 + 0 \cdot 184.5}{100}$$

$$= 12720/100 = 127.2$$

Note that $\bar{x}_g$ calculated by using the 11 grouped intervals comes out within one unit of the value of $\bar{x}$ obtained by using all 100 numbers (example 2.6), and the calculation is much shorter. Grouped data formulas are necessary when the data are obtained from frequency tables such as Table 2.3. Since this is the usual method for displaying data in published articles, it is important to know the formula for $\bar{x}_g$.

Measures of Central Tendency in Populations

Finite Populations

The measures of central tendency—the mode, median, and mean—discussed for samples, are defined exactly as before for histograms of finite populations of numbers; namely, the hump in the histogram, the 50% point, and the arithmetic average, respectively. Of these three measures, the mean is the most often used and so we employ it to illustrate some important points.

Suppose there is an important election coming up in a community of 10,000 voters and we wish to ascertain the proportion of voters who favor a particular proposition on the ballot. If we enumerate the voters in the community and record a 1 if the person is in favor and a 0 if the person is against the proposition, then we can regard our population of voters as represented by a population of 10,000 zeroes and ones. We calculate the mean of these numbers in the same manner as we did for samples, namely, add all 10,000 numbers and divide by 10,000. The resulting number is called the mean value of the population and is denoted by the symbol μ (the Greek letter mu). Thus

---

The population mean for a finite population

$$\mu = \frac{\text{sum of all numbers in population}}{\text{size of the population}}$$

In the election situation previously described the sum of all numbers in the population is the sum of all 1's in the population that is just the number of voters in the community in favor of the proposition. Hence *in this example* μ is precisely the proportion of voters in the population who favor the proposition.

An Important Version of the Formula for the Population Mean

Suppose we have a population consisting of the values 0.0, 0.1, 0.2, 0.3, 0.4, 0.5, 0.6, 0.7, 0.8, 0.9 and suppose each of these values appears 20 times in the population, yielding 200 members in the population. Then the population mean μ is

$$\mu = \frac{\text{sum of all the values}}{200}$$

$$= \frac{0.0 + 0.0 + \cdots + 0.0 + 0.1 + 0.1 + \cdots + 0.9}{200}$$

$$= 0.0 \frac{20}{200} + 0.1 \frac{20}{200} + 0.2 \frac{20}{200} + \cdots + 0.9 \frac{20}{200}$$

$$= 0.45$$

We note that $\frac{20}{200}$ or $\frac{1}{10}$ is the relative frequency of each number. Let $P(i)$ represent the relative frequency of the value i. We can rewrite the formula for μ as follows:

$$\mu = 0.0 \times (\text{relative frequency of } 0.0) + 0.1 \times (\text{relative frequency of } 0.1)$$

$$+ \cdots + 0.9 \times (\text{relative frequency of } 0.9)$$

$$= 0.0P(0.0) + 0.1P(0.1) + 0.2P(0.2) + \cdots + 0.9P(0.9)$$

In general, we have

An alternative formula for the population mean μ: If a population consists of n values $x_1, x_2, \ldots, x_n$ and $P(x_i)$ represents the relative frequency of x_i, then $\mu = x_1 P(x_1) + x_2 P(x_2) + \cdots + x_n P(x_n)$

Example 2 ▪ 8   The uniform distribution of the numbers (Fig. 2.4a) 0.00, 0.01, 0.02, 0.03, . . . , 0.99 has mean

$$\mu_1 = 0.00 \frac{1}{100} + 0.01 \frac{1}{100} + \cdots + 0.99 \frac{1}{100} = 0.495$$

The uniform distribution of the numbers (Fig. 2.4b) 0.000, 0.001, 0.002, . . . , 0.999 has mean

$$\mu_2 = 0.000 \frac{1}{1000} + 0.001 \frac{1}{1000} + \cdots + 0.999 \frac{1}{1000} = 0.4995$$

30 ∎ 2 Describing Samples and Populations

The uniform distribution of the numbers 0.0000, 0.0001, . . . , 0.9999 has mean

$$\mu_3 = 0.0000 \, \frac{1}{10,000} + 0.0001 \, \frac{1}{10,000} + \, \cdots \, + 0.9999 \, \frac{1}{10,000}$$

$$= 0.49995$$

Continuous Distributions

Example 2.8 suggests how to calculate the mean for continuous distributions. The uniform distribution of the numbers 0.00, 0.01, up to 0.99 looks like the uniform distribution over all the numbers 0 to 1. Compare Figures 2.4a and 2.4c. The uniform distribution of the numbers 0.000 to 0.999 (Fig. 2.4b) even more closely approximates Figure 2.4c. We would get even a closer approximation using the uniform distribution of numbers from 0.0000 to 0.9999. Notice from Example 2.8 that the mean goes from 0.495 to 0.4995 to 0.49995; as the approximation becomes better to Figure 2.4c, the mean gets closer to 0.5. We say that the mean of the uniform distribution over 0 to 1 (Fig. 2.4c) is 0.5.

The preceding paragraph outlines how the mean for a continuous distribution could be calculated by using finite approximations. We shall not cover this matter in any more detail but provide you with the means of continuous distributions as they are needed.

The mode and median are defined for continuous distributions analogously to finite histograms.

> The *mode* (or modes) of a continuous distribution is the value (or values) at which the distribution takes on its maximum value. It is at the hump of the density curve.

Example 2 ∎ 9 Figure 2.7 shows a particular continuous distribution of a population that takes on *all* numbers. The mode of this distribution is 0.

Example 2 ∎ 10 The uniform distribution, Figure 2.4c, does not have a single mode since it does not have a value that occurs more often than all the others.

The median of a finite collection of numbers is defined as the central number in the collection; that is, the number such that half the numbers in the population are smaller or equal to it. The median is defined similarly for continuous distributions.

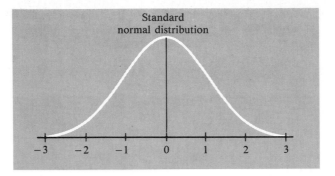

Standard
normal distribution

Figure 2 ▪ 7
A continuous distribution
called the "normal
distribution," with mean,
μ, equal to 0 and standard
deviation, σ, equal to 1.

The *median* of a continuous distribution is the value such that half the population takes on values smaller or equal to it. It is the 50% point.

Example 2 ▪ 11   The median of the uniform distribution, Figure 2.4c, is 0.5.

Parameters and Statistics

We should emphasize here the distinction between the numerical characteristics of a population and those of a sample. These characteristics are given different names to focus our attention on this distinction.

The numerical characteristics of a population are called *parameters* of the population and are certain fixed numbers that describe the population. There is no variability in the value of a population parameter. It is a single fixed number.
The numerical characteristics of a sample are called sample *statistics* and their values can change from sample to sample.

Throughout the book, we usually use Greek letters to denote population parameters (numerical characteristics of the populations) and Roman letters to denote the sample statistics. This differentiation is important to keep in mind since a population parameter is a single fixed number that is characteristic of the population (although usually unknown to us), whereas the sample statistic is a number that is calculated from the sample and that could conceivably change from sample to sample. Thus μ is a parameter and $\bar{x}$ is a statistic.

The difference between the population mean μ and the sample mean $\bar{x}$ is discussed further in the next example.

Example 2 ▪ 12 An ecologist wants to investigate the reproductive behavior of a certain species of water fowl located on an island off the California coast. It is suspected that DDT in the environment results in infertile eggs, endangering the species. There are a total of 150 bird nests on the island and *if* the number of eggs in each nest that hatch could be recorded the following population of numbers would be obtained:

1	2	1	4	0	0	3	4	2	1
0	4	3	2	1	0	1	0	2	3
3	3	4	1	0	1	0	2	2	1
1	0	0	1	5	0	2	0	0	1
3	1	2	0	4	1	0	2	3	3
2	0	3	4	0	0	1	0	0	0
2	2	0	1	2	1	0	2	1	0
1	3	1	2	0	1	3	5	1	3
0	1	0	1	3	3	1	2	1	0
4	0	2	1	2	0	1	3	2	0
4	0	0	2	1	1	0	1	0	4
1	1	1	3	2	0	1	1	2	0
0	2	0	0	1	2	1	0	2	1
2	3	2	1	2	2	1	4	3	1
1	1	0	0	3	4	2	2	1	2

The mean of this population of numbers is

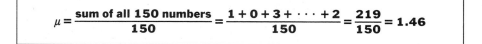

$$\mu = \frac{\text{sum of all 150 numbers}}{150} = \frac{1 + 0 + 3 + \cdots + 2}{150} = \frac{219}{150} = 1.46$$

This number $\mu = 1.46$, which is a parameter of the population and is fixed, says that this population of nests averages 1.46 fertile eggs per nest.

Suppose, however, the scientist cannot realistically observe all 150 nests but instead selects 10 nests to observe and obtains the numbers 0, 1, 3, 0, 0, 1, 4, 2, 1, 1. From this sample the scientist calculates the sample mean $\bar{x} = (0 + 1 + 3 + 0 + 0 + 1 + 4 + 2 + 1 + 1)/10 = 1.3$. Note that this value can change from sample to sample and is not necessarily the same number as μ. Indeed, if by chance a different 10 nests were observed and the number of fertile eggs counted, the scientist might obtain the 10 numbers 0, 2, 1, 3, 4, 1, 2, 2, 1, 0 and calculated $\bar{x} = (0 + 2 + \cdots + 0)/10 = 1.6$. Thus μ is a fixed number, always the same, whereas $\bar{x}$, the sample mean can change from sample to sample.

Often the population under study is too large or too intangible to measure every unit in order to calculate the population mean μ. For example, suppose an agricultural researcher develops a new hybrid of wheat and wishes to determine the mean yield μ in bushels per acre for this hybrid. The population involved in this experiment is the set of all acres that will ever be planted with this hybrid. Clearly one cannot enumerate this population since the population is to some degree theoretical. The entire population does not exist simultaneously; so, even if desired, the researcher cannot calculate μ. The researcher can, however, sample from this population by planting a small field, say, 40 acres, with the new hybrid and measuring the yield from each acre. The sample mean $\bar{x}$ calculated from these 40 numbers gives an estimate of the population mean μ, even though μ cannot itself be directly calculated. In this situation (as in most real statistical problems) the mean μ will not be known precisely and the statistician will have to estimate μ from a sample.

2 ▪ 5   MEASURES OF DISPERSION OR VARIATION

Variation in a Sample

In order to picture a histogram in your mind, you should know not only the center of the distribution of scores but also have some idea of how spread out the observations are around this center. For example, the histograms in Figure 2.8 have the same mean, median, mode, and midrange; however, they are very dissimilar in shape. The histogram in Figure 2.8b is much more spread out or disperse than that shown in Figure 2.8a.

Thus, in order to describe a histogram adequately, we should have not only a measure of central tendency but also a measure of how spread out the histogram is about this center.

As with measures of central tendency, there are several different "reasonable" measures of dispersion of a distribution. We mention four: *range, absolute deviation, variance, and standard deviation.*

The *range* of a sample is the largest value in the sample minus the smallest value in the sample. Clearly, the more spread out the histogram is, the larger the range is, and vice versa. An advantage of using the range as a measure of spread is that it is quickly calculated. A disadvantage is that it only uses two numbers in the sample and hence wastes a lot of information. The *absolute deviation* avoids this disadvantage by utilizing all the sample in its calculation. The absolute deviation is calculated by taking the arithmetic average of the distances of the sample observations from their median. For example, consider the following litter sizes for a sample of five dogs: 5, 3, 2, 3, 4. The sample arranged in order is 2, 3, 3, 4, 5, and the median is 3. The absolute deviation $= (|2 - 3| + |3 - 3| + |3 - 3| + |4 - 3| + |5 - 3|)/5 = \frac{4}{5}$. The absolute deviation is interpreted as the "average distance" from the center of a point selected by chance in the sample.

The variance and standard deviation are the most commonly used measures

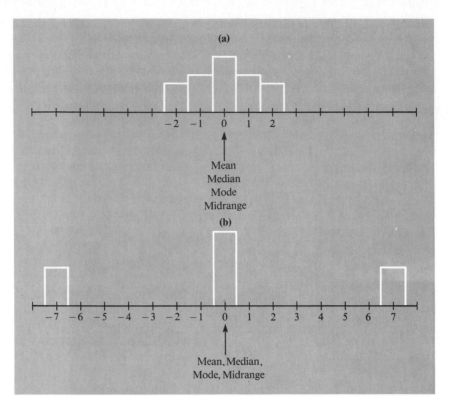

Figure 2 ▪ 8
Histograms with the same
mean, median, mode, and
midrange but different
amounts of dispersion.

of dispersion. These have essentially the same intuitive interpretation as the
absolute deviation, as being the average distance of the sample points from their
center. There are reasons for preferring the variance and standard deviation and
they shall be revealed as our study unfolds.

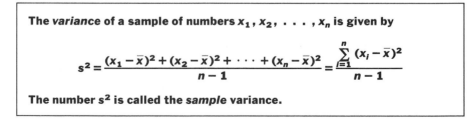

The *variance* of a sample of numbers $x_1, x_2, \ldots , x_n$ is given by

$$s^2 = \frac{(x_1 - \bar{x})^2 + (x_2 - \bar{x})^2 + \cdots + (x_n - \bar{x})^2}{n-1} = \frac{\sum\limits_{i=1}^{n}(x_i - \bar{x})^2}{n-1}$$

The number s^2 is called the *sample* variance.

Thus the sample variance is the average squared distances of the sample points
from their mean. We divide by $n - 1$ rather than by the more obvious number n

for reasons discussed later. For now we merely note that for *n* large there is very little difference between dividing by $n - 1$ and dividing by n. For smaller *n*, dividing by $n - 1$ will turn out to be preferable.

A disadvantage of variance as a measure of dispersion is that if the numbers in the sample are measured in units of feet, say, then the variance is in units of square feet (ft^2). By taking square roots, we eliminate this problem and arrive at the measure known as the standard deviation.

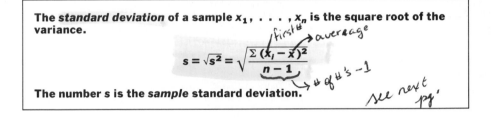

The **standard deviation** of a sample $x_1, \ldots, x_n$ is the square root of the variance.

$$s = \sqrt{s^2} = \sqrt{\frac{\Sigma (x_i - \bar{x})^2}{n - 1}}$$

The number *s* is the **sample** standard deviation.

By expanding $(x_i - \bar{x})^2$ and using $\bar{x} = \Sigma x_i / n$, we may collect terms to obtain the following shortcut formulas:*

Computational formulas for the sample variance and standard deviation:

$$s^2 = \frac{n \Sigma x_i^2 - (\Sigma x_i)^2}{n(n - 1)}$$

$$s = \sqrt{s^2} = \sqrt{\frac{n \Sigma x_i^2 - (\Sigma x_i)^2}{n(n - 1)}}$$

Example 2 ▪ 13   An entymologist studying the morphological variation in one species of mosquito recorded the following data on body length:

Body Length x (cm)	1.2	1.4	1.3	1.6	1.0	1.5	1.7	1.1	1.2	1.3	$\Sigma x_i = 13.3$
x^2 (cm²)	1.44	1.96	1.69	2.56	1.0	2.25	2.89	1.21	1.44	1.69	$\Sigma x_i^2 = 18.13$

* See problem 49 at the end of the chapter for the derivation of these results.

The mean body length is $\bar{x} = 13.3/10 = 1.33$ cm and the variance is calculated by either formula,

$$s^2 = \frac{(1.2 - 1.33)^2 + (1.4 - 1.33)^2 + (1.3 - 1.33)^2 + \cdot \cdot + (1.3 - 1.33)^2}{10 - 1}$$

$$= \frac{0.441}{9} = 0.049$$

or by the shortcut formula,

$$s^2 = \frac{10(18.13) - (13.3)^2}{10(9)} = 0.049$$

The sample standard deviation is

$$s = \sqrt{s^2} = \sqrt{0.049} = 0.221.$$

Grouped Data

When the data in the sample are grouped into intervals as in the frequency distribution of Table 2.3, we can still calculate the variance (and consequently, the standard deviation) by assuming all the observations in the ith interval fall exactly at the midpoint m_i of the interval as we did in calculating the mean by using grouped data. The formula then becomes

---

Calculation of the variance of a grouped sample of numbers:

$$s_g^2 = \frac{\Sigma \, (m_i - \bar{x}_g)^2 f_i}{n - 1} = \frac{n \, \Sigma \, m_i^2 f_i - (\Sigma m_i f_i)^2}{n(n - 1)}$$

where f_i is the frequency and m_i is the midpoint of the ith interval.

Example 2 ▪ 14   Calculating s_g^2 and s_g from the systolic blood pressure data in Table 2.3, we obtain

$$s_g^2 = \frac{1}{100(99)} [100((84.5)^2 \cdot 0 + (94.5)^2 \cdot 5 + \cdot \cdot \cdot + (174.5)^2 \cdot 2$$

$$+ (184.5)^2 \cdot 0) - ((84.5) \cdot 0 + (94.5) \cdot 5 + \cdot \cdot \cdot + (174.5) \cdot 2$$

$$+ (184.5) \cdot 0)^2]$$

$$= \frac{100(1649954.8) - (12720)^2}{100(99)} = 322.94$$

and

$$s_g = \sqrt{322.94} = 17.97$$

Variation in Populations

Finite Populations

The most common measures of the variability of a population distribution are the variance and standard deviation. The variance of a finite population is defined by analogy to the sample variance. It is a population parameter and is denoted by the symbol σ^2 (σ is the Greek letter sigma).

> The variance of a population:
> σ^2 = average squared distance from the center μ of the population $= \dfrac{1}{N} \sum\limits_{i=1}^{N} (x_i - \mu)^2$, where N is the number of units in the population.

The analogy to the sample variance is clear since the sample variance is essentially the average of the squared distances of the numbers in the sample from the center $\bar{x}$ of the sample. Notice that the value of the sample variance s^2 changes from sample to sample and is a statistic, whereas σ^2 remains the same and is a fixed parameter of the population. Similarly, we can define the standard deviation σ of the population to be the square root of the variance σ^2, that is

> The standard deviation of a population: $\sigma = \sqrt{\sigma^2}$ = square root of the average squared distance from the center μ.

Note that when defining the population variance σ^2, we divided by N, the *population* size, whereas when calculating the sample variance s^2, we divided by $n - 1$, where n is the *sample* size. We shall discuss this matter subsequently. For now we merely note that although the population variance is centered about a fixed number μ, the sample variance is centered about $\bar{x}$.

Example 2 ▪ 15   Returning to Example 2.12, we can calculate the population variance and standard deviation of the number of fertile eggs by using the 150 numbers constituting the population. Thus we have $\mu = 1.46$, so

$$\sigma^2 = \frac{1}{150} [(1 - 1.46)^2 + (0 - 1.46)^2 + \cdots + (2 - 1.46)^2]$$

$$= 1.66$$

and

$$\sigma = \sqrt{1.66} = 1.29$$

Another Version of the Formula for Population Variance

Consider a population consisting of the values 0.0, 0.1, 0.2, 0.3, 0.4, 0.5, 0.6, 0.7, 0.8, and 0.9 and suppose each of these values occurs 20 times. We have found that the mean of this population is

$$\mu = 0.0 \times (\text{relative frequency of } 0.0) + \cdots + 0.9$$

$$\times (\text{relative frequency of } 0.9)$$

$$= 0.0 \frac{1}{10} + 0.1 \frac{1}{10} + \cdots + 0.9 \frac{1}{10} = 0.45$$

The variance is

$$\sigma^2 = \frac{(0.0 - 0.45)^2 + (0.0 - 0.45)^2 + \cdots + (0.9 - 0.45)^2}{200}$$

However, combining terms in the numerator, we have

$$\sigma^2 = (0.0 - 0.45)^2 \frac{20}{200} + (0.1 - 0.45)^2 \frac{20}{200} + \cdots + (0.9 - 0.45)^2 \frac{20}{200}$$

$$= (0.0 - 0.45)^2 \times (\text{relative frequency of } 0.0) + \cdots$$

$$+ (0.9 - 0.45)^2 \times (\text{relative frequency of } 0.9)$$

$$= 0.0825$$

In general, another version of the formula for σ^2 is

Another formula for σ^2: Suppose a population takes on the value x_i with the relative frequency $P(x_i)$ and μ is the population mean. Then

$$\sigma^2 = (x_1 - \mu)^2 P(x_1) + (x_2 - \mu)^2 P(x_2) + \cdots + (x_n - \mu)^2 P(x_n)$$

Continuous Distributions

We shall find that the standard deviations of continuous distributions will also be of interest, but we shall not discuss their calculation in this text. Suffice it to say, the method of calculating these standard deviations can be motivated in the same way as the calculation of the mean of a continuous distribution. For all cases of interest the values of the standard deviation for continuous distributions will be given.

Example 2 ▪ 16   Figure 2.7 displays a continuous distribution that we use repeatedly in our work. The distribution is called the normal distribution with mean 0 and standard deviation 1 or the standard normal distribution. Notice that the distribution is symmetrical about 0; hence one half the population lies to the left of 0. Consequently, both its median and mode are 0.

2 ▪ 6   TECHNIQUES OF DATA PRESENTATION

The purpose of presenting data graphically, whether in the form of tables, figures, or charts, is to communicate the results of a study in a form that can be readily understood by even those who are unfamiliar with the study. A satisfactory presentation of the results is essential. The efforts of investigators and the expenditure of time and money are all wasted if the results of the study cannot be effectively communicated.

We now present a few of the techniques that are used in organizing statistical material that goes into graphs and tables.

Descriptions of Samples, Histograms, Relative Frequency, and Cumulative Frequency Polygons

Histograms

As we have seen, the histogram yields some information, such as the dispersion of data, that is not readily available from raw data. We see that the histogram is particularly useful for purposes of presentation when the sample size is large.

We shall employ the density histogram throughout this text, but there are also other types of histograms. For example, we could have used a vertical scale consisting of the frequencies (rather than the relative frequencies) in the construction of the histogram in Figure 2.1. The resulting picture would have looked the same as Figure 2.1, but with only the vertical scale changed. Another common practice is to multiply the relative frequencies by 100 to obtain a vertical scale going from 0 to 100. The vertical scale then consists of percentages, and again the resulting histogram looks identical to Figure 2.1. All these histogram constructions are used occasionally in studies.

Frequency Polygon

Another way to express the same data graphically is to use the *relative frequency polygon.* To make this graph, we utilize the modified frequency distribution given in Table 2.3, only this time we represent the relative frequency of an interval by a dot above the midpoint of that interval instead of by a line over the entire interval. As before, the midpoint of each interval is obtained by adding the upper and lower limits of the interval and dividing by 2. The height of the dot corresponds to the relative frequency of that interval. For example, the midpoint of the interval 89.5 to 99.5 is (89.5 + 99.5)/2 = 94.5, so we put a dot of height 0.05 above the point 94.5. The dots above the intervals are then connected with a straight edge to give a picture like Figure 2.9. Note that the lines are extended at either end of the polygon so that they touch the horizontal axis.

Cumulative Distribution Polygon

In some situations, one wants to know the proportion of the sample that falls below some value. For example, in the doctor's search for what constitutes "abnormal" or "normal" systolic blood pressure, he or she wants to know the proportion of individuals with systolic pressure below, say, 150 or 160. It would be desirable to graph this information so that it can be read visually. This is the purpose of the *cumulative distribution polygon* or *ogive.*

To construct the cumulative distribution polygon, we must add another column to the (modified) frequency distribution of Table 2.3 to obtain Table 2.4.

The cumulative relative frequency for an interval is obtained by adding (accumulating) all the relative frequencies before that particular interval and then adding the relative frequency of that interval. In practice, to calculate the entry for the *i*th interval, we take the cumulative relative frequency for the

Figure 2 ■ 9
Relative frequency polygon for systolic blood pressure of 100 males age 29 to 59. (Inclusion of the apparent class intervals instead of the real class intervals sometimes makes the horizontal scale easier to read. Such is the case here, where the ticks indicate the apparent interval limits).

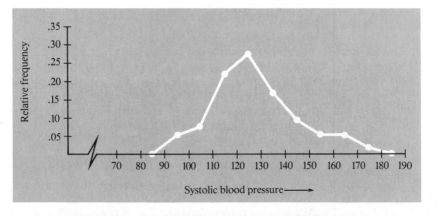

Table 2 ■ 4 Frequency Distribution of Systolic Blood Pressure with Intervals Modified for Construction of Histogram					
Apparent Interval Limits	*Real Interval Limits*	*Midpoint*	*Frequency*	*Relative Frequency*	*Cumulative Relative Frequency*
	79.5–89.5	84.5	0	0	0
90–99	89.5–99.5	94.5	5	5/100 = 0.05	0 + 0.05 = 0.05
100–109	99.5–109.5	104.5	8	8/100 = 0.08	0.05 + 0.08 = 0.13
110–119	109.5–119.5	114.5	22	22/100 = 0.22	0.13 + 0.22 = 0.35
120–129	119.5–129.5	124.5	27	27/100 = 0.27	0.35 + 0.27 = 0.62
130–139	129.5–139.5	134.5	17	17/100 = 0.17	0.62 + 0.17 = 0.79
140–149	139.5–149.5	144.5	9	9/100 = 0.09	0.79 + 0.09 = 0.88
150–159	149.5–159.5	154.5	5	5/100 = 0.05	0.88 + 0.05 = 0.93
160–169	159.5–169.5	164.5	5	5/100 = 0.05	0.93 + 0.05 = 0.98
170–179	169.5–179.5	174.5	2	2/100 = 0.02	0.98 + 0.02 = 1
	179.5–189.5	184.5	0	0	1 + 0 = 1

$(i - 1)$st interval, and add to that number the relative frequency of the ith interval. For example, to find the cumulative relative frequency for the blood pressure interval 119.5 to 129.5 in Table 2.4, we add the cumulative frequency for the previous interval 109.5 to 119.5 and the relative frequency for the interval 119.5 to 129.5. Thus the desired frequency is $0.35 + 0.27 = 0.62$.

The frequency distribution of Table 2.4 is usually simplified to just two columns as in Table 2.5. The resulting frequency table is called the *cumulative frequency distribution*. The first column consists of the right-hand end points of the real intervals in Table 2.4.

The process of graphing the cumulative distribution polygon from the cumulative frequency distribution is as follows. Put a dot over the number in the first column of Table 2.5 of a height equal to the cumulative relative frequency listed in the second column. Thus, for example, a dot of height 0.13 is put over the score 109.5. These dots are then connected with a straight edge to form the cumulative distribution polygon as shown in Figure 2.10.

Percentiles

From the cumulative distribution polygon we can obtain the various percentiles. The 20th percentile is that number with exactly $20\% = 0.20$ of the scores below it when the scores are ordered in increasing order; the 30th percentile is the number with $30\% = 0.3$ of the scores below it when the scores are ordered. In general, the *Pth percentile is that score with exactly P% of the scores below it when the scores are ordered from the smallest to largest*. These percentiles are easily read off the cumulative distribution polygon as follows. To find the *P*th percentile score, convert *P* to the decimal equivalent by moving the decimal

Table 2 ▪ 5 Frequency Distribution Needed to Construct Cumulative Distribution Polygon	
Point	Cumulative Relative Frequency
89.5	0
99.5	$0 + 0.05 = 0.05$
109.5	$0.05 + 0.08 = 0.13$
119.5	$0.13 + 0.22 = 0.35$
129.5	$0.35 + 0.27 = 0.62$
139.5	$0.62 + 0.17 = 0.79$
149.5	$0.79 + 0.09 = 0.88$
159.5	$0.88 + 0.05 = 0.93$
169.5	$0.93 + 0.05 = 0.98$
179.5	$0.98 + 0.02 = 1$
189.5	$1 + 0 = 1$

point two places to the left (20% = 0.2, 50% = 0.5, and so on). The decimal equivalent is looked up on the vertical cumulative relative frequency axis and a horizontal line is drawn from the vertical axis to the cumulative distribution polygon. From this point on the graph, drop down to the horizontal score axis and read the score at the point where the dropped line meets the horizontal axis. This score is called the Pth percentile. For example, the 50th percentile is obtained from the graph by reading over and down from the number 0.5 as

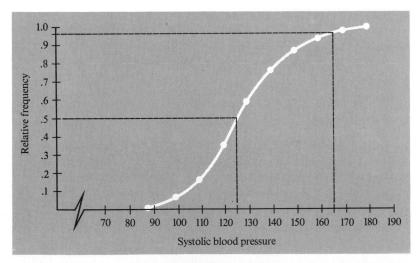

Figure 2 ▪ 10
Cumulative distribution polygon for systolic blood pressure of 100 males age 29 to 59. (For ease in reading the apparent interval limits are shown.)

shown in Figure 2.10. In this case the 50th percentile corresponds to approximately 123; that is, about 50% of the scores will be less than 123. Similarly, one reads the 20th percentile as approximately 115.

Percentile Rank

It is also of interest to determine to what percentile a given score corresponds. This is called the *percentile rank* of the given score and is calculated by retracing the steps previously outlined for finding percentile scores. Start on the horizontal score axis at the score for which you want to find the rank and read up until you hit the cumulative distribution polygon. Then read directly to the left to find the corresponding relative frequency. The percentile rank is the relative frequency converted to a percentage (multiplied by 100). For example, the score 165 has percentile rank of 96, whereas the score of 136 has a percentile rank of 73; that is 96% of the men in the sample have blood pressure scores that are less than or equal to 165 and 73% have scores less than or equal to 136. These percentile ranks may also be used to determine the percentage of people whose blood pressure falls between two scores a and b. This is done simply by subtracting the percentile rank of a from b. Thus the percentage of people with blood pressure between 136 and 165 is $96 - 73 = 23\%$, whereas the percentage of people with blood pressure scores between 123 and 165 is $96 - 50 = 46\%$, and between 123 and 136 the percentage is $73 - 50 = 23\%$.

Description of Populations

Cumulative Distribution

In some situations it is important to present in readily understandable form the proportion of the *population* that falls below some value. For example, in examining the size of eggs of chickens in a given population, the proportion of those eggs under a certain size may be indicative of the health of those chickens. In such instances first draw a graph so that this information can be read off visually. This is the purpose of the *cumulative population distribution.*

The cumulative distribution for an interval is obtained by adding (accumulating) all the areas under the population distribution for those intervals up to and including that particular interval. To graph the cumulative population distribution for the uniform distribution shown in Figure 2.3, place a dot over each population value; the height of the dot represents the accumulated area up to and including that area under the bar that contains the population value. For example, Table 2.6 shows the cumulative distribution for this population, whereas Figure 2.11 displays the distribution.

The cumulative distribution of continuous populations can also be graphed.

Table 2 ■ 6 Cumulative Distribution for Uniform Distribution of Numbers 0.0 through 0.9

x	Less than 0	0.0	0.1	0.2	0.3	0.4	0.5	0.6	0.7	0.8	0.9
Cumulative Probability up to x	0	0.1	0.2	0.3	0.4	0.5	0.6	0.7	0.8	0.9	1

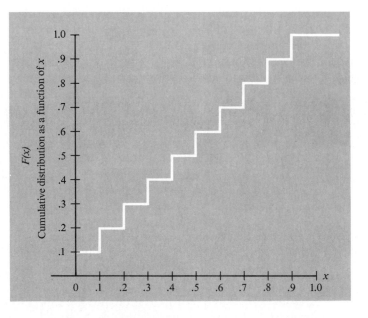

Figure 2 ■ 11
Cumulative distribution for the uniform distribution of numbers .0, .1, .2, .3, .4, .5, .6, .7, .8, .9.

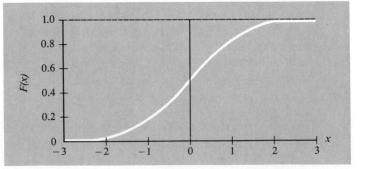

Figure 2 ■ 12
Cumulative distribution function of the normal distribution with mean = 0 and standard deviation = 1.

Figure 2.12 shows the cumulative distribution of the distribution shown in Figure 2.7, the normal distribution with mean of 0 and standard deviation of 1. The height of the cumulative distribution over any value represents the area under the normal distribution up to that value. For example, we see that the area up to 0 is 0.5 or 50%; up to 1.65, the area is about 95% of the total. These observations can be put in terms of population *percentiles,* a measure defined for population distributions in a way analogous to the definition used for percentiles of distributions of samples. Observations can also be described by population percentile ranks, which are defined analogously to sample percentile ranks. Population percentiles and percentile ranks are discussed in detail in Chapter 5.

2 ▪ 7   STATISTICAL COMPUTATION; CROSS-TABULATIONS

We have discussed several ways of summarizing data by using graphs and statistics such as $\bar{x}$ and s^2. As one can experience by doing the end-of-chapter exercises, burdensome organizational and computational effort is required to develop these summaries, even for moderate sample sizes. When the observations number in the thousands, the work may be impossible to perform manually. However use of electronic data processing computers makes these onerous tasks possible. Computers are especially efficient when the summarization methods are routine: Many techniques, such as the presentation of histograms, have become standard for displaying data. Program packages have been developed so that one needs only to organize and enter the data, with the proper instructions, into the computer in order to obtain the desired summaries quickly. Programs have even been devised for some of the microcomputers.

We discuss some of the program packages in this text and their applications to data analysis as well as to data presentation. In fact, the application of the computer to statistical analysis has become so extensive and profound that a new field, "statistical computation," has developed.

To begin, we discuss briefly an important type of data presentation table, the "cross-tab," which is easily generated by the computer.

The Cross-Tab

A type of table used frequently in social science and medical investigations is called a *cross-tabulation* (*cross-tab* for short). This table exhibits one variable against another. For example, the following tabulation shows the starting job (first variable) of all employees of a given chain of small grocery stores who were hired in 1976 versus their position in 1979 (second variable).

Cross Tabs.

		Present Job, 1979				
		Part-time Employee	Full-time Employee	Trainee	Assistant Manager	Manager
Job at Hire 1976	Part-time Employee	8	15	9	11	8
	Full-time Employee	12	26	14	14	12
	Trainee	0	11	1	18	20
	Assistant Manager	1	1	0	17	30
	Manager	1	1	0	4	10

This cross-tabulation indicates the progress employees made over time from various starting positions.

The following cross-tab shows the progress by race for all full-time employees who were hired in 1976.

		Present Job, 1979					
		Part-time Employee	Full-time Employee	Trainee	Assistant Manager	Manager	Total
Race of Full-time Employees Hired in 1976	White	8	19	9	10	6	52
	Black	4	7	5	4	6	26
	Total	12	26	14	14	12	78

Cross-tabs are popular because they are easily generated on the computer. Such computer packages as the "Statistical Package for the Social Sciences (SPSS)" (see references at the end of this chapter) facilitate the production of cross-tabs. They are popular for other reasons that will become apparent subsequently.

More complicated cross-tabs than those previously presented can also be developed. In fact, it is possible to make the two preceding cross-tabs into a *three*-variable cross-tab, showing by race the present job versus job at hire.

Present Job, 1979

		Part-time Employee		Full-time Employee		Trainee		Assistant Manager		Manager	
		W	B	W	B	W	B	W	B	W	B
Job at Hire 1976	*Part-time Employee*	5	3	9	6	5	4	5	6	4	4
	Full-time Employee	8	4	19	7	9	5	10	4	6	6
	Trainee	0	0	7	4	1	0	8	10	12	8
	Assistant Manager	1	0	1	0	0	0	9	8	17	13
	Manager	1	0	1	0	0	0	2	2	7	3

B = Black
W = White

The three-variable cross-tabs are more difficult to interpret than are several two-variable cross-tabs. Three-variable cross-tabs should only be used when there is a clear purpose. For example, if one is interested in comparing the promotions by race, the three-variable cross-tab can be confusing because the last line "Managers hired in 1976" shows *demotions,* not promotions. Of 11 whites and 5 blacks hired as managers in 1976, 4 whites and 2 blacks were demoted. It would be misleading to compare one row with other rows without noting which jobs were promotions and which demotions.

Summary

Distributions of Samples

In this chapter we have described the distributions of samples. By its very nature, most data contain a certain amount of variation, and we presented two types of methodologies for recognizing patterns in the data: graphical and numerical. As graphical techniques, we discussed the histogram, frequency polygon, and cumulative distribution polygon. The numerical techniques included measures of central tendency—the mean, median, mode, and midrange—and measures of dispersion—the range, absolute deviation, variance, and standard deviation. Computational formulas and grouped data formulas were given for the mean, variance, and standard deviation. We shall see that the mean, variance, and standard deviation play a very important role in the inferential statistics we discuss subsequently.

Population Distributions

A population distribution of a finite population can be regarded as a histogram (using a density scale) that shows the proportion of times each value occurs in the population. Each proportion is given by the area in the bar that lies above the value. Two properties of this distribution, those of central tendency and variation, can be described by measures that are analogous to those that describe histograms of samples. Central tendency can be depicted by such measures as the population mean, median, and mode. Variation can be described by the population variance and standard deviation. Measures that describe the characteristics of a population distribution are called *population parameters.* They are fixed values as contrasted to those measures that describe a sample and that can vary from sample to sample.

The distinction was made in this chapter between *finite* populations, those populations containing a fixed number of members, and *infinite* continuous populations, those populations that take on all values in a given interval. An important example of a distribution of a continuous population is the normal distribution. This distribution has many applications in this text.

Key Terms

The following basic terms were introduced in this chapter. Define these in your own words or by formulas.

Terms related to samples:

sample unit
sample value
sample size
relative frequency
    histogram
density scale
frequency distribution
frequency polygon
cumulative frequency
    polygon
percentile
percentile rank
variance
grouped data

apparent interval limits
real interval limits
mode
median
midrange
mean
dispersion
range
absolute deviation
standard deviation
skewed distribution
midpoint of an interval
frequency

Terms related to populations:

population distribution
population mean
population median
population mode
population variance
cumulative population
    distribution
population standard
    deviation

population parameter
finite population
continuous population
uniform distribution
normal distribution with
    mean 0 and standard
    deviation 1

References

Colton, T.: *Statistics in Medicine.* Boston, Little, Brown, 1974. This book has several interesting sections on the construction of graphs and tables, illustrated with medical data.

Huff, D.: *How to Lie with Statistics.* New York, Norton, 1954. Presents examples on how *not* to construct graphs and numerical characterizations of data. A very amusing short book.

Lutz, R. R.: *Graphic Presentation Simplified.* New York, Funk & Wagnalls, 1949.

Nie, N., et al.: *Statistical Package for the Social Sciences,* 2nd ed. New York, McGraw-Hill, 1975. This manual outlines computer programs for organizing data into histograms and tables, computing the values of statistics such as $\bar{x}$ and s^2, and performing data analyses, many of which we will learn in this text.

Tukey, J.: *Exploratory Data Analysis.* Reading, MA, Addison-Wesley, 1977. A marvelous book addressed to anyone who must try to make sense of or construct hypotheses from masses of data. Shortcut methods of graphically and numerically exploring and examining data are presented. This book is difficult for beginners but can be read with some effort.

Exercises

Section 2.1

1. *Variability of Data:* A tennis ball was dropped 10 times and the height to which it bounced each time is recorded in the following table:

Height of Bounce of Tennis Ball				

	Bounce Number			
	1	2	3	4	5
Height (inches)	10	9	11.5	12	10

	Bounce Number			
	6	7	8	9	10
Height (inches)	9.2	13	7.5	11	10.8

a. Give reasons why you do or do not find it surprising that there is variability in the data.

b. Would you expect to find more, less, or the same variability if you bounced 10 different tennis balls (produced by the same manufacturer) once instead of 1 tennis ball 10 times?

c. In part b suppose each of the 10 tennis balls had been produced by a different manufacturer, would you expect more, less, or the same variability in the results?

Sections 2.1 through 2.3

2. The following list is a sample of 50 hospital bills submitted to patients at a private hospital in New Orleans for 1 week of care (July 2 to July 9, 1979):

Hospital Bills for 1 Week ($)				
765.30	1070.10	990.20	1071.64	1261.89
1253.40	988.00	1041.81	1872.11	1282.91
624.28	1100.00	1362.11	1135.21	1867.58
1105.22	1736.34	1222.68	2080.29	1340.35
1065.36	1107.39	1289.73	1402.30	1071.02
1362.40	1250.56	1091.10	1028.73	1235.90
1115.10	1219.76	1100.00	1031.83	1212.11
1244.41	1190.30	1363.75	1279.36	1035.40
1275.35	920.11	1240.62	1250.00	1087.32
1580.27	1340.10	1220.00	1211.00	1191.01

a. Tally the numbers between 600 and 700, 700 and 800, 800 and 900, 900 and 1000, 1000 and 1100, 1100 and 1200, 1200 and 1300, 1300 and 1400, 1400 and 1500, and 1500 and 1600 in such a way that the first number is included in the interval but not the second number.

b. What useful information does this tally contain for an administrator of the hospital? Are there any numbers missing from this tally? If so, what would be the advantage of including them in the tally?

c. Calculate the relative frequency in each interval in part a. Into which interval does the number 1100 go?

d. Draw a relative frequency histogram for the preceding data.

e. What advantage is there in defining the final interval as all numbers that are larger than 1500 and less than or equal to 2100?

f. Choose intervals one half the size of those chosen in part a and redraw the histogram. Now choose intervals twice the size of those in part a and redraw the histogram. Which interval size do you feel presents the most information?

3. In order to ascertain the demand for an expansion of their maternity ward, a hospital administrator recorded the number of births per day in the hospital for 91 days. The following data were obtained:

Number of births per day	0	1	2	3	4	5	6	7	8	9	10	11	12	13	14	15
Number of days births observed	0	0	1	0	2	5	4	6	0	6	8	0	8	9	12	7

Number of births per day	16	17	18	19	20	21	22	23	24	25	26	27	28	29	30
Number of days births observed	0	6	4	0	0	4	0	3	0	2	2	0	1	0	1

a. Construct a frequency distribution for these data.

b. Draw a relative frequency histogram for these data.

c. Usually, about how many babies are born in a day at this hospital?

d. Suppose the hospital administrator decided to hire more staff in the maternity ward to accommodate the birth of 30 children per day. Would this be a sound decision in the light of the information provided by the histogram in part b?

4. A paint manufacturer wishes to determine how much area may be painted with a gallon of his paint. One hundred and fifty gallons were selected and the square footage covered by each gallon was as follows:

Coverage in sq ft	Number of times 1 gal of paint produced this coverage
400–450	15
450–500	19
500–550	37
550–600	41
600–650	26
650–700	12
	150

a. Draw a relative frequency histogram for these data.

b. Would the paint manufacturer be safe in assuring a customer that 3 gal of his paint would cover 1200 sq ft?

5. The Environmental Protection Agency must estimate the gasoline mileage of new cars. For one particular model the agency ran 25 tests and obtained the following data in miles per gallon (mi/gal):

31.3	28.6	35.1	39.1	29.3
27.6	29.9	38.6	36.1	28.1
40.1	34.2	31.7	32.7	40.8
37.2	41.2	34.3	35.6	42.1
33.4	44.3	37.0	30.0	36.1

a. Select intervals and group the data to obtain a frequency distribution.

b. Draw a relative frequency histogram for these data.

c. State two conclusions you could immediately draw from the histogram.

6. A manufacturer of metal bolts for aircraft construction must maintain close quality control on the breaking strengths of his product. The quality control analyst selects 25 bolts from the day's production and obtains the following frequency distribution:

Breaking strengths	Midpoint	Frequency
1550–1575	1562.5	1
1575–1600	1587.5	3
1600–1625	1612.5	5
1625–1650	1637.5	7
1650–1675	1662.5	7
1675–1700	1687.5	2
		25

To get a better idea for the variability of the production process, draw a relative frequency histogram.

Table 2 ▪ 7 Systolic Blood Pressure of 100 Male Patients by Age Groups 21–44 and 45–59

a. Age Group 21–44			b. Age Group 45–59	
100	110	107	160	150
130	138	117	105	125
120	124	121	120	118
138	122	130	168	115
110	126	119	151	129
110	120	121	130	90
115	130	132	120	152
134	142	118	140	120
120	110	126	110	128
122	128	117	106	94
110	120	98	140	134
120	124	115	165	174
115	110	123	137	115
162	119	141	107	165
130	132	129	136	138
130	125	140	174	126
110	131	120	108	158
147	117	96	149	99
122	112	141	158	
120	148	106		
131	108	114		

7. Table 2.7 displays the same systolic blood pressure data as those shown in Table 2.1 but divides the systolic blood pressures into two groupings; those belonging to male patients from 21 to 44 and 45 to 59 years of age.
   a. Draw relative frequency histograms of the data in each of parts a and b of Table 2.7. Use the same class intervals in both.
   b. Suppose 120 is approximately "normal" systolic blood pressure. From the histograms, which group is "closer" to the normal?
   c. Which group displays more variation in the data?
   d. What inferences could you make from these histograms concerning the relationship between age and systolic blood pressure?

8. A state hospital that provides free medical care to a segment of a city requested data from the U.S. Census Bureau on the distribution of the ages of the 20,000 people living in its catchment area. The bureau provided the following information, obtained by taking a sample from its information on the population:

Age	0–5	6–13	14–17	18–20	21–24	25–29
Frequency	920	1170	740	450	660	670

Age	30–34	35–39	40–44	45–55	55–65	65–75	75+
Frequency	620	680	610	1230	1130	980	100

   a. Draw a relative frequency histogram of these data.
   b. Based upon the data displayed in this histogram, would you conclude that 45 to 55 is the largest age group?
   c. Draw a density histogram. Which age group is the largest, that is, has the highest bar over its interval in this histogram?

9. A jar contains 30 tags, each of which has one of the numbers 1, 2, 3, 4, 5, 6 written on it. Draw the population distribution when
   a. each number appears on five tags

b. the number 2 appears on 25 tags, whereas each of the other numbers appears only on 1 tag

c. the numbers 1 and 6 each appear on 10% of the tags, whereas the remaining numbers each appear on 20% of the tags.

d. the numbers 2 and 5 each appear on 30% of the tags, whereas each of the other numbers appears on 10% of the tags.

10. Examine Table 2.8 and Figure 2.13.

a. Comment on any age differences you notice in the population of the two zip code neighborhoods.

b. Is it easier to detect differences in the age distribution of the populations from Figure 2.13 than from Table 2.8?

c. Given age differences shown in the histograms; does this information suggest which population may be more affluent? State why. (This type of information is useful for businesses that are choosing where to locate).

d. In a density histogram the area of the bar located on a class interval is equal to the proportion of the population in that interval. Using the densities in Figure 2.13, find the percentage of people in each zip code area who are more than 44 years of age.

Table 2 ■ 8 Age Distribution in Zip Codes 70005 and 70037 in New Orleans, LA, 031970 U.S. Census		
Age	No. of People in Zip 70005	No. of People in Zip 70037
0–5	725	274
6–13	1159	427
14–17	583	187
18–20	355	111
21–24	520	141
25–34	1001	292
35–44	915	256
45–54	1048	161
55–64	891	101
65–94	702	40
Total	7899	1990

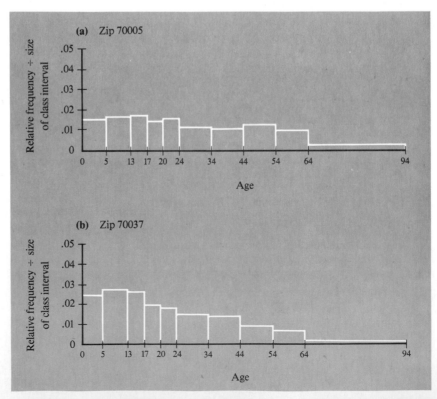

Figure 2 ■ 13
Age histogram using density scale for zip codes 70005 and 70037 in New Orleans, Louisiana, 1970 U.S. Census. (*Note:* The class limits are shown along the horizontal axes. The class boundaries, shown in Table 2.8 as 0 to 5, 6 to 13, and so on, are not shown in these histograms.)

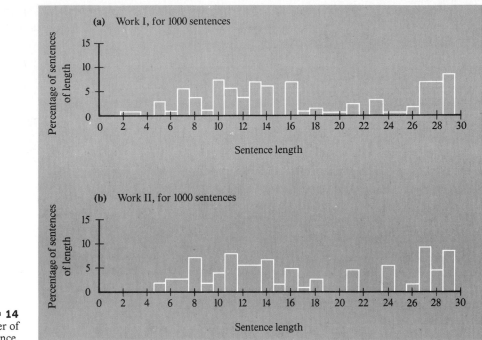

Figure 2 ▪ 14
Histogram of number of
words per sentence.

11. Literary detective work has been the fascination of many for centuries. A great deal of study has been expended in trying to discover whether or not Francis Bacon authored Shakespeare's works. Even more effort has gone into trying to determine if one or several people authored the Old Testament. Recently, much investigation has attempted to uncover the authorship of the Federalist Papers. One technique utilized in these studies is to compile the distribution of the number of words used in sentences by one author and to compare this distribution with that of an unknown author. Figure 2.14 shows the distribution of sentence length from two works. Based upon this evidence, would you say the same person authored both works?

12. John Q. is accused of typing a blackmail letter on a typewriter in his boss' office. The letter is 25 pages long. John Q. claims the secretary typed the letter on that typewriter. As evidence at his trial, his attorney shows the distribution of typing errors John Q. has made on all pages of his typing, taken from old files (Fig. 2.15a), the distribution of typing errors made by the secretary, also taken from old files (Fig. 2.15b), and the distribution of typing errors on the blackmail note (Fig. 2.15c). What is your conclusion concerning the guilt of John Q?

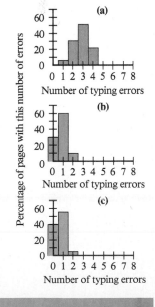

Figure 2 ▪ 15
(a) Histogram of number of John Q.'s typing errors per page (150 pages). (b) Histogram of number of typing errors per page made by secretary (300 pages). (c) Distribution of typing errors on blackmail note (25 pages).

13. You make an appointment with a friend to meet on the corner of Gauss and Euclid streets at noon. You know that your friend will show up with equal likelihood anywhere between noon and 1 PM.
    a. Draw the population distribution.
    b. What is the chance that your friend shows up between noon and 12:30 PM?

14. Figure 2.16a shows a spinner in which a pointer is mounted on a pivot. The angle the pointer makes with the horizontal line as shown in the figure is measured when the pointer comes to rest. This is the type of spinner or wheel used in carnivals for which you win a certain prize (or nothing), depending on where the pointer comes to rest. (Sometimes the wheel is spun and the pointer is at rest but the principle is the same.) It is hoped that the spinner or wheel is fair; that is, any angle between 0° and 360° is equally likely.
    a. Assuming that the wheel is fair, find the proportion of the population that lies between 90 and 180°.
    b. (For discussion) Suppose you win a teddy bear (worth $1) if the angle of the pointer when it comes to rest is between 210 and 220° and a wristwatch (worth $10) if the angle is between 221 and 222°; otherwise you lose the 50¢ you paid to spin the pointer. Determine the proportion of times you will win a teddy bear and a wristwatch and you will lose. Is it worthwhile playing the game?

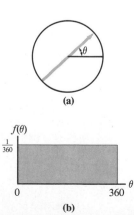

Figure 2 ▪ 16
(a) Spinner generating a uniform distribution. (b) Uniform distribution generated by spinner.

Section 2.4

15. Using the data in problem 3,
    a. calculate the sample mean $\bar{x}$
    b. calculate the sample median
    c. calculate the sample mode
    d. calculate the sample variance and standard deviation
    e. answer problem 3c again
    f. answer 3d again, using the information you now have about the value of the standard deviation.

16. Using the data from problem 5, calculate parts a, b, and c of problem 15 and state two conclusions you can immediately draw from these numbers concerning the gasoline mileage of new cars.

17. Using the data from problem 4, calculate parts a, b, and c of problem 15.

18. Using the data of problem 7,
    a. calculate the sample mean and variance for male patients from 21 to 44 years of age
    b. calculate the sample mean and variance for male patients from 45 to 59 years of age,
    c. and from the results in parts a and b, what inferences could you make concerning the relationship between age and systolic blood pressure?

19. A production process produces ball bearings that are either defective or good. A sample of 15 ball bearings is taken from a day's output of the production line. If a ball bearing is good, we assign the number "0" to it; if it is defective, we assign the number "1." Let x_i represent the value assigned to the ith ball bearing in the sample. For example, if the first ball bearing examined is good, then $x_1 = 0$. The sample of 15 yields

$$x_1 = 0, \qquad x_2 = 0, \qquad x_3 = 1,$$
$$x_4 = 0, \qquad x_5 = 0, \qquad x_6 = 0,$$
$$x_7 = 0, \qquad x_8 = 1, \qquad x_9 = 0,$$
$$x_{10} = 0, \qquad x_{11} = 0, \qquad x_{12} = 1,$$
$$x_{13} = 1, \qquad x_{14} = 1, \qquad x_{15} = 0$$

Calculate $\bar{x}$ from this sample. What estimate would you make of the proportion of defectives produced by this process?

Practice with the addition sign, Σ

20. Find the numerical values of the following summations using the numbers $x_1 = 1$, $x_2 = 7$, $x_3 = 4$, $x_4 = 2$, $x_5 = 7$, $x_6 = 2$, $x_7 = 3$, $x_8 = 10$, $x_9 = 1$, $x_{10} = 4$.

   a. $\displaystyle\sum_{i=1}^{10} x_i$ d. $\displaystyle\sum_{i=1}^{5} (2x_i + 3)$

   b. $\displaystyle\sum_{i=5}^{10} x_i$ e. $\displaystyle\sum_{i=2}^{5} x_i$

   c. $\displaystyle\sum_{i=1}^{10} x_i^2$ f. $\displaystyle\sum_{i=3}^{9} x_{i-1}$

21. Verify the following operations on summations using the numbers in problem 20 and $y_1 = 3$, $y_2 = 5$, $y_3 = 4$, $y_4 = 7$, $y_5 = 1$, $y_6 = 8$, $y_7 = 10$, $y_8 = 1$, $y_9 = 1$, and $y_{10} = 0$.

   a. $\displaystyle\sum_{i=1}^{10} ax_i = a\sum_{i=1}^{10} x_i$ for any number a

   b. $\displaystyle\sum_{i=1}^{10} (x_i + a) = \sum_{i=1}^{10} x_i + a$ for any number a

   c. $\displaystyle\sum_{i=1}^{10} (x_i + y_i) = \sum_{i=1}^{10} x_i + \sum_{i=1}^{10} y_i$

22. Show that the mean, median, and mode of the population are, respectively, in problem
   9a: 3.5, 3.5, and the mode does not exist
   9b: 2.3, 2, and 2
   9c: 3.5, 3.5, whereas 2, 3, 4, and 5 are all modes
   9d: 3.5, 3.5, whereas 2 and 5 are both modes
      A population that has two modes is called *bimodal.*

23. Calculate the mean of each part in problem 9 by multiplying each number by the proportion of times it occurs (proportion = % of times the number occurs ÷ 100) and then summing all the results.

24. Jane Eyre and her four friends, Tom, Bill, Sue, and June, have attended a dance every Friday night for 5 years. They usually wait by the punch bowl for about 30 min before someone summons the courage to ask one of them to dance. They have kept records for the past 5 years as to how often each of them was the first to be asked to dance. Each was asked first exactly 52 times.
   a. Describe the population about which this information was collected.
   b. Draw a population histogram.
   c. Is each of these five people equally likely to be chosen first to dance on a Friday night?
   d. Suppose Jane had been asked to dance first 104 times, whereas each of the others had been asked 39 times. What is the relative frequency with which each has been asked to dance?
   e. (Optional) Suppose Jane is 5'9" tall, whereas the other girls are each 5'5" tall. Using the data in part d, what is the average height of the girl who is chosen first to dance? In your society, would this information tell you something about the average height of males at the dances?

25. Is it possible that the sample average $\bar{x}$ is not equal to the population average μ? Is it possible that the sample median is not equal to the population median? Explain your answers and give examples.

26. What is the difference between a population parameter and a statistic?

27. (Optional) If the data are symmetrical about $\bar{x}$, does the sample mode necessarily equal $\bar{x}$?

28. a. Table 2.9 shows the family income distribution of each of zip codes 70005 and 70037 in New Orleans, Louisiana. Draw a population distribution of each zip code and determine if the resulting presentation of data verifies your inference from the data in Table 2.8, problem 10.

   b. At the foot of Table 2.9 we find the average income of the residents of zip code area 70005. Is the number the value of the population average (or mean) or is it the value of a sample mean $\bar{x}$?

c. True or false? The data shown in Table 2.9 are data on the *entire 1970 population* of each zip code and are not a subsample of this population.

Table 2 ▪ 9 Family Income Distribution for Zip Codes 70005 and 70037, New Orleans, LA, 1970 U.S. Census

Family Income ($)	No. of People in Zip 70005	No. of People in Zip 70037
Less than 2,000	212	153
2,000– 2,999	209	36
3,000– 4,999	512	89
5,000– 6,999	772	356
7,000– 7,999	443	173
8,000– 9,999	1,047	324
10,000–11,999	1,149	320
12,000–14,999	1,300	248
15,000–19,999	1,041	218
20,000–24,999	460	54
25,000–49,999	562	33
50,000 or more	175	0
Median ($)	11,297	9,203
Average ($)	13,874	9,920

Section 2.5

29. a. For each of the data sets in problems 3, 4, and 5, calculate the sample variance and sample standard deviation.

b. With respect to the data in problem 5, now that you know $\bar{x}$ and s, state two conclusions you can immediately draw from these numbers concerning the gasoline mileage of new cars.

30. An economist wishes to determine trends in the stock market. To do this, she compares the average price per share of 10 selected stocks from month to month. For 1 month she observes the following prices: $15.75, $17.00, $6.25, $7.25, $47.00, $32.00, $21.50, $19.00, $40.00, and $12.88. What is the average price per share for this month? How much variation in price is there?

31. A biologist investigating airborne bacteria places a petri plate containing glucose under a ventilation duct. The number of bacteria colonies that appear per square millimeter is counted, and the following data are obtained:

X = No. of colonies per cell	0	1	2	3	4	5
No. of cells		5	7	10	6	2	1

Calculate the average number of colonies per cell and the standard deviation.

32. A test given to a psychology class of 100 yields the following distribution of grades. What is the mean grade of the class? What is the variance?

Grade	60	65	67	70	75
No. of students obtaining grade	1	3	7	14	13

Grade	77	80	85	90	95	97
No. of students obtaining grade	18	20	8	10	2	4

33. Consider the following data:

1.0	3.5	7.0	11.5	8.5
3.0	14.5	12.0	10.0	18.0
16.5	11.0	18.5	15.4	11.0
12.5	7.5	19.5	8.0	2.5
9.0	15.0	9.5	8.0	16.0
4.0	10.4	7.5	2.0	13.0
5.0	9.2	6.5	7.5	15.0

a. Construct a frequency distribution for the data using the apparent limits 1 to 5, 6 to 10, 11 to 15, and 16 to 20.

b. Calculate a modified frequency distribution similar to that in Table 2.3.

c. Use the grouped data formulas and the frequency distribution from (b) to calculate $\bar{x}_g$ and s_g^2.

d. Use the raw data and the raw data formulas to calculate $\bar{x}$ and s^2.

e. Compare the results in parts c and d.

34. a. Find the mean and standard deviation of each histogram shown in Figure 2.15a and b.
    b. Knowing the population parameters in Figure 2.15a and b and the sample mean and standard deviation of the sample histogram in Figure 2.15c, what is your opinion concerning the guilt of John Q?

35. a. Calculate the mean of the population consisting of the numbers 2.5, 3, 3.5, 4, and 4.5, where each number occurs the same proportion of times as every other number; that is, the numbers are *equally likely.*
    b. Show that the mean of the population described in problem 9a is the same as that in problem 35a.
    c. Find the standard deviation of the population in problem 35a.
    d. Which population has greater variation about its mean, that in problem 9a or 35a?

36. There are 200 first year students; each of whom must buy $50 worth of books; each of the 150 sophomore students must buy $80 worth of books; and each of the 100 juniors and 75 seniors must buy $90 worth of books. What is the average cost of books per student in this school? What is the variance of this cost? the standard deviation?

37. a. A population consists of 10,000 tags marked with a 0 and 8500 tags marked with a 1. Find μ and σ.
    b. Fifty-four percent of an infinite population are "tagged" with a 0 and the remainder with a 1. Find μ and σ.

38. Is it possible that the sample standard deviation s is not equal to the population standard deviation σ? Explain your answer and give examples.

Section 2.6

39. For the data of problem 3
    a. draw a histogram using a *density scale*
    b. draw a relative frequency polygon for the data

c. draw a cumulative distribution polygon and estimate the 75th percentile number of births per day. What is the percentile rank corresponding to 13 births per day?
    d. State one conclusion that can be drawn from the cumulative distribution that cannot be drawn directly from a knowledge of the sample mean $\bar{x}$ and the sample variance s^2 of the data.

40. a. For the data of problem 4 draw a relative frequency polygon.
    b. For the data of problem 4 draw a cumulative distribution polygon and estimate the 50th percentile and the percentile rank of a gallon that covers 640 sq ft.
    c. If you were the paint manufacturer, which characteristics of your paint as observed from the information in part b would you advertise?

41. a. For the data of problem 6 draw a cumulative frequency polygon.
    b. For the data of problem 6 estimate the 25th, 50th, and 75th percentiles.
    c. The manufacturer would like to know if the breaking strength is at least 1560. Using the cumulative frequency polygon, estimate the percentile rank of 1560.

42. A city manager wishes to ascertain the monthly electricity bill of a certain segment of the population so the manager polls 40 homeowners and finds the following bills (in dollars):

83.27	96.30	79.17	69.43
76.43	75.72	86.20	47.16
44.73	61.97	74.31	55.25
59.64	53.75	69.10	71.00
102.17	97.10	67.30	63.21
91.12	110.10	61.72	57.17
80.23	57.30	93.40	49.90
67.51	61.73	79.63	76.10
72.62	71.76	84.21	73.60
61.72	83.91	83.01	70.05

a. Calculate a relative frequency distribution for these bills after grouping them into intervals.

b. Draw a histogram and a cumulative frequency polygon.

c. What is the 50th percentile as read from the cumulative frequency polygon?

d. Find the sample mean $\bar{x}$ and sample variance s^2 of these data.

43. A questionnaire given by a psychologist determines the degree of mental stress exhibited by an individual. The verbal answer to each question is assigned the numerical value 1, 2, 3, or 4, depending on the amount of stress present in the response. A sample of 50 people had a distribution of scores as follows:

Score interval	20–25	25–30	30–35
No. of people	3	4	18

Score interval	35–40	40–45	45–50	50–55
No. of people	16	6	1	2

a. Draw a relative frequency histogram for these data using a density scale.

b. Draw a cumulative frequency polygon and a relative frequency polygon based upon these data.

c. Find the grouped sample mean and variance.

44. A consumer group checked grocery prices in 100 stores in a particular state to see how much variation there was in the price of a basketful of groceries. They found the data listed in the following distribution:

Price range	18.00–18.49	18.50–18.99	19.00–19.49	19.50–19.99
No. of prices	2	4	6	18
Price range	20.00–20.49	20.50–20.99	21.00–21.49	21.50–21.99
No. of prices	31	24	9	6

a. Draw a relative frequency histogram and one using a density scale, to illustrate the variation in prices.

b. Combine the first three and last two intervals and repeat part (a). Which histogram best illustrates the variation in prices?

c. Draw a cumulative distribution polygon and use it to estimate the 50th percentile.

d. Find the grouped sample mean and variance.

45. a. Describe the type of skewness found in each of the histograms in Figure 2.17.

b. In each case, which numerical value or values best describes the central tendency of the histogram; the mean, median, or mode?

c. Can you tell at a glance which histogram has the smallest dispersion? the largest?

d. Calculate the grouped standard deviation for each histogram.

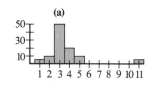

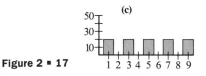

Figure 2 ▪ 17

46. Draw the cumulative distribution for the population distribution shown in problem 13a.

47. The cumulative distribution of the normal distribution with $\mu = 0$, $\sigma = 1$ is perhaps the most often used population cumulative distribution in all statistics and is known as the normal cumulative distribution. Use the following table (taken from Appendix A.1) to do parts a, b, and c.

a. Construct a cumulative distribution.

b. Note that only some of the z values in Appendix A.1 were used in the preceding table: in fact, only values from -3.00 on that differ by 0.25. Redo part a using z from -3.00 on that differ by 0.1.

c. Observe that the cumulative distribution you obtain in part b is "smoother" than that obtained in part a. If you continue to take z values that differ by smaller and smaller amounts, eventually you will obtain a cumulative distribution that looks

Value z	Cumulative proportion of z
-3.00	0.001
-2.75	0.003
-2.50	0.006
-2.00	0.023
-1.75	0.090
-1.50	0.067
-1.25	0.106
-1.00	0.159
-0.75	0.227
-0.50	0.309
-0.25	0.401
0.00	0.500
0.25	0.599
0.75	0.773
1.00	0.841
1.50	0.933
1.75	0.960
2.00	0.977
2.25	0.988
2.75	0.997
3.00	0.999

like Figure 2.12. Find the percentile ranks of $-1, 0, 1, 2, 3$, using Figure 2.12.

48. a. Table 2.10 is developed from Table 2.8. Complete the table.

b. Figure 2.18 is a plot of the cumulative percent in zip code 70005 versus the cumulative percent in zip code 70037 age group from Table 2.10. (This type of graph is called a quantile-quantile plot or Q-Q plot.) The first coordinate of the points on the Q-Q curve are the percentiles for the zip code 70037. The second coordinate is derived from the first coordinate as follows. If the first coordinate is

Table 2 ▪ 10 Percent Populations of Zip Codes 70005 and 70037 in Various Age Groups

Age	Percent in Zip 70005	Cumulative % in Zip 70005	Percent in Zip 70037	Cumulative % in Zip 70037
0–5	9.2	9.2	13.6	13.6
6–13	14.7	23.9	21.2	34.8
14–17	7.4	31.3	9.3	44.1
18–20	4.5		5.5	
21–24	6.6		7.0	
25–34	12.7		14.5	
35–44	11.6		12.7	
45–54	13.3		8.0	
55–64	11.3		5.0	
65+	8.9		3.2	

$P\%$, we find the P percentile score for the 70037 zip code area. Using this number, we then ask for the percentile rank of the number in the 70005 zip code area. This is the second coordinate. Thus the Q-Q plot for two distributions is the curve $[P, q(P)]$, where $q(P)$ is the percentile rank, using the second distribution of the pth percentile

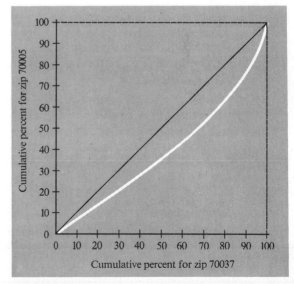

Figure 2 ▪ 18
Q-Q plot of zip code 70005 versus zip code 70037 in New Orleans, Louisiana, 1970 U.S. Census.

score obtained by using the first distribution. For example, age 5 corresponds to the 13.6 percentile in the first (70037 zip code) distribution, and 5 has percentile rank 9.2 in the second (70005 zip code) distribution. The point (13.6, 9.2) is the first point on the Q-Q curve.

If the age distributions were the same for the two populations, should the curve in the figure be a straight line? If so, what important information about these two zip code areas does the Q-Q plot demonstrate?

49. (Extra problem) Derivation of the computational formula for the sample variance. The sample variance s^2 is given by the formula

$$s^2 = \frac{\sum_{i=1}^{n} (x_i - \bar{x})^2}{n - 1}$$

Now

$$\sum_{i=1}^{n} (x_i - \bar{x})^2 = \sum_{i=1}^{n} (x_i^2 - 2x_i\bar{x} + \bar{x}^2)$$

$$= \sum_{n=1}^{n} x_i^2 - \sum_{n=1}^{n} 2x_i\bar{x} + \sum_{i=1}^{n} \bar{x}^2$$

a. Justify each step.

$$\sum_{i=1}^{n} 2x_i\bar{x} = 2\bar{x} \sum_{i=1}^{n} x_i = 2n\bar{x}\bar{x} = 2n(\bar{x})^2$$

b. Justify each step.

$$\sum_{i=1}^{n} (\bar{x})^2 = n(\bar{x})^2 = \frac{1}{n}\left(\sum_{i=1}^{n} x_i\right)^2$$

c. Justify this step. From parts a and b,

$$\sum_{i=1}^{n} (x_i - \bar{x})^2 = \sum_{i=1}^{n} x_i^2 - \frac{1}{n}\left(\sum_{i=1}^{n} x_i\right)^2$$

d. From c, take the last steps to obtain the computational formulas for s^2 and s.

The **Additive Rule** (mutally exclusive)

$$P(A \cup B) = P(A) + P(B)$$

or $P(A \cup B) = P(A) + P(B) - \boxed{P(A \cap B)}$

INDependent Proof

$$P(A \cap B) = P(A)P(B) \quad \checkmark \text{ independant.}$$

one won't efect prob of the if the other

$$P[A/B] = P[A] \qquad P[B/A] = P(B)$$

↳ Subract what is common.

more than 2 events (mutvally exclusive)

$$P(A \cup B \cup C) = P(A) + P(B) + P(C)$$

$$P(A \cup B \cup C) = P(A) + P(B) + P(C) - \left[P(A \cap B) + P(A \cap C) + P(B \cap C)\right] + P\left[A \cap B \cap C\right]$$

CONDITIONAL PROBABILITY

$$P[A/B] = \frac{P[A \cap B]}{P[B]}$$

Multiplicative Rule

$$P(A \cap B) = P(A) \cdot P(B) \quad \text{in dependant}$$

$$P(A \cap B) = P(A) \cdot P(B/A) \quad \text{general}$$

$$\left. \begin{array}{l} P[A/B] = P[A] \\ P[B/A] = P[B] \end{array} \right\} \text{independant if } P[A \cap B] = P(A) \cdot P(B)$$

CHAPTER · 3

Probability Models

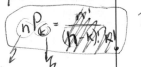

The origins of the theory of probability are to be found at the gaming tables of the Middle Ages, but this theory finds numerous applications in almost every aspect of our technological society. It is not possible to understand statistical concepts without understanding basic probability theory. We first discuss the interpretation of probability and introduce rules for the calculation of probabilities. These rules are applied to various probability models, which are approximations to actual situations and which satisfy a need or point of view, just as a road map "approximates" the nature of a given area to satisfy the needs of drivers for guidance. Finally, we show how probability and statistics are related.

Handwritten annotations:

order is not important
permutation

$$_nP_k = \frac{n!}{(n-k)! \, k!} \quad ?$$

total in group

how many groups you want.

= more

$$_nP_k = \frac{n!}{(1-k)!}$$ ✓

total · # of selections

Combination

$_nC_k$ — total / # you are selecting from total.

$$\binom{10}{3} = \frac{10!}{3! \, 7!} = \text{less}$$

eg. $\dfrac{10!}{(10-3)! \, 3!}$

(# of ways to selected red balls) (# of ways to select n-a black balls)

perm   comb

$f(x) = P[Y = x]$
prob of Funct of

$x = \binom{n}{x} p^x q^{n-x}$

Combo.

3 ■ 1 ORIGINS OF PROBABILITY

A good part of the study of statistics is devoted to the problem of analyzing samples to obtain information concerning the population from which the samples were drawn. We are about to begin those studies in probability theory that connect the information in samples with the characteristics of populations. It is this knowledge that enables us to study a sample, make statistical inferences about the population from which it comes, and test the validity of these inferences.

The origins of the theory of probability are found in the history of gambling and other games of chance. In fact, it is still possible to utilize probability theory to win substantial amounts of money at the gambling tables (witness Ed Thorp in his book, *Beat the Dealer,* listed in the References of this chapter). Nevertheless, most people view gambling with superstition. The gambler relies on the rabbit's foot, magical invocation, "lucky day," and so on. Probability theory has helped to dispel much of this superstition and to develop groundwork for the rational analysis of gambling games as well as serious applications in the social, behavioral, and physical sciences.

The first tenuous links between gambling and the "probability" we will study were established by Gerolamo Cardano (1501 – 1576), a romantic charlatan of many professions, including medicine and astrology. It is not surprising that a man who was engaged in these two professions in those days was also interested in luck, since success in both areas depended a good deal on chance. The occult was not separated far from medicine and certainly was intimately tied to astrology, as well as to darker activities. Cardano was accused of being a professional murderer. Perhaps the accusation was true, or perhaps it came from those who envied his success at the gambling tables.

Cardano analyzed the game of "dice" and was the first to formulate in a vague and primitive way the model of "equally likely" outcomes that we discuss later. It was his astute formulation and clever calculations of the odds that led to his success in gambling and the possible envy of his competitors. He was lucky to have escaped with the accusation of murder rather than the more serious one, at least in that era, of witchcraft.

The link established by Cardano between "chance" and mathematics was forged into an unbreakable chain by Blaise Pascal, the famous French mathematician and philosopher. His work was prompted by a problem posed to him by his friend, the Chevalier de Méré, an aristocratic player of the Parisian casinos (see part g of problem 19). This problem initiated, through the writings of Pascal and Fermat, the modern theory of probability.

3 ■ 2 INTERPRETATION OF PROBABILITY

The idea underlying the word "probability" is that of uncertainty. The concept is so broad that it is necessary to distinguish among various kinds of uncertainty. To avoid being sidetracked from our main theme, however, we shall briefly discuss only two kinds of expressions of uncertainty. These can be typified by the statements:

1. If I sail from east to west across the Atlantic Ocean, I shall probably fall off the edge of the earth.
2. The probability of "heads" in coin tossing is $\frac{1}{2}$.

Statements 1 and 2 both express uncertainty about the events A and B, respectively, where

A = voyage east to west across the Atlantic resulting in a

plunge off the side of the earth's surface

B = the occurrence of heads in coin tossing

Underlying statement A is the belief that the world is flat and that those sailing across it will fall off its edge. The word "probably" is used to indicate that the belief is not held with absolute conviction. Indeed, after Columbus' voyage many an ancient who had previously expressed the belief in statement 1 may well have forcefully asserted that voyagers need never again fear the consequences expressed in A.

The characteristic that distinguishes events A and B is that the question of whether or not A will occur can be cleared up once and for all simply by sailing across the Atlantic, whereas the question of the next outcome of a coin toss can never finally be settled. Once the coin is tossed and the outcome decided, the outcome of the next toss is still uncertain. To rephrase the distinction, one can think of a *sequence* of events such as B. On the other hand, one cannot consider a sequence of events such as A, for once we have traversed the Atlantic, no doubt about future crossings remains.

To understand why this difference is important, we must determine what we mean by "an event has a 50-50 chance of occurring," or "the probability of an event is $\frac{1}{2}$." In the case of coin tossing, this statement means that in a large number of tosses, the event heads should occur in "approximately" one half of the tosses. If the probability of heads were $\frac{2}{3}$, then we interpret this statement to mean that heads is more likely than tails in the sense that approximately two thirds of the tosses should be heads in a large number of tosses. Statement 2 does not concern the outcome of one particular trial but is a statement about the outcomes of a *sequence* of trials. This interpretation of the number called a probability as the relative frequency of an event in a long sequence of trials is called the *frequency interpretation*.

Frequency Interpretation of Probability: The probability of an event is the relative frequency (proportion of times) with which the event occurs in a long series of repeated experiences or trials.

Certainly we cannot interpret the probability statement 1 in the same way. To consider the number 70% as that approximate proportion of a large number of voyages in which we may expect to fall off the edge of the earth is clearly nonsense. As another example, consider the statement "the probability that the moon has another side is $\frac{2}{3}$." This statement cannot mean that in a long series of trials in which we travel to the other side of the moon, approximately two thirds of the time the moon has another side. We need only travel to the moon once to determine whether or not it has another side. Once determined, the uncertainty disappears. Thus the frequency interpretation of statements such as 1 makes no sense. How then are we to interpret probability statement 1? This probability represents the degree of belief in the occurrence of an event. This type of probability is called "subjective probability." We cannot go into this matter here, although there are several introductory references on the subject. For example, see Kyburg and Smokler listed in the references for a collection of interesting essays on the subject and an extensive bibliography.

3 ■ 3 MODEL OF EQUALLY LIKELY EVENTS FOR CALCULATING PROBABILITIES

Now that we have developed an interpretation of the "probability of an event," we can begin to develop our first probability model.

We use the symbol $P(A)$ to represent the phrase "the probability of the event A." For example, if we let H represent the event "the outcome of the coin toss is heads," then the symbols $P(H)$ represent the phrase "the probability of the outcome 'heads' on the coin toss." We state that the probability of heads for a "fair" coin is $\frac{1}{2}$ and write this statement as

$$P(H) = \frac{1}{2},$$

which reads "the probability that the outcome on the coin toss is heads equals $\frac{1}{2}$." This probability was calculated by noting that the total number of outcomes (denoted $n_{Tot.}$) was 2, whereas the number of outcomes that yield heads (denoted n_H) was 1. Since we assumed that both outcomes were equally likely, we concluded that

$$P(H) = \frac{\text{number of outcomes that yield heads}}{\text{total number of outcomes}} = \frac{n_H}{n_{Tot.}} = \frac{1}{2}$$

This formula works as well for the probability of any event we wish to calculate under the model that all events are equally likely.

If A is an event and n_A is the number of outcomes in which A occurs then

$$P(A) = \frac{\text{number of outcomes that yield } A}{\text{total number of outcomes}} = \frac{n_A}{n_{\text{Tot.}}}$$

Example 3 ▪ 1   Suppose a chemist is hired to determine the gold content of a certain coinage. He is given three coins to test; one has 50% gold; one, 75% gold; and one, 99% gold. What is the probability that the coin he chooses to assay has 99% gold? Let A denote the event "he selects the coin with 99% gold." Then $n_A = 1$ and $n_{\text{Tot.}} = 3$, so

$$P(A) = \frac{n_A}{n_{\text{Tot.}}} = \frac{1}{3}$$

Since calculating the probability of different events occurring using the model of equally likely outcomes involves calculating $n_{\text{Tot.}}$ and n_A, we must first be able to enumerate the set of all possible outcomes that are equally likely. Sometimes this is easy, but not so at other times. One way to do this is simply to write them all down.

Example 3 ▪ 2   Consider all possible outcomes of a throw of two dice, one red and one green. Each die has six sides. One possible outcome of a toss of the dice is "3 on the red die, 1 on the green" as pictured here:

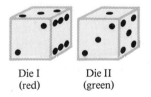

Die I Die II
(red) (green)

We may represent this outcome in more convenient notation as (3, 1), where we list in parenthesis the numbers of dots on the faces that are up. The 36 possible outcomes of a toss of the dice are listed in Figure 3.1 as the set S: The first member of each pair represents a possible outcome of the red die; the second represents a possible outcome of the green die. To illustrate this notation, we use the symbol (5, 2) to represent a "5" on the red die and a "2" on the green die. The event E that "a 7 or 11 occurs" is composed of the outcomes (6, 1), (1, 6), (5, 2), (2, 5), (4, 3), (3, 4), (5, 6), (6, 5). If any of the outcomes on this list occurs on a toss of the dice, then the event "7 or 11" occurs. Thus the event E is composed on 8 outcomes ($n_E = 8$). There are 36 total possible outcomes listed in Figure 3.1 ($n_{\text{Tot.}} = 36$). Thus

$$P(E) = \frac{8}{36} = \frac{2}{9}$$

$$\left\{ \begin{array}{cccccc} (1,1) & (1,2) & (1,3) & (1,4) & (1,5) & (1,6) \\ (2,1) & (2,2) & (2,3) & (2,4) & (2,5) & (2,6) \\ (3,1) & (3,2) & (3,3) & (3,4) & (3,5) & (3,6) \\ (4,1) & (4,2) & (4,3) & (4,4) & (4,5) & (4,6) \\ (5,1) & (5,2) & (5,3) & (5,4) & (5,5) & (5,6) \\ (6,1) & (6,2) & (6,3) & (6,4) & (6,5) & (6,6) \end{array} \right\}$$

Figure 3 ■ 1
The set S of 36 possible outcomes of a toss of two dice.

Example 3 ■ 3 Consider the event B, which is "an even number appears on the red die and the number 3 appears on the green die." The event B consists of the following outcomes: $(2,3),(4,3),(6,3)$. There are 36 possible outcomes ($n_{Tot.} = 36$) while $n_B = 3$. Thus

$$P(B) = \frac{3}{36} = \frac{1}{12}$$

The title of this section is "Model of Equally Likely Events for Calculating Probabilities." To understand the significance of this heading, let us reconsider our example of coin tossing. We listed the possible outcomes as "heads" or "tails." However, we know that other outcomes are possible: the coin may roll away and disappear down a drain, or it may stand on end. Nevertheless, in enumerating the possible outcomes, most of us would list only heads and tails. The other outcomes, we feel, do not occur frequently enough to influence the applicability of any theory built upon this idealization; namely, that there are only two possible outcomes of a toss of a coin. This is the idea behind a *mathematical model.* It abstracts the pertinent features of the occurrences and ignores the others. *The model describes only those aspects of the physical situation we consider pertinent and important.* It is hoped that the set we have abstracted from reality is a sufficiently good approximation to the physical situation that conclusions we draw concerning the "nature" of the set can be applied to physical problems.

As an analogy, consider a good road map. Such a map does not contain all the features of an area; it abstracts from that area certain pertinent features—the location of roads, types of roads, distances between points, and so on—and ignores others, such as location of trees, hills, and so on. These characteristics are presented in symbolic (graphic) form. A road map is essentially a "symbolic" (as opposed to a physical) model of an area in which only information useful for decision making is presented. Similarly, the set S (Fig. 3.1) of possible outcomes of a throw of a pair of dice is a symbolic representation of the physical situation under study.

As a further example, consider the set of outcomes of a single toss of a die. We can represent the possible outcomes as

However, this description of the outcomes gives too much irrelevant information (e.g., the shape of the die, the geometric dot pattern, the numerical relationships between adjacent sides, and so on). If we are only interested in the total number of dots that are forward, we can give a symbolic representation of the set of possible outcomes as

$$R = \{1, 2, 3, 4, 5, 6\}$$

Like the map analogy, this set contains what we consider to be the relevant information for our purposes.

Since we shall work often with the set of possible outcomes, we give it a special name. The pertinent definitions are as follows:

> The set of all possible outcomes is called the *sample space.* The individual members of the sample space are called elementary or simple events. Events that are combinations of several elementary events are called compound events.

In this section we have developed a mathematical model for some situations (coin tossing, dice throwing, and so on); this model is composed of two assumptions:

Model of Equally Likely Outcomes

Assumption 1: The list or set of $n_{Tot.}$ possible outcomes that are of interest contains all pertinent information about the outcomes.

Assumption 2: The probability of an event E is equal to

$$P(E) = \frac{\text{number of ways event } E \text{ can occur}}{\text{total number of possible outcomes}} = \frac{n_E}{n_{Tot.}}$$

Example 3 ■ 4   An obstetric nurse wishes to calculate the probability of obtaining at least one boy in a set of twin babies, assuming the applicability of the model of equally likely outcomes. The nurse might argue as follows: A male is born either first, second, or not at all, so the set of possible outcomes is listed as

$$S = (\text{male first, male second, no male at all})$$

The probability calculated for the event A, that a male child is born, might be thought to be $P(A) = (n_A/n_{\text{Tot.}}) = \frac{2}{3}$, *but this is wrong!* The problem here is that Assumption 1 in the model of equally likely outcomes is violated; thus, the events "male first," "male second," and "no male at all" are not equally likely. We enumerated the set incorrectly. To determine how to calculate $P(A)$ better, let (x_1, x_2) denote the birth of twins, where x_1 is the sex of the first child born and x_2 is the sex of the second child. The set of possible outcomes now may be listed as

$$S' = \{(M, F), (M,M), (F, F), (F, M)\}$$

where M stands for male and F stands for female. Assuming all outcomes in S' are equally likely, and noting that A is the event consisting of the three outcomes, $\{(M, F), (M, M), (F, M)\}$, we find that $P(A) = n_A/n_{\text{Tot}} = \frac{3}{4}$. The problem with the previous set S of possible outcomes is that it does not adequately account for the possibility of (M,M), a "male first *and* second."

The purpose of this example is to warn you of a common pitfall. When using the model of equally likely outcomes, you must count *all* possible outcomes. A common error is not to count all outcomes correctly. Sometimes this is difficult, especially if there are a large number of outcomes. For this reason efficient counting techniques have been developed. We have included a discussion of this subject at the end of this chapter.

One final comment: We have assigned probabilities to events by counting the components of the events. There is no reason to expect at this point that the assignment of probabilities in this way has anything to do with the long-run relative frequency of the event. We shall show the connection later in the chapter.

3 ■ 4   GENERAL MODELS FOR PROBABILITIES

The model of equally likely outcomes is the classic model used when there are only finitely many possible outcomes. However, it cannot be used if there are infinitely many possible outcomes, or if the probabilities of the elementary events are not all equal. Then we must resort to another model. Nevertheless, all probability models have something in common; namely, they all satisfy the general rules that we outline here. First, we need to discuss some nomenclature used to describe events.

The Language of Events

Let S denote the sample space, and let A and B be two events. The event "A or B" denotes the collection of elementary events that are in either A or B, or in both. For example, let S be the 36 outcomes of a dice throw shown in Figure 3.1, and let A be the event "the two faces show the same number" and B represent the event that "the sum of the faces is 3 or less." Figure 3.2 shows both events A and B. A consists of the six elementary events shown, whereas B consists of the three elementary events as shown. The event "A or B" consists of eight elementary events: namely, (1, 1), (2, 2), (3, 3), (4, 4), (5, 5), (6, 6), (1, 2), (2, 1). These eight elementary events constitute the event "two same faces *or* the sum of the faces is 3 or less." If any one of these elementary events occurs on a throw of dice, then either A or B or both occur. For example, if (6, 6) occurs, then A occurs; if (1, 2) occurs, then B occurs; if (1, 1) occurs, then both A and B occur.

We also introduce the notation "A and B" to denote the set of elementary events that are in both A and B. The notation "A and B" indicates that A and B occur together, that there is a joint occurrence of A and B. In the previous example, "A and B" consists of the elementary event (1, 1). If (1, 1) occurs, then both the events "two same faces" and "the sum of the faces is less than 3" occur. The event "A and B" is shown in Figure 3.2.

We also define the concept of two *mutually exclusive* events: A and B are *mutually exclusive* if they cannot occur together; that is, if they have no elementary events in common. The events "obtaining a 3 or less" on a dice throw and "obtaining a 4 or more" on a dice throw are mutually exclusive events. If one of them occurs, the other cannot.

Rules of Probability

We use the model of equally likely outcomes to motivate the general rules for assigning probability values to events. Recall that in this model the probability of the event A is $n_A/n_{\text{Tot.}}$, where n_A is the number of elementary events that

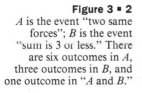

Figure 3 ▪ 2
A is the event "two same forces"; B is the event "sum is 3 or less." There are six outcomes in A, three outcomes in B, and one outcome in "A and B."

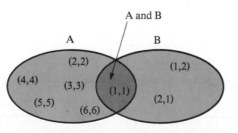

constitute A and $n_{\text{Tot.}}$ is the total number of possible elementary events in the sample space S. The probability value $n_A/n_{\text{Tot.}}$ is a number that cannot be less than 0 or greater than 1 since n_A is positive and always less than or equal to $n_{\text{Tot.}}$. This observation gives us our first rule.

Rule (i): The probability of an event cannot be less than 0 or greater than 1.

We note that the probability of the sample space must be 1 since

$$P(S) = \frac{n_{\text{Tot.}}}{n_{\text{Tot.}}} = 1$$

It is only reasonable that $P(S) = 1$ since S is the event "any outcome." Certainly the chance of "any outcome" should be 1 since it is the *certain* event.

Rule (ii): The probability of obtaining an outcome from the sample space is 1

$$P(S) = 1$$

Suppose the event A has n_A elementary events, B has n_B elementary events, whereas the event "A and B" has $n_{A \text{ and } B}$ events in common (see Fig. 3.2). In the figure, $n_A = 6$ and $n_B = 3$ while $n_{A \text{ and } B} = 1$. Hence the number in the event "A or B" is 8 or $6 + 3 - 1 = n_A + n_B - n_{A \text{ and } B}$. The numbers n_A and n_B already include the $n_{A \text{ and } B}$ elementary events of "A and B" in their count so that $n_A + n_B$ counts these elements twice. Thus we must adjust for this over counting. Hence $n_{A \text{ or } B} = n_A + n_B - n_{A \text{ and } B}$. Dividing by $n_{\text{Tot.}}$ yields

$$P(A \text{ or } B) = \frac{n_{A \text{ or } B}}{n_{\text{Tot.}}} = \frac{n_A}{n_{\text{Tot.}}} + \frac{n_B}{n_{\text{Tot.}}} - \frac{n_{A \text{ and } B}}{n_{\text{Tot.}}}$$

$$= P(A) + P(B) - P(A \text{ and } B)$$

Addition rule
Rule (iiia):

$$P(A \text{ or } B) = P(A) + P(B) - P(A \text{ and } B)$$

If A and B are mutually exclusive events, then $n_{A \, \text{and} \, B} = 0$. In this special case

> Rule (iiib): If A and B are mutually exclusive, then
>
> $$P(A \text{ or } B) = P(A) + P(B)$$

In summary, we have
A probability assignment for S is a rule that satisfies

i. The value assigned to the probability of an event is between 0 and 1; that is, $0 \le P(A) \le 1$ for any event A.
ii. The probability of "any outcome" is 1; that is, $P(S) = 1$.
iii. Addition rule:
   a. $P(A \text{ or } B) = P(A) + P(B) - P(A \text{ and } B)$.

   Special case of the Addition rule:
   b. $P(A \text{ or } B) = P(A) + P(B)$ if A, B are mutually exclusive.

Example 3 ▪ 5 When a certain illness strikes a family of five people, the number of family members who contract the disease is of interest for epidemiological reasons. The sample space is $S = \{0, 1, 2, 3, 4, 5\}$ since anywhere from 0 to 5 members could obtain the disease. Due to close contact between family members, these events are not all equally likely. Suppose it is determined that

$$P(\{0\}) = 0.168, \qquad P(\{1\}) = 0.36, \qquad P(\{2\}) = 0.309,$$
$$P(\{3\}) = 0.132, \qquad P(\{4\}) = 0.028, \qquad P(\{5\}) - 0.003.$$

This assignment of values defines a probability assignment rule over S, as can be seen by verifying rules (i), (ii), and (iii). Clearly the elementary events are not equally likely. The benefit of defining general rules for calculating probability is that we no longer need restrict our attention to equally likely elementary events.

Using rule (iii), we can obtain additional general rules for calculating probabilities. For example, let A' denote the event "A does not occur." That is, A'

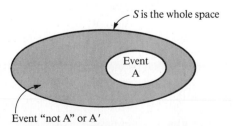

Figure 3 ▪ 3
Either A or A' occurring means the whole space S occurs. Hence probability of S = probability of A or A'.

consists of those elementary events that are *not* in A. We call A' the complement of A. Then, since A' and A have no common elements (Fig. 3.3), $P(A$ and $A') = 0$; however, since every elementary event is in either A or A', $P(A$ or $A') = P(S) = 1$. Hence $P(A$ or $A') = 1$. Since A and A' are mutually exclusive, we have by (iii) that $P(A$ or $A') = P(A) + P(A')$. Since $P(A$ or $A') = 1$, we have the following rule:

> If A is any event and A' is the event "A does not occur," then
>
> $$P(A) + P(A') = 1$$

Example 3 ▪ 6   A certain mineral compound is extracted from a mine and the probability it will contain copper, iron, and silver is 0.5, 0.3, and 0.05, respectively. Let these events be denoted by A_1, A_2, and A_3, respectively. The event $A_1' = \{$does not contain copper$\}$ has probability $1 - P(A_1) = 1 - 0.5 = 0.5$, the event $A_2' = \{$does not contain iron$\}$ has probability $1 - P(A_2) = 1 - 0.3 = 0.7$, and the event that the sample contains no silver has probability $P(A_3') = 1 - P(A_3) = 1 - 0.05 = 0.95$.

3 ▪ 5   CONDITIONAL PROBABILITY AND INDEPENDENCE

Often when trying to calculate the probability that a particular event will occur, we have some knowledge about auxiliary events having occurred that we can use in our calculation. For example, suppose a physician wishes to diagnose whether or not a particular patient has coronary heart disease. If she knows that the patient smokes, she would assess this probability differently than if she did not know he smoked. In fact, this change in probability assessment is really the purpose of taking a medical history from a patient.

To show how to incorporate this information into the calculation of the desired probability, let A and B be two events. Suppose you want to calculate the probability that event A will occur, and you are given the fact that the event B has occurred. Once you know that B has occurred, you can eliminate from consideration any elementary event that does not fall into B. In essence, B becomes your new sample space or set of possible outcomes. The only way A can occur now is in conjunction with B (since you know B occurred). Using the model of equally likely outcomes, we would arrive at the conditional probability of A given B has occurred as

$$\frac{\text{Number of ways } A \text{ can occur within } B}{\text{Number of possible outcomes in } B} = \frac{n_{A \text{ and } B}}{n_B} = \frac{n_{A \text{ and } B}/n_{\text{Tot.}}}{n_B/n_{\text{Tot.}}}$$

$$= \frac{P(A \text{ and } B)}{P(B)}$$

where n_B, of course, must be greater than 0. In the general (i.e., in the not necessarily equally likely) situation the same formula holds. For ease of writing we let the symbol $P(A \mid B)$ denote the conditional probability of the event A occurring given B has occurred.

The conditional probability of A given B:

$$P(A \mid B) = \frac{P(A \text{ and } B)}{P(B)}$$

where $P(B)$ is greater than 0.

Another useful way to write this formula is

$$P(A \text{ and } B) = P(A \mid B)P(B)$$

which allows us to calculate the probability of the joint occurrence of two events in terms of the conditional and unconditional probabilities.

Example 3 ▪ 7   The Austin Police Department wanted to stake out certain convenience stores that they suspected would be robbed during the first 15 days of November. The problem confronted by the Crime Analysis Unit was to determine during which hours, and on which days, a robbery was most likely to occur, then to assess the probability of catching the robber(s) if the police staked out these stores. They chose 10 locations based upon past robbery records and trends for the current year and decided to stake them out for 18 hr per week. Again resorting to past records, the crime unit decided to use the hours 7 PM to 4 AM, Friday through Monday.

  Let B denote the event "at least 1 of the 10 stores on the list is robbed during the surveillance period." From past records it is determined that $P(B) = 0.6$. Since there was not enough money to stake out every location on the list, each of the 10 stores was assessed for its probability of being hit. For example, one of the stores was a Seven Eleven store on First Street, which accounted for 35% of the robberies at the 10 target stores. Let A denote the event "the Seven Eleven store on First Street is robbed." Then we are given $P(A \mid B) = 0.35$ and $P(B)$ and we wish to calculate the probability that a stake out at this store will be successful. This is $P(A \text{ and } B) = P(A \mid B)P(B) = (0.35)(0.6) = 0.21$, a 21% chance.

This analysis was performed for each of the 10 stores and the top 5 were selected for stake outs. The results were significant. An average of two robbers were caught during each 15-day period by using this method of determining stake out locations, whereas previously the same number of manhours was committed to stake outs but with no success.

Example 3 ▪ 8 A certain American Indian artifact is discovered in Mexico. The probability of discovering such an artifact in Mexico is 0.2. Given that the artifact was found in Mexico, the probability it belongs to a certain tribe is 0.7. If an archeologist is trying to find an artifact from this particular tribe, what is the probability it will be found in Mexico? (If the probability is high, it might be worthwhile to make a trip to Mexico.) Let A denote the event "found an artifact from this tribe" and B denote the event "artifact was found in Mexico." The desired probability is $P(A \text{ and } B) = P(A \mid B)P(B) = (0.7)(0.2) = 0.14$.

Example 3 ▪ 9 The quality control department of a clock manufacturer uses the following technique. As parts come off the assembly line they are visually inspected and defective parts are removed. The probability that a defective part sneaks through this first visual inspection is 0.3. At the warehouse, each part is again scrutinized, this time more closely. The probability a defective part is not detected in this second inspection is 0.05. What is the probability that a defective part will make it through these two inspections? This probability can be calculated as follows: $P(\text{gets through 1 and 2}) = P(\text{gets through 2} \mid \text{gets through 1})P(\text{gets through 1}) = (0.05)(0.3) = 0.015$.

If 20% of the parts initially produced by the assembly line are defective, what is the proportion of defectives that will be shipped? For this we calculate

$$P(\text{defective shipped}) = P(\text{defective and gets through both inspections})$$
$$= P(\text{defective}) \, P(\text{gets through both inspections} \mid \text{defective})$$
$$= (0.2)(0.015) = 0.003$$

If an auxiliary event B actually has no influence on the probability of A occurring, then we call A and B independent. Put another way, A and B are said to be independent if you assess the probability of A being the same number whether or not you knew B had occurred.

Two events A and B are independent if

$$P(A \mid B) = P(A) \qquad \text{or} \qquad P(B \mid A) = P(B)$$

For calculating the joint occurrence of two independent events, we can use the preceding definition to obtain

---

Product Rule for Two Events
If A and B are two events, then

$$P(A \text{ and } B) = P(A \mid B)P(B) \qquad \text{or} \qquad P(A \text{ and } B) = P(B \mid A)P(A)$$

However, if A and B are independent, then

$$P(A \text{ and } B) = P(A)P(B)$$

Example 3 ▪ 10   A coin is tossed four times. Let H represent the outcome "heads" and T the outcome "tails." Then $HHTT$ represents the outcome "heads on the first two tosses and tails on the last two." There are 16 equally likely possible outcomes $HHHH, HHHT, HHTH, HTHH, THHH, HHTT, HTHT, HTTH$, and so on. What is the probability of the event $A =$ "exactly two heads" if it is known that event B occurred; $B =$ "at least one head appears on an even trial"? We must calculate

$$P(A \mid B) = \frac{P(A \text{ and } B)}{P(B)}.$$

Let us first calculate $P(B')$, where B' is the event "B does not occur." Thus B' occurs if a tail occurs on the second and fourth trial. That is, if one of the following outcomes occurs: $TTTT, HTTT, TTHT, HTHT$. Hence $P(B') = \frac{4}{16}$ and $P(B) = 1 - P(B') = \frac{12}{16} = \frac{3}{4}$.

[Here is a lesson to be learned. When considering the calculation of the probability of an event B, first try to determine if it is easier to calculate $P(B')$.]

The event A and B occurs if one of the following events occurs: $HHTT$, $THHT, THTH, HTTH, TTHH$. Hence

$$P(A \text{ and } B) = \frac{5}{16}$$

and

$$P(A \mid B) = \frac{5}{16} \div \frac{3}{4} = \frac{5}{12}$$

Notice that $P(A) = \frac{6}{16}$, so that $P(A \mid B) \neq P(A)$, and hence A is *not* independent of the information B.

Example 3 ■ 11   Consider set S of equally likely outcomes of a dice throw (Fig. 3.1). Let A be the event "1 or 2 on first die" and B the event "2, 3, or 4 on first die." Then

$$P(A) = \frac{12}{36}, \qquad P(B) = \frac{18}{36}, \qquad P(A \text{ and } B) = \frac{6}{36}$$

Hence

$$P(A \mid B) = \frac{6/36}{18/36} = \frac{6}{18}$$

which is equal to $P(A) = \frac{12}{36} = \frac{6}{18}$. Hence A is independent of B.

3 ■ 6   MUTUAL INDEPENDENCE AND THE PRODUCT RULE

From the preceding discussion we see that if two events are independent, then $P(A \text{ and } B) = P(A)P(B)$. Suppose also the events A and B are independent of C, and suppose also that the joint event "A and B," is independent of C, then $P(A$ and B and $C) = P(A$ and $B \mid C) \, P(C) = P(A$ and $B) \, P(C) = P(A)P(B)P(C)$. If this is so, we can show that "A and C" is independent of B and "B and C" is independent of A. This leads us to state an important definition and result.

Mutually Independent Events: The n events $A_1, A_2, \ldots, A_n$ are said to be *mutually independent* if every joint event *such as* "A_1 and A_6 and A_n" is independent of any other distinct joint event such as "A_5 and A_7 and A_{10} and A_{n-1}"

The underlying concept behind mutually independent events is that information about the occurrence of any joint event does not change the probability of any other joint event.

Product Rule: If $A_1, A_2, \ldots, A_n$ are mutually independent, then

$$P(A_1 \text{ and } A_2 \text{ and } \ldots \text{ and } A_n) = P(A_1)P(A_2) \ldots P(A_n)$$

Application: *People* vs. *Collins* (Supreme Court of California, 1968, 68 Cal. 2d 319, 66 Cal. Rptr. 497, 438 P. 2d 33.)

A jury found a black man Collins and his white wife guilty of second-degree robbery in San Pedro, California. Collins had a beard and mustache, his wife had blond hair tied in a ponytail. They owned a partly yellow automobile.

At the 7-day trial the prosecution experienced some difficulty in establishing the identities of the perpetrators of the crime. The victim could not identify Mrs. Collins and had never seen Mr. Collins. Another witness was able only to testify that he saw a white girl with a blond ponytail get into a partially yellow getaway car with a bearded black man.

In an attempt to add credibility to the identifications, the prosecutor called an instructor of mathematics at a state college. The prosecutor assigned probabilities as follows:

Characteristic	*Individual Probability*
A. Partly yellow automobile	1/10
B. Man with mustache	1/4
C. Girl with ponytail	1/10
D. Girl with blond hair	1/3
E. Negro man with beard	1/10
F. Interracial couple in car	1/1000

The prosecutor asked the instructor to calculate the probability of A and B and C and . . . and F using the product rule. Applying the rule, the instructor arrived at the probability that there was 1 chance in 12 million that any couple possessed the distinctive characteristics of the defendants. The prosecutor concluded that there could be but 1 chance in 12 million that the defendants were innocent and that another equally distinctive couple committed the robbery. The jury drew the same conclusion and convicted the couple.

The verdict was reversed. The appeals court correctly noted three errors in the previous reasoning.

1. No data had been presented to substantiate the individual probabilities used.
2. More important from our standpoint was the fact that no proof was presented that the characteristics selected were *mutually independent* "even though the witness himself acknowledged that *this condition was essential to the proper application of the product rule.*" To the extent that the traits or characteristics were not mutually independent (e.g. Negroes with beards and men with mustaches obviously represent nonindependent events), the "product rule" would yield an erroneous result even if all the individual components had been determined with precision.

3. The entire calculation was irrelevant to the determination of the guilt of the couple. To understand this point, we let

> B = robbery was committed by a bearded black man and blond white woman with a ponytail, who have a partly yellow car.
>
> A = the Collins couple is guilty.

The pertinent probability is $P(A \mid B)$.

Assuming all couples described in B are equally likely to have committed the robbery, the chance the Collins couple is the guilty one *given* that the description in B is accurate is

$$P(A \mid B) = \frac{1}{n_B}$$

where n_B is the number of couples in the population B. The appeals court made a calculation (which we repeat in the next chapter) to show that $P(A \mid B)$ could be less than 0.8; that is, the probability of the Collins couple *not* performing the robbery even given that the description of the guilty couple was correct could be greater than 20%. The court found this probability too high to constitute a conviction "beyond a reasonable doubt" and so reversed the decision.

3 ▪ 7   BRIDGES TO STATISTICS FROM PROBABILITY

Our primary aim is to study statistics. It is important then to understand how probability theory aids us in this study. As we go along, we shall discover links between probability and statistics. In this section we discuss two of them, but first emphasize the difference between probability and statistics.

Contrast between Probability and Statistics

Suppose we are playing a game of dice. With knowledge of the physical situation or "state of nature" (i.e., the construction of the dice, the possible outcomes), our task is to construct a model that will permit us to predict outcomes. Thus, we might assume the "Model of Equally Likely Outcomes" and the sample space S of Figure 3.1 so that the probability of an outcome such as "a 1 on each face" is $\frac{1}{36}$. Probability theory deals with the determination of such probabilities based on models of the state of nature.

On the other hand, suppose we go to a gambling casino and are not permitted to watch the game. All we are told is that two objects with numbers on the faces are being tossed; we are also told the outcomes of each toss and are asked to try to determine the numbers on the faces of each object. This is a typical problem in statistics; namely, to try to reconstruct the state of nature from the given information or data.

To summarize, the role of probability theory in applications is to construct models and predict outcomes of experiments. In "statistical theory" one major problem is to construct the physical situation from the outcomes of experiments, that is, using the outcomes, try to infer which model was correct.

Law of Large Numbers; Odds

An important bridge that connects the concept of probability with statistical applications is the Law of Large Numbers. This law is commonly referred to as the "law of averages" and is one of those mysterious popular scientific or mathematical phrases that hold an important place in the lexicon of gamblers, sports announcers, stockmarket brokers, newscasters, soldiers in combat, and the average person aware of, but not schooled in, the "fickleness of chance." This "law" can be observed in action by tossing a balanced coin a large number of times. To illustrate, we actually tossed a coin 30 times and recorded the following results:

$$H, H, H, T, T, H, H, T, T, T, H, T, H, H, T, T, H, T, H, H, T, T, T, T, H,$$

$$T, H, H, H, T$$

where H = heads and T = tails. Let n_H/n represent the relative frequency of heads; where n_H represents the number of heads in n trials. Figure 3.4 shows the value of n_H/n as n increases, that is, as we tossed the coin more and more often.

As we can see, n_H/n does appear to approach the probability of heads (namely, $\frac{1}{2}$) as n increases. The reader might perform this experiment in order to obtain some experience in this matter.

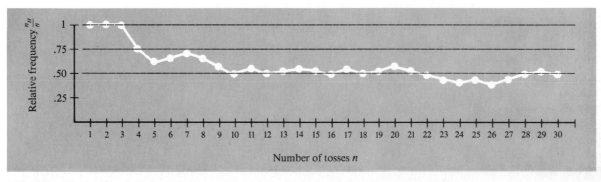

Figure 3 ■ 4
Relative frequency of the outcome "heads" vs. "the number of tosses" in a specific set of 30 tosses.

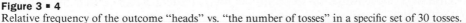

We know that there exist unfair coins: those coins that are weighted so that heads, say, has a higher probability than tails. However, whatever this probability is, the relative frequency of heads in an actual sequence of coin tosses will approach this probability. We can demonstrate this statement by simulating an unfair coin by using a die; a die has six sides, each side has given a number from 1 to 6, indicated by the number of dots on the face of the side. If the die is fair, each side is equally likely to face up on a throw of the die; that is to say, each of the six numbers is equally likely to face up or occur on a throw. Hence each number has the probability of occurring of $\frac{1}{6}$. Let us say "heads occurs if a '1' occurs on a toss of the die." In this case the probability of heads, $P(H)$ is $\frac{1}{6}$. In a large number of tosses, about one sixth of the outcomes should be "heads."

The reader who performs this experiment will observe an important natural regularity in operation. To state this natural regularity as a mathematical law, we first note that for any event A the outcome of a trial can be classified into one of the two groups: "A occurs," "A does not occur." For example, in the toss of a die as previously described, let the outcome 1 on a toss be classified as the event "A occurs"; then the occurrence of any of the other five possibilities forms the group "A does not occur." We may summarize the mathematical law as follows:

> The Law of Large Numbers: In a long sequence of repeated experiences or trials, the relative frequency of the occurrence of an event A will tend to approach the probability of the occurrence of the event A on an individual trial.

The Law of Large Numbers is just a mathematical way of quantifying the regularity we have observed in nature when examining repeated trials.

It is this law that forms the basis for the prognosis of an illness. Person J. is about to have a common kidney operation. Past records show that 90% of those people who have the same physical profile as person J. recover completely. Thus person J. is told by his doctor that there is a 90% chance of complete recovery.

One must be cautious in applying this law, for it calls for *repeated experiences,* which means that the *exact* experience or experiment is repeated again and again. We can never exactly repeat any experience, although sometimes, as in coin tossing, we can do so for all practical purposes. In the case of person J., there may not have been enough past cases that were sufficiently similar to person J.'s to permit even an approximate application of the law.

The Law of Large Numbers does, however, mathematically demonstrate an important phenomenon in nature; namely, under certain conditions, random phenomenon can yield nonrandom regularity. We shall have occasion to study other examples of this phenomenon and to find that it is such regularity that enables our study of statistics to be so useful.

Application: Odds

Let p represent the probability that an event A occurs and q the probability that it does not occur. Recall that $p + q = 1$. Hence if $p = \frac{4}{7}$, then $q = \frac{3}{7}$. The interpretation that the Law of Large Numbers permits is that in a long sequence of repeated experiences A will occur about 4 out of 7 times and will not occur about 3 out of 7 times. The odds that A occurs is defined as the ratio p/q, which in this example is $\frac{4}{3}$.

> Odds: If the probability of an event is p, the odds that it occurs is equal to $p/(1 - p)$.

Suppose we win if the event A occurs. The interpretation of the odds permitted by the Law of Large Numbers is that the odds gives the ratio of "wins" to "losses" in a long sequence of trials.

There is another meaning of odds in the lexicon of gamblers. If a gambler offers odds in favor of an event A of 3 to 2, that means he will bet $3 that A will occur for every $2 you bet that A will not occur. The ratio of his bet to yours, $\frac{3}{2}$, is called the *betting odds*. If the betting odds equal the odds that A occurs, the betting odds are said to be fair.

Example 3 ■ 12   The probability of "7 or 11" on a dice throw is $\frac{8}{36}$. The odds of this event is $\frac{8}{28}$ or $\frac{2}{7}$. These are the betting odds you should accept in fairness if you wish to bet in favor of the event; that is, for every $2 you bet in favor of the event, your opponent should offer $7.

Bayes Theorem (Optional)

A problem that arises often in statistics is the determination of the probability that a particular state of nature exists given an observed sample. For example, a person takes a screening test for cancer and the result is positive (the disease is present). This result does not necessarily mean the person has cancer. Given a positive result, what is the probability the person has cancer? We seek the probability of the state of nature given some data.

To illustrate how this type of probability is calculated, we solve the following problem:

Urn I contains two red and two black balls, whereas Urn II contains one red and five black balls. One of these urns is chosen at random and a ball is sampled from the urn. Find the probability that Urn I was chosen given that a black ball was obtained on the draw.

Let E = event "black on the draw." There are two possibilities: Black came from I or II. We identify the possibilities as B_1 and B_2, respectively. We wish to determine $P(B_1 \mid E)$. Note that

$$P(B_1 \mid E) = \frac{P(B_1 \text{ and } E)}{P(E)}$$

$$= \frac{P(B_1)P(E \mid B_1)}{P(E)}$$

Since the two urns are equally likely to be chosen, we have that $P(B_1) = \frac{1}{2}$. The value $P(E \mid B_1)$ is obtained by noting that there are two red and two black balls in Urn I so that the chance of obtaining a black ball (event E given B_1) is $\frac{1}{2}$. It remains to find $P(E)$. The event E consists of the mutually exclusive events

E and B_1 = choosing Urn I and then selecting a black ball

E and B_2 = choosing Urn II and then selecting a black ball

Hence

$$E = (B_1 \text{ and } E) \text{ or } (B_2 \text{ and } E)$$

Thus from rule (iiib) of the rules of probabilities

$$P(E) = P(B_1 \text{ and } E) + P(B_2 \text{ and } E)$$

Since $P(B_1 \text{ and } E) = P(B_1)PE \mid B_1)$ and $P(B_2 \text{ and } E) = P(B_2)P(E \mid B_1)$ we can write $P(E)$ as

$$P(E) = P(B_1)P(E \mid B_1) + P(B_2)P(E \mid B_2)$$

Substituting this expression for $P(B_1 \mid E)$, we have

$$P(B_1 \mid E) = \frac{P(B_1)P(E \mid B_1)}{P(B_1)P(E \mid B_1) + P(B_2)P(E \mid B_2)}$$

$P(B_2) = \frac{1}{2}$ since the urns are equally likely and $P(E \mid B_2) = \frac{5}{6}$. Hence

$$P(B_1 \mid E) = \frac{P(B_1)P(E \mid B_1)}{P(B_1)P(E \mid B_1) + P(B_2)P(E \mid B_2)}$$

$$= \frac{(\frac{1}{2})(\frac{2}{4})}{(\frac{1}{2})(\frac{2}{4}) + (\frac{1}{2})(\frac{5}{6})} = \frac{3}{8}$$

The preceding example illustrates the use of an important formula in probability; this formula has the name "Bayes Theorem" (named after the clergyman Thomas Bayes, 1702–1761). It can be generally statcd as follows:

Let $B_1, B_2, \ldots, B_k$ be *mutually exclusive* events of *which one must occur* and E be another event; then

$$E = (E \text{ and } B_1) \text{ or } (E \text{ and } B_2) \text{ or } \cdots \text{ or } (E \text{ and } B_k)$$

and from rule (iiib)

$$P(E) = P(E \text{ and } B_1) + P(E \text{ and } B_2) + \cdots + P(E \text{ and } B_k)$$

Using

$$P(E \text{ and } B_i) = P(B_i)P(E \mid B_i)$$

we have

$$P(E) = P(B_1)P(E \mid B_1) + P(B_2)P(E \mid B_2) + \cdots + P(B_k)P(E \mid B_k) \qquad (1)$$

Note that

$$P(B_i \mid E) = \frac{P(B_i \text{ and } E)}{P(E)} = \frac{P(B_i)P(E \mid B_i)}{P(E)} \qquad (2)$$

Substituting equation (1) in equation (2), we have

Bayes Theorem: If $B_1, B_2, \ldots, B_k$ are mutually exclusive events of which one must occur and E another event, then

$$P(B_i \mid E) = \frac{P(B_i)P(E \mid B_i)}{P(B_1)P(E \mid B_1) + P(B_2)P(E \mid B_2) + \cdots + P(B_k)P(E \mid B_k)}$$

for $i = 1, 2, \ldots, k$

Example 3 ■ 13   The population is divided into three groups according to their smoking habits: $B_1 = \{$smoker$\}$, $B_2 = \{$never smoked$\}$, $B_3 = \{$ex-smoker$\}$. Let E denote the event $\{$died of lung cancer$\}$. Suppose 50% of the smokers die of lung cancer, 10% of the people who never smoked die of lung cancer, and 20% of those who are ex-smokers die of lung cancer, and suppose it is known that 61% of the population smokes, 24% never smoked, and 15% are ex-smokers. If we observe that a person died of lung cancer, what is the probability that he never smoked? To calculate this we note that

$$P(B_1) = 0.61, \qquad P(B_2) = 0.24, \qquad P(B_3) = 0.15, \qquad P(E \mid B_1) = 0.5,$$

$$P(E \mid B_2) = 0.1, \qquad P(E \mid B_3) = 0.2$$

Then

$$P\{\text{never smoked} \mid \text{died of lung cancer}\} = P(B_2 \mid E)$$

$$= \frac{P(E \mid B_2)P(B_2)}{P(E \mid B_1)P(B_1) + P(E \mid B_2)P(B_2) + P(E \mid B_3)P(B_3)}$$

$$= \frac{(0.1)(0.24)}{(0.5)(0.61) + (0.1)(0.24) + (0.2)(0.15)}$$

$$= 0.067$$

that is, there is only a 6.7% chance he never smoked.

In the preceding discussion, the probability, $P(E \mid B_i)$, is the probability that an event E will occur given the "state of nature" B_i. This probability is called a prior probability since it is the probability about an event prior to any observation. The probability $P(B_i \mid E)$ is the probability of a given state of nature given that E occurred. This probability is called a posterior probability because it is calculated on what has been observed to occur; that is, it is calculated after observation.

Application: Medicine, Epidemiology

Test for a rare disease. Suppose that 0.1% of the population has a rare disease, and a medical test for the disease turns out either "positive" (the disease is present) or "negative" (the disease is absent). Suppose such a test makes a mistake: 1% of the time for those who do not have the disease and 2% of the time for those who have the disease. What is the probability that a person with a positive test result actually has the disease?

Solution. Let "+" represent the event "test positive"; let d represent the event "person has the disease" and $d' = $ not "d." We wish to determine $P(d \mid +)$.

Using Bayes Theorem and the previous figures, we find

$$P(d\,|+) = \frac{P(d)P(+\,|\,d)}{P(d)P(+\,|\,d) + P(d')P(+\,|\,d')}$$

$$= \frac{(0.001)(0.98)}{(0.001)(0.98) + (0.999)(0.01)}$$

$$= \frac{0.00098}{0.01097} = 0.09$$

which is equal to $\frac{1}{11}$.

This result states that of 100 people who have positive test results, only about 9 will actually have the disease. In ordinary medical terminology, there will be about 91 cases where the test will be positive and the patient will not have the disease.

It may appear intuitively that this result is incorrect because the test is so accurate. After all, the probability of a negative test result, $+'$, *given* that a person is ill is only 2% whereas the probability of a positive result for a well person is only 1%. Yet the chance that a person is *well* given a positive test result is 91%! As in many instances in the study of probability and statistics, intuition is misleading: In this case it does not take into account that the disease is *rare;* a *false* positive result will occur frequently because *so few people have the disease.*

A false positive is serious in that it can lead to unnecessary treatment and expense as well as cause the patient to suffer mental discomfort from the belief that he has the disease when he does not. However, many medical people believe *false negatives* to be much more damaging. A false negative is the event that the test is not positive but the person is ill. The probability of this event, $P(d\,|+')$, is the probability that a person for whom the test shows no disease does in fact have the disease. This event is dangerous since someone with negative test results will not be given treatment. In many cases, especially in rare diseases, early treatment can mean the difference between life and death. In this example, $P(d\,|+') = 0.00002$.

Table 3.1 is presented to help clarify the definitions of "false positive," "false negative," and other standard definitions that appear frequently in the medical and epidemiological literature.

We note that there is usually an instinctive reaction against a test that is subject to false negatives, since presumably these individuals may die if treatment is deferred until it is too late. On the other hand, false positives also exact a price, aside from the unpleasant shock to the persons concerned. If, as a result of false positives, the *limited facilities* for treatment are expended on those who do not need treatment, then the efficiency of the use of the facilities is lowered. Expressed differently, a test that never gives a false positive means that available facilities are used with maximum efficiency and therefore with best overall consequences to the population as a whole. It is clear, then, that a knowledge of

Table 3 ■ 1 Results of a Diagnostic Test			
Classified by Tests			
Individuals	*Positive*	*Negative*	*Totals*
Known diseased	a	b	$n_1 = a + b$
Known well	c	d	$n_2 = c + d$
Totals	$a + c$	$b + d$	$n_1 + n_2$

FN: False negative $= b/(b + d)$; FP: False positive $= c/(c + a)$

the probabilities of false positives and negatives is of great value in planning health delivery facilities.

3 ■ 8 SOME IMPORTANT COUNTING FORMULAS (OPTIONAL)

In applying the model of equally likely outcomes to the evaluation of probabilities, we found it necessary to count the number of elements in various sets. When the number of elements is small there are no difficulties; we simply list out the sample space. When the number is large, however, we need to develop efficient counting methods. We digress here to develop some of these methods.

An important principle of counting that we use to derive our various formulas is

---

The Multiplication Principle
If there are n_1 ways for some first event to occur, and for each of these n_1 ways there are n_2 ways for some second event to occur, then there are $n_1 \cdot n_2$ ways for both events to occur together.

For example, when flipping a coin twice, there are 2 possibilities for the outcome of the first flip, and each of these first outcomes could be paired with 2 possible outcomes for the second flip, hence there are $2 \cdot 2 = 4$ possible outcomes for the pair of flips. As another example, consider drawing 2 cards from a deck of 52 cards. There are 52 possibilities for the first draw, and each of these can be coupled with 51 remaining choices for the second draw, so there are altogether $52 \cdot 51 = 2652$ possible ways to draw 2 cards from the deck.

Permutations

Consider the number of different arrangements or orderings that can be formed by placing three coins marked separately 0, 1, and 2 into three slots. Any one of the three coins can fill the first slot or place. Thus there are three ways to fill the first place. There are two coins remaining to fill the second place. Hence there are 3 · 2 different possible arrangements among the first two places. Since there is only one coin remaining to fill the last place, there are 3 · 2 · 1 distinguishable orderings among the three places.

   To illustrate the preceding procedure, suppose the coin chosen to fill slot 1 is coin 2:

For this choice there are two possible choices for slot two.

After the second choice there is only one choice remaining.

We can repeat this diagram for the other possible initial choices.

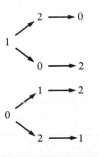

We see there are six pathways, indicating the six possible arrangements shown here.

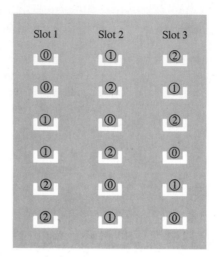

The number of arrangements is $6 = 3 \cdot 2 \cdot 1$. For purposes of brevity, we write

$$3! = 3 \cdot 2 \cdot 1$$

or, in the general case in which there are n objects

The number of ways to arrange n distinct objects in n distinct places is

$$n! = n(n-1)(n-2) \cdot \cdots \cdot 2 \cdot 1$$

where "$n!$" is read "n factorial."

In general, if there are n coins and only r places ($r \leq n$), the number of orderings is $n(n-1)(n-2) \cdots (n-r+1)$, since there are n possibilities for the first place, $n-1$ remaining choices for the second place, and so on until we come to the rth place. For this last slot there are $n-(r-1) = n-r+1$ possible choices left. Any of the choices for the first place may be coupled with any of the choices for the second place, and so on, so we multiply the numbers n, $n-1, \ldots, n-r+1$ together to obtain the total possible number.

We call each different ordering of the objects a *permutation* and denote the number of permutations of n objects among r places by the symbol $_nP_r$. Thus

> **The number of ways to arrange *n* distinct objects in *r* distinct places is**
>
> $$_nP_r = n(n-1)(n-2) \cdot \cdots \cdot (n-r+1) = \frac{n!}{(n-r)!}$$

The number $_nP_r$ is also often referred to as "the number of distinct ways to pick *r* objects from a set of *n* objects when the order of selection is important." For example, there are $_5P_3 = 5!/(5-3)! = 5 \cdot 4 \cdot 3 = 60$ distinct ways to select a president, vice-president, and treasurer for a committee from a list of five candidates since order is important.

Note that if $r = n$, then $(n-r)! = 0!$ and so, since we know there are $n!$ distinct ways to arrange *n* objects in *n* distinct places, we must have $n! = {}_nP_n = n!/0!$. Hence we set $0! = 1$ for purposes of consistency. Thus $0! = 1$, $1! = 1$, $2! = 2$, $3! = 6$, $4! = 24$, $5! = 120$, and so on.

Example 3 ■ 14   A certain chemical reaction requires that five compounds be combined in the correct order. How many ways are there to combine the compounds?

The easiest way to calculate this is to note that there are five places to fill (first compound added, second compound added, and so on) and five compounds to fill them. Thus there are five choices for the first place, four remaining for the second, three for the third, and so on. All in all, there are $5 \cdot 4 \cdot 3 \cdot 2 \cdot 1 = 5! = 120$ ways to combine the compounds.

If we wish to determine how many ways we could select and combine three of the five compounds, we would use the same procedure, except with three slots to fill. There are still five ways to fill the first slot, four to fill the second, and three to fill the third, for a total of $5 \cdot 4 \cdot 3 = {}_5P_3 = 60$ ways to combine three of the five compounds.

Combinations

In many problems we are not interested in the number of distinguishable *ordered* arrangements but just in the number of different collections without regard to order. For example, there are six permutations of the letters *A, B, C*; that is,

ABC   BAC   CAB

ACB   BCA   CBA

However, suppose order is *not* important. Then there is only *one* distinct collection, namely, the set {*A, B, C*}. Different collections contain different elements, and we do not consider a rearrangement of the same collection to be any different from the original collection. The *number of distinct collections of r objects that may be formed from a set of n objects is referred to as the number of combinations of n objects taken r at a time.*

In order to better understand the distinction between permutations and combinations, consider first the type of poker game called "open" poker. A player first obtains two cards face down and then three face up. The consequences of receiving a "6" and a "9," and then three aces in this order are much different from receiving the three aces first and then a 6, 9. In the former case the other players might not bet because they can see the three aces face up. In the latter case they only see one ace, a 6 and a 9; they would then be more inclined to bet, for they would not know the player has three aces. In this instance the *order* or *permutation* of the appearance of the cards is important. On the other hand, in five-card "closed" poker, none of the five cards are seen by the other players. Thus as the cards are dealt, the appearance of three aces in any order has the same effect. It does not matter in this game in what *order* the cards are dealt to the player; only the *combination* of the cards he or she receives is important.

To develop a formula for counting the number of combinations, we first consider the problem of how many different three-digit combinations can be formed by using the digits 1, 2, 3, and 4.

We shall let the symbol $\binom{4}{3}$ represent this number of combinations. Now consider one possible combination, such as the one composed of the numbers 1, 4, and 3. These three numbers can be arranged in 3! different ways, each arrangement being a different permutation. Hence, associated with each *combination,* there are 3! *permutations* that are identical in composition and different only in order. Thus

$$\binom{4}{3} 3! = \text{number of possible permutations of four things}$$

$$\text{among the three places} = {}_4P_3$$

But we know that

$$_4P_3 = \frac{4!}{1!}$$

Thus

$$\text{Number of combinations} = \binom{4}{3} = \frac{4!}{3! \; 1!} = 4$$

Note that the four combinations are

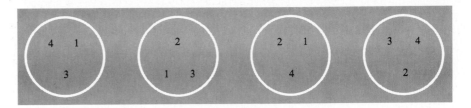

We now use the same reasoning to derive the general case when we have n things and r places. We let $\binom{n}{r}$ represent the number of combinations of n things taken r at a time, that is, the number of distinct collections of r objects that can be formed from an original collection of n objects. We now derive a formula to compute $\binom{n}{r}$. Associated with *each* combination there are $r!$ permutations. Thus

$$\binom{n}{r} r! = \text{number of possible permutations of } n \text{ things among the } r \text{ places}$$
$$= {}_nP_r$$

We know that

$$_nP_r = \frac{n!}{(n-r)!}$$

Hence

> **The number, $\binom{n}{r}$, of distinct combinations of r things that can be formed from a set of n distinct things is**
>
> $$\binom{n}{r} = \frac{n!}{r!(n-r)!}$$

Example 3 ▪ 15   A restaurant owner in Austin, Texas, has 15 different vegetables available for the dinner menu and wishes to try all the combinations of three vegetables that are possible. How many are there? Since the order in which the vegetables are put into the pan is unimportant for the final dish, we must consider the number of combinations of 15 things taken 3 at a time. This number is

$$\binom{15}{3} = \frac{15!}{3!(15-3)!} = \frac{15!}{3!\,12!}$$

$$= \frac{15 \cdot 14 \cdot 13 \cdot 12 \cdot 11 \cdot 10 \cdot 9 \cdot 8 \cdot 7 \cdot 6 \cdot 5 \cdot 4 \cdot 3 \cdot 2 \cdot 1}{3 \cdot 2 \cdot 1 \cdot 12 \cdot 11 \cdot 10 \cdot 9 \cdot 8 \cdot 7 \cdot 6 \cdot 5 \cdot 4 \cdot 3 \cdot 2 \cdot 1}$$

$$= \frac{15 \cdot 14 \cdot 13}{3 \cdot 2} = 5 \cdot 7 \cdot 13 = 455$$

Note the method we used to calculate this combination. We first canceled the 12! from the top and bottom of the fraction, and then canceled as much of the remaining fraction as possible. This left us with just 3 multiplications to do instead of the 14 multiplications for the top, 13 multiplications for the bottom, and a division. Also, if we had not canceled out before multiplying, we probably would have run over the capacity of our hand calculator.

Example 3 ■ 16 What is the total number of combinations of five cards taken from a deck of 52 cards? The number is

$$\binom{52}{5} = \frac{52!}{47!\,5!} = 2,598,960$$

Note that this is the number of possible hands in five-card closed poker. It is this large number of hands that makes this game interesting.

Example 3 ■ 17 How many possible hands are there in bridge? There are 13 cards in a bridge hand, so this number is

$$\binom{52}{13} = \frac{52!}{39!\,13!} = 635,013,559,601$$

This game is even more interesting than poker!

Now that we have these formulas for counting the number of objects in various sets, we may use them to calculate probabilities by using the model of equally likely outcomes.

Example 3 ■ 18 Assume that each of the $\binom{52}{5}$ possible hands in five-card closed poker are equally likely. What is the probability of being dealt a five-card hand with four aces? The total number of combinations or outcomes is $\binom{52}{5}$. We now derive the number of combinations that contain four aces. Given that four aces are in the hand, there are 48 cards, each of which could be the remaining card in the hand. Thus there are 48 different hands containing four aces.

$$P(\text{four aces}) = \frac{\text{number of hands containing four aces}}{\text{total number of hands}}$$

$$= \frac{48}{\binom{52}{5}} = \frac{48}{2,598,960} = 0.000018$$

The probability is about 1 out of 55,500 hands.

For your convenience in calculations, a table of combination values $\binom{n}{r}$ is given in Appendix A.2, for $n = 1, \ldots, 20$.

Application to the Law:

The application of probability and statistics to the enforcement of the law and prosecution is one of the oldest, most important and least studied applications of these fields. As far back as 1785, the French mathematician Condorcet published his "Essai sur l'application de l'analyse à la probabilité des decisions rendues à la pluralité des voix." In this work Condorcet studied the probability of jury decisions being in error. Other French mathematicians became highly involved in this and related problems. The works of LaPlace and Poisson remain classics in demonstrating the connection between mathematics and the law. This connection has continued to the present day. An important example in enforcing the law is given in "Reapportionment and Redistricting" by R. C. Silva in the November 1965 *Scientific American*. The problem discussed in this paper is that of district reapportionment in accordance with court decisions to enforce the concept of "one man, one vote." We have already discussed one case involving criminal justice. Other important areas we shall discuss in detail are those of "jury discrimination" and Title VII cases involving discrimination by sex or race, or both, in employment. These are just a few areas of applications to the law. There are more profound ones that require extensive research; for example, the question of how the legal concept of materiality is related to the concept of statistical significance.

   Considering all the rich and fruitful applications of probability and statistics to the law, it is remarkable that the law itself (and the law schools*) has re-

* One of the authors (A. L.) suggested to a Dean of a well-known law school that a seminar be offered the law students on the application of statistics to the law. His offer was rejected on the basis that such applications were a "fad." Two hundred years of history apparently prove nothing.

mained virtually immune to its influence. To this day lawyers have made only slight use of probability and statistical theories, and then only when these applications are so obvious that they have no choice but to delve into them.

We present another application to the law with the case of *Leah, Bernadette, Hodges, et al.* vs. *State of Louisiana #216-624 Sec. 1 of the Criminal District Court of the Parish of Orleans.* It turns out that this case demonstrates the utility of developing simple counting methods. It also demonstrates a counting method that we have not previously considered but that is useful to master. The case resulted from an armed confrontation between the police and the defendants at one of the headquarters of the Black Panthers party. No one was injured. Twelve people were arrested in September 1970. Each was charged with five counts of attempted murder. The possible verdicts on each count were "guilty of attempted murder," "guilty of attempted manslaughter," and "not guilty." All 12 defendants were to be tried together. The jurors were to consider all possible verdicts and to bring in verdicts on each of the defendants. How many possible different arrangements of verdicts were possible? Each defendant was charged with five counts, there were 12 defendants. Thus there were $5 \times 12 = 60$ charges for the jury to consider. For each charge there were three possible verdicts. There were 3(possible verdicts)$^{5(\text{counts}) \times 12(\text{defendants})} = 3^{60} = 42,391,158,247,900,000,000,000,000,000$ different possible arrangements of the verdicts. Do you believe a jury could give serious consideration to each of these possibilities? A psychologist at the trial testified that no jury could possibly seriously consider so many different possibilities. The purpose of presenting this testimony was to convince the court to break the defendants up and to try them separately.

Summary

In this chapter we presented the general background in probability theory that is necessary for our subsequent study of statistics. The model of equally likely outcomes is the classical probability model in which $P(A) = n_A/n_{\text{Tot.}}$. We developed some general rules of probability to be used for cases other than that of "equally likely outcomes."

We learned the "frequency interpretation" of probabilities and found that this interpretation was supported by the Law of Large Numbers. This law forms one of the bridges from probability to statistics, because it permits us to check the probability of an event through sampling; that is, in a long sequence of identical experiments the proportion of times (the relative frequency) with which an event occurs should be close to the probability of that event.

Another bridge to statistics was given by Bayes Theorem which related the posterior probability of an event to its prior probability.

In order to count n_A and $n_{\text{Tot.}}$ better in solving problems involving equally likely outcomes, we derived formulas for the number of permutations and the number of combinations that can be formed by taking k objects from among n

objects. If the order of presentation is important, we use the permutation formula; if order is not important, we use the combination formula. In both permutations and combinations, the items are selected without replacing the previously selected items.

Key Terms

probability	conditional probability
frequency interpretation	independence of events
sample space	Product rule
elementary event	Law of Large Numbers
compound event	odds
model of equally likely	*Optional*
outcomes	Bayes Theorem
mathematical model	factorial
Addition rule	permutation
Probability assignment rule	combinations
complement of *A*, and its	
probability relationship	
to *P(A)*	

References

Kyburg, H. E., and H. E. Smolker: *Studies in Subjective Probability.* New York, John Wiley & Sons, 1963. This text contains essays on interpretations other than the frequency interpretation of probability by pioneers in this field of study.

Levinson, H. C.: *Chance, Luck and Statistics.* New York, Dover, 1950. An elementary discussion of probability with many simple examples.

Mosteller, F., et al. (eds.): *Statistics by Example: Weighing Chances.* Reading, MA, Addison-Wesley, 1973. An elementary treatment of statistics with many interesting applications.

Thorp, E.: *Beat the Dealer.* New York, Random House, 1966. Discusses how to win at blackjack, a practical application of a specific mathematical model and of the calculations of conditional probability.

Exercises

Section 3.2

1. Which of the following "probability statements" permit a frequency interpretation and which do not? Explain.
   a. According to the best available evidence, the earth is probably flat. (Statement made by a courtier in Queen Isabela's court, circa 1490.) *No*
   b. The probability of obtaining the ace of spades on the draw of a card from a deck of 52 cards is $\frac{1}{52}$. *True*

c. The economic policies of the present U.S. government will probably reduce inflation in the next 4 years. No.
d. Based on the data displayed in today's weather map, the probability of rain tomorrow is $\frac{1}{20}$. TRUE

Section 3.3

2. Write the number 0 on each of 10 slips of paper and the number 1 on 30 slips of paper. Mix the slips and select one of them. (The probability of selecting a slip with a "1" on it is $30/(30+10)=0.75$.) Note whether or not you have selected a 1. Put the slip back in the pile and mix the pile again. Repeat this experiment 20 times. What proportion of 1's did you obtain? Repeat the experiment another 60 times. What proportion of 1's did you obtain of the 80 trials? Is this value close to 0.75? Do you believe there is a relationship between the probability of obtaining a 1 in a single trial and the proportion of 1's you obtained in 80 trials?

3. The integers 0, 1, 2, up to 9 are equally likely in the Table of Random Numbers, Appendix A.3; that is, each of these values appears equally as often.
   a. About what proportion of 0's would you expect to obtain if you selected 50 of these numbers at random? $.1 = \frac{4}{50} = .08$ .1
   b. Use the following procedure to pick 50 numbers at random: Flip a coin; if you obtain

   heads, pick page 1 of the table
   tails, pick page 2 of the table

   Close your eyes and pick a spot on the page selected. Read the 50 numbers down the column. These are your 50 random choices. How often does 0 occur? What is the proportion of 0's you have obtained?"

4. Let S be the set of outcomes of a throw of dice (Fig. 3.1). For every event A, let $P(A)=$

$n_A/36$, where n_A is the number of elements in A. Find the probability of each of the following events:
   a. a double six
   b. the sum of the faces is 7
   c. at least one 6
   d. no face has a 1
   e. one of the faces has a 1 whereas the other does not
   f. not obtaining the sum 3

5. a. A fair coin is tossed twice. Assume the model of equally likely outcomes. What is the probability of two heads in a row?
   b. The fair coin is tossed three times. What is the probability of three heads in a row?
   c. Suppose the fair coin is tossed 10 times. What is the probability of 10 heads in a row?

6. Three balls are distributed among two boxes. Assume for each ball that each box is equally likely. Find the probability of exactly one empty box.

7. Let S be the set of outcomes of a throw of dice (Fig. 3.1). How many members of S are in each of the following events?
   a. A is the event "two same faces." 6
   b. B is the event "sum of the faces is even." 18
   c. The event A and B, which is the event "two same faces and the sum of the faces is even." 6
   d. C is the event "sum of the faces is odd." 18
   e. The event A and C 0
   f. D is the event "a 1 or 2 on the first die while any number appears on the second die." 12
   g. E is the event "a 2, 3, or 4 on the first die and any number on the second die." 18
   h. The event D and E 6
   i. The event D or E. 24

8. Find the probability of each event (a) to (i) in problem 7.

9. In a given group of 100 people 60 were smokers, 30 alcoholics, and 25 beat their spouses. Every one of the 100 people is in at

least one of these categories. Further, it is known that 80 smoked or drank.

a. How many people smoked and drank? 10

b. Must every person who is a spouse beater also be a smoker or an alcoholic? Give reasons. No if everyone who beat drank smoke neg.

10. In the following figure we see that the event D, which is the event "the outcome 1 or 2 on the first die and any outcome on the second die," is that set of boxes with horizontal shading. The event E, which is "the outcome 2, 3, or 4 on the first die and any outcome on the second die," is represented by that set of boxes with vertical shading. The event D and E is that set of boxes with both shadings. Draw the same type of representations for the events A, B, "A and B," C, "D or E" defined in problem 7.

would only total a high of 80 people

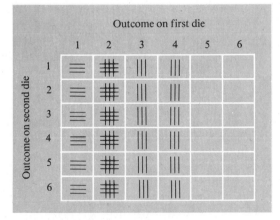

Outcome on first die

11. There are three people at a table; each has the initials A. B. C. Assuming every person has a middle name, and assuming all names have equally likely starting letters, the probability of this event is

a. 26^{-3};

b. 26^{-9};

c. $4(26)^{-3}$;

d. $(\frac{1}{3})^3$

12. Suppose you have 10 gum balls in your pocket, 3 of which are licorice flavored. If you put your hand in your pocket and pull out 2 gum balls, what is the probability that neither is licorice flavored? What is the probability that at least 1 is licorice flavored?

13. A jar of cookies contains three fudge cookies and four chocolate creme cookies. If three are selected at random, what is the probability that exactly two of them are fudge cookies?

14. Let $n_{Tot.}$ represent the total number of members of a population and the number in the event "event A" be equal to n_A, where $n_{Tot.} = 10n_A$. Assuming the model of equally likely outcomes, show that $P(A) = \frac{1}{10}$.

15. In the table of random numbers, Appendix A.3, there are 10 numbers, 0, 1, 2, 3, 4, 5, 6, 7, 8, 9. Each of these numbers appears as often as every other number; that is, number of times 0 appears in the table = number of times 1 appears in the table = . . . = number of times 9 appears in the table. Assume the model of equally likely outcomes; that is, each number has the same probability as every other number. Probability of number $i = P(i) = n_i/n_{Tot.}$; $i = 0$, 1, . . . 9. Since $n_{Tot.}$ is 10 times the number of times n_i any number i appears, we have $P(i) = n_i/10n_i = \frac{1}{10}$. If you select a number from the table at random, find

a. the probability that it is odd.

b. the probability that it is greater than 0.

c. the probability that it is less than 9.

d. the probability that it is greater than 0 and less than 9.

e. the probability that it is not 0 or 9.

f. the probability that it is equal to 0 or 9.

16. Suppose that the population of a certain city is 40% male and 60% female. Assume the model of equally likely outcomes. What is the probability of "selecting a male" if one selection is made from the population?

17. If two coins are tossed, there are four results that may occur; these are HH, HT, TH, TT, where H represents heads and T represents

tails. Assume the model of equally likely outcomes. Find
a. the probability of two heads
b. the probability of one head and one tail
c. the probability of no heads appearing
18. If three coins are tossed, find the probability of the following events (assuming the model of equally likely outcomes):
a. getting at least two heads
b. getting an odd number of heads
c. getting no heads
d. getting three tails
19. Let two dice be rolled. The possible outcomes are listed in Figure 3.1. Assuming the model of equally likely outcomes, find the probability of each of the following events:
a. getting (3, 1)
b. getting (1, 1)
c. getting one of the following outcomes {(1, 1), (3, 2), (4, 6)}
d. getting one of the following outcomes {(4, 6), (5, 5), (6, 4), (5, 6) (6, 5), (6, 6)}
e. getting an outcome in which the sum of the faces is ≥ 10
f. getting an outcome in which the sum of the faces is < 10
g. (Optional) The Chevalier de Méré had been losing steadily at the gaming tables. His misfortunes had led him to test his conclusions about what the frequency of outcomes should be by more closely observing what they actually were. For example, he thought that the probability of throwing a 6 in 4 rolls of one die should be the same as the probability of throwing two 6's in 24 rolls of a pair of dice. His observations led him to conclude that he was wrong and so he wrote to Pascal asking him to calculate the two probabilities. Find the two probabilities.

Now that you have calculated the probabilities and checked your result with the answer given at the back of this book, you should be able to find the flaw in the Che-

valier's reasoning that almost led to his financial ruin. He reasoned that the probability of a 6 on one throw of the die is $\frac{1}{6}$ and so in four throws, the chance of exactly one appearance of 6 is $\frac{4}{6}$. The chance of a double 6 on a throw of dice is $\frac{1}{36}$, and so the chance of one such event in 24 throws is $\frac{24}{36} = \frac{4}{6}$. What is wrong with this reasoning?

20. Two balls are drawn without replacement from an urn containing two red balls and three white ones. Assuming equally likely outcomes, what is the probability of
a. getting a red ball on the first draw and a white one on the second? Multiplicative rule.
b. getting a ball of each color?
21. Two dice are rolled. Show that the probability that the sum of the faces equals 2 is the same as the probability that the sum of the faces is 12. Show that the probability that the sum of the faces equals 3 is the same as the probability that the sum is 11. Show that $P(\text{sum} = 4) = P(\text{sum} = 10)$; $P(\text{sum} = 5) = P(\text{sum} = 9)$; $P(\text{sum} = 6) = P(\text{sum} = 8)$.
22. Four coins are tossed. Consider the events: "four heads," "3 heads and 1 tail," "2 heads and 2 tails," "1 head and 3 tails," "four tails." Which of these events has the highest probability?
23. (Optional) Three drugs A, B, and C are to be administered to six patients. Three vials of each drug are placed in a box and each patient selects one. What is the probability that no patient selects drug A? What is the probability that two of each of the drugs are selected?
24. Three of the sides of a die are painted red. What is the probability that when the die is thrown a "red" side will show face up? Another die has three faces painted green. What is the probability that a red side and a green side will show face up when the two dice are thrown?
25. A coin is tossed four times. For every head

that appears a gambler receives $5. For every tail, $6 is lost. What is the probability of being ahead after four tosses?

26. A person with an inferiority complex goes to lunch with a friend and is served after the friend. Should it be believed that this person is being discriminated against? The next day the person goes to lunch with four friends and again is served last. Ignoring the previous day's event, should the person feel discriminated against? Suppose the person goes to lunch with four friends for n days in a row, what is the probability $p(n)$ that the person will be served last every time, assuming no discrimination on the part of the waiter? What is the value n_0 such that $p(n_0) \leq 10^{-5}$? (For discussion) Would you say that if the person were served last while going to lunch with four friends for n_0 days in a row that the person was being discriminated against?

Section 3.4 $P(D) + P(E) - P(D \cup E) = \dfrac{24}{36}$

27. Use the "Addition Principle" to calculate the probability of "D or E" in problem 7(i).
28. Let B' represent the event "B does not occur." Suppose $P(B) = 0.5$. What is $P(B')$?
29. If $P(A) = 0.8$, $P(B) = 0.5$, $P(A$ or $B) = 0.9$, find $P(A$ and $B)$. Also, find the probability that either A does not occur or B does not occur or both do not occur.

Section 3.5

$P(A)=.8$ $P(A \cap B)=?$ $P(A \cup B)$
$P(B)=.5$ $P(A')=?\cdot2$
$P(A \cup B)=.9$ $P(B')=.5$

30. a. If $P(A) = 0.8$, $P(B) = 0.5$, and $P(A$ or $B) = 0.9$, find $P(A \mid B)$. Are A and B independent?
   b. (Optional)
      i. Show that if $P(A \mid B) = P(A)$ then $P(B \mid A) = P(B)$.
      ii. If $P(A$ and $B) = P(A)P(B)$ show that $P(A \mid B) = P(A)$.
      This last result taken with the fact that $P(A \mid B) = P(A)$ implies $P(A$ and $B) = P(A)P(B)$ shows that we can take the con-

dition $P(A$ and $B) = P(A)P(B)$ as a definition of independence of A and B, which is equivalent to that definition given in the text.

31. Show that the events D and E defined in problem 7(f) and (g) are independent events.
32. Suppose that the population of a certain city is 40% male and 60% female. Suppose also that 50% of the males and 30% of the females smoke. Find the probability that a person known to be a smoker is male. *Hint:* Probability of a person being a smoker = $P(S)$ = probability (person is male and smoker *or* female and smoker) = P(male and smoker) + P(female and smoker) = 0.2 + 0.18 = 0.38.
33. Suppose 5 out of every 100 men and 25 out of every 10,000 women are colorblind. A colorblind person is chosen at random in the United States. What is the probability that this person is male? (Assume males and females to be equal in number.)
34. The probability that a certain movie will get an award for good acting is 0.16; for good directing is 0.30; and for both is 0.09. What is the probability that the movie will get
   a. either or both awards?
   b. only one of the two awards?
   c. neither of the awards?
   d. the award for good acting given that it gets the award for good directing?
35. An urn initially contains 10 red balls and 6 black balls. At each trial a ball is selected from the urn at random, its color is noted, and then it is returned to the urn along with 2 more balls of the same color.
   a. Compute the probability of obtaining a red ball on each of the first two trials.
   b. Show that the events "red ball on the first trial" and "red ball on the second trial" are not independent.
   c. (Optional) Explain how the model presented in this problem might be applicable to the study of spread of diseases.

#29. $P(A), P(B)$; $P(A \cup B)$
$= P(A \cap B) = -.9 + .8 + .5$
$= 0.4$

the prob of complements must equal 1
∴ if prob $(A \cap B) = .4$
the $P(A' \cap B)'$ must $= .6$

36. In the senate rules committee of a certain state there are 3 democrats and 2 republicans. A subcommittee of 3 senators is to be selected (ignore the order of selection and consider each senator as equally likely to be selected).
   a. How many ways can this be done?
   b. If it is required that the subcommittee contain exactly 2 democrats, how many possible committees are there?
   c. What is the probability of obtaining a subcommittee with a majority of republicans? What is this probability if we are given that at least 1 democrat has been selected?

Section 3.6

37. A man goes shopping in a well-known department store and selects at random one light bulb from a box containing 10% defective light bulbs, one fuse from a box containing 5% defective fuses, and one extension cord from a stock shelf containing 1% defective cords. After arriving home, the man installs the fuse, screws the light bulb into the reading lamp, and plugs the lamp into a wall socket using the extension cord. What is the probability that the lamp will work when switched on? Use the notion of mutually independent events in making this calculation.

38. In what is known as a "triangle taste test" the taster is presented with three samples, two of which are alike, and asked to pick out the odd one by tasting. If a taster has no well-developed sense of taste, and picks at random, what is the probability that in six triangle taste tests the taster will make
   a. all correct decisions.
   b. no correct decisions.
   c. at least one correct decision.

39. Two baseball teams are playing in the World Series. The series consists of at most seven games, the first team to win four games is declared the winner. (Thus the series could last four, five, six, or seven games). Suppose when team A plays team B there is a 60% chance of team A winning, and assume the games are independent.
   a. What is the probability that team A (the better team) loses in four straight games? in five games?
   b. What is the probability that team A loses the series?

Section 3.7

40. (Optional) A resort in Florida decides to issue insurance to its patrons in case it rains while they are vacationing there. The resort knows that, if it is a clear day, then the probability of rain the next day is $\frac{2}{10}$, and if it is a rainy day, the probability of rain the next day is $\frac{4}{10}$. How much of the time will the resort have to pay the rain insurance? *Hint:* Assume the probability of rain is the same each day.

41. (Review problems 3 and 15.) Select 100 numbers from the Table of Random Numbers, Appendix A.3. What proportion is
   a. odd?
   b. greater than 0?
   c. less than 9?
   d. greater than 0 and less than 9?
   Compare these proportions to the corresponding probabilities (a), (b), (c), (d) in problem 15.

42. Breaking codes (how samples approximate populations). Up until World War II, when modern enciphering machines were first developed, cryptographic methods remained essentially the same since Caesar's time. One ancient principle for breaking codes is explained in Edgar Allen Poe's "The Gold Bug." A less literary explanation is given through this problem.

   Table 3.2 gives the percentage of times each letter in the English language occurs over many pages of ordinary writing.

Table 3 ▪ 2 Approximate Percent of Occurrence of Letters in English Writing												
A	B	C	D	E	F	G	H	I	J	K	L	M
11.0	1.5	1.9	3.2	13.0	1.8	1.6	5.0	6.5	0.7	2.1	5.1	1.7
N	O	P	Q	R	S	T	U	V	W	X	Y	Z
6.4	5.6	3.6	0.3	6.0	6.3	12.0	2.1	0.4	2.1	0.1	0.1	0.1

The table yields equivalently the approximate probability of each letter.

a. Select one page from an ordinary novel and verify that the percentage of times each letter occurs is approximately that given in the table.

b. Decode the following message:

FWFSZPOFDBOMFBSOTUBUJ-
TUJDT

XJUIUIJTUFYU

Hint: In the code used, each letter in the message is replaced by another letter. Assume that the letters occur in the actual message with about the same frequency as they do in Table 3.2. Hence the letter that replaces the actual letter in the message occurs about as frequently in the message as does the actual letter in Table 3.2. For example, one of the most frequently occurring letters in the coded message (F) corresponds to the most frequently occurring letter in the table (which is E).

43. What should the betting odds be on the outcome
   a. "7 or 11" in a toss of dice?
   b. "snake eyes" [outcome (1,1)] on a toss of dice?
   c. "1 ace from a 52-card deck"?
   d. "10 heads in 10 tosses of a fair coin"?

44. I have two coins; one is a fair coin, the other is weighted so that the probability of heads is 0.2. I choose one of the coins at random, toss five times, and obtain five heads. What is the chance that it is the fair coin?

45. (Bluffing) Part of the intelligent playing of closed poker is to evaluate the strength of your opponent's hand from a knowledge of his betting pattern. If a player never bluffs, then the size of his bet will indicate the evaluation of the strength of his hand with respect to the other players. If this evaluation is good (i.e., he is a "good" player), the other players will learn this after a relatively small number of hands and will learn not to respond to his large bets. Thus a good player is obliged to bluff occasionally (bet a large sum with a weak hand) if he is to win substantial sums. The other players must try to detect a pattern in the good player's bluffing in order to counter the bluffing strategy. In a six-person game the probability that an individual has a hand that beats all the others is $\frac{1}{6}$. Given that good player A bets big whenever he has a hand that beats all the others and bluffs once for every five times he bets a hand that does not beat all others, what is the chance that A is bluffing given that he bets big on a hand?

46. (Lying) Bluffing is a form of lying. The principle in detecting lying is the same as detecting bluffing: Listen to what is said but watch carefully what is done. The "what is done" corresponds to the data, the posterior information.

In Orleans Parish, Louisiana, the law requires that people be selected by "jury commissioners" for a jury by random selection (i.e., all people equally likely) from a list compiled from the population. One such list in 1958 consisted of 30 blacks and 40 whites. Of the 12 people chosen, none were black.

The commissioners insisted they selected the jury according to the law. Given that the commissioners were telling the truth, what was the probability of selecting 12 white jurors from the list?

(It was a case of this kind, *Willie B. Brooks* vs. *Dr. George J. Beto,* 1966, that prompted the Fifth Circuit Court of Appeals to state in its decision ". . . the courts have consistently held that statistics speak louder than jury commissioners.").

47. a. In the application of Section 3.7 (Bayes Theorem), verify that the probability of a false negative is 0.00002 as claimed.

b. Two other probabilities that appear frequently in the medical literature are *Sensitivity* (*Sn*) and *Specificity* (*Sp*). To define these terms, let A represent the event "a diseased person is classified by the test as diseased" and B the event "a well person is classified as well." Then assuming all outcomes in the sample space are equally likely, we have from Table 3.1,

$$Sn = \text{probability of event } A$$

$$= P(A) = a/n_1 \qquad (1)$$

$$Sp = \text{probability of event } B$$

$$= P(B) = d/n_2 \qquad (2)$$

Thus if we select a diseased person "at random" from the group of n_1 diseased people, the probability that the test will classify this person correctly is given by expression (1). Similarly, expression (2) gives the probability of correctly classifying a person from the normal group. In the application of Section 3.7 (Medicine, Epidemiology), what is the *Sn, Sp* of the rare disease?

c. Other terms that appear frequently in the medical literature and that are probabili-

ties or combinations of probabilities are

$$\text{Incidence} = \frac{n_1}{n_1 + n_2};$$

n_1, n_2 are counted over time T.

i. What is the incidence of the rare disease discussed in the application?

$$\text{Relative Risk} = \frac{a}{a + c} \div \frac{b}{b + d}$$

ii. Show that the relative risk $=$ $(1 - FP)/FN$.

Section 3.8

48. Explain the difference between "permutations" and "combinations."

49. I have 20 books, each of which has different colored covers and a bookshelf that holds only 10 books. How many different arrangements of colors on the shelf can I have if
a. I consider the order of the books on the shelf?
b. I do not consider the order?

50. A batch of 50 fuses contains 4 defective fuses. If 10 are tested, what is the probability that none of them are defective?

51. If there are two possible outcomes on the toss of a coin, how many different results can be obtained in 10 tosses of the coin?

52. Three cards are selected at random from a 52-card deck. What is the probability that all 3 are spades?

53. An IRS inspector randomly checks 5 income tax returns out of a batch of 12. If 6 of these 12 have illegal deductions what is the probability that the inspector finds 2 of the illegal returns?

54. Suppose there are five married couples from England and three married couples from Ireland. They are all in the same room. Six people in this room are chosen at random.

a. What is the chance of obtaining exactly four English people out of the six choices?

b. What is the chance that exactly two English men will be chosen?

c. What is the chance that the six people are the married couples from Ireland?

55. (Genetics) A cell contains a gene; each gene contains N components. Of these components, let us suppose the i are "healthy" and $N - i$ are "mutants." When a cell divides, each component also divides so that $2N$ components result. Mutant components produce mutants. The cell now splits and the gene in each new cell contains N components. The assumption is made that all distributions of the $2N$ components among the two new genes are equally likely. What is the probability that one gene contains j healthy components?

56. If there are 4 defective fuses in a box of 100 fuses, what is the chance of selecting a defective fuse?

57. In problem 56, how many ways can two fuses be selected if

a. sampling is done with replacement; that is, the first fuse selected is returned to the box before the second selection.

b. sampling is done without replacement; that is, the first fuse selected is not returned to the box.

58. Referring to problem 56, suppose two fuses are sampled without replacement. What is the probability that two good fuses are selected?

59. Ten defendants are tried together before a jury. Each defendant has two different charges against him and each defendant can be found guilty or innocent on each charge separately from all other findings. How many different verdicts can the jury bring in on all the defendants and all the charges?

(This type of problem faced a judge in the trial of 12 Black Panthers held in 1970 in New Orleans, Louisiana. The judge was asked by defense lawyers to separate the trials of the 12 defendants because the jury members would be faced with too many possible decisions to be able to remember all of them.)

60. A travel agency is asked by a client to price tours of South America that include 5 stopovers selected from a list of 10 cities. How many different tours must be priced

a. if the order of the stopovers influences the price?

b. if the order of the stopovers does not matter with respect to price?

61. I plan to hike up one of four different trails to the top of a small mountain, arrive at noon, eat lunch, and then hike down again. Mr. Z (whom I detest) plans to follow the same schedule. What is the probability that I will select the same trail up the mountain as Mr. Z? What is the probability that I will select the same downward trail as Mr. Z? What is the probability that neither trail I select will be the same as Mr. Z's choice (so that I will have a day free of Mr. Z)?

62. The Olympic committee has bids from three countries for the 1992 and 1996 Olympics. How many ways can they select sites if

a. the same country cannot have the Olympic games twice in a row?

b. they may be held in the same location twice?

c. the same as part a but the order of selection does not matter?

63. The National Center for Education Statistics was legislatively mandated in the Bilingual Education Act to count the number of school-age children and adults "with limited English-speaking ability" in each state and report the findings to Congress by July 1, 1977. If each person was classified into 1 of the 50 states, and into 1 of 3 categories of English-speaking ability, how many different ways could a person be cross-classified?

64. A statistics test has a true-false section of 10 questions. How many ways can you
    a. get exactly three answers correct?
    b. get eight right and two wrong?
    c. get five right and five wrong?
    d. get seven right and three wrong?

65. Quality control procedures for a batch of manufactured goods often are as follows: If there are n objects in the batch, a sample of k objects are selected and tested. If too many of the tested products are defective, the entire batch is rejected for shipment.

    If a batch contains 25 objects, of which 10 are defective, and a sample of 6 are selected for inspection,
    a. how many ways are there total to select these 6 objects?
    b. how many ways are there to select exactly 3 defectives in this sample of 6?
    c. what is the probability of selecting exactly 3 defectives in the sample of 6 from the batch?

66. Consider a slot machine with three windows. When you pull the handle, the objects showing in each window change in a random way. Suppose there are four objects that could possibly be shown in each window.
    a. how many distinct arrangements of objects are possible?
    b. How many ways can all three windows show the same object (the player wins in this case)?
    c. The player can also win if exactly two "cherries" appear. How many ways can this happen?
    d. What is the probability of a player winning?

67. A student planning a schedule for next semester wishes to attend three classes, one at 10 AM, at 11 AM, and at 12 PM, chosen from among Music 200, History 150, Math 112, Spanish 201, and PE 125. Consulting the schedule of classes, the student finds that not all these courses are offered at each of the times 10, 11, and 12.

10 AM	11 AM	12 PM
Music	History	Music
History	Spanish	Spanish
Math	Math	Math
Spanish	PE	PE
PE		

   a. How many possible schedules are there (in which the same course is not attended at two different times)?
   b. If one of these schedules is chosen at random, what is the probability that it contains both the history and music courses?
   c. If the student decides first to reduce the list of courses by randomly selecting three from the original list of five, what is the probability that PE will not be among the three?

68. A man who claims to have extrasensory perception (ESP) concentrates on the 2, 3, 4, and 5 of diamonds. The cards are then taken from him, shuffled, and placed face down on a table. He then guesses the value of two cards. Suppose the man does not have ESP. What is the probability
    a. both guesses are correct?
    b. the first one is correct and the second one is incorrect?
    c. the first one is incorrect and the second one is correct?
    d. at least one is correct?
    e. both guesses are wrong?

69. In problem 68, suppose the man points to one card, then the other, clasps his brow and says, "One card is the 5 of diamonds, the other is the 2 of diamonds" (i.e., he does not specify the order or which card is which). What is his chance of being correct?

70. The forecaster on KTVV television station in Austin, Texas, says the probability of rain tomorrow is $\frac{1}{3}$. Let A denote the event "rain tomorrow." Find $P(A')$. The *odds* in favor of rain tomorrow are $P(A)/P(A')$. What are the odds of rain tomorrow?

71. A radio or television station must have four call letters as an identification code, the first letters of which must be a "K" or a "W." How many possible distinct identification codes are possible? Those stations on the West Coast must start their call letters with a K. How many possible call letter choices are there for a West Coast station?

72. A California car license plate has three letters followed by three numerals. How many possible plates can be made? There are 198 "unacceptable" three letter combinations (swear words in various languages, and so on). How many plates are now possible if these 198 are eliminated?

73. A graduating senior with a major in liberal arts has four interviews scheduled. Based upon his assessment of the job demands, his qualifications, and so on, he assesses the probability of being offered a job by each company as follows:

Company	Probability of Being Hired
Megalithic International	1/10
Goodtime Rubber and Tires	2/5
A. R. and Sons Fire Insurance	1/2
Robust Estimator Service Co.	1/20

a. What is the probability all the companies offer him a job?
b. What is the probability none of them will offer him a job?
c. What is the probability he is offered a job by at least one of the four companies?

74. Red-Green color-blind people see these two colors the same. This type of color blindness occurs with a probability of 0.049. The chance that a person is red-green color-blind, however, is quite dependent upon that person's sex, in fact. The proportion of people who are males and red-green color-blind is 0.042.

a. Define the events M = "male," R = "red-green color-blind." What is $P(R)$, $P(M$ and $R)$, and $P(M \mid R)$?
b. If a person is known to have red-green color blindness, what is the chance the person is female?
c. Assuming $P(M) = \frac{1}{2}$, what is the proportion of red-green color-blind men?

75. A local computer firm has an annual customer-employee convention in which certain employees are chosen to demonstrate the computer capabilities to their customers. Each year the convention is located in a different location. Last year it was in New Orleans, the year before that, in Dallas, and in general, always in some real nice place. All the employees like to attend this fun-filled, work-free (almost), expense paid convention, but only one expert on each computer model is sent. To be fair, the supervisor randomly selects qualified employees to attend.

a. If the company has five models, and there are 10 employees who are qualified on all the models (i.e., each of the 10 employees is qualified on each of the five models), how many ways can the group of employees be selected?
b. How many ways can the group be selected if
   i. five qualify on model I only?
   ii. three qualify on models I, II, and III only?
   iii. two qualify on models III, IV, and V only?
c. How many ways can the group be selected if two qualify only on each model?

76. Evaluate the following formulas:

a. $\binom{5}{3}$

b. $_5P_3$

c. $4!$

d. $\binom{10}{4}$

e. $_{15}P_4$

f. $\binom{12}{10}$

g. $\binom{7}{3}$

h. $\binom{13}{2}$ and $\binom{13}{11}$

77. How many seven-digit telephone numbers are possible (a telephone number cannot start with a 0)? All the numbers at the University of Texas start with 471-. How many numbers can the University of Texas have?

$q \cdot 33$ $5/100$ = colour blind male ie $P(M \cap C)$

$25/20,000$ = colour blind female

equal chance of being male | female ∴

$P(M) = .5$

$P(M') = P(F) = .5$

find $P[M/C]$

FoemuLa = $P(M/C) = \dfrac{P(M \cap C)}{P(C)}$ ⟶ prob of being male.

$P = \dfrac{\left(5/100\right)\left(\frac{1}{2}\right)}{\dfrac{5}{200} + \dfrac{25}{20,000}}$

ans. = $\dfrac{20}{21}$

$E(x) = X_1\left(f_{x_1}\right) + X_2\left(f_{x_2}\right) \dots X_n\left(f_{x_n}\right)$

$E(2) = \mu x$ ie $\mu =$

CHAPTER · 4

Random Variables, Probability Distributions, and Probability Models

Everything should be made as simple as possible, but not more so.

Albert Einstein

We are often interested in events that have a certain value to us, rather than in the occurrence of just any event. If we find only high gains or losses in a gambling game important, then our interest settles on those events that yield such results. We introduce the *value* of random events through the concept of the random variable and the expectation of a random variable.

We also study the construction of *mathematical models* of nature, for in the context of such models the concepts of probability find their most useful applications.

4 ■ 1   RANDOM VARIABLES

Let us suppose that we are playing a game in which we roll a pair of dice and count the number of dots that appear on the uppermost side. We let Y denote the number of dots facing up on a particular roll of the dice. For example, the roll

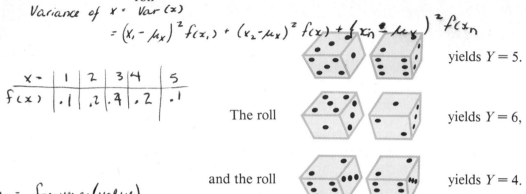

yields $Y = 5$.

The roll yields $Y = 6$,

and the roll yields $Y = 4$.

[handwritten annotations:]

Variance of $x = Var(x)$

$= (x_1 - \mu_x)^2 f(x_1) + (x_2 - \mu_x)^2 f(x) + (x_n - \mu_x)^2 f(x_n)$

$x =$	1	2	3	4	5
$f(x)$	.1	.2	.4	.2	.1

$\mu_x = \dfrac{\text{frequency (value)}}{n_{\text{total}}}$

$\sigma = \sqrt{npq}$

failure / sucess

\# trials / repititions.

Since each die has six sides, there are $6 \times 6 = 36$ possible outcomes for the game. We list these outcomes together with the associated values of Y (Figure 4.1). Thus we see that Y gives a number for each possible outcome. Y is just a rule that tells us the number assigned to each outcome. The set of numbers associated with the outcomes is called the range of Y. Y is called a variable since the values in the range of Y vary with the outcomes. Since the outcome of a roll of the dice is controlled by chance, and each value of Y has a probability of occurring, we call Y a *random variable*.

Figure 4 ■ 1
Possible outcomes of dice throw together with the values of the random variable Y that represents the sum of the faces.

Outcome of dice throw	Y value	Outcome of dice throw	Y value	Outcome of dice throw	Y value	Outcome of dice throw	Y value
1st 2nd die die		1st 2nd die die		1st 2nd die die		1st 2nd die die	
(1, 1)	2	(2, 4)	6	(4, 1)	5	(5, 4)	9
(1, 2)	3	(2, 5)	7	(4, 2)	6	(5, 5)	10
(1, 3)	4	(2, 6)	8	(4, 3)	7	(5, 6)	11
(1, 4)	5	(3, 1)	4	(4, 4)	8	(6, 1)	7
(1, 5)	6	(3, 2)	5	(4, 5)	9	(6, 2)	8
(1, 6)	7	(3, 3)	6	(4, 6)	10	(6, 3)	9
(2, 1)	3	(3, 4)	7	(5, 1)	6	(6, 4)	10
(2, 2)	4	(3, 5)	8	(5, 2)	7	(6, 5)	11
(2, 3)	5	(3, 6)	9	(5, 3)	8	(6, 6)	12

Standard deviation = ~~ka~~
σ

~~variance~~ = μ
~~median~~ *mean* = ~~mid~~

ie ~~midpoint~~. ~~ave~~

If an experiment, game, or social study is performed in which the outcomes are determined by chance, then the rule that assigns a numerical value to each outcome of the experiment is called a random variable. The set of assigned values is called the *range* of the random variable. A particular assigned value is called here a value of the random variable.

$\sigma_x = \sqrt{(x-\mu)^2 (f_x)} \, \Sigma$ It should be noted that once the experiment or study has been performed and the outcome determined, then the value of the random variable is specified. In our previous example, once the dice have been rolled yielding (3, 4), then the value of Y is no longer up to chance but is known to be $3 + 4 = 7$. In a social study, once a person is selected to participate and is given the required tests or measurements, the score is no longer a random variable but is determined. The range of the random variable represents the set of possible numerical outcomes of an experiment *before* the actual performance of the experiment. After performing the experiment, we obtain a *particular value* of the random variable, which is just a number.

Example 4 ▪ 1 Suppose we roll a pair of dice and let X denote the number of 3's that appear on the upper side of the dice. The possible outcomes of the dice are the 36 pairs listed previously. The values of X for each outcome are easily calculated. For example, (5, 4) yields $X = 0$, (3, 1) yields $X = 1$, and (3, 3) yields $X = 2$.

Example 4 ▪ 2 A doctor is interested in the reflex reaction rate of men who have had 5 oz of alcohol to drink. The experiment is to give a man 5 oz of alcohol and test his reflex reaction time. The random variable is R, the reflex reaction time, which varies from person to person. However, once a person is picked and the experiment performed, we obtain a specific value of R for that person at that time.

Example 4 ▪ 3 A marketing specialist desires to know the effectiveness of a certain type of packaging. The experiment she performs is to show the package to a prospective buyer and ask if he likes, dislikes, or is neutral to the package. A random variable to describe this experiment could be $W = +1$ if the response is "likes," $W = -1$ if the response is "dislikes," and $W = 0$ if the response is "neutral." Each value has a probability of occurrence.

We should also note here that if X and Y are two random variables, then $X + Y, X - Y$, and XY are also random variables.

4 ▪ 2 PROBABILITY DISTRIBUTIONS

Let us return to example 4.1. We roll a pair of dice and let X denote the number of 3's that appear. The possible values that X can assume are {0, 1, 2}, and by looking at the 36 previously listed possible outcomes of the dice, we see that

$X = 0$ occurs in 25 cases, $X = 1$ in 10 cases, and $X = 2$, once. We write $P(X = 0) = \frac{25}{36}$, where $P(X = 0)$ represents the probability of the set of elementary events for which X is equal to zero. We also find $P(X = 1) = \frac{10}{36}$ and $P(X = 2) = \frac{1}{36}$. The entire probability structure of the experiment is described by the knowledge of the preceding probabilities. This set of probabilities is called the distribution of the random variable X.

The distribution of a random variable tells us how the probabilities are distributed among the values that X takes on (i.e., the values in the range of X).

> Probability Distribution: The probability distribution of the random variable X is the set of values of X together with their respective probabilities.

For example, if X is a random variable that takes on the values $\{x_1, x_2, \ldots\}$ with probabilities $P(X = x_1)$, $P(X = x_2)$, . . . , respectively, then this collection of probabilities over $x_1, x_2, \ldots$ is called the distribution of the random variable X. Note particularly that

$$P(X = x_1) + P(X = x_2) + P(X = x_3) + \cdots = 1$$

Example 4 ▪ 4   Consider the random variable described in Figure 4.1; that is, Y denotes the sum of the numbers showing on a toss of a pair of dice. We have $P(Y = 2) = \frac{1}{36}$ since there is only one outcome that yields the value 2 for Y while there are 36 total outcomes. The probability that $Y = 3$, $P(Y = 3)$ is $\frac{2}{36}$ since there are two ways that yield $Y = 3$. The remaining probabilities are: $P(Y = 4) = \frac{3}{36}$, $P(Y = 5) = \frac{4}{36}$, $P(Y = 6) = \frac{5}{36}$, $P(Y = 7) = \frac{6}{36}$, $P(Y = 8) = \frac{5}{36}$, $P(Y = 9) = \frac{4}{36}$, $P(Y = 10) = \frac{3}{36}$, $P(Y = 11) = \frac{2}{36}$, $P(Y = 12) = \frac{1}{36}$.

Notice that since Y must have one of the values 2, 3, . . . , 12,

$$P(Y = 2) + P(Y = 3) + \cdots + P(Y = 12) = 1$$

We can draw a *density* histogram of the probability $P(Y = 2)$, $P(Y = 3)$, and so on. Figure 4.2 displays this histogram. For example, one can read $P(Y = 7)$ as the height of the bar over the interval 6.5 to 7.5. The height is $\frac{6}{36} = 0.167$. Note also that the area of the bar is equal to 0.167 as well since the base of the bar is 1. In a density histogram, the area of the bar always equals the probability.

Example 4 ▪ 5   The total number of outcomes or configurations for a throw of three dice is $6 \cdot 6 \cdot 6 = 216$; however, there are only 16 different values for the sum of the faces. Let X represent the random variable that represents this sum of faces, and let $f(x) = P(X = x)$. Figure 4.3 displays the probability distribution of X as the graph of $f(x)$ vs. x.

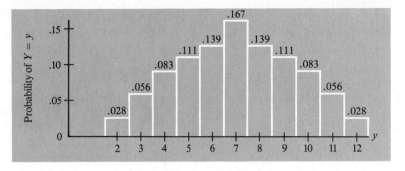

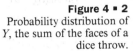

Figure 4 ▪ 2
Probability distribution of
Y, the sum of the faces of a
dice throw.

From Figure 4.3 we can calculate various probabilities by using the Addition Rule from Chapter 3. For example, $P(X = 3$ or $X = 6) = P(X = 3) + P(X = 6)$ = area in bar over $(x = 3)$ + area in bar over $(x = 6) = 0.005 + 0.046 = 0.051$. The probability that X is less than or equal to 6 is denoted by $P(X \le 6)$ and is equal to

$$P(X \le 6) = P(X = 3 \text{ or } X = 4 \text{ or } X = 5 \text{ or } X = 6)$$
$$= P(X = 3) + P(X = 4) + P(X = 5) + P(X = 6)$$

which we can find by adding the areas under the bars.

$$P(X \le 6) = 0.005 + 0.014 + 0.028 + 0.046 = 0.093$$

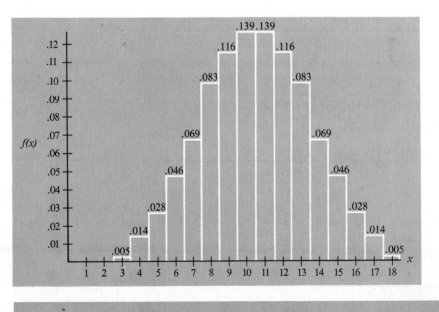

Figure 4 ▪ 3
Probability distribution of
the sum (x) of the faces for
a throw of three dice.

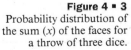

PROBABILITY DISTRIBUTION AND POPULATION DISTRIBUTION

Using the "frequency interpretation," we see that a probability distribution displays how frequently a random variable will take on various values in repeated enactments of an experiment or trial study.

We have seen that a *population distribution* (defined in Section 2.2) displays how frequently members of a population that possess a certain measured value will appear in that population. The Law of Large Numbers tells us that this particular value occurs about as frequently in a very large sample as it does in the population. Hence a *population* distribution displays the approximate frequency with which population values appear in a large sample. A *probability* distribution displays the approximate frequency of values of a random variable in a large sample. We see, then, that both distributions yield the same information; in other words, the population distribution and the probability distribution are different names for the same thing.

CONTINUOUS RANDOM VARIABLES: DENSITY FUNCTIONS AND DISTRIBUTIONS

Thus far we have considered in our examples mainly random variables that take on only a finite number of values. It is not difficult, however, to construct experiments for which *every* number in an interval constitutes a possible outcome. For example, suppose we are employed by a consumer affairs agency to test a manufacturer's claim that his light bulbs last at least 500 hr. We put a bulb in the socket and let X denote the lifetime of the bulb or 500 hr, whichever comes first. Certainly, any number from 0 to 500 is a possible value for X. Random variables for which entire intervals constitute the range of possible values are called *continuous random variables*.

> Continuous Random Variables: Suppose X is a random variable for which the range of possible values for X constitutes an entire interval. Then X is called a continuous random variable.

Example 4 ■ 6   Suppose we have a pointer that we spin on a wheel whose circumference is

marked with numbers greater than 0 and up to 2. If we let X denote the number that the pointer lands on, then X is a continuous random variable. All values of X are equally likely. The density histogram of X is shown in Figure 4.4a; it is called the continuous uniform distribution.

As before, the probability of obtaining a value of the random variable X between two values a and b is given by the area under that portion of the histogram that lies over the interval (see Fig. 4.4a).

The density histogram for continuous random variables is given a special name in the literature. It is called the *density function* of X. We denote it by $f(x)$.

> Density Function: The curve $f(x)$ is called the density function of the random variable X if the probability, $P(a < X \le b)$, of obtaining a value of X between a and up to and including b is the area under the curve between the points a and b.

Note that since the area under a single point is zero, we can loosely say things like "X between a and b" to denote either $a < X \le b$ or $a \le X \le b$, provided that we are dealing with continuous random variables.

Just as we previously defined a cumulative population distribution, we define the cumulative distribution for a random variable.

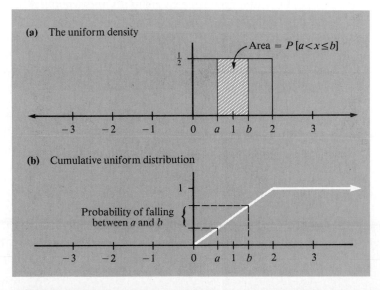

(a) The uniform density

Area $= P[a < x \le b]$

(b) Cumulative uniform distribution

Probability of falling between a and b

Figure 4 ▪ 4

> Cumulative Distribution: Suppose that X is a random variable. Then the cumulative distribution function of X (denoted by F_X) is defined for any number t by $F_X(t) = P(X \leq t)$. That is, the value of the cumulative distribution function at the number t is the probability of picking a value of X that is no larger than t.

Example 4 ▪ 7   Figure 4.4b shows the cumulative distribution of the density function of Example 4.6.

4 ▪ 5   MEAN AND VARIANCE OF A DISCRETE RANDOM VARIABLE

The mean of a population was defined (see Section 2.4) as

$$\mu = x_1 P(x_1) + x_2 P(x_2) + \cdots + x_n P(x_n)$$

where $P(x_i)$ is the relative frequency of the number x_i. From the preceding discussion, we see that a *population* distribution is the same as a *probability* distribution of a random variable X that takes on each population value x_i with probability $P(X = x_i) = P(x_i)$, $i = 1, \ldots, n$. The mean of a random variable with n values $x_1, x_2, \ldots, x_n$ in its range of values is then the same as the population mean μ as defined in Chapter 2, namely, $\mu = x_1 P(x_1) + \cdots + x_n P(x_n)$. We now present a slightly more general definition.

We consider here a random variable having in its range a *discrete* number of values $x_1, x_2, \ldots$. A discrete number of values is either a finite number of values or an infinite number of values such that you cannot make a continuous interval by putting any of them together. A random variable that has a discrete number of values in its range is called a *discrete random variable*.

> Mean of a Discrete Random Variable: If X is a random variable with values $x_1, x_2, \ldots$ that it takes on with probabilities $p_1 = P(X = x_1)$, $p_2 = P(X = x_2), \ldots$, then we define the mean value for X to be $\mu = x_1 p_1 + x_2 p_2 + \cdots$.

Example 4 ▪ 8   Let Y be a random variable that has distribution $P(Y = 0) = P(Y = 0.1) = \cdots = P(Y = 0.9) = \frac{1}{10}$. Such a distribution is called a uniform distribution over the set $(0, 0.1, \ldots, 0.9)$. Then

$$\text{The mean of the random variable } Y = 0 \cdot \frac{1}{10} + 0.1 \cdot \frac{1}{10} + \cdots$$

$$+ 0.9 \cdot \frac{1}{10} = \frac{0 + 0.1 + 0.2 + \cdots + 0.9}{10} = 0.45$$

Thus we see that if we are given a collection of numbers, and a random variable that is uniformly distributed over them, then the mean value of the random variable is precisely the arithmetic average of the numbers.

Example 4 ▪ 9   Let $X = 1$ if heads occurs on a toss of a coin and $X = 0$ if tails occurs. Let $p = P(X = 1)$. Then $(1 - p) = P(X = 0)$ and

$$\mu = 1 \cdot p + 0 \cdot (1 - p) = p$$

Variance of a Random Variable

Recall that the variance σ^2 of a population that contains the values x_i with relative frequency $P(x_i)$ was defined (Chapter 2, Section 5) as

$$\sigma^2 = (x_1 - \mu)^2 P(x_1) + \cdots + (x_n - \mu)^2 P(x_n)$$

where μ is the mean of the population. Hence we define the variance σ^2 of a random variable X the same way.

> Variance and Standard Deviation of a Discrete Random Variable: If X is a random variable with the range of values $x_1, x_2, \ldots$ that it takes on with probabilities $p_1 = P(X = x_1)$, $p_2 = P(X = x_2)$, $\ldots$, then the variance of X is
>
> $$\sigma^2 = (x_1 - \mu)^2 p_1 + (x_2 - \mu)^2 p_2 + \cdots$$
>
> where μ is the mean of X.
>   The standard deviation of $X = \sqrt{\sigma^2} = \sigma$.

Example 4 ▪ 10   Let $X = 1$ if heads occurs on a toss of a coin and $X = 0$ if tails occurs. Let $p = P(X = 1)$. Then the mean μ of X is equal to p (see example 4.9) and

$$\sigma^2 = (1 - p)^2 p + (0 - p)^2 (1 - p) = p(1 - p)$$
$$\sigma = \sqrt{p(1 - p)}$$

4 ▪ 6   EXPECTATIONS

We now develop another interpretation of the mean and variance of a random variable that is useful in a wide variety of problems and will enable us to generalize our notion of the "mean."

The early origins of probability theory were mostly concerned with gambling and games of chance. A gambler is primarily interested in increasing his fortune, and hence would like to determine (before wagering) what the expected return on his investment will be. He would specifically like to know if he plays a particular game over and over, what he could expect his "long-run" earnings to average *per game*. Suppose, for example, he plays a game in which there are k possible outcomes each time he plays. (Think of a "game" as being the toss of a die; there are six possible outcomes on each game.) Suppose, also, he wins x_1 dollars if outcome 1 occurs, x_2 dollars if outcome 2 occurs, and so forth. (In most games it is possible to lose; a loss means that the "win" is negative.) Since each outcome is random, each gain is random also.

To rephrase the preceding description, we have a random variable X that can assume the range of values $x_1, x_2, \ldots, x_k$, with probabilities $p_1 = P(X = x_1), p_2 = P(X = x_2), p_3 = P(X = x_3), p_4 = P(X = x_4)$, and so forth up to $p_k = P(X = x_k)$. Suppose we repeat the game n times under identical conditions. Let n_1 represent the number of times we obtain the value x_1, n_2 represent the number of times we obtain value x_2, and so on. After n trials or repeated experiments the

$$\text{Total gain} = x_1 n_1 + x_2 n_2 + \cdots + x_k n_k$$

The gain *per game* is the arithmetic average.

$$\text{Gain per game} = \frac{\text{total gain}}{n} = \frac{x_1 n_1 + x_2 n_2 + \cdots + x_k n_k}{n}$$

$$= x_1 \frac{n_1}{n} + x_2 \frac{n_2}{n} + \cdots + x_k \frac{n_k}{n}$$

By the Law of Large Numbers, we recall that if n is large, p_1 should be about equal to n_1/n, p_2 to n_2/n, and so on so that the *expected gain per game played* in the long run is

$$x_1 p_1 + x_2 p_2 + \cdots + x_k p_k$$

which is just the definition of the mean μ of a random variable; thus the mean μ can be interpreted as the gain *per game* a gambler can expect if he plays a very long time.

It should be remembered that μ is to be "expected" only in the sense that *in the long run* the average gain will converge to μ. It does not necessarily mean that half the observed gains will be above μ and half below, or that μ is the value that occurs most often.

Example 4 ▪ 11   A gambler enters a lottery. If his ticket wins he receives a $100 prize; however, if his ticket loses, he must pay the cost of a ticket (a negative win of $2). If he knows there is a probability of $\frac{1}{50}$ of his ticket being a winner, how much money can he expect to win? That is, if he could make the same wager over and over, how much would he expect to average per game in the long run?

X takes on the value $100 and $-$2, with probabilities $\frac{1}{50}$ and $\frac{49}{50}$, respectively. Hence

$$\mu = \$100 \, \frac{1}{50} + (-\$2) \, \frac{49}{50} = \$0.04$$

His average winning per game is $0.04. Note that the value of μ, which is $0.04, is not even a possible value for X to have; but rather, it is a long-run average value.

Example 4 ▪ 12   A stockbroker has $500,000 to invest. If we let X denote the value of her investment after 6 months, then it is known that the distribution of X is given by

X:	$0	$100,000	$300,000	$500,000	$800,000	$900,000	$1,000,000
$P(X = x)$:	0.01	0.09	0.30	0.10	0.40	0.05	0.05

What return should she expect to average on her investment? That is, what is μ? We calculate this quantity as follows:

$$\mu = (\$0)(0.01) + (\$100,000)(0.09) + (\$300,000)(0.30) + (\$500,000)(0.10)$$
$$+ (\$800,000)(0.40) + (\$900,000)(0.05) + (\$1,000,000)(0.05)$$
$$= \$564,000$$

Thus her expected return is $564,000 $-$ $500,000 $=$ $64,000.

Example 4 ▪ 13   An automobile dealer has four types of cars selling for $6000, $5000, $3500, and $2900. The probability a buyer will buy a $6000 car is $\frac{1}{8}$; a $5000 car, $\frac{1}{4}$; a $3500 car, $\frac{3}{8}$; a $2900 car, $\frac{1}{4}$. What is the average amount the dealer will receive on a sale? The random variable involved in this problem is X equal to the number of dollars earned on a sale. It depends on chance since chance determines which type of buyer enters the dealership.

To calculate μ we have

$$\mu = (\$6000)\left(\frac{1}{8}\right) + (\$5000)\left(\frac{1}{4}\right) + (\$3500)\left(\frac{3}{8}\right) + (\$2900)\left(\frac{1}{4}\right)$$

$$= \$750 + \$1250 + \$1312.50 + \$725 = \$4037.50$$

He should expect an average gain of \$4037.50 on the sale of each car.

Variance and Standard Deviation

When we discussed the mean value for a random variable we said that a gambler would like to find the average value of his winnings in order to determine whether or not to play the game in question. Since he only has a certain amount of money with which to wager, he also would be interested in the amount his fortune would vary from this average value from game to game. If there is a large amount of variation, then his chance of going broke is fairly large no matter what his expected winnings may average in the long run. To measure this variation, we use the variance σ^2 as previously defined. In financial analysis the mean is often used as a measure of expected return, and the variance as a measure of "expected risk" for an investment.

Example 4 ▪ 14   To compare two hybrid grains, a farmer would like to choose the hybrid with the highest yield and smallest variance. Suppose, for example, we have hybrids type A and type B. Let X denote the yield in bushels from type A and Y the yield in bushels of type B. Suppose we know that X and Y have the following distributions:

x	$P(X = x)$	y	$P(Y = y)$
1	0.2	2	0.4
2	0.1	3	0.4
3	0.4	4	0.1
4	0.2	5	0.1
5	0.1		

Then $\mu_X = 1 \cdot (0.2) + 2 \cdot (0.1) + 3 \cdot (0.4) + 4 \cdot (0.2) + 5 \cdot (0.1) = 2.9$ and $\mu_Y = 2 \cdot (0.4) + 3 \cdot (0.4) + 4 \cdot (0.1) + 5 \cdot (0.1) = 2.9$ so the mean yields are the same. Which hybrid is more consistent? It is the one with the smaller variance.

$$\sigma_X^2 = (1 - 2.9)^2(0.2) + (2 - 2.9)^2(0.1) + (3 - 2.9)^2(0.4) + (4 - 2.9)^2(0.2)$$

$$+ (5 - 2.9)^2(0.1) = 1.49$$

$$\sigma_Y^2 = (2 - 2.9)^2(0.4) + (3 - 2.9)^2(0.4) + (4 - 2.9)^2(0.1)$$

$$+ (5 - 2.9)^2(0.1) = 0.89$$

Thus the farmer should plant hybrid type B since

$$\sigma_X = \sqrt{1.49} = 1.22 \quad \text{and} \quad \sigma_Y = \sqrt{0.89} = 0.943$$

4 ▪ 7   THE INTRODUCTION OF EXPECTED VALUE NOTATION: $E(X)$ AND $V(X)$

One interpretation of the mean μ of a random variable X is the expected total gain per game or trial. Many texts use the notation $E(X)$ for μ to make this idea explicit.

> $E(X)$ is read "expected value of X" and denotes the mean of X.

We adopt this notation here as well for the following reason: We have occasion to discuss more than one random variable at a time, say, X and Y, and it is convenient to distinguish their expected values by the notation $E(X)$, $E(Y)$.

For the same reason we often denote the variance of a random variable X by the notation $V(X)$.

4 ▪ 8   EXPECTATION OF THE SUM OF RANDOM VARIABLES

If X and Y are any two random variables, then $E(X + Y) = E(X) + E(Y)$. We see that expected value is additive. If we expect to receive $E(X)$ dollars by playing "game one" and $E(Y)$ dollars by playing "game two," then combining the two we expect to receive $E(X) + E(Y)$ dollars by playing both games. In general, we have the following rule for the expected value of the sum of n random variables $X_1, X_2, \ldots, X_n$.

> Property of the Sum of Random Variables:
>
> $$E(X_1 + X_2 + \cdots + X_n) = E(X_1) + E(X_2) + \cdots + E(X_n)$$

Example 4 ▪ 15   Let X_i represent the possible outcomes of the ith toss of a coin; then the outcomes are 0 and 1, with $P(X_i = 1) = p$, $P(X_i = 0) = 1 - p$. Let $S = \sum_{i=1}^{n} X_i$ represent the total number of times the outcome 1 occurs in n tosses. The expected number of times is $E(S)$; and hence by the preceding property, $E(S) =$

$E(X_1) + \cdots + E(X_n)$. We note that for each $i = 1, 2, \ldots, n$, $E(X_i) = 1 \cdot p + 0 \cdot (1 - p) = p$. Thus the expected value of the sum is equal to $p + p + \cdots + p = np$.

Example 4 ■ 16 I play two games. My expected winnings in the first game is $10; in the second game, $-$16. By the previous property, my expected combined winnings are $-$6.

Application to Gambling: The St. Petersburg Paradox

The following game illustrates the fact that an expected value may be infinite; in fact, a gambler playing this game may expect to win an infinite amount of money. Legend has it that this game was first proposed by a poor mathematics student who saw a chance to gain a fortune by breaking a St. Petersburg casino. The legend relates that the student succeeded and that, ever since, the St. Petersburg (Leningrad) casinos employ a resident mathematician to analyze prospective games. Let us examine this student's proposal and his reasoning (assuming he existed).

A coin was to be tossed until heads appeared. The student offered to pay a fixed fee for each trial, say, 5 rubles. If heads occurred on trial n, the student would win 2^n rubles minus the entrance fee. If heads appeared on the first trial, the total gain would be 2 rubles minus the entrance fee of 5 rubles; that is, the gain would be $2 - 5 = -3$ rubles. If heads appeared for the first time on trial 2, the student would gain 2^2 rubles minus the total entrance fee paid for the two tosses; that is, the gain would be $2^2 - 5 \cdot 2 = -6$ rubles. If heads appeared for the first time on the third trial, the gain to the student would be $2^3 - 5 \cdot 3 = -7$ rubles. In general, the student agreed to play a game such that if heads appeared for the first time on trial n, the gain would be $2^n - 5 \cdot n$. The probability of this gain is $(\frac{1}{2})^n$.

$$\text{The expected gain} = (2 - 5)\frac{1}{2} + (2^2 - 5 \cdot 2)\left(\frac{1}{2}\right)^2 + (2^3 - 5 \cdot 3)\left(\frac{1}{2}\right)^3$$

$$+ (2^4 - 5 \cdot 4)\left(\frac{1}{2}\right)^4 + (2^5 - 5 \cdot 5)\left(\frac{1}{2}\right)^5 + \cdots$$

$$+ (2^{30} - 5 \cdot 30)\left(\frac{1}{2}\right)^{30} + \cdots = \infty$$

To understand why the sum is infinite, note that the first term is $(1 - 5 \cdot 1/2^1)$; the second, $(1 - 5 \cdot 2/2^2)$; the third, $(1 - 5 \cdot 3/2^3)$; and so on. The second term in each parenthesis gets smaller and smaller as the number of trials increases; for example, $(1 - 5 \cdot 30/2^{30}) = 1 - 0.000000139 = 0.999999961$; the second term in the parenthesis is very small. Thus the terms get very close to 1 as the number of terms becomes large so that the sum up to the nth trial is about equal to n for n large; for example, the sum of the first 30 terms is about 20; for 100 terms, about 90; for 2000 terms, about 1990; and so forth. Hence as the number of trials goes toward infinity, so does the sum.

Note that the expectation is infinite only if the casino has an infinite amount of money. No casino has this quantity of rubles. Thus, in any real situation, the expectation is not infinite. Nevertheless, if the casino had a very large amount of money as compared to the entrance fees, and if the student had the stamina, he could have come out quite a winner after playing many, many such games.

4 ▪ 9   OTHER IMPORTANT PROPERTIES OF EXPECTATION

There are two other important properties of the expectation. These are

> *Property 1.* If a is any number, X any random variable, then $E(aX) = aE(X)$
> *Property 2.* If a is any constant, then $E(a) = a$.

The proof of these properties is given in problem 21.

Example 4 ▪ 17   Let $X = \$2$ if a head appears on the toss of a fair coin and $-\$2$ otherwise; then $E(X) = 0$. Suppose the stakes are doubled and there is an entrance fee of $\$1$. Then the outcome of the new game is given by $Y = 2X - 1$, where $Y = \$3$ if a head appears and $-\$5$ otherwise.

$$E(Y) = \$3 \cdot \frac{1}{2} - \$5 \cdot \frac{1}{2} = -\$1$$

We can also calculate $E(Y)$ by using the properties of the expected value. Since $Y = 2X - 1$ (justify each step),

$$E(Y) = E(2X - 1) = E(2X) + E(-1)$$
$$= 2E(X) - 1 = 2 \cdot 0 - 1 = -1$$

4 ▪ 10   AN IMPORTANT PROPERTY OF $V(X)$

In Section 4.7, we introduced the notation $V(X)$ for the variance of the random variable X. Let a and b be any numbers, X a random variable, and let

$Y = aX + b$. The following property is proven in problem 22.

$$V(Y) = V(aX + b) = a^2 V(X)$$

Note particularly that multiplying constants come outside the parenthesis as squared, whereas added constants are ignored.

Example 4 ▪ 18 Suppose $P(X = 2) = P(X = -2) = \frac{1}{2}$ and $Y = 2X - 1$. Then $E(X) = 0$, $E(Y) = -1$ and

$$V(X) = 4 \cdot \frac{1}{2} + 4 \cdot \frac{1}{2} = 4$$

By definition,

$$V(Y) = (3 + 1)^2 \cdot \frac{1}{2} + (-5 + 1)^2 \cdot \frac{1}{2} = 16$$

We can also use the preceding formula to obtain

$$V(Y) = V(2X - 1) = 4V(X) = 16$$

4 ▪ 11   DEPENDENCE AND INDEPENDENCE OF RANDOM VARIABLES: INDEPENDENT TRIALS

The idea of independent random variables is an integral part of all statistical theory. Indeed, the mathematical models necessary to make statistical inferences would be virtually unmanageable without the knowledge of independence and dependence of random variables. To understand these concepts, we must first understand that when we write the symbol, $(X = 3)$, this symbol stands for an *event*, namely, the event that the random variable X takes on the value 3. Similarly, $(Y = 7)$ represents the event that the random variable Y takes on the value 7. If these two events are independent, we know that $P(X = 3 \mid Y = 7) = P(X = 3)$. This leads us to the following definition.

> Independent Random Variables: X and Y are independent if $P(X = x \mid Y = y) = P(X = x)$ for every value x and y that X and Y can take on respectively. If random variables are not independent, they are said to be dependent.

Example 4 ▪ 19   a. Let X be the outcome of a flip of a coin and Y the outcome of a roll of dice. Then X and Y are independent since $P(X = \text{heads} \mid Y = \text{dice outcome}) = P(X = \text{heads})$; also, $P(X = \text{tails} \mid Y = \text{dice outcome}) = P(X = \text{tails})$.

b. Let X be the weight of a person before going on a particular diet for 2 weeks, and let Y be the weight of the same person after the diet. Then X and Y are dependent random variables since the chance that $Y = 70$ lb is certainly different for $X = 200$ lb from that for $X = 80$ lb.

Suppose we have n random variables $X_1, X_2, \ldots, X_n$. We can generalize the previous concept of independence (i.e., that one variable conveys no information about another).

> Mutually Independent Random Variables: If $P(X_i = x \mid$ any specific values of the other random variables$) = P(X_i = x)$ holds for each x that X_i can take on, then the random variables are said to be *mutually independent*. Thus no configuration of observed values for the other random variable can change our assessment of the chances that $X_i = x$.

Just as for events, the Product Rule holds for mutually independent random variables.

> Product Rule: If $X_1, X_2, \ldots, X_n$ are mutually independent, then
>
> $$P(X_1 = x_1 \text{ and } X_2 = x_2 \text{ and } \cdots \text{ and } X_n = x_n)$$
> $$= P(X_1 = x_1)P(X_2 = x_2) \cdots P(X_n = x_n)$$
>
> In addition, if the Product Rule holds, then the random variables are mutually independent.

Example 4 ▪ 20   Let $X_1, X_2, \ldots, X_n$ represent n mutually independent random variables where for each X_i

$$P(X_i = 1) = p$$
$$P(X_i = 0) = 1 - p$$

Then, $P(\text{all } X_i = 1) = P(X_1 = 1 \text{ and } X_2 = 1 \text{ and } \cdots \text{ and } X_n = 1)$

$$= P(X_1 = 1)P(X_2 = 1) \cdots P(X_n = 1)$$

$$= \overbrace{p \cdot p \cdot p \cdots p}^{n} - p^n$$

Example 4 ■ 21   A coin is tossed until heads appears for the first time and then the game stops. Let $X_i = 1$ if heads appears on the ith toss and 0 otherwise; and $P(X_i = 1) = p$, $P(X_i = 0) = 1 - p$ for each toss i. Assume the tosses are mutually independent; that is, assume the random variables are mutually independent. What is the probability that the first head will occur on the fifth (5th) toss?

This event is "$X_1 = 0$ and $X_2 = 0$ and $X_3 = 0$ and $X_4 = 0$ and $X_5 = 1$." Hence

$$P(\text{first head on 5th toss}) = P(X_1 = 0)P(X_2 = 0)P(X_3 = 0)P(X_4 = 0)P(X_5 = 1)$$

$$= (1 - p)(1 - p)(1 - p)(1 - p)p$$

$$= (1 - p)^4 p$$

Independent Trials

Consider a sequence of coin tosses and assume that a knowledge of any of the outcomes does not give us information about any of the other outcomes. Under these conditions, we feel that the trials are "independent." We can precisely articulate this concept in terms of mutually independent random variables.

Independent Trials: Let the random variable X_i represent the outcome of the ith trial. If, for any number of trials, n, the n random variables $X_1, X_2, \ldots, X_n$ are mutually independent, the trials are said to be *independent*.

Example 4 ■ 22   Example 4.21 illustrates a problem involving independent trials. Namely, the tosses of a coin constitute independent trials.

Example 4 ■ 23   One hundred people are randomly selected for a heart disease study. Each person has a physical examination and their blood pressure is recorded. The 100 readings constitute 100 independent trials.

4 ▪ 12   VARIANCE OF THE SUM OF MUTUALLY INDEPENDENT RANDOM VARIABLES

We have the following important property of the variance:

> The variance of the sum $X_1 + X_2 + \cdots + X_n$, where the n random variables are mutually independent, is equal to the sum of the variances that is $V(X_1) + V(X_2) + \cdots + V(X_n)$.

Example 4 ▪ 24   Consider n independent trials, where for each X_i

$$P(X_i = 1) = p$$
$$P(X_i = 0) = 1 - p$$

We know from example 4.10 that $V(X_i) = p(1 - p)$ for each i. Then

$$V(X_1 + X_2 + \cdots + X_n) = V(X_1) + V(X_2) + \cdots V(X_n)$$
$$= p(1 - p) + p(1 - p) + \cdots + p(1 - p)$$
$$= np(1 - p)$$

4 ▪ 13   REPEATED INDEPENDENT TRIALS— THE BINOMIAL MODEL

Coin-Tossing Model

One probability model that occurs quite often is an analog of the model for the outcomes of coin flippings. Rather than stating this model in terms of heads and tails on a coin flip, we state it in terms of obtaining a success or a failure in an experiment or study. The assumptions of this model are

> **Coin-tossing or Binomial Model**
> 1. The experiment, study, or observation is performed n times under identical conditions. The sample space consists of the set of all sequences of outcomes on the individual experiment or study. Each individual experiment, study, or observation is called a *trial.*
> 2. There are exactly two possibilities (or possible outcomes) on each individual trial, which we label as success and failure, or 1 and 0.
> 3. Let $X_i = 1$ if a success occurs on trial i and $X_i = 0$ if a failure occurs on trial i. The probability of obtaining a success is the same for each of the n trials; that is, $P(X_1 = 1) = P(X_2 = 1) = \cdots = P(X_n = 1)$.
> 4. Trials are independent.

Some examples of phenomena that have been described by this model in the scientific literature are

1. a sequence of births in a hospital in which each birth is classified as being either a boy or a girl
2. a sequence of items (say, machine parts) being produced on an assembly line in which each item is classified as acceptable or defective depending on whether or not it is in specification
3. a sequence of missile firings in which each firing is rated as a success or failure
4. an examination of a group of students in which each is judged to have passed or failed.

It is important to emphasize that once we have decided that a particular phenomenon such as example 1 can be described by independent trials, then we may apply all conclusions drawn from the general model to the particular phenomenon in question. For example, we soon describe a formula based on this model that will permit us to calculate $P(k$ successes in n trials). Since each of the preceding four phenomena can be described by the binomial model, we immediately will have a formula for finding the probability of k boys in a sequence of n births, or the probability of k "acceptable" items in a production of n items off a production line, and so on. Thus once having decided that a given phenomenon satisfies the description given by the preceding model, we can automatically answer many questions concerning the phenomenon. It is not surprising that a scientist or engineer encountering a new phenomenon at once begins to look about for a thoroughly investigated model such as the previous one to describe this phenomenon. If one can be found, then many results can be immediately derived. However, before using a given model, one must ascertain whether it is applicable to the given phenomenon. The phenomenon must first be tested to determine if it satisfies the underlying postulates of the model. As an example, suppose we are treating the phenomenon of weather. In particular, suppose we wish to know the probability there will be three rainy days in September. Each day is a trial, and there are only two possible outcomes on each trial, "rain" and "no rain." One might be inclined to use the formula for $P(3$ successes in 30 days), based on a sequence of 30 independent trials, where success stands for rain. To see whether or not this model is applicable, we test the phenomenon in question to determine if the assumptions hold. First, characteristic 2 of the model is satisfied since there are only two possible outcomes on each trial. We may suppose that the probability of rain on any given day is the same as on any other day; that is, we may suppose 3 holds. However, certainly 4 is not satisfied since bad weather tends to last several days in a row, and hence the trials or daily events are not independent. Thus the coin-tossing model is not applicable to this particular phenomenon.

One further comment: Regarding characteristic 4 of the binomial model, there has been much discussion concerning its applicability to such phenomena as coin tossing, roulette, and gambling games, in general. We emphasize that the binomial model may or may not be suitable as a description of a given phenomenon. A model must be tested by determining from *experience* whether the conclusions one draws from this model are valid. The suitability of a model is determined from experimental evidence. This is a matter we shall study in statistics.

We are interested in a formula for the probability of obtaining exactly k successes out of n independent trials. For example, an assemble-it-yourself bookcase requires 10 screws. If the production of screws by the manufacturer is represented as a sequence of independent trials, where success on trial i means the ith screw produced is nondefective, then we are interested in P(box of 10 screws contains exactly k nondefectives) $= P(k$ successes in 10 independent trials). Appendix B presents a derivation of the formula for calculating this probability. This formula is as follows:

The probability of obtaining exactly k successes in a sequence of n independent trials is

$$P(k \text{ successes}) = \frac{n!}{(n-k)!k!}\, p^k (1-p)^{n-k}$$

where the symbol $n!$ is read "n factorial" and represents the product $n(n-1)(n-2) \cdots 1$. We write $\dfrac{n!}{(n-k)!k!}$ in the shorthand $\dbinom{n}{k}$. The values $\dbinom{n}{k}$ for $n \leq 20$ are given in Appendix A.2.

These probabilities are called *binomial probabilities.* The binomial probabilities for various values of n and p are given in Appendix A.8.

Example 4 ■ 25   If an assemble-it-yourself bookcase requires 10 screws and each screw has a probability of 0.05 of being defective, what is the probability a box of 11 screws will have less than 10 good screws? This is an application of the binomial model, with $n = 11$ and where success corresponds to getting a good screw; $p = P$(success) $= 0.95$. The desired probability is

$$P(\text{less than 10 good screws}) = 1 - P(10 \text{ or } 11 \text{ good screws})$$

$$= 1 - P(10 \text{ good}) - P(11 \text{ good})$$

$$= 1 - \binom{11}{10}(0.95)^{10}(0.05)^1 - \binom{11}{11}(0.95)^{11}(0.05)^0$$

$$= 1 - (11)(0.599)(0.05) - 0.569$$

$$= 1 - .329 - 0.569 = 0.102$$

There is approximately a 10.2% chance of obtaining less than 10 good screws.

Example 4 ■ 26   A social scientist has determined that approximately 60% of the people who received a questionnaire will respond on the first mailing. If 20 questionnaires are mailed out, what is the probability less than three people respond? This probability is calculated by using the binomial model, with $n = 20$ and $p = P(\text{returning a questionnaire}) = 0.6$. Thus

$P(\text{less than 3 responses}) = P(0, 1, \text{ or } 2 \text{ responses})$

$$= \binom{20}{0}(0.6)^0(0.4)^{20} + \binom{20}{1}(0.6)^1(0.4)^{19} + \binom{20}{2}(0.6)^2(0.4)^{18}$$

$$= 0.00000001098 + 0.0000003 + 0.0000047 = 0.00000501.$$

There is less than 1/1000% chance of less than three responses. What is the probability that 16 of the questionnaires will be returned? This probability is

$$P(16 \text{ responses}) = \binom{20}{16}(0.6)^{16}(0.4)^4$$

$$= 0.035, \text{ a } 3\frac{1}{2}\% \text{ chance}$$

Example 4 ■ 27   A medical test yields a positive result on 20% of the people who do not actually have the disease. If 15 people who do not have the disease are given the test, what is the probability 3 of them will show a positive result on the test? This can be answered by the binomial model with $n = 15$ and $p = P(\text{positive result}) = 0.2$. We have

$$P(3 \text{ positive results}) = \binom{15}{3}(0.2)^3(0.8)^{12} = (455)(0.008)(0.0687) = 0.25$$

Example 4 ■ 28   Assume the coin-tossing model with n trials. The probability of *at least* one success is $1 - \text{probability of 0 successes} = 1 - (1 - p)^n$. The probability of *greater than* one success is

$1 - \text{probability of exactly 0 } or \text{ exactly 1 success}$

$\quad = 1 - \text{probability of 0 successes} - \text{probability of exactly one success}$

$\quad = 1 - (1 - p)^n - np(1 - p)^{n-1}$

Given that *at least* one success was *observed* in n trials, what is the probability of *more than* one success? Let A be the observed event of "at least one success," and U be the event of "more than one success."

$$P(U|A) = \frac{P(U \text{ and } A)}{P(A)}$$

$$= \frac{P(U)P(A|U)}{P(A)}$$

Note that $P(A|U)$ is the probability of at least one success, given more than one success. But if you know there is *more than one success,* there must be at least one; hence $P(A|U) = 1$. Thus $P(U|A) = P(U)/P(A)$ or

$$P(U|A) = \frac{1 - (1-p)^n - np(1-p)^{n-1}}{1 - (1-p)^n}$$

Suppose $p = 1/12{,}000{,}000$ and $n = 12{,}000{,}000$, then the probability of more than one success, given at least one success, is about 0.41.

This calculation served as the basis for the reversal of a criminal conviction in San Pedro, California. This matter is discussed in the application section that follows

Application to Law: *People* vs. *Collins* (review example 4.28.)

This case was discussed in Section 3.6. Witnesses claimed that a robbery was committed by a black man with a beard accompanied by a white woman who had a blond ponytail and who drove a partly yellow automobile. Police found a married couple, Collins, that fit this description. The prosecution claimed that the chance of the existence of such a couple is $1/12{,}000{,}000$. The prosecutor concluded, and the jury agreed, that since the characteristics of the "identified" couple were so rare, the Collins couple must be the one that committed the crime.

The appeals court used the boxed formula of example 4.28 to refute this conclusion. The event, success, was defined by the court as "finding a couple such as Collins." It denoted this event by C. After deriving the preceding formula, the court argued as follows:

> Turning to the case in which C represents the characteristics which distinguish a bearded Negro accompanied by a ponytailed blond in a yellow car, the prosecution sought to establish that the probability of C occurring in a random couple was $1/12{,}000{,}000$ — i.e., that $p = 1/12{,}000{,}000$. Treating this conclusion as accurate, it follows that, in a population of n random couples, the probability of C occurring *exactly once* is $\{(n) \times (1/12{,}000{,}000) \times (1 - 1/12{,}000{,}000)^{n-1}\}$. Subtracting this product from $\{1 - (1 - 1/12{,}000{,}000)^n\}$, the probability of C occurring in *at least*

one couple, and dividing the resulting difference by $\{1 - (1 - 1/12{,}000{,}000)^n\}$, the probability that C will occur in at least one couple, yields the probability that C will occur more than once in a group of n random couples of which at least one couple (namely, the *one seen by the witnesses*) possesses characteristics C. In other words, the probability of *another* such couple in a population of n is the quotient A/B, where A designates the numerator $\{1 - (1 - 1/12{,}000{,}000)^n - (n) \times 1/12{,}000{,}000 \times (1 - 1/12{,}000{,}000)^{n-1}\}$, and B designates the denominator $\{1 - (1 - 1/12{,}000{,}000)^n\}$.

n, which represents the total number of all couples who might conceivably have been at the scene of the San Pedro robbery, is not determinable, a fact which suggests yet another basic difficulty with the use of probability theory in establishing identity. One of the imponderables in determining n may well be the number of n couples having easy access to this San Pedro bank. Such considerations make it evident that n, in the area adjoining the robbery, is in excess of several million; as n assumes values of such magnitude, the quotient A/B computed as above, representing the probability of a second couple as distinctive as the one described by the prosecution's witnesses, soon exceeds $\frac{4}{10}$. Indeed as n approaches 12 million, this probability quotient rises to approximately 41 percent. . . .

Hence, even if we should accept the prosecution's figures without question, we would derive a probability of over 40 percent that the couple observed by the witnesses could be "duplicated" by at least one other equally distinctive interracial couple in the area, including a Negro with a beard and mustache, driving a partly yellow car in the company of a blond with a ponytail. Thus the prosecution's computations, far from establishing beyond a reasonable doubt that the Collins were the couple described by the prosecution's witnesses, imply a very substantial likelihood that the area contained *more than one* such couple, and that a couple *other* than the Collins' was the one observed at the scene of the robbery.

To complete the argument the judge should have written: Suppose there was at least one other Collins-type couple; in the absence of other evidence, suppose the chance that another couple did the crime is *at least* 0.5. The chance that there is another couple is 0.4. Hence the chance another couple committed the crime is *at least* $0.4 \times 0.5 = 0.2$ or 20%, which is too great a chance to convict the Collins.

Binomial Distribution

The density histogram shown in Figure 4.5 displays the probabilities derived from binomial formula; namely, the probabilities.

$$\binom{n}{x} p^x (1 - p)^{n-x} = \text{probability of } x \text{ successes in } n \text{ trials}$$

with $n = 10$, $p = 0.2$, and $x = 0, 1, \ldots, 10$. The histogram of the binomial probabilities is called the binomial distribution.

Since the base of each bar is of length equal to 1, both the height of each bar and the area of each bar are equal to the probability of the integer under the bar.

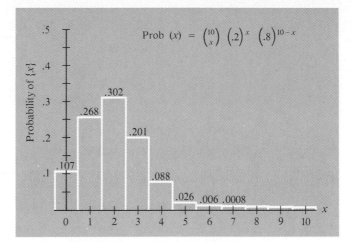

Figure 4 ■ 5
Probability of x successes
in n trials for the binomial.

From the figure we can calculate the probability of various events. For example, we can calculate the probability that the number of successes is 2, 3, 4, or 5; that is, we can calculate the probability that the number of successes is greater than 1 and less than 6 by adding the areas of the bars over 2, 3, 4, and 5.

We now derive the two important descriptive parameters of this distribution, namely, the mean and the variance. These actually have been derived before in examples 4.15 and 4.24, but at the time we did not explicitly state the connection with the binomial distribution.

Expected Number of Successes; Mean of the Binomial Distribution

Consider again the binomial model. The random variable $X_i = 1$ if a success occurs on the ith trial, whereas $X_i = 0$ if a failure occurs on the ith trial. Let us call the sum of the random variables $X_1 + X_2 + \cdots + X_n$ by the name "S." If all the random variables $X_i = 0$, then S $= 0$; hence S $= 0$ represents the event "0 success in n trials." If only one of the $X_i = 1$, then S $= 1$, which the event "1 success in n trials." If exactly k of the $X_i = 1$, then S $= k$, which is the event "k successes in n trials." Hence

$$P(S = k) = P(k \text{ successes in } n \text{ trials})$$

$$= \binom{n}{k} p^k (1 - p)^{n-k} \qquad k = 0, 1, \ldots, n$$

We see, then, that the random variable S has a distribution given by the binomial probabilities. The distribution is called the binomial distribution. We can find the expected value of S by noting the fact that $S = \sum\limits_{i=1}^{n} X_i$ and recalling the property that the expected value of a sum of random variables is the sum of the expected values of the random variables. Then

$$E(S) = E(X_1 + X_2 + \cdots X_n)$$
$$= E(X_1) + E(X_2) + \cdots + E(X_n)$$

But for each i,

$$E(X_i) = 1 P(X_i = 1) + 0 P(X_i = 0)$$
$$= P(X_i = 1) = p$$

Hence

$$E(S) = np$$

Since the distribution of S is the binomial distribution, we have the result

Mean μ of the Binomial Distribution: $\mu = np$

Example 4 ▪ 29   Suppose the probability is 0.5 that the sex of a newborn baby is female. What is the mean or average number of females in 100 births? In this case, $n = 100$, $p = 0.5$ so that $\mu = 50$.

Variance of Binomial Distribution

Since S is the sum of mutually independent random variables, we have

$$V(S) = V(X_1) + V(X_2) + \cdots + V(X_n)$$

where $V(X_i)$ represents the "variance of X_i." Recall that

$$V(X_i) = (1 - p)^2 P(X_i = 1) + (0 - p)^2 P(X_i = 0)$$

where $p = E(X_i)$. Hence

$$V(X_i) = p(1 - p) \text{ for each } i$$

so that

$$V(S) = np(1 - p)$$

Consequently,

> Variance, σ^2, of the Binomial Distribution:
>
> $$\sigma^2 = np(1 - p)$$

Example 4 ■ 30   The binomial with $p = 0.5$ and $n = 100$ has $\mu = 100(0.5) = 50$ and $\sigma^2 = 100(0.5)(0.5) = 25$ and $\sigma = \sqrt{25} = 5$.

4 ■ 14   OTHER MODELS (OPTIONAL)

Independent Random Sampling; Sampling With and Without Replacement; Simple Random Sampling

Independent Random Sampling and Sampling with Replacement

An important assumption central to many probability models concerns the type of sampling used. For example, in the coin-tossing model we have assumed *independent trials.* In terms of sampling procedures, *independent trials* means that on each trial a population is sampled and the outcome is independent of those on any other trial. The term "independent random sampling" is often used in place of "independent trials."

> When we state the assumptions, "independent trials" or "independent random sampling" we mean the assumption that the observations are obtained in a sequence of "experiments" that are mutually independent.

The term "independent random sampling" is used in the field of sampling theory, which we discuss subsequently.

Example 4 ▪ 31   **Independent random sampling.** A sample of size *n* is a collection of *n* observations from a sample space or *population*. Suppose our sample space is a box filled with 90 tags, one third of which have a 0 written on them, whereas two thirds have a 1 written on them. We mix the tags and select one of them. After noting the number on the tag, we return it to the box, mix the tags again, and then select another tag. We repeat this procedure, which is called *sampling with replacement,* until we obtain 10 observations. Note that the samples are *independent;* that is, the result of one sample gives no information as to the result of any other sample. Sampling with replacement such that the samples are independent yields an *independent random sample.*

The probability of obtaining seven 1's in the preceding situation by using independent random sampling is

$$\binom{10}{7}\left(\frac{2}{3}\right)^7\left(\frac{1}{3}\right)^3 = 0.26$$

that is, this probability is given by the binomial formula you have previously studied.

Sampling Without Replacement

A typical problem we have encountered is the following: Two balls are drawn from an urn containing six red balls and three white balls. What is the probability of obtaining a red ball on the first trial and a red ball on the second trial? Assume in each trial that all balls in the urn are equally likely.

The probability of selecting a red ball on the first trial is $\frac{6}{9}$ and of selecting a white ball on the first trial is $\frac{3}{9}$. (Why?) If sampling were done with replacement; that is, if the selected ball is returned to the urn after selection, the situation would remain the same as at the start of sampling and the probability that a red ball is again chosen would still be $\frac{6}{9}$, no matter which ball was first selected and the sampling would be independent random sampling. If the red ball is not returned after selection (called sampling without replacement), then only five red balls remain and the probability of obtaining a red ball on the second trial, given a red ball on the first draw, would be $\frac{5}{8}$. The probability of obtaining a red ball on the second trial, given a white ball on the first trial, is $\frac{6}{8}$. The probability of obtaining a red ball on the second trial is equal to

P(red ball on first *and* second trial or white on first *and* red on second)

$$= \left(\frac{6}{9}\right)\left(\frac{5}{8}\right) + \left(\frac{3}{9}\right)\left(\frac{6}{8}\right) = \frac{6}{9}$$

Note that the trials are not independent since the probability of a red ball on the second trial given a red ball on the first trial $= \frac{5}{8} \neq \frac{6}{9}$.

*Simple Random Sampling**

Suppose we have N members in the population and we select n *without replacement*. The number of different samples without regard to the order in which the observations occur is given by the expression $\binom{N}{n}$. (The derivation of this formula is given in Chapter 3, Section 8.) If these $\binom{N}{n}$ samples are all equally likely, the sampling is said to be *simple random sampling*.

Example 4 ▪ 31   Given an urn containing six red balls and three white ones and sampling without replacement, what is the probability of obtaining one red and one white ball (without respect to the order of the drawing)? Assume "equally likely outcomes."

There are $\binom{9}{2}$ possible samples. By the assumption of "equally likely outcomes," these samples are all equally likely. The number of ways to get one red ball is $\binom{6}{1}$, and for one white ball is $\binom{3}{1}$, so that the number of samples containing one red and white ball is

$$\binom{6}{1}\binom{3}{1}$$

Thus this probability is

$$\frac{\binom{6}{1}\binom{3}{1}}{\binom{9}{2}} = \frac{1}{2}$$

This example leads to our next model.

---

* This material is used in Chapters 13 and 14 and may be reviewed just before those chapters are studied.

Hypergeometric Distribution and Its Expectation

An important model is the following:

Hypergeometric Model.
a. The sample space (population) consists of *N* objects; *m* of them are of one type and *k* of another type.

$$N = m + k$$

b. A sample of *n* objects is selected, using *simple random sampling*; that is, all possible samples of size *n* are equally likely. Let *X* = *c* if exactly *c* of the *m* things of type 1 are obtained in the sample of size *n*; then

$$P(X = c) = \frac{\dbinom{m}{c}\dbinom{k}{n-c}}{\dbinom{N}{n}} \qquad c = 0, 1, \ldots, h$$

where *h* = *m* or *n*, whichever is smaller

This set of probabilities is called the "hypergeometric" probability distribution.

The formula for the mean of the hypergeometric distribution is

Expectation of Hypergeometric:

$$E(X) = nm/N$$

Example 4 ▪ 33   *Gambling:* The game of Keno is now regularly played in Las Vegas casinos. Twenty numbers are selected at random by the casino from the numbers 1 through 80. The player selects from 1 to 15 numbers from the 80. Suppose the player chooses the numbers 21, 22, 60, and 71. The player wins a certain amount of money if one fourth these numbers match with any of the 20 numbers selected by the casino; she wins more if one half the numbers match, and so forth. The chance that one fourth of the four will match is

$$P(1) = \frac{\dbinom{20}{1}\dbinom{80-20}{4-1}}{\dbinom{80}{4}} = 0.433$$

The probability 2 out of 4 will match is

$$P(2) = \frac{\binom{20}{2}\binom{80-20}{4-2}}{\binom{80}{4}} = 0.213$$

In general, if the player selects n numbers, the probability $P(k)$ that $k(k \leq n)$ will match is

$$P(k) = \frac{\binom{20}{k}\binom{80-20}{n-k}}{\binom{80}{n}}$$

If $n = k = 1$, then $P(1) = 0.25$.

Application to Ecology: Capture – Recapture

It is standard practice in ecological research to trap and capture animals of a certain species, mark them with a dye or a tag, then let them go, and after a while, to attempt their recapture. The number recaptured is an indication of how well the species is surviving. For example, suppose 100 beavers are caught and marked with red dye; then after a month, we hunt until 100 beavers are again caught. Suppose all those caught have the mark of red dye. We would then suspect that the beaver population is quite small, perhaps near 100, for if there were many beavers, say, 1000, the chance of recapturing all the beavers previously captured would be small. Thus a result in which all 100 marked beavers are recaptured would indicate a very small beaver population. (This might suggest that the beavers were near extinction and we would attempt to determine a method to ensure their future existence. For example, we might check for contamination of their food supply or limit hunting.)

As an example, what is the probability of capturing 10 of the 100 marked animals if the total population is 500 beavers and 100 animals are caught the second time? Denote this event by $X = 10$. Using the hypergeometric model, we find the probability of capturing exactly 10 marked beavers is

$$P(X = 10) = \frac{\binom{100}{10}\binom{400}{90}}{\binom{500}{100}} = 0.00196$$

which is about 20 out of 10,000. The number of captured marked beavers we would expect to catch is $100(10)/50 = 20$. The result makes sense, for since 20% of the population was examined (100 out of 500) we should expect to find 20% of the marked beavers, namely, 20 out of 100. The question now is whether the captured number of 10 is too small as compared to the 20 expected to be attributable to chance? We investigate this question in our study of statistics.

Application to Quality Control: Acceptance Sampling

A government agency purchases electronic tubes from a manufacturer in batches of 1000 and, as invariably happens, requires the manufacturer to inspect the batches before delivery. Testing every tube (100% inspection) is sometimes impractical because of the expense in both time and money. Consequently, a subset of each batch is tested. The number to be tested is an important matter for negotiation between manufacturer and customer. One aid in deciding this matter and others associated with sampling inspection is the set of *operating characteristic (OC) curves.* To demonstrate how these curves are constructed, consider the following generalization of the preceding case.

We are to inspect batches containing K objects. From these K objects we test samples *without replacement.* If there are d or more defective objects in the sample, we reject the batch; that is, we do not sell this batch to the customer. We wish to calculate the probability of rejecting a batch.

The first step in this calculation is as follows: Given K objects of which c are defective and $K - c$ are not defective, we calculate the probability of (j), of obtaining exactly j defectives in a sample of size n by using the hypergeometric formula

$$P(j) = \frac{\binom{K-c}{n-j}\binom{c}{j}}{\binom{K}{n}}$$

We may now calculate the probability of rejecting a batch. We use the criterion of rejecting a batch if there are d or more defectives. Let R_d represent the event; rejecting the batch if there are d or more defectives in the sample. Noting that

$$R_d = \text{event } (d) \text{ or } (d+1) \text{ or } \cdots \text{ or } (n)$$

and that all the events (d), $(d+1)$, and so on are mutually exclusive, we have

$$P(R_d) = P(d) + P(d+1) + \cdots + P(n)$$

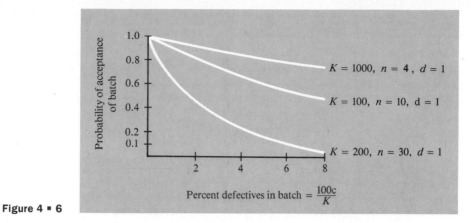

Figure 4 ▪ 6

Consequently, knowing K, c, and n, we may now calculate $P(R_d)$. Thus since the probability of acceptance equals $1 - P(R_d)$, we are able to draw curves of the type shown in Figure 4.6, plotting the chance of acceptance versus the percentage that is defective in the batch.

Now if we interpret $P(R_d)$ as the proportion of times we reject a batch that contains c or more defectives in a long sequence of such inspections, then these curves, called operating characteristic *(OC)* curves, tell us the frequency with which we accept or reject batches containing a certain percent of defective items.

Poisson Model (optional)

Approximation to Binomial: Model of Rare Events

The calculation of binomial probabilities can be tedious and, even for a computer, time consuming. This task can be reduced significantly if the probability of success p is small, whereas the number of trials is large. In this case the binomial probabilities, $P(X = x)$, can be approximated by using the formula

$$P(X = x) \doteq \frac{(np)^x e^{-np}}{x!} \qquad x = 0, 1, \ldots$$

where the symbol $\doteq$ means "is approximated by" and e is the base of the natural logarithm, a number equal to about 2.71828. The values of e^{-np} may be obtained from Appendix A.4.

Example 4 ▪ 34   A computer system has 1550 elements such that if any of these elements are defective the system will not work properly. The probability that any element is defective during a year is 1/500. What is the probability that the system will work properly during the entire year?

The probability that the system will have no failures during the year is approximately $e^{-np} = e^{-1550/500} = e^{-3.1} = 0.045$.

Let $\lambda = np$. The set of probabilities

$$P(X = k) = \frac{e^{-\lambda}\lambda^k}{k!} \qquad k = 0, 1, 2, \ldots$$

is called the Poisson distribution. It has the interesting property that its mean $E(X)$ and variance $V(X)$ both equal λ: $E(X) = V(X) = \lambda$. This distribution has many important applications other than to the binomial. However, because of its application to approximating the binomial for small p (which means that success is a rare event), it is sometimes called the "model of rare events." In the next section we shall see that there is additional motivation for this name.

Application to Law: "Odds Trip Up Pair in New York Case"*

MINEOLA, L. I.: A Queens woman and a Baldwin, L. I., man have been convicted in an unusual case over disputed claims to a winning pari-mutuel ticket worth $5050.

Key elements in the Nassau County Court trial were a Belmont Park racetrack computer printout of more than 350 bets made at a particular betting window on the day in question, and a Hofstra University mathematics professor's testimony that the odds against two different people independently choosing the same combination of nine $2 bets in nine races on the same day were a decillion to one.

The jury trial stemmed from a complaint to racetrack security personnel and later to the Nassau County Police and district attorney's office by a Brooklyn nurse, Rose Grant. She said she had been cheated out of her winning ticket on the track's ninth-race triple bet for May 21, 1974.

The triple is a high-return betting arrangement in which the bettor must pick the first-, second-, and third-place finishers in exact order. At other tracks it is known as the trifecta.

According to Grant's complaint, she went to the track that day but had to leave early. She said she wrote down her picks in each of the nine races on a piece of lavatory paper before leaving. She said she gave the paper and an old, wrinkled $20 bill to a track lavatory matron, Evelyn Jones, with a request that Jones place the $2 bets as listed for her. She said Jones consented.

The next day or so she learned that her triple bet paid off and was worth $5050, Grant said. She went to the track to get her winning ticket from the matron. But she said that Jones told her she had not been able to place the bets after all and that Jones handed her back her paper with the choices and a $20 bill. The bill, Grant

---

* Copyright, 1976 by *Newsday*, Inc. Reprinted by permission.

said, was a new one and not the one she had had originally. Grant filed a complaint with the track.

Within about a week, Howard R. Graham, a retired restauranter, tried to cash a winning ticket on that triple. Graham said he had placed his own bet. That ticket was the only winning triple ticket sold at Window 18.

Jones and Graham were indicted in January on charges of second-degree grand larceny and first-degree criminal possession of stolen property.

A computer printout of bets placed at Window 18 on May 21, 1974, showed that the precise series of 27 betting choices noted on Grant's paper had indeed been made that day. Track personnel also testified that the ticket produced by Graham came from Window 18. Assistant District Attorney James Boland produced an expert witness, Hofstra mathematics Professor Sylvia Pines, who said the chances of two strangers deciding on their own to bet the same 27 choices in sequence at the same window the same day were 1,000,000,000,000,000,000,-000,000,000,000,000 to 1.

The jury found Jones guilty of grand larceny and Graham guilty of criminal possession.

The conviction hinges on the belief that if "one person winning" is a rare event, then two people winning independently of one another is unbelievable. We can obtain an approximation to the probability of two people winning the trifecta from the Poisson approximation to the binomial. The value of p is obtained as follows: The chance of getting three choices correct *in order* in, say, a 10-horse race is $1/(10 \cdot 9 \cdot 8) = 1/720$. There are nine races; assume the outcomes are mutually independent. By the product rule, the chance of winning all nine is $(1/720)^9$, which is the value of p; we have $n = 350$ since 350 people made bets. Hence the chance of two people winning is (using $\lambda = 350/(720)^9$)

$$\frac{e^{-\lambda}\lambda^2}{2!}$$

which yields about the same result as that described in the preceding article.

Poisson Model: Model of Catastrophic Events

The story goes (see the Reference section) that Prussian cavalry soldiers in the late 19th century refused to wear their helmets unless it was absolutely necessary. Prussian horses, being high-spirited animals, kicked a great deal. These two facts came together to produce a great many fatalities among the soldiers from being kicked in the head while shoeing or cleaning their horses. The story continues that a mathematician decided to construct a mathematical model of this situation. Of course, he generalized the model by calling a "kick in the head" an "event." He stated the following assumptions:

1. *Independent intervals* A knowledge of the number of events (kicks in the head) in one interval of time does not supply information about the number in another nonoverlapping interval of time.
2. *Time is homogeneous* The probability of k events in one interval time is equal to the probability of k events in another nonoverlapping interval of time of the same length.
3. *Events happen infrequently* For a small enough time interval, the probability of more than one event is zero in this interval and the probability of one event in this interval $= \lambda \times$ (size of the interval), $\lambda > 0$.

Let $(X = k)$ be the event "k kicks in time t." Using the preceding assumptions, it can be shown that

$$P(X = k) = \frac{e^{-\lambda t}(\lambda t)^k}{k!} \qquad k = 0, 1, \ldots$$

There are other phenomena described by this model.

a. the number of telephone calls originating in an exchange between certain hours
b. the number of atoms disintegrating per second in quantity of radioactive material
c. the number of typing errors per page
d. the number of electrons emitted by a heated cathode in a given time interval
e. the number of light bulbs burning out in a large factory per week
f. the number of tire blowouts per week in a large fleet of trucks

So many applications are to failures that this model, developed by the French mathematician Poisson, is often called the "model of catastrophic events." As noted, it is also called the model of rare events. Assumption 3 motivates this name.

Note that

$$E(X) = \lambda t$$

hence

$$\lambda = \frac{E(X)}{t}$$

and so λ is called the average failure rate. The value $T = 1/\lambda$ is called the "time to failure."

Example 4 ▪ 35   A typist makes an average of six typing errors per page. What is the probability that on a random page there will be no typing errors?

We may regard one page as a "unit of time." Then

$$P(X = 0) = \frac{e^{-\lambda}\lambda^0}{0!} = e^{-6} = 0.0025$$

We had better proofread carefully!

4 ▪ 15   THE IMPORTANCE OF THE MEAN AND VARIANCE: TCHEBYCHEV'S APPROXIMATION (OPTIONAL)*

We shall encounter a number of reasons why the mean and variance of a random variable play an important part in statistics. One such reason is the fact that large departures from the mean have a small probability. The variance measures the variation of a random variable from its mean; we feel intuitively that when the variance is small, the probability also is small that a random variable deviates greatly from its mean. This intuitive feeling was shown to be true by the mathematician Tchebychev and is expressed as follows:

> Tchebychev's Approximation: Let $E(X) = \mu$. The probability that a random variable X will take on a value greater than k standard deviations from its mean μ is less than $1/k^2$; that is,
>
> $$P(|X - \mu| \geq k\sigma) \leq 1/k^2$$

Example 4 ▪ 36   Assume the coin-tossing model with $p =$ probability of heads $= 0.5$. For 100 tosses, $\mu = 100(0.5) = 50$, heads, $\sigma^2 = 100(0.5)(0.5) = 25$, and $\sigma = 5$ heads. Suppose you obtained 20 heads, would you feel that this number is so far from the expected number that it constitutes evidence that the coin is not fair?

Note $50 - 20 = 6(5) = 6\sigma$. The chance of getting an observation six standard deviations or further from the mean $\mu = 50$ is less than $1/6^2 = 0.0278$. The chance is so small of obtaining the deviation that was observed, that one would be justified in doubting that the coin is fair.

---

* This material is intended to give insight into the importance of the mean and variance. It is referred to in a footnote in Section 5.5.

| 4 ▪ 16 | USING THE COMPUTER TO SIMULATE RANDOM EVENTS |

We have developed some mathematical models of random events. Using these models, we have been able to find formulas that enable us to calculate various probabilities of interest. However, there are many important models in which the formulas for probabilities of interest are simply too difficult to derive or, if found, too complicated to use for easy computation. One then must consider finding the probabilities from observation. This might be done by carrying out an experiment or survey and analyzing the results. Unfortunately, even the simplest experiments are time-consuming and expensive. For example, if we had no idea what the value is of the probability of a 7 or 11 on a throw of two dice, we might decide to roll the dice 1000 times and observe the proportion of the trials that result in a 7 or 11. This observed proportion would be an *estimate* of the probability of a 7 or 11. Such an effort takes time. To save time, we can have a computer *simulate* the experiment and print out the results we seek.

Simulation is a technique used in statistics for obtaining results that are just too difficult to obtain by using standard mathematical tools.

Every statistical programming package, such as SPSS discussed in Chapter 2 and referenced there, provides a program for simulating experiments in which you can sample from a binomial distribution. Some of the smaller packages, such as MINITAB (see References), provide simple programs in which you can easily simulate coin tossing.

Summary

We have introduced the idea of a random variable, a variable that describes the outcomes of an experiment, study, survey, or game whose exact numerical value is determined by chance. The probabilistic description of a random variable via the probabilities and values $P(X = x_1)$, $P(X = x_2)$, . . . , $P(X = x_n)$ was called the distribution of the random variable. We found that random variables are a convenient way to describe how values are distributed in a population as well. Namely, the distribution of a population was defined as the distribution of the random variable. Knowing this distribution allows us to calculate the probability of various events of interest.

The mean and variance of a probability distribution were defined similarly as those of a population distribution. The cumulative distribution was defined as well.

The number that represents the probability of a particular event is given a frequency interpretation; that is, this number is interpreted as representing the value of the relative frequency of occurrence of the event in a large number of trials.

Using this frequency interpretation of probability, we developed another interpretation of the mean μ of a probability distribution as the expected value, $E(X)$, of a random variable X. The expected value represents the gain to a gambler per game played as the number of games he plays increases indefinitely.

The importance of μ and σ^2 was demonstrated through a discussion of Tchebychev's approximation.

The important concept of the independence of random variables was presented, which led to the definition of *independent trials* and then to the coin-tossing model. The main result of this model is: If there are only two possible outcomes on each of n independent trials, then the probability of getting k of type 1, and $n - k$ of the other type is given by the formula $\binom{n}{k} p^k(1 - p)^{n-k}$, where p is the probability of getting a type 1 outcome on any trial. This model is also called the binomial model and is very important.

Other models were introduced: The Poisson model is an approximation to the binomial model when the probability of success is very small in this latter model and the number of trials is large. The Poisson model is also applicable when time is homogeneous, intervals are independent, and events are infrequent. The hypergeometric model arises when you have *simple random sampling* from a finite population.

Key Terms

The following new terms were introduced in this chapter. Define them in words and symbols.

random variable
distribution of a random
  variable
probability density
probability distribution
cumulative probability distribution
mean and variance of a
  probability distribution
expectation
mean of a random variable
variance of a random
  variable
standard deviation of a
  random variable

mutually independent
  random variables
independent trials
coin-tossing model
Optional
  Tchebychev's approximation
  independent random
    sampling
  simple random sampling
  hypergeometric distribution
  Poisson model
  Poisson distribution

References

The following references contain discussions of mathematical models involving elementary probability theory as applied to various fields. The field of application is indicated. The discussions are readable at an elementary mathematical level; however, because of the sophisticated writing style, the asterisked references require considerable patience to read.

Sociology and Business
Blumen, I.: *The Industrial Mobility of Labor as a Probability Process.* New York State School of Industrial Relations, Cornell University Press, 1955.
Rogoff, N.: *Recent Trends in Occupation Mobility.* New York, The Free Press, 1953.

Anthropology*
Buchler, R. R., and H. A. Selby: *Kinship and Social Organization.* New York, Macmillan, 1968.

Learning
Bush, R. R., and F. Mosteller: *Stochastic Models for Learning.* New York, John Wiley & Sons, 1955.

Sociology*
Glass, D. V. (ed.): *Social Mobility in Britain.* New York, The Free Press, 1954.

Dentistry
Lee, K. H.: On the Theoretical Distribution of Surface Survival Times, and Caries Occurrence in Dental Caries Process. Abstracts of the Sixth Annual Symposium on Biomathematics and Computer Science on the Life Sciences, March 14–16, 1968, Houston. University of Texas, p. 19.

Accidents
Bortkewitsch, Ladislaus von: *Das Gesetz der Kleinen Zahlen.* Leipzig, Teubner, 1898. This reference contains applications of the Poisson distribution to suicide and accident data, as well as to the notorious Prussian horse kicks.

Computer Package
Ryan T. A., B. L. Joiner, and B. F. Ryan: *MINITAB Student Handbook.* Duxbury Press, 1971. This manual is designed to be used with MINITAB, a computer package developed for student use of computer applications in studying statistical problems.

Exercises

Sections 4.1–4.2

1. Define a random variable that describes
   a. the outcomes of a toss of a coin
   b. the number of defective fuses in the output of 100 fuses off a production line
   c. the possible outcomes of a toss of one die
   d. an item off a production line as being either defective or nondefective
   Remember: The range of a random variable consists of numerical values.

2. If you wish to study the proportion of people

in college who are classified as overachievers, you may give them a standardized psychological test. Define a random variable pertinent to this investigation.

3. I paint three faces of a die red. Let $X = 1$ if a red face comes up on a throw of the die and $X = 0$ otherwise. Assume equally likely events. Draw the probability distribution of the random variable X.

4. Three balls are distributed among two boxes. Assume all arrangements are equally likely. Let $X =$ number of balls in the first box. Draw the probability distribution of X. Find $P(X = 1$ or $X = 2)$.

Section 4.4

5. Classify the following random variables as discrete or continuous.
   a. the number of boys among 50 births at a city hospital
   b. the time until failure of a certain brand of light bulb
   c. the number of defective watches among a batch of 25 received by a jeweler
   d. the time spent in jail by a person convicted of robbery
   e. the height of a man selected at random
   f. the number of alpha particles emitted by a radioactive source

g. the weight of a tomato grown by using a certain fertilizer
h. the length of a baby born to a mother who smokes
i. the length of time spent in a mental hospital by a neurotic patient
j. the return on the investment of $100,000 by using a certain investment scheme

6. The uniform distribution (or density) over 0 to 2 is shown in Figure 4.4a.
   a. The area under the curve between any two points a, b represents the probability of falling between a and b. Calculate this probability if $a = 0.1$, $b = 0.2$; $a = 1.3$, $b = 1.7$; and $a = b$.
   b. Verify that the cumulative probability for this uniform is that shown in Figure 4.4b. Convince yourself that the probability of falling between a and b is given by the length shown on the vertical axis. Use this information to calculate the probability of falling between a and b from the graph of the cumulative probability when $a = 0.1$, $b = 0.2$; $a = 1.3$, $b = 1.7$; and $a = b$.

7. a. Draw the cumulative probability distribution for the distribution of Figure 4.2; show that it looks like that drawn in Figure 4.7.
   b. Draw the cumulative probability distribution for the distribution of Figure 4.3;

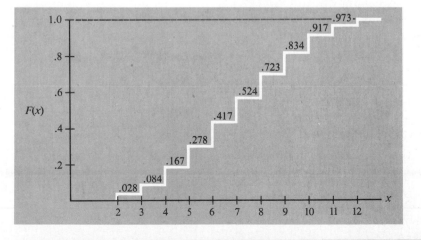

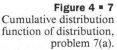

Figure 4 ■ 7
Cumulative distribution function of distribution, problem 7(a).

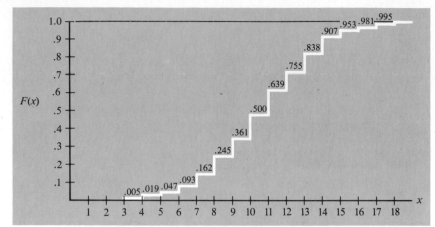

Figure 4 ▪ 8
Cumulative distribution
function of distribution,
problem 7(b).

show that it looks like that drawn in Figure 4.8.

8. There are two equally likely outcomes (0, 1) of a toss of a coin. The coin is tossed twice. The possible outcomes are 00, 01, 10, and 11.
   a. What is the probability of each outcome?
   b. The *score* is defined as the sum of the face values. The possible scores are 0, 1, and 2. What is the probability of each score?
   c. The average of the observations is the score divided by the number of tosses. There are three possible averages; $\frac{0}{2}, \frac{1}{2}$, and $\frac{2}{2}$. Find the probability of each average.
   d. Define the random variable $\overline{X}_2$ as having those possible values listed in part c. Draw the probability distribution of $\overline{X}_2$.
   e. The coin is tossed three times. The average is the sum of the outcomes divided by 3.

   Average $= \overline{X}_3 =$

   $$\frac{\text{1st outcome} + \text{2nd outcome} + \text{3rd outcome}}{3}$$

   Draw the probability distribution of the average of the three observations.
   f. Let X_i represent the outcome of the ith toss where X_i equals 0 or 1. The average of

n observations is

$$\overline{X}_n = \frac{X_1 + X_2 + \cdots + X_n}{n}$$

Draw the probability distribution of $\overline{X}_6$; $\overline{X}_8$.

9. Suppose a random variable Y has a cumulative distribution function $F_Y(t) = P(Y \le t)$. We define $F_Y(t)$ as follows:

$$F_Y(t) = \begin{cases} 0 & \text{if} \quad t < -2 \\[2mm] \dfrac{1}{4} & \text{if} -2 \le t < 0 \\[2mm] \dfrac{3}{4} & \text{if} \quad 0 \le t < 2 \\[2mm] \dfrac{7}{8} & \text{if} \quad 2 \le t < 6 \\[2mm] 1 & \text{if} \quad t \ge 6 \end{cases}$$

a. Calculate $P(-3 < Y \le -1)$.
b. Calculate $P(Y = -2)$.
c. Calculate $P(-1 < Y \le 6)$.
d. Graph the cumulative distribution function F_Y.

Section 4.5

10. Find the mean and standard deviation of $\bar{X}_n$ for $n = 2, 3,$ and 8 in problem 8.

11. Median of a random variable. The median of a random variable X is a number m such that $P(X \geq m) \geq 0.5$ and $P(X \leq m) \geq 0.5$; that is, the median is a number for which one half the probability distribution is around and on either side of m.

    a. What is the median of $\bar{X}_2$ in problem 8?
    b. Let $P(X = 0) = 0.1,$ $P(X = 2) = 0.1,$ $P(X = 3) = 0.6, P(X = 5) = 0.2.$ What is the median of X?

Section 4.6

12. Players A and B play the following game: A bridge deck (52 cards) is shuffled and A picks a card. If the card is a Jack, player B pays A \$10. If player A does not pick a Jack, what should A pay B in order that this game be fair? A fair game is defined here as one in which the expected value of each person's winnings is zero.

13. A game is played where you toss a pair of dice and lose \$3 if a 7 or 11 sum appears, win \$6 if both dice show the same number, and no money is exchanged in any other case. Define a random variable describing this game and calculate your expected winnings. If you had to pay a certain fee to enter the game, what would you consider a "fair" price?

14. A student taking a certain exam has probability 0.75 of scoring an 80%; 0.15 of scoring an 85%; 0.05 of scoring a 90%; and 0.05 of scoring 100%. What is his expected score on the test? How would you interpret this number?

15. Suppose you play the game outlined in problem 13 with triple the stakes and an entrance fee of \$2. What are your expected winnings?

16. Calculate the variance of each of the random variables given in problems 13 and 15.

17. Draw the distribution function for the random variables given in problems 13 and 15.

Section 4.7

18. If a random variable X has the distribution $P(X = -3) = \frac{1}{3},$ $P(X = 0) = \frac{1}{3},$ $P(X = 2) = \frac{1}{4},$ and $P(X = 6) = \frac{1}{12},$ find the expected value of X, $E(X)$, and the variance of X, $V(X)$.

19. Find $E(X)$ in each of the following cases:

    a. $P(X = i) = 0.1;$ $i = 0, 1, 2, \ldots, 9.$
    b. $P(X = -1) = \frac{1}{4},$ $P(X = 2) = \frac{1}{4},$ $P(X = 5) = \frac{1}{4},$ $P(X = 10) = \frac{1}{4}.$
    c. $P(X = 1) = 0.2,$ $P(X = -1) = 0.2,$ $P(X = 4) = 0.2,$ $P(X = -27) = 0.2,$ $P(X = 0) = 0.2.$

Section 4.8

20. a. Let S_2 represent the random variable of scores resulting from the toss of two dice. Verify that the probability distribution of S_2 is that shown in Figure 4.2. Find the expected value, $E(S_2)$, of S_2.
    b. Let S_3 represent the random variable of scores resulting from the toss of three dice. Verify that the probability distribution of S_3 is that shown in Figure 4.3. Find the expected value of S_3.

Section 4.9

21. *Other Properties of E(X)*

> *Property 1.* If a is any number and X is any random variable, then aX is also a random variable and $E(aX) = aE(X).$

To prove this assertion, suppose that X can assume only the values $x_1, x_2, \ldots, x_n.$ Then aX assumes the values $ax_1, ax_2,$ $\ldots, ax_n.$ And, $P(X = x_i) = P(aX = ax_i)$ so that

$$E(aX) = \sum_{i=1}^{n} ax_i P(aX = ax_i)$$

equals

$$\sum_{i=1}^{n} ax_i P(X = x_i) = a \sum_{i=1}^{n} x_i P(X = x_i)$$
$$= aE(X)$$

> *Property 2.* If a is any constant, then $E(a) = a$.

Indeed, a can be considered as a random variable X such that $P(X = a) = 1$. Then $E(a) = a \cdot 1 = a$.

Use properties 1 and 2 and the property of expected value of sums for the following:
a. Find $E(2X)$.
b. Find $E(2X + 3)$ for those X given in problems 13 and 18.
c. Suppose you play the game outlined in problem 13 with triple the stakes and an entrance fee of $2. What are your expected winnings?

Section 4.10

22. Let a and b be any constants and X a random variable; then, $V(aX + b) = a^2 V(X)$.
    We now prove this result. Give reasons for each step.
    1. $E(aX + b) = aE(X) + b$
    2. $V(aX + b) = E\{(aX + b) - aE(X) - b\}^2$
       $= E\{(aX) - aE(X)\}^2$
       $= E\{a^2[X - E(X)]^2\}$
       $= a^2 E\{X - E(X)\}^2 = a^2 V(X)$

Section 4.11

23. Classify the following random variables X and Y as being either independent or dependent.
    a. A person is selected at random; X denotes her height and Y denotes her weight.

b. A college student is selected at random; X represents the student's grade point average and Y represents the yearly average temperature of New Orleans.
c. X equals the dosage of a drug given and Y equals the number of side effects observed.
d. X equals a person's IQ score and Y equals the person's height.
e. X represents a college student's entrance test score and Y represents the student's major in college.
f. X equals the number of miles from a worker's home to the factory in which he works; Y represents the worker's income.

24. a. I go to a restaurant every day for lunch with four friends. This last week we had a new waiter for our favorite table. For the last 5 days, he has served me last. Should I suspect he doesn't like me? (Calculate the chance of my being served last 5 days in a row and assume each day (trial) is independent of every other day).
    b. Twelve men are selected at random from the population to form a jury. No member of social group G has served on the last 18 juries, yet this group constitutes 20% of the population. Would you attribute the event to chance that no member of G served on any of the last 18 juries? Assume independent trials; that is, juries are selected independently of each other.

Section 4.13

25. Fatigue is the cause of 20% of all industrial accidents. Last year there were 10 accidents at E. Chemical Co. What is the probability exactly 3 of these were caused by fatigue?
26. Use the binomial to calculate the probability of $X = x$ in each of the following situations:
    a. $p = 0.5$, $n = 4$, $x = 2$
    b. $p = 0.1$, $n = 5$, $x = 3$
    c. $p = 0.9$, $n = 5$, $x = 3$

27. Sketch the graph of the density histogram for the binomial with $n = 8$, $p = 0.1$ and also with $n = 10$, $p = 0.4$.

28. If we assume that the probability of a boy on a birth is equal to 0.5 and that on each birth the only possibilities are a boy or a girl, compute the probability that exactly 20 out of 32 babies born will be boys.

29. A game that is fun to play with children is to take them for a walk and at each corner decide whether to go north or south by tossing a coin. Suppose the children tire after walking six blocks. What is the probability that you return home before walking more than six blocks? The game ends only when home again.

30. The probability is 0.05 that a patient will cancel a dental appointment. Assume that such an event is statistically independent from patient to patient.
    a. What is the probability that exactly 2 out of 10 appointments will be canceled?
    b. Find the probability that fewer than 4 will be canceled.
    c. What is the expected number of patients who will cancel their appointments? What is the standard deviation of the number of cancellations?

31. I have nine coins in my pocket: one is a worthless counterfeit; two are dimes, two are nickels, one is a quarter, three are pennies. I select two coins at random. Their sum is less than $0.02. What is the probability that one of them is counterfeit? Suppose the sum was greater than $0.10. What would the probability be that one of the coins is counterfeit?

32. I have two pennies and a nickel. I toss all three coins. For each coin, $p =$ probability of a head.
    a. What is the probability of obtaining three heads? two heads and a tail?
    b. What is the probability that both pennies will come up heads, and the nickel will come up tails?

c. What is the probability that more heads will come up on the pennies than on the nickel? Repeat this problem assuming five pennies and two nickels.

33. (Optional) Let A, B, C each toss a coin in order, but there are no more than 9 successive tosses. The first to obtain heads wins. Let 0 represent tails and 1 represent heads. The sample space, Ω, of all possible outcomes is

$$\Omega = \{1, 01, 001, 0001, 00001, 000001,$$
$$0000001, 00000001, 000000001,$$
$$000000000\}.$$

Find the probability of the event "B wins"; "no one wins."

34. Assume the "coin-tossing model" with probability p for success. Compute
    a. the probability of obtaining at least one success in n trials
    b. the probability of obtaining i successes followed by a failure given at least one success in the first i trials $(i > 0)$

35. a. In tossing a coin n times, what is the probability that on 10 tosses there are an equal number of heads and tails?
    b. Suppose you are given that in 10 tosses a head appears on tosses 1, 5, and 7. What is the probability of obtaining exactly 4 tails in the 10 tosses?

36. Using the binomial distribution, Figure 4.5, find the probability of obtaining 3 or more successes in 10 trials.

37. What is the expected gain for a person who is to receive $80 if he obtains three heads in a single toss of three coins and $0 otherwise? What is the variance of this game?

38. To which of the following phenomena is the binomial formula applicable for calculating the probability of "x successes in n trials."
    a. A success is a rainy day; a trial is a day.
    b. A success is a winning hand in blackjack; a trial is a single deal.

c. A success is an increase on a given day in the value of my stocks; a trial is a day.

d. Each week I take an airplane trip home and back to my university; a success is no air accident; a trial is each trip.

e. A success is winning a game of Bingo; a trial is a Bingo game.

Section 4.14

39. a. A man wants to open his door. He has n keys, of which one fits his door. The gentleman is not very clever so he samples the keys from his pocket with replacement. What is the probability that he chooses the proper key at the rth trial and no sooner?

    b. Suppose he does not sample as previously described but remembers each key he has tried. What is the probability of the event described in part a if r is less than or equal to n? if r is greater than n?

40. A lake contains N fish, 100 of which are tagged with a red tag, and the rest are untagged. A fisherman catches 15 fish. Find the probability in terms of N that he catches exactly 5 tagged fish?

41. The Equal Employment Opportunity Commission (EEOC) wishes to determine if a company is discriminating against women in employment. There are supposed to be 50 men and 50 women in the company. A commission member randomly selects 10 records from the 100 employee files and finds that the employees selected are all men. What is the probability of obtaining 10 men in a sample of 10 files if there are really 50 men and 50 women each in the company?

42. Twelve grand jurors are selected by using simple random sampling from a list of 100 people, six of whom belong to a minority group. Let X represent the random variable that represents the number of the minority group chosen for the jury.

    a. Verify that Figure 4.9 is the distribution of X.

    b. Find the probability that two or fewer members of the minority group will appear on the grand jury.

    c. Find $E(X)$.

43. Among 14 qualified applicants for a job, 9 are female. If four applicants are randomly selected for interviews, find the probability that

    a. no female is interviewed

    b. two females are interviewed

    c. four females are interviewed

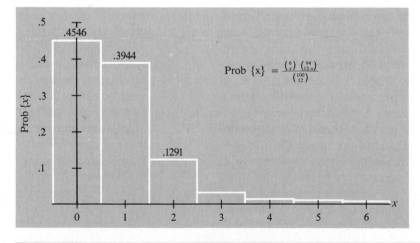

Figure 4 ▪ 9
Probability that members of a minority will be on a grand jury.

$$\text{Prob } \{x\} = \frac{\binom{6}{x}\binom{94}{12-x}}{\binom{100}{12}}$$

Section 4.14: Poisson Model

44. A population of a rural town is 12,500. The chance of being struck by lightning in a storm is 0.000001. There are approximately eight electrical storms per year. What is the chance that at least one person is struck by lightning in a 1-year period?

45. One million cars produced by Company X were bought last year. The engineers have discovered a defect in the steering mechanism and have determined that the probability of the defect causing an accident is 0.000001. What is the chance that the defect caused at least one accident?

46. The Swiss government estimated that 1 out of every 100 foreign workers does not have a working permit. If the authorities randomly checked 1000 foreign workers, what is the expected number that would be found without a permit? What is the chance of finding exactly this number?

47. Records show that the average number of tire blowouts per year in a large fleet of trucks is 50. What is the probability of obtaining half this number of blowouts in a given year?

48. The expected number of arrivals at a toll booth per minute is two. What is the probability of four arrivals in 2 minutes?

Section 4.15

49. The average number m of accidental automobile deaths in a city during the summer months, June through August, is 30, with a standard deviation of 6. Suppose 50 accidents occurred one summer; would you suspect some breakdown in the traffic system?

50. The average number of tourists going to the top of Mt. Blanc on a weekend is 120, with a standard deviation of 18. One person on the staff of tourist guides can handle 20 people; there are 10 such guides. Is there a good chance that there will be too few guides on a weekend?

Other Models (optional)

51. (Geometric Distribution) Assume the coin-tossing model with $P(X_i = 1) = p$; $P(X_i = 0) = 1 - p$ for each trial i. Let $Y = m$ be the event m zeros followed by a one; show that $P(Y = m) = q^m p$; $q = 1 - p$ and $m = 0, 1, 2, \dots$. The distribution of Y is called the *geometric distribution,* and it counts the number of failures before the first success in repeated trials.

52. (Geometric Distribution and Reliability Theory) Reliability theory has wide applications to engineering problems. One of the primary investigations of reliability theory is the determination of the distribution of the random variable, T, the time until failure (i.e., the lifetime) of a device. This distribution is usually called the "failure law," and it depends, of course, on the model of the physical situation. We discuss one model here.

   Assume that time is measured in discrete intervals called cycles. For example, suppose a machine is designed to run for 2 hours and is then shut down, then runs for two more hours, and so on. Each 2-hour period is called a cycle. Let p represent the probability that the machine will fail in a given cycle. We assume that P (failure in cycle k| none in $k - 1$) $= p$. (This assumption is analogous to the assumption that the length of time the machine has been used does not influence its future life.)

   a. Let X be the random variable that gives the time of the first failures, that is, X equals n if the first failure occurs on cycle n. Then $P(X = n) = pq^{n-1}$, where $q = 1 - p$. Show that

$$P(X > n) = P(\text{no failure in } n \text{ cycles})$$
$$= q^n$$

b. Let $P(X = n|X > r)$ (where $n \geq r$) represent the conditional probability of the first failure on the nth trial given that no failure has occurred up to the rth cycle. In other words, we wish to know the probability of the event that, after finding the system operable for r cycles, it fails for the first time on the nth cycle $(n > r)$. Show that

$$P(X = n|X > r) = pq^{n-r-1}$$

53. (Negative Binomial Distribution) Consider again the model described in problem 51. Our model here is the same as the coin-tossing model in which heads replaces failures and trials replaces cycles. Hence

$$P(1 \text{ failure in 3 cycles}) = \binom{3}{1} p^1 q^2$$

In general, show that

$$P(r - 1 \text{ failures in } t - 1 \text{ cycles})$$
$$= \binom{t-1}{r-1} p^{r-1} q^{t-r} \qquad (t > r)$$

Hence for any $t \geq r$, the probability that the rth failure occurs in cycle t is

$P(r - 1 \text{ failures in } t - 1 \text{ cycles and then a failure on the } t\text{th cycle})$

$$= p\binom{t-1}{r-1} p^{r-1} q^{t-r} = \binom{t-1}{r-1} p^r q^{t-r} \quad (1)$$

$$= P[r\text{th failure occurs on the } t\text{th cycle}].$$

Expression (1) is called the "negative binomial." We emphasize the difference between the binomial and negative binomial in the following table.

Binomial
$P(S_n = x) = \binom{n}{x} p^x q^{n-x};$
$x = 0, 1, 2, \ldots, n$

Negative binomial
$P(S_x = n) = \binom{n-1}{x-1} p^x q^{n-x};$
$n = x, x+1, \ldots$

In the binomial, n is fixed and x takes on the integer values from 0 to n. In the negative binomial, x is fixed and n takes on the values $x, x + 1, \ldots$. If we are interested in how long it will take for the rth failure to occur, then we employ the negative binomial.

Problem: the negative binomial and dental caries. Let p represent the probability that a carie lesion (cavity) develops on a given tooth. Suppose the time between dental checkups to the dentist is defined as one cycle. If we affirm that if the tooth needs treatment the dentist successfully treats it during the visit, and if we affirm that the probability of a caries lesion developing on the tooth during the next cycle is p, independent of the number of preceding cycles, and that at most one carie developed per cycle, then it is of interest to determine the number of cycles needed to produce f caries. Show that

$$Pr\,(f \text{ caries in } t \text{ cycles}) = \binom{t}{f} p^f q^{t-f}$$

where $f \geq 0$, $t = f, f + 1, \ldots$

(For an interesting application of the negative binomial distributions to dentistry, see the reference by K. H. Lu.)

54. (Statistical Mechanics) The following problems are associated with the field of statistical mechanics in physics.

   a. If we have M urns and N distinguishable balls, $M > N$, show that the probability of the event S that the first N urns each contain exactly one ball is $N!/M^N$. Assume equally likely occurrences.

   b. Suppose we assert that two balls cannot occupy the same urn; then Ω, the set of all possible arrangements of the balls in the urns, contains only those outcomes in which at most one ball occupies an urn. If we assume that the outcomes in this set are equally likely, show that the probability of the event S described in (a) is

   $$\frac{N!}{M(M-1)\cdots(M-N+1)} = \frac{1}{\binom{M}{N}}$$

   c. Suppose the balls are indistinguishable and not more than one can fit in an urn; also, assume the sample space Ω' consists of all such distinguishable arrangements and all outcomes in Ω' are equally likely.

Show that the probability of S is

$$\frac{1}{\binom{M}{N}}$$

   d. (Discussion question). One model of small particles such as atoms used in physics is that they are balls in "urns" where the urns are classifications of the balls according to their positions and energy level. The urns in physics are called "states." In the theory called Maxwell–Boltzman statistics* after the physicists who developed the ideas, the assumption is made that the particles (balls) are distinguishable. Fermi–Deriac statistics, another theory of small particles, assume the particles are indistinguishable and not more than one can be in a given state. Experimental evidence indicates that this theory is applicable to electrons, protons, and neutrons. Bose–Einstein statistics assume indistinguishable particles and more than one can be in any state; this theory applies to protons and pimesons. In all the aforementioned theories, equally likely outcomes are assumed for the respective sample spaces.

   Take some boxes and balls and try to construct physical situations under which each of the statistics previously described would be applicable.

---

* The word "statistics" as used in physics has a peculiar meaning, not the same as that used by statisticians.

$P(E)$ = probability of Event E

$= \dfrac{\text{no. of out comes in E}}{\text{total \# of out comes}}$
~~base~~

$= \dfrac{n(E)}{N}$

ADDITIVE RULE
(mutually Exclusive Event)

$$P[A \cup B] = P[A] + P[B]$$

(general Rule

$$P[A \cup B] = P[A] + P[B] - P[A \cap B]$$

Multiplicative Rule
(mutually exclusive Event)

$$P[A \cap B] = P[A]\, P[B]$$

(general rule)

$$P[A \cap B] = P[A]\, P\!\left(\dfrac{B}{A}\right)$$

Counting Techniques

n_1 = \# of ways for Event 1

n_2 = " " " " " 2
     Aft E1 has occured.

N = total \# of ways to perform
the operation.

$$N = n_1 \cdot n_2$$

$$P[A/B] = \dfrac{P[A \cap B]}{P[B]}$$

PERMUTATIONS   ORDER NB

$$nP_k = \dfrac{n!}{(1-k!)}$$

Where   n = \# of objects / LARGER \#
     k = \# " ways to select / sm.\#

COMBINATIONS

$$nC_k = \dfrac{n!}{(n-k)!\,k!} \quad \text{select } k \text{ from } n$$

RANDOM VARIABLES

Discrete = isolated out comes
     definite X

Continuous = no difinite limit.
     continuous interal

IN A Success / FAILURE SITUATION

$$f(x) = \binom{n}{x}(p)^x (q)^{n-x}$$

Where   X = \# of desired outcomes
     n = \# of trials
     p = sucess
     q = failure

STANDARD DEVIATION σ_x of $x = \sqrt{Var(x)}$

VARIANCE = $Var(x)$ =

$$(X_1 - \mu_x)^2 f(x)_1 + (X_n - \mu_x)^2 f(x)_n$$

$\underline{\mu_x \cdot Ex}$ = Mean or Average
     = balance pt. of graph.

$$= \left(f_{x_1}\right)\left(x_1\right) + \left(f_{x_n}\right)(x_n)$$

IN A SUCCESS / FAILURE SITUATION

$\mu_x = np$ RANGE of σ =

$Var(x) = npq$ $\mu_x - \sigma_x$ &

$\sigma(x) = \sqrt{npq}$ $\mu_x + \sigma_x$

CONTINUOUS R.V

SEE CH.6

CHAPTER · 5

The Normal Distribution

I know of scarcely anything so apt to impress the imagination as the wonderful form of cosmic order expressed by the "Law of Frequency of Error." The law would have been personified by the Greeks and deified, if they had known of it. It reigns with serenity and in complete self-effacement amidst the wildest confusion. The huger the mob and the greater the apparent anarchy the more perfect its sway. It is the supreme law of Unreason. . . .

Francis Galton
Order in Apparent Chaos
concerning the normal distribution

As odd as it may seem, much random phenomena exhibit a regular and consistent behavior. One expression of this regularity is that the added effect of a large number of random values leads to a random variable that has a normal distribution. This result occurs so often in nature that the normal distribution has taken a central role in the study of statistics. We now study this distribution and some of its applications.

5 ■ 1 INTRODUCTION: CALCULATION OF PROBABILITIES

Without exaggeration, the continuous density distribution known as the *normal distribution* is the most important distribution that arises in the study of statistical inference. Many of the statistical techniques we encounter in this and subsequent chapters require that the population under consideration have a normal distribution. It is important that we become familiar with this distribution and learn how to use it. We begin with the following:

A random variable **X** is said to have a *normal distribution* with mean μ and variance σ^2, if it is a continuous variable with the density given in Figure 5.1. The equation of the density curve is

$$f(x) = \frac{1}{\sqrt{2\pi\sigma^2}}\, e^{-(x-\mu)^2/2\sigma^2} \qquad \text{for } -\infty < x < \infty$$

where e is a constant approximately equal to 2.7183.

The probability of obtaining a value between two given numbers is the area under the density curve between these points.

To see how the normal distribution changes shape as μ and σ^2 change, refer to Figure 5.2. We see that the mean μ is really the "center" of the curve and tells where to put the hump in the bell shape, whereas σ^2, the variance of the normal distribution, controls the spread of the curve. The smaller σ^2 is, the more peaked the density appears.

The center of a normal distribution is determined by the value of the mean μ, and the shape is determined by the value of the variance σ^2. Thus these two values completely describe what a normal distribution looks like.

Note that a normal random variable can take on any value between $-\infty$ and ∞. In reality there are few phenomena in which all values between $-\infty$ and ∞ are possible values. As with many mathematical models, the normal distribution model is an idealization of nature. Nevertheless, many populations such as the

Figure 5 ■ 1
The normal density curve with mean μ and variance σ^2 is the familiar bell-shaped curve. If a population has this distribution, then the probability of selecting a population member with a value between the numbers a and b is the area under the curve between a and b.

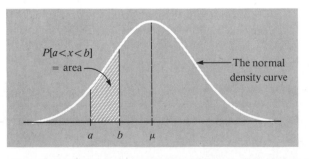

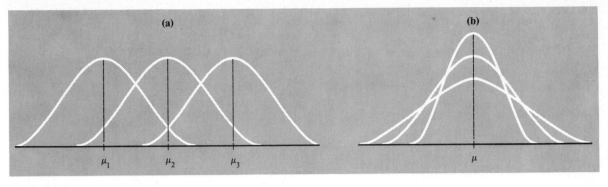

Figure 5 ▪ 2
How the shape of the normal curve changes with μ and σ^2. (a) Normal distributions with the same variance σ^2 but different means. (b) Normal distributions with the same mean μ but different variances. The larger the variance σ^2 is, the fatter and flatter the curve will be.

population of heights of all people in America, the lifetimes in miles of a brand of tires, the scores on exams, and so on, follow this distribution quite closely. In working problems in statistics, we often assume that the population under consideration follows a normal distribution and proceed to use results known for the normal distribution. It is important in developing statistical habits to always justify this assumption (or any assumption) in practical work. Be aware that often normality is only an approximation to the actual population distribution.

The Normal Distribution with Mean 0 and Variance 1

The cumulative distribution function $F(z) = P(Z \leq z)$ for a normal population with mean 0 and variance equal to 1 is tabulated in Appendix A.1. The shaded area $F(z)$ shown in Figure 5.3 is the quantity tabulated in the table for various values of z. It is important that we learn how to use this table so that we are able to find the probability that a random variable Z with this distribution has a value between two points a and b. To find this probability, namely,

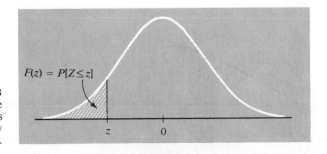

Figure 5 ▪ 3
The area under the curve up to and including z is equal to the probability $P[Z \leq z]$.

$F(z) = P[Z \leq z]$

$P(a < Z \leq b)$, we look up a and b in the left-hand column (marked z) of the table and read $F(a)$ and $F(b)$ in the corresponding right-hand column (marked $F(z)$). Then $P(a < Z \leq b) = F(b) - F(a)$. For example, $F(-2.95) = 0.0016$, $F(-1.25) = 0.1056$, and $F(1.30) = 0.9032$. Thus we calculate $P(-2.95 < Z \leq -1.25) = 0.1056 - 0.0016 = 0.1040$, and $P(-2.95 < Z \leq 1.30) = 0.9032 - 0.0016 = 0.9016$, and $P(-1.25 < Z \leq 1.30) = 0.9032 - 0.1056 = 0.7976$.

Example 5 ▪ 1 A certain population is known to have a normal distribution with mean 0 and variance 1. Find the proportion of the population that lies below or equal to 1 and above -1: To solve this problem, we find from Appendix A.1 that $F(1) = 0.8413$ and $F(-1) = 0.1587$ so that $P(-1 < Z \leq 1) = 0.8413 - 0.1587 = 0.6826$. Thus 68.26% of the population lies between $+1$ and -1.

To calculate $P(a < X \leq b)$ in the case where X has a normal distribution with mean μ not zero and variance σ^2 not 1, we could also construct a table similar to Appendix A.1 for this distribution; however, this would entail constructing a new table for each possible choice of μ and σ^2. To avoid working with thousands of tables, we use a "standardization" technique so that we can use Appendix A.1 for any normal distribution.

The Standardization Technique

To see how this standardization is accomplished, note that if X is normally distributed with mean μ and variance σ^2, then the "hump" of the density is above the number μ. The first thing to do is to move the center of the distribution of X from μ to 0; that is, we want to get the hump over 0. This can be done by subtracting μ from X (which just shifts the curve over until the hump is in the right place). The mean of $X - \mu$ is now 0. We have not changed the shape of the distribution by shifting it. Hence the variance of $X - \mu$ is still σ^2, the same as the variance of X. We must now adjust the width of the curve to get a variance of 1. Since variance is a squared measurement, the variance of Y/a equals $1/a^2$ times the variance of Y. Since $X - \mu$ has a variance of σ^2, the variable $Z = (X - \mu)/\sigma$ has a variance of $\sigma^2/\sigma^2 = 1$.*

---

* By using the properties of expected value and variance, we can show rigorously that $E(Z) = 0$ and $V(Z) = 1$ since

$$E(Z) = E\left(\frac{X}{\sigma} - \frac{\mu}{\sigma}\right) = \frac{1}{\sigma} E(X) - \frac{\mu}{\sigma} = \frac{\mu}{\sigma} - \frac{\mu}{\sigma} = 0$$

$$V(Z) = V\left(\frac{X}{\sigma} - \frac{\mu}{\sigma}\right) = \frac{1}{\sigma^2} V(X) = \frac{\sigma^2}{\sigma^2} = 1$$

> If X has mean μ and variance σ^2, then $Z = (X - \mu)/\sigma$ has mean 0 and variance 1. *This is true no matter what distribution X has!* The variable Z is called the standardization of X.

To calculate $P(a < X \le b)$, where X is normal with mean μ and variance σ^2, we note that

$$P(a < X \le b) = P\left(\frac{a-\mu}{\sigma} < \frac{X-\mu}{\sigma} \le \frac{b-\mu}{\sigma}\right) = P\left(\frac{a-\mu}{\sigma} < Z \le \frac{b-\mu}{\sigma}\right)$$

where Z is the standardization of X, and Z is normally distributed with mean 0 and variance 1. The latter probability statement involving Z can easily be calculated by using Appendix A.1 and looking up the numbers $(b - \mu)/\sigma$ and $(a - \mu)/\sigma$.

Also notice that for the normal distribution (as with all continuous distributions) we have

$$P(a < X \le b) = P(a < X < b) = P(a \le X < b) = P(a \le X \le b)$$

since the probability is 0 of obtaining any specific pregiven number. (The area under a point is 0). Thus we may speak of "the probability of a value falling between a and b" without ambiguity in the value of the probability.

Example 5 ▪ 2   The scores on a particular statistics exam are assumed to have a normal distribution with $\mu = 80$ and $\sigma^2 = 25$. What is the probability of obtaining a score from 70 and up to and including 85 on the exam? To find this probability, we standardize X by using $\mu = 80$ and $\sigma = 5$ to obtain $Z = (X - 80)/5$ which has a normal distribution with mean 0 and variance 1. Then

$$P(70 < X \le 85) = P(70 - 80 < X - 80 \le 85 - 80)$$

$$= P\left(\frac{70-80}{5} < \frac{X-80}{5} \le \frac{85-80}{5}\right)$$

$$= P(-2 < Z \le 1) = F(1) - F(-2) = 0.8413 - 0.0228$$

$$= 0.8185$$

where the last set of numbers comes from Appendix A.1. Thus the probability is 0.82 of falling in this score range.

Example 5 ▪ 3   If X has a distribution that is normal with $\mu = 53$ and $\sigma^2 = 16$, find two values a and b that are equidistant from μ and such that $P(a < X \le b) = 0.95$ (see

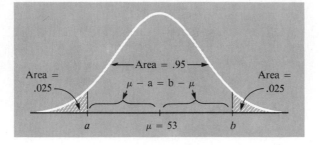

Figure 5 ▪ 4
The normal density with $\mu = 53$, $\sigma = 4$ showing two points on opposite sides of μ and equidistant from μ such that $P(a < X \le b) = 0.95.$

Fig. 5.4). To solve this problem, we note that $Z = (X - 53)/4$ has a normal distribution with $\mu = 0$ and $\sigma^2 = 1$. To find the required value of a, we look up 0.025 in the $F(z)$ column and find that $P(Z \le -1.96) = 0.025$. By the symmetry of the normal curve, $P(Z > 1.96) = 0.025$ also, so $P(-1.96 < Z \le 1.96) = 0.95$. Since $Z = (X - 53)/4$, we have

$$0.95 = P[-1.96 < (X - 53)/4 \le 1.96] = P[(-1.96)(4) < X - 53 \le (1.96)4]$$

$$= P[53 - (1.96)(4) < X \le 53 + (1.96)4] = P[45.16 < X \le 60.64]$$

Thus $a = 45.16$ and $b = 60.84$.

Example 5 ▪ 4   The height of students at a Texas University is normally distributed with mean height, μ, equal to 70 in. and a standard deviation, σ, of 5 in. What proportion of students are taller than 6 ft?

Let X represent the random variable "student height." Recall that the proportion of the student population that is taller than 6 ft is given by the probability $P(X > 72$ in.$)$, which is equal to $1 - P(X \le 72)$. We standardize

$$z = \frac{x - \mu}{\sigma} = \frac{72 - 70}{5} = 0.4$$

and by using Appendix A.1 find that $P(Z \le 0.4) = 0.6554$. As shown by the explanation in example 5.2, the value of $P(X \le 72)$ is also equal to 0.6554. Hence $P(X > 72) = 1 - P(X \le 72) = 1 - 0.6554 = 0.3446$.

The type of problem considered in the previous example, where we must find $P(X > x)$ for a specified x, occurs quite often. Another problem that occurs frequently is the following: For a given probability value α (the Greek letter alpha), determine a value x_α such that $P[X > x_\alpha] = \alpha$. This type of problem is of fundamental importance in statistical analysis. We give a special name to the value x_α associated with the probability α:

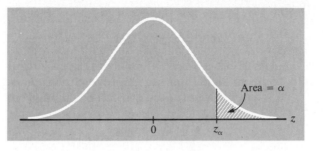

Figure 5 ▪ 5
Normal density curve with
$\mu = 0$ and $\sigma = 1$.

If X is a random variable, and if $P(X > x_\alpha) = \alpha$, then x_α is called the
upper α centile for X.

As an example, consider a normal random variable, Z, with mean 0 and
variance 1. The value z_α satisfies $P(Z > z_\alpha) = \alpha$; that is, z_α is the value such that
the area under the normal curve to the right of z_α is α (see Fig. 5.5). The value z_α
can be found by looking up the number $1 - \alpha$ in the column $F(z)$ of Appendix
A.1 and then reading in the z column to find the value z_α. The last column gives
various upper centile values directly. For example, $P(Z > 1.282) = 0.10$ and so
$z_{0.10} = 1.282$. Other examples are $z_{0.50} = 0$, $z_{0.20} = 0.842$, $z_{0.05} = 1.645$,
$z_{0.025} = 1.96$.

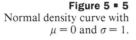

CENTRAL LIMIT THEOREM[1]

We now come to another example of that remarkable regularity of behavior
that random phenomena often exhibit and that explains the importance of the
normal distribution.

Let $X_1, X_2, \ldots, X_n$ be n independent random variables, each with mean μ
and variance σ^2. Let "Sum" represent the sum of these random variables:

$$\text{Sum} = X_1 + X_2 + \cdots + X_n$$

Recall that the mean for the sum is

$$E(\text{Sum}) = E(X_1) + E(X_2) + \cdots + E(X_n)$$

[1] This section presents a concise discussion of the central limit theorem. The reader may wish to
review problems 8 and 10 of chapter 4 before studying this section or may instead wish to read
section 5.5, which presents a more detailed, slower paced discussion. See the "Outline of Courses"
at the beginning of this text for comments concerning the order of sections 5.2 to 5.5.

so that

$$E(\text{Sum}) = \overbrace{\mu + \mu + \cdots + \mu}^{n} = n\mu$$

because each random variable has the same mean μ. Also the variance for the sum is

$$V(\text{Sum}) = V(X_1) + V(X_2) + \cdots + V(X_n)$$

$$= \overbrace{\sigma^2 + \sigma^2 + \cdots + \sigma^2}^{n} = n\sigma^2$$

because each random variable has the same variance σ^2, and they are independent. We define a new random variable $\overline{X}$;

$$\overline{X} = \sum_{i=1}^{n} X_i/n \quad \text{(sample mean)}$$

where $\overline{X} = \text{Sum}/n$ is the sample mean, which has expectation $E(\overline{X}) = n\mu/n = \mu$ and variance $V(\overline{X}) = n\sigma^2/n^2 = \sigma^2/n$.

The mean and variance of $\overline{X}$ are respectively μ and σ^2/n, but what is the shape of the distribution of $\overline{X}$? The following statement gives us insight into this question: Although for small samples, $\overline{X}$ can have almost any distribution, for large enough n, the distribution resembles the normal distribution. It is this property of sums that gives the normal distribution its fundamental place in statistics, and that motivated Galton's quote at the beginning of the chapter.

> Central Limit Theorem: Assume that we have n independent random variables $X_1, X_2, \ldots, X_n$, each having the same probability distribution with mean μ and variance σ^2. Then for large n, the average of these random variables, $\overline{X} = \Sigma_{i=1}^{n} X_i/n$ is approximately normally distributed with mean μ and variance σ^2/n.

How large n must be in order for the normal distribution to be a good approximation to the distribution of $\overline{X}$ will be discussed in Section 5.4. However, we do not need to worry about how large n must be in the special case where each X_i is normally distributed, for in this case $\overline{X}$ is exactly normally distributed with mean μ and variance σ^2/n for *any* n.

There are a number of ways to state the central limit theorem. Another frequent way of stating the conclusion of this theorem is to multiply $\overline{X}$ by n and

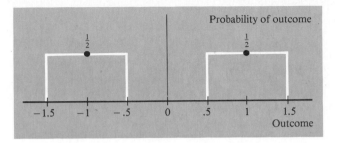

Figure 5 ▪ 6

speak of the sum $S = \Sigma_{i=1}^{n} X_i$. Thus: S is approximately normally distributed with a mean of $n\mu$ and a variance of $n\sigma^2$.* We illustrate both these versions in the examples that follow.

Example 5 ▪ 5   To observe the central limit theorem in action, consider the following game: A coin is tossed; if heads comes up, your gain is \$1; if not, your gain is $-$\$1 (you lose \$1). The two outcomes are -1 and 1, each with probability $\frac{1}{2}$. This probability distribution is shown in Figure 5.6.

This distribution certainly does not resemble the normal distribution. Now let us consider the random variable S_2 that represents the sum of two outcomes, where X_1 represents the possible outcomes of the first toss and X_2 the possible outcomes of the second toss.

$$S_2 = X_1 + X_2$$

The possible outcomes of S_2 are

$$S_2 = \begin{cases} -1 - 1 = -2 \\ -1 + 1 = 0 \\ 1 - 1 = 0 \\ 1 + 1 = 2 \end{cases}$$

The chance that S_2 takes on the value -2 is $\frac{1}{4}$, 0 is $\frac{1}{2}$, and 2 is $\frac{1}{4}$. The probability distribution for S_2 is shown in Figure 5.7.

Consider now the sum S_4 of four tosses. If you write down all the possibilities, you will find that the distribution of S_4 is as shown in Figure 5.8.

Notice that for only four trials the distribution of the sum begins to take on a shape that is reminiscent of the normal distribution (like beginning to see the shape of the baby through the developing embryo). This is especially so if we

---

* There are some subtle pitfalls to avoid when using this formulation (see the Brockett reference for details). In all the applications in this text, both versions of the central limit theorem apply.

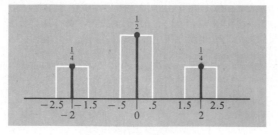

Figure 5 ▪ 7

connect the heights of the bars as shown in Figure 5.8. The distribution of S_{30}, the sum of 30 tosses shown in Figure 5.9 is almost indistinguishable from the normal distribution superimposed on it.

Application: Loss Ratios in Automobile Insurance Underwriting

An area of interest in casualty actuarial insurance research involves the probability distribution of the loss ratio which is the total losses by an insurance company divided by its total premiums. The standard deviation of the loss ratio can be interpreted as a measure of the risk involved in underwriting insurance in a particular state, since the more variable (unpredictable) the ratio of losses to premiums are, the higher is the standard deviation of the loss ratio and the riskier is the writing of insurance. In order to adequately address such issues as "is it riskier to underwrite insurance in states in which insurance premium charges are regulated by state agencies?" we must first determine the distribution of the loss ratio in both regulated and unregulated states. The following argument based on the central limit theorem shows that the loss ratio distribution should be approximately normally distributed.

As previously mentioned, the loss ratio in a state is defined as the total insured losses in that state divided by the total premiums collected. Let us number from 1 to n the policies issued during the year and let X_i denote the loss claimed on policy i during the year, and P_i the premium collected to insure policy i. The total losses for the state are given by $\sum_{i=1}^{n} X_i$, and the total premium collected is

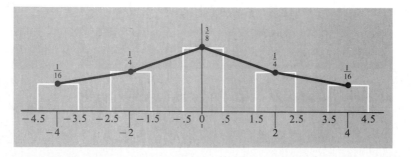

Figure 5 ▪ 8

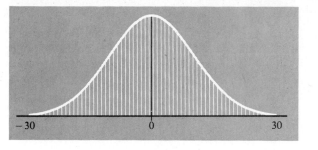

Figure 5 ▪ 9

$\sum_{i=1}^{n} P_i$. Thus the loss ratio is

$$L = \frac{\sum_{i=1}^{n} X_i}{\sum_{i=1}^{n} P_i}$$

Due to the competitive nature of the insurance underwriting business, the premiums per $1000 of insurance may be taken as approximately the same for each policy; that is, $P_i = P$. Accordingly, the loss ratio now becomes

$$L = \frac{\sum_{i=1}^{n} X_i}{nP} = \frac{1}{P} \overline{X}$$

By the central limit theorem the distribution of $\overline{X}$ is approximately normal, and since a constant times a normal random variable is also normally distributed, the loss ratio $L = \left(\frac{1}{P}\right) \overline{X}$ is approximately normally distributed. This knowledge is important because it allows us to use the statistical methods we develop in subsequent chapters to analyze the underwriting risk and return relationship for various states and various insurance lines (e.g., automobile insurance, fire insurance, homeowners insurance, and so on). For a detailed discussion of the central limit theorem and the loss ratio see the Brockett and Witt reference given at the end of this chapter.

5 ▪ 3   NORMAL APPROXIMATION OF THE BINOMIAL

Assume the coin-tossing model; that is, $X_1, X_2, \ldots, X_n$ are mutually independent with $P(X_i = 1) = p$, $P(X_i = 0) = 1 - p$ for each trial i. Hence $E(X_i) = p$, $V(X_i) = p(1 - p)$, leading to the result that $E(\text{Sum}) = np$,

$V(\text{Sum}) = np(1 - p)$, where $\text{Sum} = \Sigma_{i=1}^{n} X_i$. Moreover, we know that

$$P(\text{Sum} = x) = \binom{n}{x} p^x (1 - p)^{n-x}$$

the binomial distribution.

The central limit theorem tells us that for large n the binomial distribution looks like the normal distribution with mean p and variance $np(1 - p)$.

Example 5 ▪ 6   Let S have the binomial distribution where $S = \Sigma_{i=1}^{n} X_i$. Figure 5.10 shows for various values of n and p the distribution of

$$\frac{S - np}{\sqrt{np(1 - p)}} = \frac{\overline{X} - p}{\sqrt{\dfrac{p(1 - p)}{n}}}$$

which is the standardized binomial. Notice how closely the normal distribution with $\mu = 0$ and $\sigma = 1$ fits these distributions even for small n. Also, the closer p is to $\frac{1}{2}$, the better is the normal distribution fit.

The preceding example suggests an important technique: We can approximate the probability of the event $a < S \le b$ for the binomial by calculating $P(a < X \le b)$, where X is normal with mean np and variance $np(1 - p)$.

Example 5 ▪ 7   Assume that we have a binomial distribution with $n = 20$ and $p = 0.5$. Suppose we wish to calculate $P(8 < \text{Sum} \le 13)$; that is, we wish to calculate the probability that the number of successes is greater than 8 and less than or equal to 13. This probability is

$$P(8 < \text{Sum} \le 13) = \sum_{x=9}^{13} \binom{20}{x} (0.5)^x (0.5)^{20-x}$$

$$= \binom{20}{9} (0.5)^9 (0.5)^{11} + \binom{20}{10} (0.5)^{10} (0.5)^{10}$$

$$+ \cdots + \binom{20}{13} (0.5)^7 (0.5)^{13} = 0.69$$

This is a tedious calculation. We could approximate this probability by finding

$$P(8 < X \le 13)$$

where X is normal with $\mu = np = 10$, $\sigma^2 = np(1 - p) = 5$. To find this probabil-

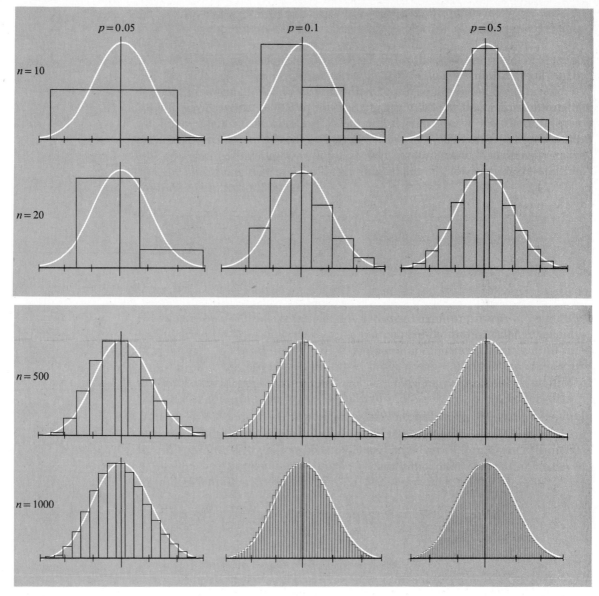

Figure 5 ▪ 10
The fit of the standard normal distribution to the standardized binomial for various values of *p* and the sample size *n*.

ity we note that by standardization

$$Z = \frac{X - 10}{\sqrt{5}}$$

is normal with mean 0 and variance 1. Hence we find

$$P\left(\frac{8 - 10}{\sqrt{5}} < Z \le \frac{13 - 10}{\sqrt{5}}\right) = P(-0.89 < Z \le 1.34) = F(1.34) - F(-0.89)$$

which we find from Appendix A.1 to be equal to 0.7232. Thus the chance of obtaining greater than 8 successes and less than or equal to 13 is *approximately* 0.7232.

The preceding example illustrates the fact that one can find an approximation to the binomial probability $P(a < \text{Sum} \le b)$ by finding

$$P\left(\frac{a - np}{\sqrt{np(1 - p)}} < Z \le \frac{b - np}{\sqrt{np(1 - p)}}\right)$$

The following discussion indicates that we can obtain a better approximation. Figure 5.11 shows a picture (density histogram) of the binomial distribution with $n = 20$ and $p = 0.5$. Since the binomial histogram is a density histogram, the *area* under each bar represents the probability of the value x shown under the bar. The probability of the value (the size of the area contained in the bar) is shown on top of the bar. Thus the probability (rounded to three places) of the value 9 is 0.160; the probability of 10 is 0.176, and so forth. Hence the probability of obtaining a number greater than 8 and less than or equal to 13 is the sum $0.160 + 0.176 + 0.160 + 0.12 + 0.074 = 0.69$. Therefore the probability of obtaining a number greater than 8 and less than or equal to 13 is given by the sum of the areas of the bars extending from 8.5 to 13.5. Now in Figure 5.11 we superimpose on the binomial histogram the normal distribution having the same mean (10) and same standard deviation ($\sqrt{5}$) as the binomial. We are trying to estimate the binomial area from 8.5 to 13.5 by the area under the normal from 8.5 to 13.5. Thus we wish to approximate the binomial areas by $P(8.5 < X \le 13.5)$, where X is normal with $\mu = 10$ and $\sigma = \sqrt{5} = 2.24$. Standardizing, we wish to find

$$P\left(\frac{8.5 - 10}{\sqrt{5}} < Z \le \frac{13.5 - 10}{\sqrt{5}}\right) = P(-0.671 < Z \le 1.57)$$

where Z is normal with $\mu = 0$ and $\sigma = 1$. From Appendix A.1 we find this area

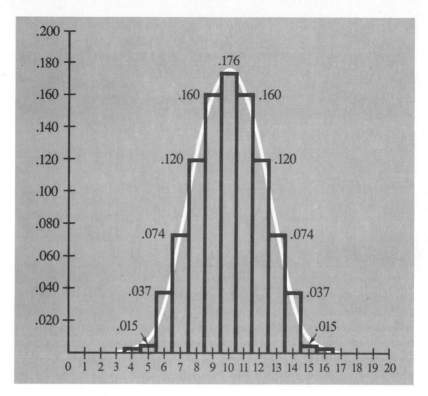

Figure 5 ■ 11
Binomial distribution;
$p = 0.5$, $n = 20$ with
normal distribution superimposed on it.

to be 0.6904.* Notice that the exact value (to three places) is 0.69 and that 0.6904 is closer to it than the approximation of 0.7232 previously obtained.

Normal Approximation to Binomial
Suppose we have n independent trials with the probability of success on each trial being p. Then the binomial distribution has mean np and variance $np(1 - p)$. The probability of obtaining the number of successes strictly greater than a and less than or equal to b is approximated by finding the area under a normal curve with mean np and variance $np(1 - p)$ between $a + \frac{1}{2}$ and $b + \frac{1}{2}$. Standardizing yields

$$P\left(\frac{a + 0.5 - np}{\sqrt{np(1 - p)}} < Z \le \frac{b + 0.5 - np}{\sqrt{np(1 - p)}}\right)$$

which may be found from Appendix A.1.

---

* The values of z equal to -0.671 and 1.57 do not appear in Appendix A.1. To find the values of the area corresponding to these values of z, one must use interpolation methods. However, even if one does not use "interpolation" but uses those values of z that are closest to -0.671 and 1.57, one can still see the improvement in the approximation.

Example 5 ■ 8 Let $p = 0.5$ and $n = 20$ as before. Suppose now we wish to calculate the probability of a number *greater than or equal to* 8 and *strictly less than* 13. This is equal to the probability of a number that is strictly *greater than* 7 and *less than or equal to* 12. We obtain an approximation to the probability by finding

$$P\left(\frac{7.5 - 10}{\sqrt{5}} < Z \leq \frac{12.5 - 10}{\sqrt{5}}\right) = P(-1.12 < Z \leq 1.12)$$

from Appendix A.1. The answer is 0.7372.

5 ■ 4 HOW LARGE MUST *n* BE TO APPLY THE CENTRAL LIMIT THEOREM TO THE BINOMIAL?

The central limit theorem states that the sum (or average) of a large number *n* of independent random variables is approximated by the normal distribution. Just exactly how large *n* must be in order to feel secure in using this approximation is a very complex problem. By observing the binomial density histograms of Figure 5.10, we can observe some fundamental facts. The more skewed the distribution is (the further *p* is from $\frac{1}{2}$), the longer it takes the histogram to resemble the bell-shaped normal curve. Symmetric distributions (like the $p = \frac{1}{2}$ binomial) appear normal quite quickly.

A rule of thumb that we personally have found useful in practical situations is that derived by W. G. Cochran in the reference given at the end of this chapter. This rule is

---

If $S = X_1 + X_2 + \cdots + X_n$ is the sum of *n* independent *X*'s, each having the same probability distribution, then in order to use the normal approximation for *S*, we should have $n \geq 30$ or

$$n \geq 25 \left[\frac{E(X - \mu)^3}{\sigma^3}\right]^2,$$

whichever is larger.

Many books use only the rule $n \geq 30$ for the normal approximation. If the distribution is highly skewed this *n* may be too small.

To implement these criteria in practice, we generally use some sample data or data on a subject related to the study under consideration. We then estimate $E(X - \mu)^3$ (usually by $\sum_{i=1}^{k} (x_i - \bar{x})^3/k$) and use this estimate and the preceding criteria to get a rough idea about the size of n required.

For the binomial random variable we have $P(X_i = 1) = p$ and $P(X_1 = 0) = (1 - p)$ and calculate $E(X) = p$, $V(X) = \sigma^2 = p(1 - p)$, and $E(X - \mu)^3 = E(X - p)^3 = (0 - p)^3(1 - p) + (1 - p)^3 p = p(1 - p)(1 - 2p)$. Continuing our calculation, we find

$$25 \left[\frac{p(1 - p)(1 - 2p)}{(\sqrt{p(1 - p)})^3} \right]^2 = \frac{25(1 - 2p)^2}{p(1 - p)}$$

Summarizing, we have

If $S = X_1 + X_2 + \cdots + X_n$ is a binomial random variable with $p = P(X = 1)$, then in order to use a normal approximation for S, we should at least have $n \geq 5/p$ or $n \geq 5/(1 - p)$ or

$$n \geq \frac{25(1 - 2p)^2}{p(1 - p)}$$

whichever is larger.

Notice that as p gets farther from $\frac{1}{2}$, the required sample size increases just as predicted from Figure 5.10.

As an illustration, consider determining how large a sample to take to use the normal approximation if we are investigating the probability of a defective part in a certain production process. Suppose we feel secure that whatever this probability is, it is definitely neither smaller than 0.1 nor bigger than 0.4. Using the preceding formula for the binomial with $p = 0.1$ first, and then with $p = 0.4$, we obtain the bounds

$$p = 0.1; \quad n \geq 25 \frac{(1 - 2(0.1))^2}{(0.1)(1 - 0.1)} = 25 \frac{(0.8)^2}{0.09} = 177.7 \approx 178$$

$$p = 0.4; \quad n \geq \frac{5}{0.4} = 12.5 \approx 13$$

Thus to use the normal approximation, we should take $n \geq 178$.

5 ■ 5   SAMPLING DISTRIBUTION OF $\overline{X}$

Let $X_1, X_2, \ldots, X_n$ represent n independent random variables and consider the sum $X_1 + X_2 + \cdots + X_n$ divided by n. We have denoted this random variable by $\overline{X}$;

$$\overline{X} = \frac{X_1 + X_2 + \cdots + X_n}{n}$$

It is the same sample average that we introduced in our discussion of descriptive numerical characteristics of samples back in Chapter 2. Now we are considering $\overline{X}$ as a random variable. In Chapter 2 we only considered one possible value of $\overline{X}$ (namely, $\overline{x}$, the average of the sample actually obtained). Here we consider how $\overline{x}$ changes as the sample changes.

Example 5 ■ 9   Suppose we sample a population of tags. On each tag, either the value 0 or the value 1 is written. An equal number of tags have 0 and 1. Hence the 0's and 1's are equally likely. The chance of obtaining a 0 equals the chance of obtaining a 1, and is $\frac{1}{2}$. We sample with replacement and take a sample of size 2. The possible values in the range of $\overline{X}$ are 0, $\frac{1}{2}$, and 1. The value of $\frac{1}{2}$ can be obtained if we obtain a 0 and then a 1, or if we obtain a 1 first, then a 0. The value 0 is obtained if there are two 0's in a row; a 1 is obtained if there are two 1's in a row.

$$\overline{X} = \begin{cases} \dfrac{0+0}{2} = 0 \\[2mm] \dfrac{0+1}{2} = \dfrac{1+0}{2} = 0.5 \\[2mm] \dfrac{1+1}{2} = 1 \end{cases}$$

The probability that $\overline{X} = 0$ is $\frac{1}{4}$; that $\overline{X} = 0.5$ is $\frac{1}{2}$, and that $\overline{X} = 1$ is $\frac{1}{4}$. This is the probability distribution of $\overline{X}$.

The distribution of $\overline{X}$ is called the *sampling distribution* of the sample mean $\overline{X}$.

Once we have obtained a *particular* sample, then we obtain a *particular* value for $\overline{X}$, namely, *one* of the values 0, $\frac{1}{2}$, or 1. We denote the actual value we obtain by a lowercase letter $\overline{x}$. Recall that $\overline{x}$ is the mean of a sample and that we previously called it the *sample mean.*

To summarize, the range of values $\overline{X}$ is the set of possible values we can obtain for the sample mean. We denote the value we actually get from sampling by the small letter $\overline{x}$. Unfortunately, the observed value $\overline{x}$ is also called the *sample mean.* It should be called the *observed* sample mean but this terminology is cumbersome and not often used. We should always keep in mind the difference between $\overline{X}$ and $\overline{x}$. One of the major sources of confusion in the study of statistics is not distinguishing between the random variable, $\overline{X}$, and one of its observed values, $\overline{x}$. The next example explains the differences between the population mean μ, the sample mean $\overline{X}$, and the observed sample mean $\overline{x}$.

Example 5 ▪ 10 Let us suppose we have a population $\mathcal{P}_1 = \{1, 3, 4, 5, 7\}$ where the numbers are equally likely to be selected into the sample. Using the formulas from previous chapters, we calculate the population mean

$$\mu = \frac{1 + 3 + 4 + 5 + 7}{5} = \frac{20}{5} = 4$$

and the population variance

$$\sigma^2 = \frac{(1-4)^2 + (3-4)^2 + (4-4)^2 + (5-4)^2 + (7-4)^2}{5} = 4$$

If we now sample *with replacement* twice from this population, then the possible samples we can obtain along with their corresponding observed sample means are

(Choice 1, Choice 2)	$\overline{x}$	*(Choice 1, Choice 2)*	$\overline{x}$
(1, 1)	1	(4, 5)	4.5
(1, 3)	2	(4, 7)	5.5
(1, 4)	2.5	(5, 1)	3
(1, 5)	3	(5, 3)	4
(1, 7)	4	(5, 4)	4.5
(3, 1)	2	(5, 5)	5
(3, 3)	3	(5, 7)	6
(3, 4)	3.5	(7, 1)	4
(3, 5)	4	(7, 3)	5
(3, 7)	5	(7, 4)	5.5
(4, 1)	2.5	(7, 5)	6
(4, 3)	3.5	(7, 7)	7
(4, 4)	4		

so that $\mathcal{P}_2 = \{1, 2, 2.5, 3, 4, 2, 3, 3.5, 4, 5, 2.5, 3.5, 4, 4.5, 5.5, 3, 4, 4.5, 5, 6, 4, 5, 5.5, 6, 7\}$ is the population of possible values of $\overline{X}$ for samples of size 2. The mean of this population is 4, which is exactly equal to μ, the mean of the original population. The variance of this population is 2, which is equal to the variance of the original population divided by the sample size. Hence $E(\overline{X}) = 4$ and

$V(\overline{X}) = 2 = \sigma^2/2$. Note that the range of $\overline{X}$ is the *set* of possible values for $\overline{x}$, where $\overline{x}$ denotes the actual value obtained after sampling.

The question is, If we take n independent observations from the same population, do we need to go through the laborious process of constructing the sample distribution $\mathcal{P}_n$ of $\overline{X}$ to calculate the mean and variance of $\overline{X}$, or can we easily calculate these quantities from a knowledge of μ and σ^2, the mean and variance, respectively, of the original population? The answer is, we can calculate the mean and variance of $\overline{X}$ from a knowledge of μ and σ^2. In fact, by the properties of expected values (see sections 4.8, 4.9),

$$E(\overline{X}) = E\left(\frac{X_1 + X_2 + \cdots + X_n}{n}\right)$$

$$= \frac{E(X_1) + E(X_2) + \cdots + E(X_n)}{n} = \frac{\mu + \mu + \mu + \cdots + \mu}{n} = \frac{n\mu}{n} = \mu$$

On the average, $\overline{X}$ gives the right value for the population mean μ regardless of the sample size.

We may also calculate the variance of $\overline{X}$. Using the fact that each X_i has the same variance σ^2, we have by properties of the variance (see sections 4.10, 4.12),

$$V(\overline{X}) = V\left(\frac{1}{n}\sum_{i=1}^{n} X_i\right) = \frac{1}{n^2} V\left(\sum_{i=1}^{n} X_i\right)$$

$$= \frac{1}{n^2}\sum_{i=1}^{n} V(X_i) = \frac{1}{n^2}\sum_{i=1}^{n} \sigma^2 = \frac{n\sigma^2}{n^2} = \frac{\sigma^2}{n}$$

We summarize these results as follows.

> If a population has a mean μ and a variance σ^2, and if we take n independent observations from the population and calculate $\overline{X} = (1/n)\sum_{i=1}^{n} X_i$, then $\overline{X}$ has a mean equal to μ and a variance equal to σ^2/n. The standard deviation of $\overline{X}$ is $\sigma/\sqrt{n}$ and is called the *standard error of the mean.*

The formula $V(\overline{X}) = \sigma^2/n$ becomes clearer if you think of σ^2 as a measurement of how difficult it is for a value extremely far away from the center to appear. The smaller σ^2 is, the harder it is to obtain extreme values. Now, in order for an average of two values X_1 and X_2 to be extreme, both values must be extreme. Thus it should be twice as hard to get the *average* of two samples to be extreme, or equivalently, the variance of the average should be half the variance of the X's; that is, $V(\overline{X}) = \sigma^2/2$. The same argument works for samples of any size n.*

Now that we know that the mean of $\overline{X}$ is μ and the variance is σ^2/n, we wish to describe its exact distribution or shape. Since $\overline{X}$ is the sum of the independent random variables X_1/n, X_2/n, . . . , X_n/n, from the central limit theorem we have seen from the discussion of section 5.2 that

---

Approximate Sampling Distribution of $\overline{X}$
Assume that we have a population with mean μ and variance σ^2. Suppose we take n independent observations from this population and calculate
$\overline{X} = \sum\limits_{i=1}^{n} X_i/n$; then for n large, the distribution of $\overline{X}$ is approximately normal with mean μ and variance σ^2/n. Moreover, if the population is itself normally distributed, then $\overline{X}$ is exactly normally distributed with this mean and variance.

If the distribution of $\overline{X}$ is either approximately or exactly normal, we say that $\overline{X}$ is *at least* approximately normally distributed.

The distribution of $\overline{X}$ (called the sampling distribution of $\overline{X}$) describes exactly how $\overline{X}$ changes from sample to sample. The beauty of the central limit theorem (section 5.2) is that it tells us that no matter how the original population is distributed, for n large, the sampling distribution of $\overline{X}$ is at least approximately normal. Thus if we wish to calculate $P(a < \overline{X} \leq b)$, we use the standardization procedure introduced in the first section to find $Z = (\overline{X} - \mu)/(\sigma/\sqrt{n})$, where by the central limit theorem, Z is approximately normal with $\mu = 0$ and $\sigma = 1$. Thus,

$$P(a < \overline{X} \leq b) = P\left(\frac{a-\mu}{\sigma/\sqrt{n}} < \frac{\overline{X}-\mu}{\sigma/\sqrt{n}} \leq \frac{b-\mu}{\sigma/\sqrt{n}}\right) = P\left(\frac{a-\mu}{\sigma/\sqrt{n}} < Z \leq \frac{b-\mu}{\sigma/\sqrt{n}}\right)$$

and we can easily determine the value of this last probability from Appendix A.1.

Example 5 ▪ 11 We wish to calculate the probability that $\overline{X}$, the mean of a sample of size 30 from a population with mean 20 and variance 120, lies between 16 and 22; that is, to calculate $P(16 < \overline{X} \leq 22)$.

---

* We also know from Tchebychev's approximation that $\overline{X}$ will not deviate greatly from the mean μ, especially if σ^2 is small and n is large, since

$$P(|\overline{X} - \mu| \geq k) \leq \frac{V(\overline{X})}{k^2} = \frac{\sigma^2}{nk^2}$$

See Chapter 4, Section 15.

To perform this calculation, we use the central limit theorem to conclude that $(\overline{X} - 20)/\sqrt{120/30}$ is approximately normal with $\mu = 0$ and $\sigma^2 = 1$; we assume that $n = 30$ is sufficiently large so that the approximation to the normal yields an answer with acceptable accuracy.* Thus

$$P(16 < \overline{X} \leq 22) = P\left(\frac{16 - 20}{\sqrt{\frac{120}{30}}} < \frac{\overline{X} - 20}{\sqrt{\frac{120}{30}}} \leq \frac{22 - 20}{\sqrt{\frac{120}{30}}}\right) = P\left(\frac{-4}{2} < Z \leq \frac{2}{2}\right)$$

$$= P(-2 < Z \leq 1) = 0.8413 - 0.0228 = 0.8185$$

from Appendix A.1.

5 ■ 6 THE IMPORTANCE OF THE MEAN AND VARIANCE

Our knowledge of the normal distribution and its importance in nature yields a good reason for studying ways to determine the mean μ and variance σ^2 of a population. If X is normally distributed, then a knowledge of μ and σ^2 completely determines the probability distribution of X. Since the normal distribution (or a close approximation to it) occurs frequently in problems, we can often evaluate all the probabilities of interest simply by determining μ and σ^2 and then using the normal table, Appendix A.1. Thus we shall spend much time in subsequent chapters on methods for investigating the value of μ and σ^2 in various populations.

5 ■ 7 SAMPLING EXPERIMENTS: SIMULATION OF SAMPLING FROM NORMAL POPULATIONS

In Section 4.16, we briefly discussed the use of the computer in simulation experiments, particularly its use in simulating sampling from the binomial distribution. Clearly, because many things in nature, such as people's height, errors of measurement, the average yield of wheat from year to year, tend to be approximately normally distributed, the ability to simulate sampling experiments from the normal distribution would be useful. This fact has been recognized for some time and has resulted in the production of programming subroutines that enable such simulation experiments to be performed on the

* Just as in the binomial case (see Section 5.4) where the issue was of how large n should be for the normal approximation to be acceptable, this issue is also important in deciding whether the sampling distribution of $\overline{X}$ can be approximated by the normal distribution. A general rule is given at the beginning of Section 5.4. This matter is discussed in detail in the reference by Cochran given at the end of this chapter.

computer. All large statistical computer packages have such capability as do many small ones such as MINITAB, referenced in Chapter 4.

Independent Random Sampling and Sampling without Replacement

We have repeatedly stressed that a major assumption underlying the central limit theorem is that the random variables in the sum $X_1 + X_2 + \cdots + X_n$ are independent. Thus in our subsequent discussion we emphasized that the sampling used was independent random sampling. If one samples with replacement and independently on each trial, then one attains such a sequence of random variables. Such a sampling procedure is called *independent random sampling.* One can obtain an approximation to independent random sampling even if one samples *without* replacement as long as the number of samples n is small as compared to the population size N. For example, suppose $N = 1{,}000{,}000$ balls, half of which are green and half are red. If $n = 2$, what is the probability of obtaining two green balls when (1) sampling with replacement or (2) sampling without replacement? (Assume the model of equally likely events.) The answer to (1) is $(0.5)(0.5) = 0.25$, whereas to (2), it is $(0.5)(499{,}999)/999{,}999 = 0.2499997$. The difference between the two answers is 0.0000003; for all practical purposes, the answers are the same. The two methods of sampling will produce essentially the same results if N is so large that when sampling with replacement, the chance of getting the same member of the population more than once in the sample is negligible. Hence the central limit theorem should hold when sampling without replacement if N is sufficiently larger than n. (If a computer is available, the class should develop a simulation experiment to illustrate this fact.)

SUMMARY

In this chapter we introduced the most important distribution found in statistics, the normal distribution. We described how to calculate probabilities for the normal distribution when μ and σ^2 are given, using Appendix A.1, which is a tabulation of the normal distribution with $\mu = 0$ and $\sigma = 1$.

The importance of the normal distribution was underlined in the discussion of the central limit theorem. Two important applications of this theorem were

1. An application to the approximation of the binomial distribution for large n by the normal
2. An application to determine the sampling distribution of the sample mean $\overline{X}$, where $\overline{X} = (X_1 + X_2 + \cdots + X_n)/n$ and each random variable X_i has the same distribution with mean μ and variance σ^2. Then $E(\overline{X}) = \mu$. If independent random sampling is used, $V(\overline{X}) = \sigma^2/n$ and the standard deviation of $\overline{X}$ (called the standard error of the mean) is $\sigma/\sqrt{n}$. For independent

random sampling, the sampling distribution of $\overline{X}$ is, according to the central limit theorem, approximately normally distributed with mean μ and variance σ^2/n. If each X_i is normal, then $\overline{X}$ is exactly normally distributed.

We also discussed a method for determining approximately how large n must be to use the central limit theorem.

The distinction between the random variable $\overline{X}$ and one of the values it can take on, $\overline{x}$, was strongly emphasized.

Key Terms

normal distributions

cumulative normal distribution

standardization of a random variable

upper centile values

central limit theorem

sampling distribution of $\overline{X}$

mean of $\overline{X}$ as it relates to the original population

variance of $\overline{X}$ as it relates to the original population

standard error of the mean

sample mean

References

Galton, F.: *Natural Inheritance.* 1889, p. 63ff. Reprinted by the American Mathematical Society Press. In this classic work, Galton described an apparatus known today as Galton's *quincunx.* It consists of a board on which nails are arranged in rows, the nails in a given row being placed below the midpoints of the intervals between the nails in the row above. Small steel balls (ball bearings) of equal size are poured into the apparatus through a funnel located opposite the central pin of the first row. As they run down the board, the balls are likely to strike some of the nails; the final distribution of the balls is approximately normal about the position below the central pin of the first row. Galton explains why in a rather colorful description:

> The shot passes through the funnel and issuing from its narrow end, scampers deviously down through the pins in a curious and interesting way; each of them darting a step to the right or left, as the case may be, every time it strikes a pin. . . . The principle on which the action of the apparatus depends is, that a number of small and independent accidents befall each shot in its career. . . .

Applications to Psychology can be found in

DuBois, P. H.: *An Introduction to Psychological Statistics.* New York, McGraw-Hill, 1965, Chapters 11 and 12.

McNemar Quinn: *Psychological Studies.* New York, John Wiley & Sons, 1965, Chapters 6 and 11.

Gnedenko, B. V., and A. N. Kolmogorov: *Limit Distributions for Sums of Independent Random Variables.* Reading, MA, Addison-Wesley, 1968. The following quote from this book by the founders of modern probability theory emphasizes the importance of limit theorems such as the law of large numbers and the central limit theorem. Notice that they regard such theorems not only as derivations from mathematical models of nature but also as reflecting properties found in nature. The quote from Francis Galton at the beginning of this chapter shows that he concurs with Gnedenko and Kolmogorov.

> The epistemological value of the theory of probability is revealed only by limit theorems. Moreover, without limit theorems it is impossible to understand the real content of the primary concept of all our sciences and the concept of probability. . . .

> The epistemological value of the theory of probability is based on this: that large-scale random phenomenon in their collective action create strict, non-random regularity. The very concept of mathematical probability would be fruitless if it did not find its realization in the frequency of occurrence of events under large-scale repetition of uniform conditions (a realization which is always approximate and not wholly reliable, but that becomes, in principle, arbitrarily precise and reliable as the number of repetitions increases.)

Detailed tables of the normal with $\mu = 0$ and $\sigma = 1$ and detailed versions of other tables we will use can be found in

Pearson, E. S., and H. O. Hartley: *Biometrika Tables for Statisticians,* 3rd ed., New York, Cambridge University Press, 1966.

The statistical 'rule of thumb' we presented for determining the sample size n central limit theorem may be found in

Cochran, W. G.: *Sampling Techniques,* 3rd ed. New York, John Wiley & Sons, 1977, p. 42.

A discussion of some pitfalls in using the central limit theorem for estimating probabilities for sums is found in

Brockett, P. L.: On some misuses of the central limit theorem in risk management and ruin calculations. *Journal of Risk and Insurance,* Vol. 50, 1983, pp. 727–731.

Further discussion of the central limit theorem as used in insurance loss ratio calculations can be found in

Brockett, P. L. and R. C. Witt: The underwriting risk and return paradox revisited. *Journal of Risk and Insurance,* Vol. 49, 1982, pp. 621–627.

Exercises

Section 5.1

1. a. Find the area under the normal distribution with mean 0 and variance 1 that lies from (i) 0 to 1.65, (ii) 0 to 1.95, (iii) −1.65 to 1.65, (iv) −1.95 to 1.95, (v) 1.65 to 1.95.
   b. Find the area to the left of (i) 1.65, (ii) 1.95, (iii) −1.65, (iv) −1.95.
   c. Find the area to the right of (i) 1.65, (ii) 1.95, (iii) −1.65, (iv) −1.95.

2. Using Appendix A.1,
   a. find the values b (b positive) if the area under the standard normal curve between

$-b$ and b is (i) 0.5, (ii) 0.95, (iii) 0.9, (iv) 0.99.

b. find the value of c if (i) the area between 0 and c is 0.45 and c is greater than 0, (ii) the area between $-c$ and c is 0.95, (iii) the area to the left of c is 0.1, (iv) the area to the right of c is 0.99.

3. (Centiles) Often it is of interest to find the upper *centile values* of a distribution. If F is a cumulative distribution function, then the upper α centile value of F is defined to be that value x_α such that $1 - F(x_\alpha) = \alpha$, that is, if X is a random variable with distribution function F, then x_α is the number for which $P(X > x_\alpha) = \alpha$ (i.e., α of the area is to the right of x_α).

a. For the normal distribution with mean 0 and variance 1, find $z_{0.99}$ and $z_{0.975}$.

b. For the normal distribution with mean 6 and variance 25, find the upper 0.90 centile and the upper 0.99 centile.

4. Suppose Z has a normal distribution with $\mu = 0$ and $\sigma = 1$. Using Appendix A.1, find

a. $P(Z \le -2.80)$ b. $P(Z \le 0)$
c. $P(Z \le 0.25)$ d. $P(Z \le 2.80)$
e. $P(Z \ge 2.80)$ f. $P(Z \ge -0.25)$
g. $P(0.25 < Z \le 2.8)$
h. $P(-1.96 < Z \le 1.96)$
i. $P(|Z| \ge 2) = P(Z \ge 2 \text{ or } Z \le -2)$
j. $P(|Z| \ge 1.6) = P(Z \ge 1.6 \text{ or } Z \le -1.6)$
k. $P(|Z| < 0.3) = P(-0.3 < Z < 0.3)$

5. Find the upper centile values for Z corresponding to

a. $z_{0.05}$, b. $z_{0.80}$, c. $z_{0.975}$, d. $z_{0.10}$, e. $z_{0.95}$, f. $z_{0.99}$, g. $z_{0.995}$, h. $z_{0.005}$.

6. Suppose X has a normal distribution with $\mu = 25$ and $\sigma^2 = 16$. Find

a. $P(X \le 31)$ b. $P(X \le 16)$
c. $P(16 < X \le 31)$ d. $P(X \ge -3.5)$
e. $P(|X - 25| \ge 3)$ f. $P(|X - 25| < 4)$
g. $P(19 < X < 23)$
h. $x_{0.05}$, $x_{0.80}$, $x_{0.975}$, $x_{0.95}$, $x_{0.99}$

7. The following letter to Ann Landers appeared in the *New Orleans Times Picayune*, March 23, 1977 (Copyright Field Newspaper Syndicated):

Dear Ann Landers: When my son first married his wife, she would spend a couple of weeks with him and then go visit her parents across the state for a week or so. This pattern kept up for more than a year. My son knew his wife dated other men while on her visits home.
Last June she went for a visit with her folks and stayed almost six weeks. 253 days after the night of her return, she gave birth to a full-term healthy son. According to my information, 270 days must elapse after insemination, or 275 to 280 days after the first day of monthly period. Taking all this into consideration, she would have been impregnated while on her visit at her home.
I don't think my son is aware of these facts. Should I mention to him that even though he is the legal father of the baby he may not be the biological father? Or would it be better not to say anything?
—His Mom.

Ann Landers' reply was decidedly nonstatistical (she told his mom to mind her own business). We, on the other hand, are in a position to use our statistical knowledge to give a much better answer. Assume that X represents the gestation period required for a randomly selected woman, and assume that X has a normal distribution with a mean $\mu = 270$ days, and a standard deviation of $\sigma = 16$ days.

a. The wife could have been pregnant before she left, in which case her pregnancy lasted 253 days + 6 weeks = 295 days. Calculate $P(X > 295)$.

b. The wife could have gotten pregnant after returning home. Calculate $P(X \le 253)$.

c. Calculate $P(X > 295 \text{ or } X \le 253)$.

d. If there are 100,000 births, how many would you expect to last ≤ 253 or > 295 days gestation?

What do you conclude about the chances that the wife was impregnated by the husband?

8. Insurance companies must often calculate the potential outlay of cash that might occur due to insured loss. This is, of course, a random variable each year and frequently is modeled by a normal distribution. In order to ensure that they have a sufficient amount of liquid funds (i.e., cash) on hand, they calculate the maximum probable yearly loss (MPYL) for the year, which is just the upper 0.05 centile of the assumed total loss variable X.
   a. Calculate the MPYL if X has a normal distribution with $\mu = \$25,000,000$ and $\sigma = \$3,000,000$.
   b. Calculate the MPYL if the expected loss is $16,000,000 and the standard deviation is $1,000,000.

9. The distribution of the height of a given population is normal with mean 5 ft, 5 in. and variance 41.4 in. Find the probability of someone being
   a. over 7 ft tall
   b. under 4 ft tall

10. One of the serious risks in a blood transfusion is that the receiver could get anicteric hepatitis virus from the donor. Consequently, serum enzymes are used to screen hospital blood bank donors for hepatitis, since if you could tell which donors were carriers, the risk could be substantially reduced.

    Prince and Gershon discuss an enzyme called serum glutamic pyruvic transaminase (SGPT) and conclude that the logarithm of SGPT is normally distributed in both healthy and diseased (i.e., hepatitis carrying) individuals. Let Y be the logarithm of SGPT level in the blood. For normal individuals Y has a mean 1.25 and standard deviation 0.12. In the diseased group the mean is 1.55 and the standard deviation is 0.13.
    a. Find the upper tenth centile, $y_{0.1}$ for the *healthy* group.
    b. Find the proportion (i.e., probability) of *diseased* individuals with a log SGPT

level less than the value $y_{0.1}$ found in part a.
    c. What level of log SGPT could be used to screen out 95% of the diseased donors?
    (Prince, A. M. and R. K. Gershon: The use of serum enzyme determinations to detect anicteric hepatitis. *Transfusion,* 5:130, 1965.)

Section 5.2

11. The normal distribution is found throughout nature because many random variables are the sums of a large number of random variables, each having a small variation. For example, if one looks at the pattern of shots along the horizontal axis of a target, one will see the normal distribution. This result comes as no surprise to those who know the central limit theorem, for the deviation of the bullet from the center is a random variable that is the sum of the deviations due to the variation in (1) the bore of the bullets fired, (2) the slightly different position the firearm is held and pointed on each shot, (3) the slight variation in the wind (even on a calm day), and many other variables.
    a. Suppose the deviation from the target center is normally distributed with $\mu = 0$ and $\sigma = 4$, where the deviation is measured in terms of inches. What is the probability that the next bullet will land within 2 in. of the target's center?
    b. Every Sunday morning there is a jogger's race at Audubon Park in New Orleans. The runners run five laps (each of 2 mi) around the park. All types of people participate, from oil rig workers to college students to fine uptown ladies whose servants wait at the finish line with towels and freshening drinks. About 128 people participate. They are divided into four groups according to "ability" as determined by age, performance in past races,

and so on. The groups are staggered at the start, the least "able" starting first, and so forth. The keeper of the records last year was a student of statistics. He was idly watching the runners pass the starting line after the first lap when he jumped up in surprise and shouted to the astonishment of those around him, "Look! The normal distribution is running by!" When pressed by the spectators to explain, he said "Notice that only a few runners lead the pack but the density of runners increases until you find a tight bunching, and then there is a trailing off of runners again until there are only a few stragglers left in the tail of the distribution. However, now that I think about it, I shouldn't have been surprised by this phenomenon because, after all, it is a natural result of the central limit theorem." The spectators' reaction was mystification. The student was too busy to explain further. He set out to check whether his observation of "normality" was correct. We shall learn in Chapter 10 how he did this. After he verified this observation, he looked at the records of old races and found that the average time it takes a person to finish is 1.4 hours, with $\sigma = 0.2$ hours. The student was interested in the probability that a person would finish in less than 2 hours and, for 128 people, the chance that everyone would finish in less than 2 hours. (The student had to stay until everyone finished. Since he had an exam the next day, he was interested in the amount of study time he might lose. He figured he could safely lose 2 hours.) Calculate each of these numbers. Calculate the chance that everyone finishes within 2 hours, assuming that each jogger runs his or her own pace independently of the pace of every other jogger.

12. Refer to problem 8 and give an explanation involving the central limit theorem as to why a large insurance company with many small individual policyholders might assume that its yearly loss follows (approximately) a normal distribution.

Section 5.3

13. Assume that S has a binomial distribution. Use the normal approximation to find $P(S \leq 15)$ if $n = 30$, $p = 0.3$. What is the approximate upper 0.10 centile of this distribution?

14. One hundred light bulbs are selected from a large production lot. Suppose the probability of a light bulb being defective is 0.1. What is the probability that fewer than five defective light bulbs are in the lot?

15. It is found that 10% of all people who take a drug have severe side effects: loss of weight, rash, depression.
    a. If 120,000 people used this drug last year, what is the probability that more than 10,000 people suffered side effects?
    b. It is found that of those who suffer side effects only 5% see a doctor about them. What is the probability that doctors will see more than 10,000 people who suffer these side effects? more than 1000?
    c. The World Health Organization (WHO) collects information on drug side effects from 22 countries. Doctors in these countries have forms on which they record information concerning the side effects patients suffer from drugs; these forms are to be sent to a national center that forwards them to WHO in Geneva. Suppose doctors only report 2% of the side effects they observe. What is the probability that WHO will receive more than 10,000 reports on the drug? more than 1000? more than 100? Assume the data in parts a and b are valid for each country.

16. The experience of an airline indicates that the probability of a person not showing up for his

or her reservation is 5%. Suppose the airline sells 500 seats on an international flight that has 450 seats. What is the probability that more than 450 people will show up? What is the maximum number of seats the airline should sell so that there is a 99% chance that everyone who shows up will get a seat?

17. Cards are drawn 30 times with replacement from a set of four cards. On each draw a "clairvoyant" guesses the value of the drawn card.

   a. What is the probability of guessing more than one card correctly?
   b. What is the probability of guessing more than two cards?
   c. What is the probability of guessing more than eight cards?
   d. Let N denote the number of cards in 30 such draws that are correctly identified by an alleged clairvoyant. What number must N be in order that there is less than a 5% chance that he or she randomly made at least this number of correct choices?

Section 5.4

18. A medical study is being contemplated concerning the effect of a certain drug on a blood enzyme of interest. The question of how large a sample of patients to test is very important, and the researchers wish to use the central limit theorem to facilitate their analysis. Studies on other biochemical experiments lead them to believe that the proportion of patients who will respond to this drug is between 0.7 and 0.8. To use the normal approximation to the binomial, how large should n be?

19. A production supervisor needs to estimate the proportion of defective parts produced by a certain machinist. It is believed that this proportion is $0.05 \le p \le 0.1$. How large should n be to use the central limit theorem in this situation?

Section 5.5

20. Let the random variable X_1 represent the outcomes of the toss of the first die and X_2 the outcomes of the second die tossed. The possible outcomes for each random variable together with the associated probabilities are listed in Table 5.1.

   a. Show that $E(X_1) = E(X_2) = 3.5; V(X_1) = V(X_2) = 2.917$.
   b. Now consider another random variable $\overline{X}_2$ that is the average of the two outcomes.

Table 5 ▪ 1 List of Outcomes on Each Die and Associated Probabilities

x_1		x_2	
Outcome of die 1	Probability of outcome	Outcome of die II	Probability of outcome
1	$\frac{1}{6}$	1	$\frac{1}{6}$
2	$\frac{1}{6}$	2	$\frac{1}{6}$
3	$\frac{1}{6}$	3	$\frac{1}{6}$
4	$\frac{1}{6}$	4	$\frac{1}{6}$
5	$\frac{1}{6}$	5	$\frac{1}{6}$
6	$\frac{1}{6}$	6	$\frac{1}{6}$

$\overline{X}_2 = (X_1 + X_2)/2$. There are 36 possible outcomes of the throws of two dice but there are only 12 values possible for $\overline{X}_2$. Table 5.2 lists each possible value together with its associated probability of occurrence.

Verify each of the following calculations

$$E(\overline{X}_2) = 1.5\frac{2}{30} + 2\frac{2}{30} + \cdots$$

$$+ 5.5\frac{2}{30} = 3.5$$

$$V(\overline{X}_2) = \text{variance of } \overline{X}_2 = (1 - 3.5)^2\frac{1}{36}$$

$$+(1.5 - 3.5)^2\frac{2}{36} + \cdots$$

$$+ (6 - 3.5)^2\frac{1}{36}$$

$$= \frac{2.917}{2} = 1.459$$

c. Calculate $E(\overline{X}_2)$ by using the formula $E(\overline{X}_2) = [E(X_1) + E(X_2)]/2$.
d. Calculate $V(\overline{X}_2)$ by using the formula $V(\overline{X}_2) = \frac{1}{4}[V(X_1) + V(X_2)]$.
e. Use the facts that $E(X) = $ mean of the population and $V(X) = $ variance of the population divided by the sample size to find $E(\overline{X}_2)$, $V(\overline{X}_2)$.

f. Draw the histogram of the population X_1.
g. Draw the histogram of $\overline{X}_2$. Note that even after two samples, the histogram looks more like the normal than does the histogram of the original population.
h. Figure 4.3 shows the distribution of the sum of three dice. Use the figure to draw the histogram for $\overline{X}_3$. Note that the distribution of $\overline{X}_3$ looks even more like a normal than does that for $\overline{X}_2$. What is $E(\overline{X}_3)$, $V(\overline{X}_3)$?

21. Why is each of the following statements incorrect, misleading, or not precise?
   a. We *observe* the number $\bar{x}$ as calculated from the observation of n values. $E(\bar{x}) = \mu$, μ is the mean of the population.
   b. The distribution of the number $\bar{x}$ is normal.
   c. The random variable X is as shown in Figure 4.2. We throw the dice and the sum is 6. The random variable X is no longer random.
   d. If we observe the value of $\bar{x}$ from one sample, we must observe this value of $\bar{x}$ for every other sample of the same size that we take in the future.

22. Let $X_1, X_2, \ldots, X_n$ be n random variables such that

$$P(X_i = 1) = P(X_i = 0) = 0.5$$

   a. In example 5.9 (Section 5.5) the distribution of $\overline{X}_2$ is shown. Draw the histogram of this distribution.

Table 5 ▪ 2

Possible values of $\overline{X}_2$	1	1.5	2	2.5	3	3.5	4	4.5	5	5.5	6
Probability	$\frac{1}{36}$	$\frac{2}{36}$	$\frac{3}{36}$	$\frac{4}{36}$	$\frac{5}{36}$	$\frac{6}{36}$	$\frac{5}{36}$	$\frac{4}{36}$	$\frac{3}{36}$	$\frac{2}{36}$	$\frac{1}{36}$

b. Compare the histogram in (a) to the histogram of X_1 and X_2.
c. Draw the histogram of $\overline{X}_4, \overline{X}_{10}$. Note how the distribution appears to approach the normal. Notice also that the mean of each distribution is 0.5, whereas the variance is $0.25/n$, where n is the number of observations.

23. Use $V(\overline{X}) = \sigma^2/n$ to explain why $\overline{X}$ will not deviate far from μ for very large n. (This might lead you to conclude correctly that $\overline{X}$ can be used to estimate μ if μ is unknown).

24. The grades of our students have approximately a normal distribution with $\mu = 80$ and $\sigma^2 = 3$. What is the expected value of the class average and the standard deviation of this average for a class of 16 students? What is the probability that for this class of 16 students the class average will be greater than 80? Greater than 85?

25. Fifty men are selected from a very large population without replacement. Assume the population is so large that this sample is essentially just like sampling with replacement. The distribution of the population of the men's heights has a standard deviation $\sigma = 6$ in. Find the standard error of the mean height for the sample of 50 men.

26. How much of an increase in sample size is needed to decrease the standard error of the mean by 10%? by 50%? by 75%?

27. Gregor Mendel's mathematical formulation for the existence of genes for inheritance (1865) ranks as one of the milestones in biology. How well did his data fit his model? As an example, in one experiment he planted 8023 second-generation pea seed hybrids. His model predicted one fourth would be green flowered. He observed 2001 were green. Was this too close to the expected ($\frac{1}{4}$) $(8023) = 2006$ to be due to chance? We now discuss this matter.

a. Assume the number of green plants is binomial with $n = 8023$ and $p = \frac{1}{4}$. What are the mean and standard deviation of the number of green plants?
b. The observed deviation of 5 between what was expected and what was obtained must be compared in terms of standard deviations. How many standard deviations from the expected value was his result?
c. Using the normal approximation, calculate the probability that Mendel would find a discrepancy of no more than 5 from the expected, that is, $P(2001 \leq X \leq 2011)$.
d. If 1000 scientists were to perform Mendel's experiment independently, each examining 8023 plants, how many would you expect to get as a good or better fit than did Mendel?
e. Mendel was this close on numerous experiments. What does this say about the likelihood that he "adjusted" his data? (R. A. Fisher suggests Mendel was "deceived by his gardening assistant who knew only too well" what Mendel expected.)

CONTINUOUS R.V

- described on probability
density curves.

50th %ile : Median are
the same thing ie P_{50} = Median.

$Q_U = P_{25}$ upper

$Q_L = P_{75}$ Lower

$\mu - 2\sigma$ $(\mu - \sigma)$ μ $(\mu + \sigma)$ $\mu + 2\sigma$

POINTS OF INFLECTION.

NORMAL CURVES
NOT "   " = KURTIC

MODE HEIGHT = $\dfrac{1}{\sqrt{2\pi}\ \sigma}$

AREA BENEATH CURVE "

AREA BET PTS OF INFLECTION = .95
= IQR - intra quartile RANGE

X_α = upper critical range

$P\left[\dfrac{a-\mu}{\sigma} \leqslant z \leqslant \dfrac{b-\mu}{\sigma}\right]$

IF $P\left[a \leqslant X \leqslant b\right]$

TO FIND: P bet a & b or
$P[a \leqslant X \leqslant b]$

A
z
.5   1.7

$A_{.5\ to\ 1.7} = A_{z_b} - A_{z_b}$

= .9554 - .6915

= .2639

Find P_{90}

= $P\left[z \leqslant \dfrac{b-\mu}{\sigma}\right]$

$z = 1.30$

THEN:

$(\sigma z) + \mu$ = % of items in the
90% ile.

FIND
X is at least .70

$P\left[X > .70\right]$

$P\left[z > \dfrac{.7 - .84}{.12}\right]$

$P[z > 1.17]$ find Area of
1.17

$1 - 0.1251 = 84.49\%$ chance of
being X at least .70.

* * *

X - has a binomial distn.
X^* = has a normal distn $\mu = np$ $\sigma = \sqrt{npq}$

$P\left[30 \leqslant X \leqslant 40\right]$ or $P\left[29.5 \leqslant X^* \leqslant 40.5\right]$

$(30+\frac{1}{2})(31+\frac{1}{2})(32+\frac{1}{2})(40+\frac{1}{2})3n$ $P\left[\dfrac{29.5-36}{3.79} \leqslant z \leqslant \dfrac{40.5-36}{3.79}\right]$

$P[30 \leqslant z \leqslant 40]$ $P\left[-1.70 \leqslant z \leqslant 1.20\right]$

$n = 60$ $p = 0.60$ $q = 0.4$ $\cong 0.8849 - .0446$

$f_{(30)} + f_{(31)} + f_{(32)} \cdots f_{(40)}$ $= .8403$

$= \binom{60}{30}(.6)^{30}(.4)^{30} + \binom{60}{31}(.6)^{34}(.4)^{29} \cdots$

* * *

Stem-leaf diag . , t, f, s, *

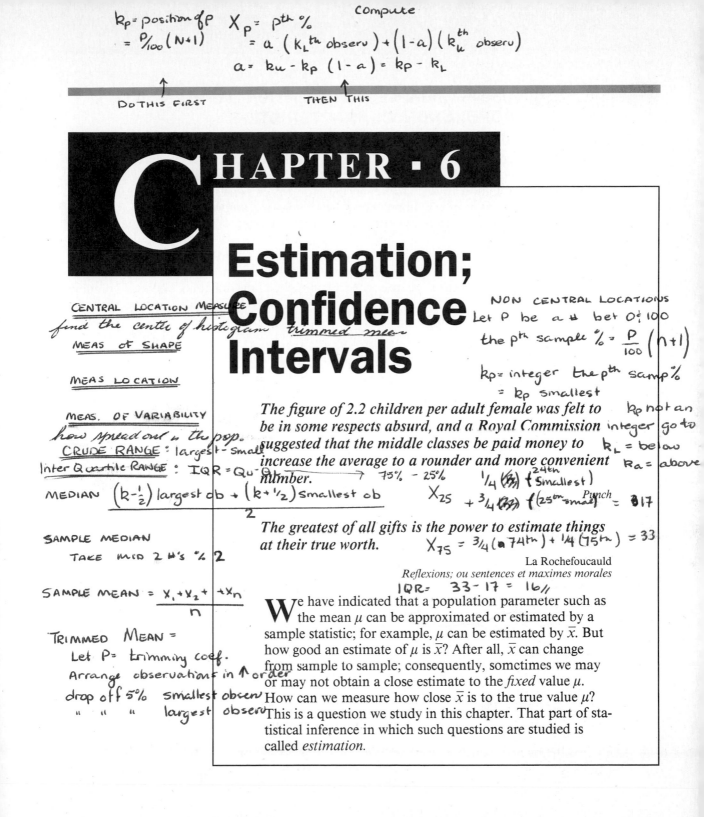

CHAPTER · 6

Estimation; Confidence Intervals

CENTRAL LOCATION MEASURE
find the center of histogram   trimmed mean
MEAS OF SHAPE

MEAS LOCATION

MEAS. OF VARIABILITY
how spread out is the pop.
CRUDE RANGE: largest - small
Inter Quartile RANGE : IQR = Qu - Q
MEDIAN $\dfrac{\left(k-\frac{1}{2}\right)\text{largest ob} + \left(k+\frac{1}{2}\right)\text{smallest ob}}{2}$

SAMPLE MEDIAN
   TAKE MID 2 #'s % 2

SAMPLE MEAN = $\dfrac{X_1 + X_2 + \cdots + X_n}{n}$

TRIMMED MEAN =
   Let P= trimming coef.
   Arrange observations in ↑ order
   drop off 5% smallest observ
   "   "   "   largest observ

NON CENTRAL LOCATIONS
Let P be a # bet 0 ¢ 100
the pth sample % = $\frac{P}{100}(n+1)$
k_p = integer the pth samp %
   = k_p smallest
k_p not an integer go to
k_L = below
R_a = above

75% - 25%   ¼ (3rd) ↑ smallest)
X_{25} + ¾ (3rd) ↑ (25th smallest Punch = $17

X_{75} = ¾ (a 74th) + ¼ (75th) = 33

IQR= 33 - 17 = 16 //

The figure of 2.2 children per adult female was felt to be in some respects absurd, and a Royal Commission suggested that the middle classes be paid money to increase the average to a rounder and more convenient number.

The greatest of all gifts is the power to estimate things at their true worth.

La Rochefoucauld
Reflexions; ou sentences et maximes morales

We have indicated that a population parameter such as the mean μ can be approximated or estimated by a sample statistic; for example, μ can be estimated by $\bar{x}$. But how good an estimate of μ is $\bar{x}$? After all, $\bar{x}$ can change from sample to sample; consequently, sometimes we may or may not obtain a close estimate to the *fixed* value μ. How can we measure how close $\bar{x}$ is to the true value μ? This is a question we study in this chapter. That part of statistical inference in which such questions are studied is called *estimation*.

6 ▪ 1 USING SAMPLE INFORMATION TO ESTIMATE POPULATION CHARACTERISTICS

We now begin to attack the problem of using the information contained in a sample to develop conclusions about the population from which the sample was drawn. This is the general problem of inferential statistics. We restrict our attention in this chapter to the problem of estimating the mean μ and the variance σ^2 of a population when employing *independent random sampling,* that is, independent sampling with replacement. We shall subsequently relax this condition.

There is one matter we must understand before we can make progress in solving these estimation problems. Recall that a random variable has a range of possible values; we must clearly comprehend the difference between a random variable and one of the values in its range. To review this, consider again the sample mean $\overline{X}$.

$$\overline{X} = \frac{X_1 + X_2 + \cdots + X_n}{n}$$

where X_i is the random variable that represents the ith observation. The sample mean $\overline{X}$ is also a random variable. It might be thought of as the potential average before taking any sample. If one actually obtains the value x_1 for the first observation, x_2 for the second, and so forth, until x_n for the nth observation, then

$$\overline{x} = \frac{x_1 + x_2 + \cdots + x_n}{n}$$

is the *observed* sample mean.

Similarly, the value

$$s^2 = \frac{(x_1 - \overline{x})^2 + (x_2 - \overline{x})^2 + \cdots + (x_n - \overline{x})^2}{n - 1}$$

is the observed value of the random variable

$$S^2 = \frac{(X_1 - \overline{X})^2 + (X_2 - \overline{X})^2 + \cdots + (X_n - \overline{X})^2}{n - 1}$$

Example 6 ▪ 1   Let us suppose we have a population $\mathcal{P} = \{1, 3, 4, 5, 7\}$, where the numbers are equally likely to be selected. We sample with replacement and independently

Wt. gain for cows. Compute 30th %ile

$n = 10$

$k_{30} = \dfrac{30}{100}(10+1) = 3.3$

$k_L = 3^{rd} \text{ small} = .7$

$k_a = 4^{th} \text{ "} = .3$

$30\% = (.7)(3^{rd} \text{ sm}) + (.3)(4^{th} \text{ small})$

$= (.7)(87) + (.3)(100$

(independent random sampling) take a sample of size 2.* The possible observations are

(Choice 1, Choice 2)	$\bar{x}$	s^2	(Choice 1, Choice 2)	$\bar{x}$	s^2
(1, 1)	1	0	(4, 5)	4.5	0.5
(1, 3)	2	2	(4, 7)	5.5	4.5
(1, 4)	2.5	4.5	(5, 1)	3	8
(1, 5)	3	8	(5, 3)	4	2
(1, 7)	4	18	(5, 4)	4.5	0.5
(3, 1)	2	2	(5, 5)	5	0
(3, 3)	3	0	(5, 7)	6	2
(3, 4)	3.5	0.5	(7, 1)	4	18
(3, 5)	4	2	(7, 3)	5	8
(3, 7)	5	8	(7, 4)	5.5	4.5
(4, 1)	2.5	4.5	(7, 5)	6	2
(4, 3)	3.5	0.5	(7, 7)	7	0
(4, 4)	4	0			

pth %ile → low group   p% of data
          → upper group (1-p) of data

$k_p = k^{th}\text{%ile} = \dfrac{p}{100}(n+1)$

↳ not an integer

> To summarize, $\bar{X}$ has a range of possible values; we denote the value we *actually* obtain from a sample by the small letter $\bar{x}$. Similarly, the sample variance S^2 has a range of values; we denote the actual value obtained by s^2.

The general problem we face is how to estimate certain numerical characteristics of the population such as the mean μ and the variance σ^2. These numerical characteristics of the population are called parameters of the population. Accordingly, this portion of statistical inference we are studying is called *parametric* statistical inference.

As noted previously, the numerical parameters of the population, such as μ and σ^2, are fixed, nonrandom numbers. The corresponding sample statistics, such as $\bar{x}$ and s^2, however, can change from sample to sample. Based on the information in the sample, it seems natural to estimate the nonrandom parameters μ and σ^2 by their sample counterparts $\bar{x}$ and s^2. We call these values *estimates.* We call the corresponding random variables $\bar{X}$ and S^2 *estimators.* Before assessing how good an estimator $\bar{X}$ is of μ, or S^2 is of σ^2, we must first see how the estimate can change from sample to sample. Remember, $\bar{X}$ and S^2 are random variables and random variables have a whole distribution of values. It may happen that there is a large chance that $\bar{X}$ may vary far from μ. In such a case $\bar{X}$ would not be much use as an estimator of μ. To determine whether $\bar{X}$ varies a great deal from μ, we can look at the sampling distribution of $\bar{X}$. The

---

* In practice, of course, sample sizes are usually much larger than 2. We use small sample sizes here for illustrative purposes only.

sampling distribution of $\overline{X}$ tells us the extent of the "randomness" of $\overline{X}$ and the tendency for $\overline{X}$ to take on values far from μ. Similarly, the sampling distribution of S^2 indicates what the tendency is for S^2 to take on values far from σ^2.

We now apply these ideas to the estimation of μ when σ^2 is a known value. Although it will not often happen in your statistical work in real applications that you know σ^2 but not μ, the type of analysis we perform in this situation is really the prototype for the general case and is easier to understand as a beginning.

6 ▪ 2 UNBIASED ESTIMATES: CONFIDENCE INTERVALS ON μ WHEN σ^2 IS KNOWN

We use $\overline{X}$ as an estimator of μ. Recall that we have previously calculated

$$E(\overline{X}) = \mu$$

Hence, on the average, $\overline{X}$ is equal to μ. This is a very desirable property for an estimator to have. The property is so important that an estimator that has it is given a special name, an *unbiased estimator.*

> An estimator is said to be *unbiased* if its expected value is always equal to the population parameter it is intended to estimate. An observed value of an unbiased estimator is called an *unbiased estimate.*

Thus $\overline{x}$ is an unbiased estimate of μ.

Even though $\overline{X}$ is, *on the average,* equal to μ, in general, the observed value $\overline{x}$ does not equal μ so all we know is that μ should be "somewhat close" to $\overline{x}$, or equivalently, μ should lie in an interval around $\overline{x}$. We are therefore led to the idea of an *interval estimate* of μ. For example, we might estimate that μ lies between 10 and 12. In this case we have estimated μ not by a number, but rather, by an *interval* in which we feel it can be found. Based on a sample of size n, we can find such an interval estimate of μ and express our "degree of confidence" that it actually contains the unknown value μ. Such an interval estimate for μ is called a *confidence interval,* which expresses not only the sample estimate $\overline{x}$ of μ but also the variability in the *estimator* $\overline{X}$, since the more variable $\overline{X}$ is, the wider the interval must be made in order to retain the same confidence. The calculation of such interval estimates proceeds as follows.

For an independent random sample of size n the sampling distribution of $\overline{X}$ will be normally distributed with mean μ and variance σ^2/n if the population is normal, or approximately normal even if the population is not normal but n is large. We say that $\overline{X}$ is *at least* approximately normal for large n. In Chapter 5 we have shown that for large n

$$Z = \frac{\overline{X} - \mu}{\sigma/\sqrt{n}}$$

is at least approximately normal with mean 0 and variance 1. If we desire a 90% confidence interval for μ (i.e., an interval we are "90% sure" contains μ), we first find from Appendix A.1 that $P(-1.645 < Z \le 1.645) = 0.90$. We use this result to build our interval by replacing Z with

$$\frac{\overline{X} - \mu}{\sigma/\sqrt{n}}$$

in this probability expression to obtain (see Fig. 6.1)

$$0.90 = P\left(-1.645 < \frac{\overline{X} - \mu}{\sigma/\sqrt{n}} \le 1.645\right)$$

$$= P\left(1.645 \frac{\sigma}{\sqrt{n}} < \overline{X} - \mu \le 1.645 \frac{\sigma}{\sqrt{n}}\right)$$

$$= P\left(-\overline{X} - 1.645 \frac{\sigma}{\sqrt{n}} < -\mu \le -\overline{X} + 1.645 \frac{\sigma}{\sqrt{n}}\right)$$

$$= P\left(\overline{X} - 1.645 \frac{\sigma}{\sqrt{n}} \le \mu < \overline{X} + 1.645 \frac{\sigma}{\sqrt{n}}\right) = 0.90$$

The interpretation of this probability is that if we calculate the value $\overline{x}$ for one sample of size n and calculate it again for a different sample of size n and keep repeating this for many samples, say, 100, of size n, each time using the calculated value to form the interval $\overline{x} \pm 1.65 \, \sigma/\sqrt{n}$, then we expect the true value μ to be in about 90% of the 100 intervals; that is, about 90 of these intervals should

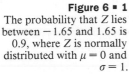

Figure 6 ▪ 1
The probability that Z lies between -1.65 and 1.65 is 0.9, where Z is normally distributed with $\mu = 0$ and $\sigma = 1$.

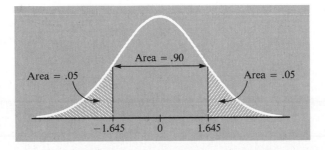

Area = .90

Area = .05

Area = .05

-1.645 0   1.645

cover μ. We express this verbally by saying we have 90% confidence that μ is in the calculated interval. If we did this over and over, we would be right 90% of the time in saying that μ is in the calculated interval. The 90% confidence interval for a particular sample is thus defined to be from $\bar{x} - 1.645\ \sigma/\sqrt{n}$ to $\bar{x} + 1.645\ \sigma/\sqrt{n}$. In general, if we wish to find the $100(1 - \alpha)$% confidence interval for μ when σ^2 is known, we go through the same calculation as previously by first finding the centile value $z_{\alpha/2}$; that is, the value such that $P(Z > z_{\alpha/2}) = \alpha/2$. We know then that

$$P(-z_{\alpha/2} < Z \leq z_{\alpha/2}) = 1 - \alpha$$

The $100(1 - \alpha)$% confidence interval can be obtained by setting $Z = (\bar{X} - \mu)/(\sigma/\sqrt{n})$ and then doing the same algebraic manipulations as were performed to get the 90% confidence interval. The result is

> The $100(1 - \alpha)$% confidence interval for μ when using independent random sampling is the interval $\bar{x} - z_{\alpha/2}\sigma/\sqrt{n}$ to $\bar{x} + z_{\alpha/2}\sigma/\sqrt{n}$. The probability $1 - \alpha$ is called the *confidence level*.

The width of the $(100)(1 - \alpha)$% confidence interval is $2z_{\alpha/2}\sigma/\sqrt{n}$ and is a measure of the *precision* of this interval estimate of μ. The number $1 - \alpha$ is called the confidence level and is a measure of the *reliability* of the interval estimator.

Example 6 ■ 2   A certain manufacturer of tires knows that the life in miles of his tires is normally distributed with a standard deviation of $\sigma = 6000$ mi. From a test sample of 144 tires he calculates $\bar{x} = 31{,}000$ mi. Find a 99% confidence interval estimate for μ, the unknown mean life of the tires.

Since $1 - \alpha = 0.99$, $\alpha/2 = 0.005$ so that from Appendix A.1 we find that $z_{0.005} = 2.576$. Thus from the confidence interval formula, the 99% confidence interval for μ is from

$$\bar{x} - 2.576\ \frac{\sigma}{\sqrt{n}} \qquad \text{to} \qquad \bar{x} + 2.576\ \frac{\sigma}{\sqrt{n}}$$

Replacing $\bar{x}$ by 31,000 and $\sigma/\sqrt{n}$ by $6000/\sqrt{144}$, we find the 99% confidence interval estimate for μ to be from

$$31{,}000 - 2.576\ \frac{6000}{\sqrt{144}} \qquad \text{to} \qquad 31{,}000 + 2.576\ \frac{6000}{\sqrt{144}}$$

or from

$$29,712 \quad \text{to} \quad 32,288$$

In this sense, therefore, we can be "99% certain" that the true mean μ lies somewhere between 29,712 and 32,288 mi. Such information is vital for setting the guaranteed mileage of the tire.

Example 6 ■ 3 An independent random sample of 36 men had their blood cholesterol measured, yielding $\bar{x} = 312$. If $\sigma = 23$, find a 95% confidence interval for the mean cholesterol level of men. Assume (as is probably the case) that the blood cholesterol levels of men form a normal distribution.

To find the confidence intervals, we note that with $1 - \alpha = 0.95$, $\alpha/2 = 0.025$ so $z_{\alpha/2} = z_{0.025} = 1.96$ and the 95% confidence interval is from

$$312 - 1.96\left(\frac{23}{\sqrt{36}}\right) \quad \text{to} \quad 312 + 1.96\left(\frac{23}{\sqrt{36}}\right)$$

or from

$$304.49 \text{ to } 319.51$$

It should be emphasized again that a confidence interval should be interpreted as an "interval estimate" for μ. We should not really say that "we are confident that μ lies in the interval with probability 0.95" because, since μ is a number, it either is or is not in the given interval. It is the *procedure for obtaining the interval in which we have 95% confidence,* and not the particular interval. Indeed, if we took 100 samples and calculated the estimate $\bar{x}$ for each of them, and then each time used these values to find a confidence interval for μ, we would obtain 100 possible different intervals. Our interpretation of "95% confidence" is that we expect 95% of the confidence intervals formed by using the various observed values $\bar{x}$ to actually contain μ. Each particular one either does or does not contain μ, but about 95% of them will contain μ.

CAUTION! The aforementioned procedure gives *exact* confidence limits when the *population* is normally distributed. If the population is not normal, the confidence limits are approximations to the actual ones; the approximation is excellent for large n.

6 ■ 3   CONFIDENCE INTERVAL ESTIMATES FOR THE MEAN OF A NORMAL DISTRIBUTION WHEN THE POPULATION VARIANCE IS UNKNOWN

We used the fact in the last section that if we are sampling from a normal population with mean μ and known variance σ^2, then the sample statistic

$$Z = \frac{\overline{X} - \mu}{\sigma/\sqrt{n}}$$

has a normal distribution with mean 0 and variance 1 no matter what the values of μ, σ, and n are. In most real situations, however, the value of σ^2 is not known and must itself be estimated by the sample variance s^2. In this case the sample statistic

$$T = \frac{\overline{X} - \mu}{S/\sqrt{n}}$$

has a distribution that is the same no matter what the values of μ and σ are but that now depends on the sample size n. This distribution is called the *t-distribution with n − 1 degrees of freedom.*

Since the t-distribution is different for each n, in order to avoid having a complete table for each n, we list in Appendix A.5 only certain upper centile values of the t-distribution that correspond to commonly used confidence levels. The upper αth centile value with $n - 1$ degrees of freedom is denoted by t_α. To read this value from the table, you look down the first column marked df (for degrees of freedom) until you find $n - 1$. You then read across that row until you find the desired upper centile value. For example, we have for df = 4, $t_{0.25} = 0.741$; for df = 9, $t_{0.025} = 2.262$; for df = 14, $t_{0.05} = 1.761$; and for df = 40, $t_{0.01} = 2.423$. Since the t-distribution is symmetric, we have $t_\alpha = -t_{1-\alpha}$; so, for example, df = 4, $t_{0.75} = -0.741$ and with df = 9, $t_{0.975} = -2.262$.

It should be noted that as n gets larger, the shape of the distribution gets closer to the normal curve (see Fig. 6.2). This result reflects the fact that as n gets larger, S^2 gets closer to σ^2 so that the distribution of $T = (\overline{X} - \mu)/(S/\sqrt{n})$ gets closer to the distribution $Z = (\overline{X} - \mu)/(\sigma/\sqrt{n})$, which is normal when we sample from a normal population.

Degrees of Freedom

The only parameter on which the shape of the t-distribution depends is the number of degrees of freedom. Since this concept occurs so often in statistics, it is important that we become familiar with it. The t-statistic has $n - 1$ degrees of freedom because we use S^2 to estimate σ^2. To understand this, recall that the

Figure 6 ▪ 2
Comparison of the normal distribution to the *t*-distribution with df = 2 and df = 9. Note that the *t*-distribution is fatter (i.e., the variance is bigger) than the normal curve because of the extra variability introduced by substituting the random variable S^2 for the number σ^2 in calculating the standardized ratio.

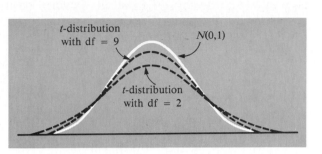

sample variance S^2 is derived from the sum of the squares of the deviations $D_i = X_i - \bar{X}$ because

$$S^2 = \{(X_1 - \bar{X})^2 + (X_2 - \bar{X})^2 + \cdots (X_n - \bar{X})^2\}/(n-1) \qquad \text{or}$$

$$S^2 = (D_1^2 + D_2^2 + \cdots + D_n^2)/(n-1)$$

The question of degrees of freedom boils down to how many of the D_i's are we really free to choose. We only have complete freedom of choice for $n - 1$ of the D_i's. Although we have n independent X's, we have only $n - 1$ independent D's, and the D's are the basis used for calculating S^2. To see this, note that

$$\sum_{i=1}^{n} D_i = \sum_{i=1}^{n} (X_i - \bar{X}) = \sum_{i=1}^{n} X_i - n\bar{X} = n\bar{X} - n\bar{X} = 0$$

Suppose, for example, that we have $n = 4$; then

$$D_1 + D_2 + D_3 + D_4 = (X_1 - \bar{X}) + \cdots + (X_4 - \bar{X})$$
$$= X_1 + X_2 + X_3 + X_4 - 4\bar{X}$$
$$= 4\,\frac{X_1 + X_2 + X_3 + X_4}{4} - 4\bar{X} = 4\bar{X} - 4\bar{X} = 0.$$

We have complete freedom to choose the values of three of the D_i's, say, $D_1 = 2$, $D_2 = -3$, $D_3 = 4$; however, *since the sum must equal zero, we have no free choice about D_4 and in fact we must have $D_4 = 0 - 2 + 3 - 4 = -3$* in order that the sum of the D's be zero. Similarly, if there are n deviations from the mean, we can only have $n - 1$ free choices in assigning values to the deviations. We express this fact by saying that the number of degrees of freedom is $n - 1$. Since S^2 is calculated from these deviations, we see that S^2 is really only calculated from $n - 1$ degrees of freedom, that is, $n - 1$ independent pieces of information. This is why the t-distribution has $n - 1$ degrees of freedom, and also why we divided by $n - 1$ instead of n when we first defined the variance s^2 of a sample in Chapter 2. We divided by $n - 1$ because we are averaging the squares of only $n - 1$ numerically independent quantities.

t-distribution

We now show how to calculate confidence interval estimates of an unknown population mean μ when the population variance is also unknown. Assuming the population is normally distributed, we know that $T = (\bar{X} - \mu)/(S/\sqrt{n})$ has a t-distribution with df $= n - 1$. If we desire a $100(1 - \alpha)\%$ confidence interval, then we may determine the value $t_{\alpha/2}$ from Appendix A.5 such that

$$1 - \alpha = P\left\{-t_{\alpha/2} < \frac{\bar{X} - \mu}{S/\sqrt{n}} \leq t_{\alpha/2}\right\}$$

$$= P\left\{-\frac{S}{\sqrt{n}}\, t_{\alpha/2} < \bar{X} - \mu \leq \frac{S}{\sqrt{n}}\, t_{\alpha/2}\right\}$$

$$= P\left\{\bar{X} - \frac{S}{\sqrt{n}}\, t_{\alpha/2} \leq \mu < \bar{X} + \frac{S}{\sqrt{n}}\, t_{\alpha/2}\right\}$$

After obtaining our observations, we can calculate the observed sample mean $\bar{x}$ and observed sample standard deviation s to obtain the confidence interval.

> The $100(1 - \alpha)\%$ confidence interval estimate of μ when σ^2 is unknown in a normal population is from $\bar{x} - t_{\alpha/2}s/\sqrt{n}$ to $\bar{x} + t_{\alpha/2}s/\sqrt{n}$, where the t-distribution has $n - 1$ degrees of freedom.

Example 6 ■ 4   An anthropologist wishes to study the brain size of a certain tribe of primitive man. Ten skulls are obtained from this tribe and the following cranial width measurements (in millimeters) are found 137, 145, 139, 126, 140, 138, 143, 149, 141, and 139. It is felt biologically that it is justified to assume that these numbers came from a normal population. The sample mean is

$$\bar{x} = \frac{137 + 145 + 139 + 126 + 140 + 138 + 143 + 149 + 141 + 139}{10}$$

$$= 139.7$$

and the sample variance is

$$s^2 = \frac{(137 - 139.7)^2 + (145 - 139.7)^2 + \cdots + (139 - 139.7)^2}{10 - 1}$$

$$= \frac{326.1}{9} = 36.23$$

The sample standard deviation is given by $s = \sqrt{s^2} = \sqrt{36.23} = 6.019$.

Based on these data we wish to find a 95% confidence interval for μ, the mean skull measurement for this tribe. Here $1 - \alpha = 0.05$, df $= 10 - 1 = 9$, and $t_{\alpha/2} = t_{0.025} = 2.262$ so that the 95% confidence interval is found by using $\bar{x} = 139.7$ and $s = 6.019$. The interval is from $139.7 - (2.262)6.019/\sqrt{10}$ to $139.7 + (2.262)6.019/\sqrt{10}$ or from 135.4 to 144. We feel 95% confident (and would give 19 to 1 odds) that the mean cranial width is between 135.41 and 143.99.

Example 6 ▪ 5   A study was made of the income of people with a bachelor's degree. Assume that this income is normally distributed. Twenty college graduates were sampled and they reported an income (in dollars) of 22,000, 21,500, 20,000, 23,000, 20,000, 22,500, 24,000, 27,000, 24,500, 19,000, 18,500, 31,000, 21,500, 22,500, 21,000 19,500, 19,600, 27,000, 30,000, and 18,000. An estimate of the mean income was desired and the estimate was to be 90% accurate in the sense of a 90% confidence interval. We calculate $\bar{x} = \$22,605$ and $s = \$3686.32$. Since $\alpha = 0.1$ and $n = 20$, we have $t_{0.05} = 1.729$ and hence the confidence interval is from

$$22,605 - 1.729 \left(\frac{3686.32}{\sqrt{20}} \right) \qquad \text{to} \qquad 22,605 + 1.729 \left(\frac{3686.32}{\sqrt{20}} \right)$$

or from

$$\$21,179.81 \qquad \text{to} \qquad \$24,030.19$$

6 ▪ 4   THE SAMPLING DISTRIBUTION OF S^2; CONFIDENCE INTERVALS ON THE VARIANCE OF A NORMAL DISTRIBUTION

Usually we not only wish to approximate the mean μ of a population but also need to assess the variability in the population. We could, of course, use S^2 as an estimator of the variance σ^2, but how close to the true value can we expect the estimate obtained after sampling to be? To answer this question we must know the sampling distribution of S^2, which we now proceed to discuss.

Let P denote the population from which we are sampling. Referring back to the population $P = \{1, 3, 4, 5, 7\}$ of example 6.1, we employed independent random sampling and calculated the observed sample variance s^2 for each of the 25 possible samples of size 2 to obtain

$$\{0, 2, 4.5, 8, 18, 2, 0, 0.5, 2, 8, 4.5, 0.5, 0, 0.5, 4.5, 8, 2,$$
$$0.5, 0, 2, 18, 8, 4.5, 2, 0\}$$

as the set of possible values of S^2. The distribution of this population describes how S^2 changes from sample to sample. Choosing a random sample and calculating the *observed* s^2 is the same as randomly picking a number from this set of values. Note that the variance of the original population P is $\sigma^2 = 4$ and that the mean for the preceding population consisting of all possible values s^2 is

$$\frac{0 + 2 + 4.5 + 8 + 18 + \cdots + 4.5 + 2 + 0}{25} = 4 = \sigma^2$$

In general, if σ^2 is the variance of the original population and if we employ independent random sampling, then the expected value of S^2 is σ^2, so that, on the average, S^2 yields a correct estimate of σ^2. S^2 is an *unbiased estimator* of σ^2. This is another reason for defining S^2 with $n - 1$ in the denominator; the equation "expected value of $S^2 = \sigma^2$" does not hold if S^2 is defined with n in the denominator, so we get a better estimator by using the denominator $n - 1$.

Suppose now we are sampling (using independent random sampling) from an original population that has a normal distribution with mean μ and variance σ^2. Then the mean or the expected value of S^2 is σ^2 and the variance of S^2 can be

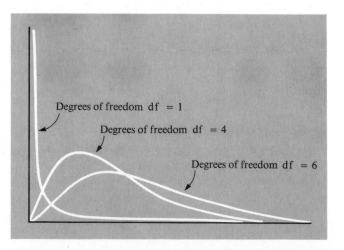

Degrees of freedom df = 1

Degrees of freedom df = 4

Degrees of freedom df = 6

Figure 6 ▪ 3
The chi-squared (χ^2) distribution for various values of the degrees of freedom.

shown to be $2\sigma^4/(n-1)$. Thus as n gets larger, the variance of S^2 gets smaller and so S^2 gets closer to the true value σ^2. Therefore S^2 is a very good estimator of σ^2.

When the population is distributed normally, the distribution of S^2 depends on the value of σ^2, and also on how large n is. It does not depend on the population mean μ. We may even remove the dependence on the value of σ^2 by dividing S^2 by $\sigma^2/(n-1)$. We find then that $(n-1)S^2/\sigma^2$ has a distribution that is independent of the value of σ^2 but still depends on $n-1$, the number of degrees of freedom involved in the calculation of S^2. This distribution is called the *chi-square distribution with n − 1 degrees of freedom*. The density curves for the chi-square distribution for various different choices of degrees of freedom are graphed in Figure 6.3.

> The random variable $(n-1)S^2/\sigma^2$ has a chi-square (χ^2) distribution with $n-1$ degrees of freedom.

We have only included the most commonly used centile values for the chi-square distribution in Appendix A.6, instead of putting in an entire table for each value of n. This table is read in a manner similar to the t-table, Appendix A.5; namely, we read down the left-hand column labeled df (degrees of freedom) until we find the correct number of degrees $n-1$, and then read across until the desired centile column is reached. We denote the upper α centile by χ_α^2. Thus, for example, $\chi_{0.9}^2 = 0.0518$ for df $= 1$; $\chi_{0.1}^2 = 9.24$ for df $= 5$; $\chi_{0.025}^2 = 20.48$ for df $= 10$; and $\chi_{0.99}^2 = 14.95$ for df $= 30$. The values $\chi_{\alpha/2}^2$ and $\chi_{1-\alpha/2}^2$ are not systematically related as were the $\alpha/2$ and $1-\alpha/2$ centiles in the normal and t-distributions since the chi-square distribution is not symmetric.

We may use the fact that $(n-1)S^2/\sigma^2$ has a chi-square distribution to derive confidence limits for an estimate of σ^2. We know that for df $= n - 1$

$$P\left\{\chi_{1-\alpha/2}^2 < \frac{(n-1)S^2}{\sigma^2} \leq \chi_{\alpha/2}^2\right\} = 1 - \alpha$$

so that rearranging to get σ^2 on top, we get

$$P\left\{\frac{1}{\chi_{\alpha/2}^2} < \frac{\sigma^2}{(n-1)S^2} \leq \frac{1}{\chi_{1-\alpha/2}^2}\right\} = 1 - \alpha$$

and hence

$$P\left\{\frac{(n-1)S^2}{\chi_{\alpha/2}^2} < \sigma^2 \leq \frac{(n-1)S^2}{\chi_{1-\alpha/2}^2}\right\} = 1 - \alpha$$

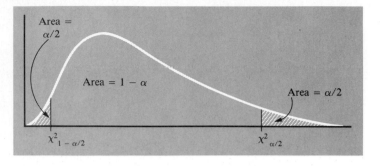

Figure 6 ▪ 4
Graph of the chi-square
density. Note that the
cut-off values $\chi^2_{1-\alpha/2}$ and
$\chi^2_{\alpha/2}$ are not symmetrically
related as they were in the
normal and *t*-distributions.

After obtaining our sample, we calculate the observed s^2 of the sample variance and so obtain

The $100(1 - \alpha)\%$ confidence interval for σ^2 for a normal distribution is

$$\frac{(n-1)s^2}{\chi^2_{\alpha/2}} \text{ to } \frac{(n-1)s^2}{\chi^2_{1-\alpha/2}} \quad \text{with df} = n - 1$$

Remember, we have assumed independent random sampling from a normal distribution. We should always check the validity of these assumptions before applying the test.

Example 6 ▪ 6   A manufacturer of chain links would like to produce chains with a breaking strength of at least 1600 lb. Since a chain is only as strong as its weakest link, he must control the variance in strength as well as the mean breaking strength of the links. He measured the breaking strength of 25 links and grouped the data as follows:

Breaking strengths midpoints	Frequency
1552.5	5
1562.5	1
1587.5	3
1637.5	7
1662.5	7
1687.5	2
	25

Calculate the 95% confidence interval on the variance of the breaking strengths.

Using grouped data formulas, we find

$$\bar{x}_g = \frac{(1552.5) \cdot 5 + \cdots + (1687.5) \cdot 2}{25} = 1622.5$$

$$s_g^2 = \frac{(1552.5 - 1622.5)^2 \cdot 5 \cdots + (1687.5 - 1622.5)^2 \cdot 2}{24}$$

$$= 2208.3$$

From Appendix A.6 with $n - 1 = 24$ and $\alpha = 0.05$ we find $\chi_{0.975}^2 = 12.401$, $\chi_{0.025}^2 = 39.364$, so that the 95% confidence interval for σ^2 is from

$$(2208.3) \left(\frac{24}{39.364} \right) \quad \text{to} \quad (2208.3) \left(\frac{24}{12.401} \right)$$

or, doing the multiplications, from 1346.39 to 4273.78 sq lb. In other words, with 95% confidence, *the standard deviation* in breaking strength falls between 36.7 and 65.4 lb.

Example 6 ▪ 7   Find a 99% confidence interval for the variance in the dollar incomes of college graduates based on the sample given in example 6.5.

To do this, we must first assume that the population of incomes is a normally distributed population. Then, using Appendix A.6 with df $= n - 1 = 19$ degrees of freedom and $\alpha = 0.01$, we find $\chi_{0.995}^2 = 6.844$ and $\chi_{0.005}^2 = 38.582$, so that since $s^2 = 13,588, 955.00$ was calculated in example 6.5, the 99% confidence interval is from

$$(13{,}588{,}955.00) \left(\frac{19}{38.582} \right) \quad \text{to} \quad (13{,}588{,}955.00) \left(\frac{19}{6.844} \right)$$

or from

$$6{,}691{,}984.3 \quad \text{to} \quad 37{,}725{,}033 \text{ sq dollars}$$

With 99% confidence, *the standard deviation* in the dollar incomes of college graduates falls between $2586.89 and $6142.07.

6 ▪ 5   THE SAMPLE PROPORTION AS AN ESTIMATE OF THE POPULATION PROPORTION

In many practical problems we are interested in the proportion of objects in the population of a particular type rather than in the total number of objects of this type. For example, consider an election involving several candidates where the victor must obtain a majority of the votes or fight out another election with the runner-up. In this case the candidates for office are not concerned with the *absolute number* of people who vote for them, but rather, with the *proportion* of people who vote for them, because it is the proportion of people voting for each candidate that determines who is elected. A drug manufacturer is interested in determining if the proportion of side effects of a certain drug is higher than some fixed acceptable level. A psychologist might be interested in determining if the proportion of people admitted to hospitals with psychoneurotic disorders is higher than 0.05. In all of these examples there is some proportion of the population that possesses a certain characteristic. The exact value of this proportion is unknown and we must utilize a sample to *estimate* the value of this proportion. Let us denote this (unknown) proportion by p. Note that p is a population parameter.

We sample n units with *replacement* and *independently* from the population. Let $X_1, X_2, \ldots, X_n$ represent the random variables associated with the first, second, up to the nth unit, respectively, and let

$$X_i = \begin{cases} 1 \text{ if the } i\text{th unit selected possesses the characteristic} \\ 0 \text{ otherwise} \end{cases}$$

Note that each X_i is a random variable where $P(X_i = 1) = p$, the proportion that has the characteristic and that the expected value of X_i is $E(X_i) = 1 \cdot p + 0 \cdot (1 - p) = p$ for each i. We can estimate p from the sample by using the estimator $\overline{X}$ which represents here the proportion of 1's in the sample.* For this case of estimating p, the sample mean $\overline{X}$ is called the *sample proportion*. It is given a special symbol $\hat{P}$ and the *estimate* is denoted $\hat{p}$ instead of $\overline{x}$. We know that $(\overline{X} - \mu)/(\sigma/\sqrt{n})$, where μ is the mean of the population and σ is the standard deviation of the population, is approximately normal with mean 0 and variance 1. Consequently, by replacing $\overline{X}$ with $\hat{P}$, μ with p, and σ with $\sqrt{p(1 - p)}$ (which is the standard deviation of each random variable X_i), we know that

$$\frac{\hat{P} - p}{\sqrt{p(1 - p)/n}}$$

is approximately normal with mean 0 and variance 1. To obtain a confidence

---

* Note that p is a population parameter of the binomial distribution. We have tried consistently to use Greek letters for population parameters (e.g., μ, σ^2). We do not use a Greek letter for this population parameter in order to be consistent with the notation found in the literature and other texts.

interval for p, we use the formula for the confidence limits on the population mean μ, namely,

$$\bar{x} \pm z_{\alpha/2} \frac{\sigma}{\sqrt{n}}$$

We replace $\bar{x}$ with $\hat{p}$ and σ with $\sqrt{p(1-p)}$. Since p is unknown, we replace p by its estimate $\hat{p}$ to obtain

> The normal approximation method for finding confidence limits on p:
> The confidence limits on p for the $100(1-\alpha)\%$ confidence interval are from
>
> $$\hat{p} - z_{\alpha/2} \sqrt{\frac{\hat{p}(1-\hat{p})}{n}} \quad \text{to} \quad \hat{p} + z_{\alpha/2} \sqrt{\frac{\hat{p}(1-\hat{p})}{n}}$$

The usefulness of the confidence limits previously developed depends on the accuracy of the normal approximation of the binomial that depends in turn on p and n. (See Section 5.4.)

When we are uncertain as to whether or not the approximation is valid we can turn to figures or tables that give *exact* results. To find an exact 95% confidence interval, we use Figure 6.5 in the following manner: Locate the value of $\hat{p}$ (which you obtain from the observed data) on the horizontal scale of the graph. Next draw a vertical line upward from the value $\hat{p}$ until you hit the

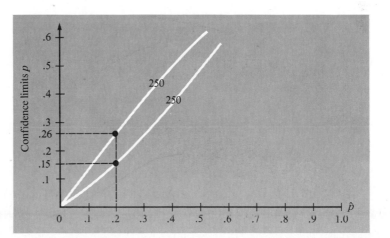

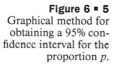

Figure 6 ▪ 5
Graphical method for obtaining a 95% confidence interval for the proportion p.

curve that corresponds to the sample size n. The lower limit of the confidence interval for p is then read on the scale directly to the left of this point on the curve. Continuing the vertical line upward still farther, you will hit a second curve corresponding to the sample size. The upper limit of the confidence interval is read on the scale to the left of this point on the curve.

Example 6 ■ 8 A manufacturing company wished to purchase a new machine to produce parts. A sample of 250 parts produced by the machine contained 50 defective parts. Find a 95% confidence interval for p, the true proportion of defective parts produced by this machine. Here $\hat{p} = 50/250 = 0.2$, and $n = 250$. From the graph in Figure 6.5 corresponding to $n = 250$, we read the 95% confidence interval as 0.15 to 0.26.

Appendix C shows figures similar to Figure 6.5 except that one can obtain confidence limits for additional values of the sample size n and for confidence levels of 0.95 and 0.99.

6 ■ 6   THE DETERMINATION OF SAMPLE SIZE

It is often desired to determine the sample size n that is necessary for estimating a particular population parameter within a certain accuracy. For this we need the width of various confidence intervals, which we now summarize.

Length *l* of Confidence Intervals

1. The length of the confidence interval on the mean, when σ is known is

$$l = 2 \frac{\sigma}{\sqrt{n}} z_{\alpha/2}$$

2. The length of the confidence interval on the mean, when σ is unknown is

$$l = 2 \frac{s}{\sqrt{n}} t_{\alpha/2} \qquad \text{with df} = n - 1$$

3. The length of the confidence interval on σ^2 is

$$l = (n - 1)s^2 \left[\frac{1}{\chi^2_{1-\alpha/2}} - \frac{1}{\chi^2_{\alpha/2}} \right] \qquad \text{with df} = n - 1$$

In each of these cases the sample size enters into the calculation of the length of the confidence interval; the larger the sample size n is, the smaller the length is. This is as it should be since the smaller the length of the confidence interval for a fixed confidence level $1 - \alpha$ is, the better the estimate under study must be. It would be odd if the length of the confidence interval did *not* decrease as we took more samples.

We can now establish a criterion by which to make a decision about how many samples we should take in sampling randomly with replacement in order to estimate a population parameter. For example, suppose we are measuring the diameters of ball bearings that are about 0.1 in. We wish our 95% confidence interval to be no more than 0.001 in. Suppose the value of σ is known to be 0.01 in. How many samples do we select?

In this case the length of the confidence interval is set at $l = 0.001$; the confidence level is $\alpha = 0.05$ so that $z_{\alpha/2} = 1.96$. Using formula 1 and solving for n, we find

$$\sqrt{n} = 2z_{\alpha/2}\left(\frac{\sigma}{l}\right) = 2(1.96)(0.01)/(0.001)$$

$$= 39.2$$

and

$$n = (39.2)^2 = 1536.6 \qquad \text{or} \qquad n \text{ should not be less than } 1537$$

Formulas 2 and 3 do not permit making comparable determinations of the sample size because s enters into the calculation of l and we only know the value of s *after* taking our sample. However, if we have some notion as to the *maximum* value of σ, we may use this value in place of s to obtain an approximate idea of what the sample size should be. This matter is illustrated in the exercises.

6 ▪ 7   COMMENT ON INDEPENDENT RANDOM SAMPLING

In this chapter we have assumed that independent random sampling was employed. However, independent random sampling involves sampling *with* replacement; in many practical situations we sample *without* replacement. As we have indicated in Chapter 5, even when random sampling is performed without replacement and where the sample size is small as compared to the population size, we may still use analyses (such as those in this chapter) based on independent random sampling. We go into this matter in more detail in the chapter on survey sampling.

Summary

In this chapter we have addressed the problem of statistical inference. In our development we assumed independent random sampling. For estimating the population mean μ based on the sample mean $\overline{X}$, we found it necessary to find the sampling distribution of $\overline{X}$. The central limit theorem states that for n large, $\overline{X}$ is at least approximately normally distributed with the mean the same as the population mean μ and the standard deviation equal to $\sigma/\sqrt{n}$ where σ is the population standard deviation.

One of the most important concepts introduced in this chapter is that of a confidence interval estimate of an unknown parameter. This is a procedure for estimating the unknown parameter with a predetermined level of confidence. We emphasize that it is the procedure and not the precise interval obtained in which we have a specified level of confidence.

To construct such an interval estimate for μ when σ^2 is known, we use the fact that $Z = (\overline{X} - \mu)/(\sigma/\sqrt{n})$ has at least approximately a normal distribution with mean 0 and variance 1 to obtain $P(\overline{X} - z_{\alpha/2}\sigma/\sqrt{n} < \mu < \overline{X} + z_{\alpha/2}\sigma/\sqrt{n}) = 1 - \alpha$ where $z_{\alpha/2}$ is the upper $\alpha/2$-centile of this normal distribution. Z will have a normal distribution if the population is normal or approximately normal if the sample size is large. If σ^2 is unknown, then we use the quantity $T = (\overline{X} - \mu)/(S/\sqrt{n})$ to build our interval estimates, if the distribution of the population is normal. The random variable T has a t-distribution with $n - 1$ degrees of freedom. Using the centile $t_{\alpha/2}$ for the t-distribution, we obtain

$$1 - \alpha = P\{\overline{X} - t_{\alpha/2}S/\sqrt{n} \leq \mu < \overline{X} + t_{\alpha/2}(1 - \alpha)S/\sqrt{n}\}$$

The interval $\overline{x} \pm z_{\alpha/2}\sigma/\sqrt{n}$ is the $100(1 - \alpha)\%$ confidence interval for μ when σ^2 is known, and $\overline{x} \pm t_{\alpha/2}s/\sqrt{n}$ is the $100(1 - \alpha)\%$ confidence interval for μ when σ^2 is unknown and the population is normal.

We also discussed estimating σ^2 by S^2 and found that when sampling from a normal distribution with variance σ^2 the variable $(n - 1)S^2/\sigma^2$ has a distribution that is not dependent on μ or σ^2. We called this distribution the chi-square distribution with $n - 1$ degrees of freedom. If χ_α^2 denotes the upper αth centile of this distribution, then we know that

$$P\left\{\frac{(n - 1)S^2}{\chi_{\alpha/2}^2} \leq \sigma^2 < \frac{(n - 1)S^2}{\chi_{1-\alpha/2}^2}\right\} = 1 - \alpha$$

and so the $100(1 - \alpha)\%$ confidence interval for σ^2 is from

$$\frac{(n - 1)s^2}{\chi_{\alpha/2}^2} \quad \text{to} \quad \frac{(n - 1)s^2}{\chi_{1-\alpha/2}^2}$$

We also discussed the sample proportion $\hat{p}$ as an estimate of the population proportion p. The approximate limits for the $100(1 - \alpha)\%$ confidence interval for p are given by

$$\hat{p} \pm z_{\alpha/2}\sqrt{\frac{\hat{p}(1 - \hat{p})}{n}}$$

The exact limits can be obtained from figures in Appendix C for levels of 0.95 and 0.99.

Finally, we examined the issue of how large a sample to take to obtain a confidence interval of a certain pregiven length.

Key Words

Define the following terms by words and formulas:

estimator

estimate

sampling distribution of a
   sample statistic

unbiased estimator

confidence interval or
   interval estimate

confidence interval for μ
   when σ^2 is known

reliability of an interval
   estimate

sampling distribution of $\overline{X}$
   when σ^2 is unknown

degrees of freedom

sampling distribution of S^2

mean of S^2 as it relates to
   the original population

variance of S^2 as it relates
   to the original population

t-distribution

chi-square distribution
   (often written as "χ^2-dis-
   tribution.")

confidence interval for μ
   when σ^2 is unknown

confidence interval for σ^2

sample proportion

confidence interval for the
   population proportion p

confidence level

References

Blackman, S., and D. Catalina: The moon and the emergency room. *Perceptual and Motor Skills,* Vol. 37, pp. 624–626, 1973. (See problem 22.)

Griffin, D. R. et al.: The echolocation of flying insects by bats. *Animal Behavior 8:* 141–154, 1960. (See problem 24.)

Tanur, J. M., et al.: *Statistics: Guide to the Unknown.* San Francisco: Holden-Day, 1978. This book has many interesting examples of statistical applications. In particular, see the article by Brown on statistics and smoking (p. 59) and by Chayes on an estimation problem using "continuous" data.

Moroney, M. J.: *Facts from Figures.* New York, Penguin Books, 1956. An interesting discussion of estimation under the title "How to be precise though vague."

Exercises

Sections 6.1 and 6.2

1. We toss a *fair* coin twice so that the outcomes are independent of each other. (This can be accomplished by having each of two people toss a coin.) If heads comes up we obtain $1 and if tails comes up we lose $1. The possible outcomes of two tosses of a coin are two

heads, two tails, a head on the first toss and a tail on the second, and a tail on the first toss and a head on the second.

a. Verify that the total possible *gains* are $0, −$2, $2.

b. What is the probability of each gain shown in (a)?

c. i. If the *average gain* is the gain divided by the number of tosses, show that the possible values of the average gain are $0, −$1, $1.

   ii. What is the probability of each of these average gains?

d. We repeat problem (c) but in different terms. Let

$$X_i = \begin{cases} 1 & \text{if heads occurs on toss } i \\ -1 & \text{if tails occurs on toss } i \end{cases}$$

Let

$$\overline{X} = \frac{X_1 + X_2}{2}$$

represent the set of possible values for the average of two tosses. The possible values of $\overline{X}$ are

$$\overline{X} = \begin{cases} \dfrac{-1-1}{2} = -1 \\ \dfrac{-1+1}{2} = 0 \\ \dfrac{1-1}{2} = 0 \\ \dfrac{1+1}{2} = 1 \end{cases}$$

Show that the probability $\overline{X} = -1$ is 0.25, that the probability $\overline{X} = 0$ is 0.5, and that the chance $\overline{X} = 1$ is 0.25.

e. Toss a coin twice. What is the observed value $\overline{x}$ you have obtained?

2. a. In problem 1d, list the possible values of the sample variance S^2. What is the probability of each value?

   b. What is the observed value s^2 you obtained in part 1(e)?

3. The two random variables X_1 and X_2 in problem 1 are independent. Show that

   a. $E(X_1) = E(X_2) = 0$

   b. $E(\overline{X}) = 0$

   c. Variance of X_1 = Variance of $X_2 = 1$

   d. Variance of $\overline{X} = 0.5$

   e. In problem 1, is independent random sampling employed? Repeat problem 1(e).

   f. Is the value $\overline{x}$ you obtained equal to the expected value of $\overline{X}$ which, in this case, is 0?

4. A production process produces screws that are either defective ($X = 1$) or good ($X = 0$). A sample of 15 yields: $x_1 = 0, x_2 = 0, x_3 = 1, x_4 = 0, x_5 = 0, x_6 = 0, x_7 = 0, x_8 = 1, x_9 = 0, x_{10} = 0, x_{11} = 0, x_{12} = 1, x_{13} = 1, x_{14} = 0, x_{15} = 0$. Calculate $\overline{x}$ from this sample. What would you estimate to be the proportion of defectives produced by this process?

5. An economist wishes to determine trends in the stock market. To do this, he finds the *average daily price (in dollars) per share of 10 selected stocks* for each month and then compares these averages from month to month. For one month he observes the following averages: 24, 28, 25, 23, 25, 27, 29, 24, 27, 23, 26, 25, 24, 25, 26, 24, 24, 26, 25, 25, 24, 24. (There were 21 working days in this month.) What would you estimate to be the mean daily price per share for the 10 stocks for the year (assuming this 1 month is a representative sample)? What should be the estimated variance of this daily average. What is the estimate of the standard deviation of this daily average?

6. A medical researcher observes a set of 16 patients with an average level of a certain serum

enzyme of 1.23 ppm. Another researcher who is skeptical of the first researcher's results duplicates the experiment on another 16 patients and finds an average level of 2.17 ppm.

a. Could they both be correct, or are their results automatically inconsistent with each other?

b. Assuming both results to be valid, what would you estimate to be the true mean level of this enzyme in a patient?

c. (Optional) What further information is needed (if any) to determine if the two researchers results are incompatible?

7. As a class project in a statistical research methods class, each of 25 students was to sample randomly 9 students and determine their grade-point average (GPA). They then reported the mean and standard deviation of the 9 numbers they obtained.

a. Although the numbers obtained by the 25 students were close together, none of them agreed. Is this surprising?

b. Suppose (from computerized student records) it is determined that for the whole school the mean GPA is 2.5 with a standard deviation of 0.5. What do you expect each of the 25 students to get for $\bar{x}$ and for s^2, the sample variance of the nine observations?

c. Consider the 25 averages reported by the students, say, $\bar{x}_1, \bar{x}_2, \ldots, \bar{x}_{25}$. What would you expect the average of these 25 numbers to be?

d. The students also reported the variance of each sample, say, $s_1^2, s_2^2, \ldots, s_{25}^2$. What would you expect the average of these 25 numbers to be?

8. A chemist determining the concentration of a compound in a colloidal suspension takes four readings and records their average. He does this seven times and records the following numbers: 17, 15, 17, 18, 14, 19, 15.

a. What is the mean of these seven numbers?

What is the variance of these seven numbers?

b. The true concentration is μ. What would you estimate μ to be?

c. There is usually quite a bit of variation between the concentration readings of individual determinations. (This is why the chemist averages four readings before recording the result.) If σ^2 is the variance of the individual readings, what should you expect the variance of the recorded readings to be? (Remember each recorded value is an average of four readings.)

d. Use (a) and (c) to estimate σ^2.

9. Find a 90% confidence interval estimate for the unknown mean μ in each of the following cases:

a. $n = 25$, $\sigma = 2$, $\bar{x} = 12$
b. $n = 36$, $\sigma = 5$, $\bar{x} = 31$
c. $n = 9$, $\sigma = 33$, $\bar{x} = 100$
d. $n = 100$, $\sigma^2 = 121$, $\bar{x} = 212$

10. A cigarette manufacturer's product is being examined by the Food and Drug Administration to ascertain the nicotine content of each cigarette. A laboratory performed tests on nine cigarettes and obtained the following nicotine contents in milligrams: 26, 24, 23, 27, 28, 22, 25, 26, 23. From past research they know this manufacturer has a standard deviation of $\sigma = 1.2$ mg nicotine per cigarette. Find a 95% confidence interval for μ, the "true" nicotine content of the cigarettes. (State your assumptions.)

11. A grand jury investigating judicial variation in sentencing for a particular crime found that a certain judge had given sentences (in months) of 123, 110, 119, 115, 102, 130, 122, 112, and 109. The standard deviation in sentences for this particular crime was calculated from the vast history of criminal cases and found to be $\sigma = 7$ months. Determine a 90% confidence interval for μ, the mean sentence imposed by this judge for this crime. (State your assumptions.)

12. In order to determine the amount of money lost to uncollectible student loans, a clerk takes a sample of 16 delinquent accounts and finds $\bar{x} = \$2312$. Assume that $\sigma = \$400$ is known. Find a 90% confidence interval for the mean value of delinquent student loans. (State your assumptions.)

13. To estimate the mileage of a certain car, the Environmental Protection Agency tested 25 on a fairly long trip and obtained $\bar{x} = 21.2$ mi/gal. Assume $\sigma = 6$ is known from past experience. Compute a 95% confidence interval for μ, the expected mileage for this car. (State your assumptions.)

14. A plastic manufacturer must control the tensile breaking strength of his plastics in order that toy manufacturers will use his product. The variability in breaking strength is known to yield $\sigma = 13$ lb due to the nature of the production machinery, chemicals, and temperature. A sample of 25 plastic chips is made from a particular production run and the following breaking strengths are recorded (in pounds pressure): 43.2, 35.3, 38.2, 46.1, 30.2, 29.9, 42.9, 51.0, 35.3, 39.1, 37.2, 41.6, 42.7, 40.3, 37.1, 40.9, 48.2, 31.2, 49.1, 40.6, 42.1, 41.2, 39.6, 41.7, 43.8. Find a 95% confidence interval for the mean breaking strength of this production run.

Section 6.3

15. (Optional) One reason for dividing by $n - 1$ instead of n in the formula for the sample variance s^2 is that the data $x_1, x_2, \ldots, x_n$ is forced to satisfy one equation, namely, $x_1 + x_2 + \cdots + x_n = n\bar{x}$ (or equivalently the sum of the deviations is 0). Consequently, we can use this equation to find one x_i in terms of the others, and hence have only $n - 1$ free choices for the x's. If the x's satisfy *two* equations, how many degrees of freedom do we have in the calculation of the sample variance?

16. (Optional) Every time we estimate a parameter in a statistical formula by the corresponding sample estimate we are forcing an equation to hold in the x's. Use the logic of problem 15 to say why the number of degrees of freedom in a problem where k parameters are estimated is $n - k$.

17. A random sample of 16 bolts is taken from a batch of bolts alleged to have a mean diameter of $\mu = \frac{3}{4}$ in. A quality control inspector examining these 16 bolts finds $s^2 = 0.0025$ sq in. What is the probability that he finds a sample mean greater than 0.7825?

18. Repeat problem 10 assuming σ^2 is unknown.

19. Repeat problem 11 assuming σ^2 is unknown.

20. In problem 12, assume the clerk found $s^2 = 136,900$ for the sample of 16 delinquent student loans. Find a 90% confidence interval for μ. (State your assumptions.)

21. Suppose the manufacturer in problem 14 does not know σ. Find a 95% confidence interval for μ. (State your assumptions.)

22. In many cultures it is believed that a full moon has an adverse affect on the mental stability of people. Indeed, the use of the word "lunatic" to describe mentally disturbed individuals is based on this supposed lunar influence. Blackman and Catalina reported the number of admissions to the emergency room of a mental health clinic during full moons from August 1971 to July 1972 as 5, 13, 14, 12, 6, 9, 13, 25, 13, 14, and 20. (Reprinted by permission of author and publisher from Blackman, S., and D. Catalina: The moon and the emergency room. *Perceptual and Motor Skills, 73*, 1973, pp. 624–626.)

   a. Construct a 90% confidence interval for μ, the true mean of the number of admissions during a full moon. (State your assumptions.)

   b. For the rest of the year the mean admission rate was 11.2 persons per night. Would you consider this to be much less than the full moon admission rate in light of your answer to (a)?

23. A study was performed to determine if monotonous tasks affected absenteeism in a large factory, and if this absenteeism could be reduced by rotating the workers' work stations daily. The first step in this study was to determine the absentee rate that is currently experienced. To do this, a sample of 16 employee records were drawn and the following absentee rate in days per last 6 months were discovered: 8, 5, 10, 12, 4, 2, 11, 7, 9, 0, 1, 13, 10, 9, 12, 7. Find a 90% confidence interval for μ, the current absentee rate for the company. (State your assumptions.)

24. To hunt flying insects, bats emit high frequency sounds and then listen for their echoes. Until an insect is located, these pulses are emitted at intervals of from 50 to 100 milliseconds (ms). When the insect is detected, the pulse-to-pulse interval will sharply decrease to about 10 ms so that the bat can pinpoint the insect's position. To determine how far apart the bat and insect are when the bat first senses that the insect is there, the researchers videotaped and sound recorded 11 catches. The following data were obtained. (Taken from Griffin et al., *Animal Behavior 8:* 141–154, 1960.)

Catch No.	Distance (cm)
1	62
2	52
3	68
4	23
5	34
6	45
7	27
8	42
9	83
10	56
11	40

Find a 90% confidence interval for the mean

distance separating a bat from an insect when the insect is first discovered ($\Sigma x_i = 532$, $\Sigma x_i^2 = 29,000$). (State your assumptions.)

Section 6.4

25. A manufacturing plant making electronic components must closely control the quality of the units produced. One symptom of an out of tune machine is that the variance in length of the produced parts becomes too large. In normal operating conditions $\sigma^2 = 1$ mm, whereas σ^2 becomes greater if the machine is out of tune. A sample of 21 parts gives $s^2 = 2$. Find a 90% confidence interval for σ^2. (State your assumptions.)

26. The octane rating of regular unleaded gasoline in California varies according to producer and according to brand. A sample of 16 gasolines yielded a variance of $s^2 = 25$. Find a 95% confidence interval for σ^2.

27. An insurance company cross-classifies clients according to age, sex, location, and other variables in order to *improve* their estimate of the expected loss from each client. The cross-classification improves the estimate by reducing the variance of the expected loss. For one cross-classified group of 36 married men aged 25 to 30 years old in Los Angeles, the company finds a standard deviation estimate of $s = \$400$ for losses experienced. Find a 95% confidence interval for σ^2, the true variation in this cross-classification.

Section 6.5

28. a. Using the following data, estimate in each case the proportion, p, of the population that possesses the characteristic in question.

Sample size n	Number in the sample with the characteristic
10	4
20	6
50	16
100	82
100	61
100	47
250	100
1000	450

b. Estimate p by a 99% and a 95% confidence interval by using Appendix C.

c. Find in each case a 99% and a 95% confidence interval for p by using the normal approximation method.

29. When determining the probability of rain on a given day, the weather forecaster compares that day's weather conditions (barometric pressure, cloud cover, and so on) to previous days with similar climactic characteristics. The National Weather Service records show that there were 50 days in the past that had climactic conditions similar to today's conditions. Of these 50 days, records show that it rained on 30 days and did not rain on 20 days. Estimate the probability p of rain today. Find a 95% confidence interval by using Appendix C and the normal approximation. (Assume that the "coin tossing model" is applicable here.)

30. The department of highway safety decided to determine the proportion of drivers who use seat belts when driving. A random sample of 250 drivers found that 100 had belts on and 150 did not. Estimate the proportion of drivers who use seat belts. Find a 99% confidence interval for this proportion by first using Appendix C and then using the normal approximation method. (State the necessary assumptions.)

31. A state prison inquiry is to ascertain the proportion of recidivists (repeat offenders) that are produced by a certain prison. A random check of 100 names taken from the files shows 65 committed offenses after release and were reincarcerated. Estimate the proportion of recidivists produced by this prison. Find a 95% confidence interval for the proportion of recidivists produced by this prison. (State your assumptions.)

32. A precision machine produces parts for aircraft motors. A sample of 256 parts found 16 defectives. Estimate the proportion of defectives produced by this machine. Find a 95% confidence interval for the proportion of defectives produced by the machine.

Section 6.6

33. A new teaching method has been introduced into the school system. It is hoped that the new method will increase the *average* grade of students; however, it is known that the standard deviation of the grades, which equals 12 points, will stay the same. If independent random sampling of grades is to be used, estimate the minimum sample size needed to obtain a 0.99 confidence interval of length 6 points on the mean. (State your assumptions.)

34. An auditor sampling the books of a large firm wishes to determine the mean dollar error per account. He will estimate this mean μ by a 95% confidence interval. If $\sigma = \$1000$ is known, how large a sample n must be taken to ensure that you are 95% confident that the error in approximating μ by the sample mean is less than $\$100$.

35. Formula 2 of Section 6 is an expression for the length l of the confidence interval on the mean when σ is unknown. This formula can yield the approximate sample size n required to obtain a specified l for a given confidence level. Suppose we know that σ must be less than some value σ_{max}; then, for sufficiently large n, say, greater than 30, we assume that s will be less than σ_{max}. In this case

$$l < 2 \frac{\sigma_{max}}{\sqrt{n}} z_{\alpha/2}$$

where we have replaced $t_{\alpha/2}$ by $z_{\alpha/2}$ because of the large sample size.* Hence

$$n \geq \left[2z_{\alpha/2} \frac{\sigma_{max}}{l} \right]^2$$

We take the expression on the right as a conservative estimate of the required sample size.

Problem: An auditor sampling the books of a large firm wishes to determine the mean dollar error per account. He will estimate this mean μ by a 95% confidence interval. He does not know σ, but from experience knows it must be less than \$4,000. How large a sample n should be taken to obtain approximately a 95% confidence interval of length less than \$200. (Compare this result to that of problem 34 and note the effect of the poorer prior information we have on σ. As a general principle, the poorer the prior information we have, the larger the sample size must be to compensate in order to obtain as "good" results when we have better information. In this case the firm must be large indeed to satisfy our sample size requirement.)

36. The determination of sample size in estimating a proportion p: The length l of the confidence interval on p is

$$l = 2z_{\alpha/2} \sqrt{\frac{\hat{p}(1-\hat{p})}{n}}$$

Solving for n, we find

$$n = \hat{p}(1-\hat{p}) \left[\frac{2z_{\alpha/2}}{l} \right]^2$$

a. Draw the graph of $\hat{p}(1-\hat{p})$ versus $\hat{p}$, and show that the maximum value of $\hat{p}(1-\hat{p})$ is equal to 0.25.
b. Show that the maximum sample size n_{max}

*Recall that the t-distribution approaches the standard normal distribution as n increases (Fig. 6.2).

required to obtain a confidence interval of width l at the confidence level $1 - \alpha$ is

$$n_{max} = 0.25 \left[\frac{2z_{\alpha/2}}{l} \right]^2$$

c. Suppose we wish to estimate the "tourist proportion" of those people walking along Royal Street in the French Quarter during a given period on Saturday night; we wish to obtain a 95% confidence interval on this estimate of size less than 0.2. How many people must we interview?
d. Suppose we have some prior information; we know that the proportion of tourists in (c) must be greater than 0.2 and less than 0.3. Instead of replacing $\hat{p}$ by 0.5 in

$$n = \hat{p}(1-\hat{p}) \left[\frac{2z_{\alpha/2}}{l} \right]^2$$

we replace it by the value from 0.2 to 0.3 which maximizes $\hat{p}(1-\hat{p})$; in this case we replace $\hat{p}$ by 0.3 (why?) to obtain

$$n_{max} \text{ is approximately } 0.3 \left[\frac{2z_{\alpha/2}}{l} \right]^2$$

Redo (c) using this formula. (Notice the effect of having better prior information. Note also that the procedure in (d) yields only an approximate value for n_{max} because $\hat{p}$ may in fact be larger than 0.3; we are assuming that the sample size is so large that $\hat{p}$ will be close to p.)

37. (Additional problem) Confidence limits when the distribution is hypergeometric: Let us assume that we are sampling from a population of which a proportion p has a certain characteristic C. We let

$$X_i = \begin{cases} 1 & \text{if the unit selected on the } i\text{th sample possesses characteristic } C \\ 0 & \text{otherwise} \end{cases}$$

and let $Y = \Sigma_{i=1}^{n} X_i$. Again, we define the sample proportion $\hat{P} = Y/n$. Let N denote the size of the population and assume the selection is by simple random sampling.* Since the sampling is without replacement, the random variables are *not* independent, and hence the distribution of Y is not binomial. To determine the distribution of Y (and hence of $\hat{P}$) in the simple random sampling case, let us suppose there are M units in the population with characteristic C (so that $p = M/N$). Then

$$P(Y = k) = \frac{\binom{M}{k}\binom{N-M}{n-k}}{\binom{N}{n}}$$

$$= \frac{\binom{Np}{k}\binom{N-Np}{n-k}}{\binom{N}{k}}$$

that is, Y has a hypergeometric distribution.* For this distribution it can be shown that the mean is $\mu_Y = E(Y) = nM/N = np$ and the variance is

$$\sigma_Y^2 = n \cdot \frac{N-n}{N-1} \cdot \frac{M}{N} \cdot \frac{N-M}{N}$$

$$= \left(\frac{N-n}{N-1}\right) np(1-p)$$

Thus for $\hat{P} = Y/n$ we have

$$\mu_{\hat{P}} = \frac{\mu_Y}{n} = \frac{np}{n} = p \quad \text{and}$$

$$\sigma_{\hat{P}}^2 = \frac{\sigma_Y^2}{n^2} = \left(\frac{N-n}{N-1}\right)\frac{p(1-p)}{n}$$

* See Section 4.4.

Notice that the mean and variance of $\hat{P}$ when using this sampling method are easily related to those of $\hat{P}$ when sampling with replacement. The only change is the inclusion of what is called the *finite population correction factor* $(N-n)/(N-1)$ in the variance.

Confidence intervals

When N and n are large we may approximate the distribution of $\hat{P}$ by the normal distribution to obtain confidence intervals. Namely, when N and n are large, the limit on the $100(1-\alpha)\%$ confidence interval for p is

$$\hat{p} \pm z_{\alpha/2}\sqrt{\left(\frac{N-n}{N-1}\right)\frac{\hat{p}(1-\hat{p})}{n}}$$

a. A certain chemical is thought to produce tumors when ingested. A group of 5000 mice are fed food containing this chemical for 6 weeks and then 500 mice are killed and autopsied to determine if they have a tumor. Fifteen of the mice are found to have a tumor. Find a 95% confidence interval for p, the true proportion of mice who will have a tumor after 6 weeks. Answer: Here $N = 5000$, $n = 500$, $\hat{p} = {}^{15}\!/_{500} = 0.03$, and $z_{0.025} = 1.96$, so the 95% confidence interval is from

$$0.03 - (1.96)\sqrt{\left(\frac{4500}{4999}\right)\frac{(0.03)(0.97)}{500}} \quad \text{to}$$

$$0.03 + (1.96)\sqrt{\left(\frac{4500}{4999}\right)\frac{(0.03)(0.97)}{500}}$$

or from

$$0.016 \quad \text{to} \quad 0.044$$

b. Place 95% confidence limits on p in "Application to Ecology," Section 4.14.

CHAPTER · 7

Hypothesis Testing About the Mean and Variance of a Population

If a man will begin with certainties, he will need end with doubts. But if he will be content to begin with doubts, he shall end in certainties.

Francis Bacon
Advancement of Learning (1605)

Statistical inference is a particular kind of inference in which one draws conclusions about a population from data obtained from sampling. The previous chapter covered some topics in one area of statistical inference called *estimation* where a statistic computed from a sample is used to estimate a population parameter.

We now begin our study of another area of statistical inference, called *hypothesis testing,* which deals with the testing of *statistical hypotheses.* Not all hypotheses are statistical hypotheses. A hypothesis is a statistical hypothesis only if it is stated in terms that are related to the distributions of populations. We shall study the methodology for developing *statistical* proofs of this kind of hypothesis.

7 ▪ 1   HYPOTHESIS TESTING

Types of Errors, Significance Level, Critical Region

Generally speaking, decision making is the process of choosing among alternatives. We usually are uncertain about the true state of nature and yet must make decisions concerning it. In order to help make decisions in the face of uncertainty, we utilize a technique known as hypothesis testing. This involves forming a *statistical hypothesis* or a tentative statement about the value of some population parameter. Thus we could hypothesize "the mean height of men is 70 in" or "the mean lifetime of a certain brand of light bulb is 500 hours." The hypothesis to be tested is stated in terms of a population parameter and is called the *null hypothesis* and labeled H_0. The competing hypothesis is called the *alternative hypothesis* and is labeled H_1. We assume that one of the two hypotheses must be true, and we treat the hypotheses as if the null hypothesis were true unless the sample data convince us otherwise. To test the null hypothesis $H_0 : \mu = 70$ versus the alternative hypothesis $H_1 : \mu = 80$, we write

$$H_0 : \mu = 70$$

$$H_1 : \mu = 80$$

We then use the observations obtained from a sample to decide if H_0 is implausible, or if the data are consistent with H_0. As another example, suppose we wish to test $H_0 : \mu \geq 3$ versus $H_1 : \mu < 3$. We would tend to reject H_0 if the observed sample mean $\bar{x}$ was much smaller than 3 since this value of $\bar{x}$ would be more in line with H_1 than with H_0. Still, because the sample was randomly chosen, there is a chance that the random variable $\bar{X}$ will take on a value smaller than 3 even if $\mu \geq 3$ is true. If we reject H_0 in this case we have erred. Similarly, because of the randomness inherent in the distribution of the random variable $\bar{X}$, we could obtain a value of $\bar{X}$ that is larger than 3 even if the alternative hypothesis $\mu < 3$ were the true state of nature. Such a result might lead us to a wrong decision. In short, because we base our decision on a sample, there is a risk of making an error. In general, there are two types of errors possible when testing a hypothesis. These are listed in Table 7.1.

We shall see that when testing a null hypothesis about a population parameter we can choose to regulate the probability of our type 1 error. The probability of this error is called the *level of significance* of the test and is determined by the statistician prior to sampling. We denote this probability by α and call it the size of the type 1 error. The probability of the type 2 error is denoted by β and is called the size of the type 2 error.

Once H_0 and H_1 are set up and the level of significance α is chosen, we must select a *test statistic*. A test statistic is a function of the sample that we utilize to quantify the intrinsic differences between the hypotheses H_0 and H_1. and it

Table 7.1 ▪ Decisions and State of Nature

		Decision	
		Do not reject H_0	Reject H_0
State of nature	H_0 *true*	Correct decision	Error (called type 1 error)
	H_1 *true*	Error (called type 2 error)	Correct decision

should alert us to which hypothesis is more likely to be true. For example, in testing

$$H_0 : \mu \geq 3$$
$$H_1 : \mu < 3$$

we would use $\overline{X}$ as our test statistic since it should have a high probability of being large when H_0 is true and small when H_1 is true. Moreover, we have already seen that $\overline{X}$ is a very good estimator of μ, and it is a good idea to use the best estimator possible in your test.

The basic question is: Once we obtain the observed value $\overline{x}$, how do we decide whether to reject, or not to reject, the hypothesis H_0. We obtain a hint on how to make this decision from the following argument: If the value of $\overline{x}$ we observe is, say, 50, we would certainly reject H_1 that the mean μ is less than 3 since the observed estimate $\overline{x}$ of μ is 50 which is much larger than 3. If $\overline{x}$ were equal to 3, we certainly would not reject the hypothesis H_0 that μ is greater than or equal to 3. If $\overline{x}$ were equal to 0, we might be inclined to accept the hypothesis H_1 that μ is less than 3. We might be so inclined if $\overline{x}$ were equal to 1. But suppose $\overline{x} = 2.999$; would we be willing to reject H_0 that $\mu \geq 3$? Where do we draw the line? At what point do we stop accepting H_0 and say "wait a minute, this value of $\overline{x}$ seems mighty unlikely to have arisen from a population with a mean bigger than 3." The question is, what is the value c such that we would reject H_0 if the observed value $\overline{x}$ were smaller than c? The set of values less than c is called the *rejection region* or the *critical region* for this hypothesis test; it is the set of values such that one rejects H_0 if $\overline{x}$ takes on a value in this region. To summarize:

Critical region (rejection region): The set of values such that if our test statistic takes on one of the values in this region, we reject H_0.

How do we determine the critical region? A critical region or rejection region is determined from the level of significance α and the test statistic. We agree to reject H_0 (and consequently, favor H_1 as the true state of nature) if the value of the test statistic falls in the rejection region. Otherwise we do not reject H_0. The rejection region is chosen so that the probability that the observed value of the test statistic (such as $\overline{X}$) falls in the rejection region is some small value, say, α. *A rejection region is a set of implausible values of the test statistic when H_0 is true.* The precise dividing line for the rejection region is selected to ensure a type 1 error whose probability is the pregiven number α. In short, the rejection region is chosen so that

$$P(\text{rejecting } H_0 \mid H_0 \text{ true}) \text{ is equal to a given value } \alpha$$

An Eight-Step Procedure for Hypothesis Testing

We summarize the preceding discussion by giving an eight-step procedure that is utilized in all our hypothesis tests.

1. State the underlying assumptions about the population being tested and decide on the method of sampling. (In this text we use independent random sampling unless otherwise stated. Usually we assume the population sampled has a normal distribution.)
2. State the null hypothesis H_0 and the alternative hypothesis H_1. (Usually the test is meant to determine if H_0 could be true, i.e., if chance variation can explain the difference between the observed data and the parameter value implied by H_0.)
3. State the level of significance α chosen for the test. (This is a subjective decision. The choices $\alpha = 0.1, \alpha = 0.05$, or $\alpha = 0.01$ are the most common.)
4. Select the test statistic to be utilized in the hypothesis test. (The test statistic measures the difference between the data and H_0.)
5. State the rejection region for H_0. (These are the "implausible" values for the test statistic if H_0 were true. We shall show how to determine it once H_0, the test statistic, and α are chosen. By implausible here we mean "has a chance less than or equal to α of occurring if H_0 is true.")
6. Sample from the population and compute the value of the test statistic.
7. If the value computed in step 6 is in the rejection region, then reject H_0; otherwise we fail to reject H_0.
8. Use the statistical conclusion arrived at in step 7 and relate this back to the original problem; that is, state your final decision concerning the problem of interest.

One fact we have not yet discussed in sufficient detail is how to select H_0. This is important since we treat H_0 and H_1 differently, our critical region being set up based on H_0 and α.

The Choice of H_0

The choice of which of two competing hypotheses to take as H_0 and which to take as H_1 should be carefully made since the test is not symmetric in the two hypotheses. Indeed, H_0 is favored, since we do not reject H_0 unless the evidence is overwhelming that H_0 is false. In general, there are two types of errors possible and we have control over the type 1 error by selecting our critical region. One type of error is usually much more critical than the other, and thus we choose H_0 so that the more critical of the two is the type 1 error. We then adjust the cutoff for the critical region so that the probability of the type 1 error, α, is at a level that is acceptable to us. For example, suppose we wish to test whether a certain nuclear power plant is safe, and based on this test, we either agree or refuse to build the plant. The possible errors are either: (1) build an unsafe plant and risk a nuclear accident with many resulting deaths or (2) do not build a safe plant and risk occasional power shortages. Certainly most of us would agree that it is much more costly to commit the first type of error than the second. We would not want to build the plant unless we have some certainty that it is safe; that is, we would insist that there is only a small chance (α) of building an unsafe plant. Consequently, we choose H_0 and H_1 as follows:

$$H_0: \text{The plant is not safe.}$$

$$H_1: \text{The plant is safe.}$$

If neither type of error is more to be avoided than the other type, then we utilize a slightly different logic to determine which hypothesis to call H_0. If we are testing whether or not a treatment of some sort has an effect, we let H_0 represent the status quo; that is, H_0 represents the hypothesis of *no effect*, whereas H_1 represents the hypothesis that there is an effect. Since H_0 is favored (it only is rejected if the evidence is clear that there is an effect) this puts the burden of proof on the experimenter who is trying to show an effect. An analogy in the legal profession is the supposition that a person is innocent (H_0) until proven guilty (H_1). Here the assumed burden of proof falls on the plaintiff or prosecution.

As an example of a situation where we would test for "no effect," suppose we are unhappy about the mean speed μ_0 with which a pill is absorbed into the bloodstream. We are doing research designed to find a coating that will either speed up or slow down the mean time of the process and believe we have discovered such a coating. We take as our hypothesis H_0 that the mean time μ, using this new coating, is the same as μ_0, the mean time of the old coating: $H_0: \mu = \mu_0$. H_0 is the hypothesis that there is *no effect*. We make this choice of H_0 in order to be careful not to make extravagant claims—we do not want to say there is an effect when there is none. This would be making a serious error. It results in placing an ineffective product on the market. Many hopeful, and perhaps desperate people, would spend their resources for nothing. In addition, our research would lose credibility and we might no longer be permitted to continue our efforts. To repeat, we wish to test the hypothesis of "no effect"— that μ is *equal to* the number μ_0.

Often only extreme deviations in one direction are to be avoided. For example, when guaranteeing a tire for 25,000 mi, the manufacturer is only concerned that the tire will not last that many miles. If it lasts much longer than 25,000 mi, the manufacturer is perfectly happy. Similarly, if a mountain climber is purchasing rope, the only concern is if the breaking strength is too low. When we are hypothesis testing in such cases we wish the hypothesis and the rejection region to reflect this concern of only avoiding one extreme. We take as the null hypothesis the inequality of most concern to us, since only if it is rejected do we feel secure that the opposite inequality is probably true. Suppose, for example, we wish to perform a test about the mean of a population and would be most concerned if μ is less than or equal to some constant μ_0. We write

$$H_0 : \mu \leq \mu_0$$

$$H_1 : \mu > \mu_0$$

This is called a *directional hypothesis test.* Only by rejecting H_0 is information about μ conveyed, since if we fail to reject H_0, we have only said the data are not inconsistent with H_0. It is possible that the data are not inconsistent with H_1 as well. If we do reject $H_0 : \mu \leq \mu_0$, we say that the mean is significantly greater than μ_0. If the hypothesis were $H_0 : \mu = \mu_0$ and we reject H_0, we say that the mean is significantly different from μ_0. If we were to reject $H_0 : \mu \geq \mu_0$, we say the mean is significantly less than μ_0. The terms "significantly less than," "significantly greater than," or "significantly different from" also enter into our language when we ask questions about tests. If we ask, Is the mean significantly greater than μ_0? we are asking whether $H_0 : \mu \leq \mu_0$ is to be rejected.

For further clues on how to select H_0, do problems 3, 4, 5, and 6.

7 ▪ 2 TESTING HYPOTHESES ABOUT THE MEAN μ WHEN THE POPULATION VARIANCE σ^2 IS KNOWN

In this section we assume that the value of the variance σ^2 of the population is known.

The Eight-Step Procedure When Testing About the Mean

Suppose we use independent random sampling to sample from a population with a known variance σ^2 but with an unknown mean μ. We wish to test the hypothesis that μ is *equal to the number* μ_0 against the hypothesis that it is not equal to μ_0. We follow our eight-step procedure.

1. Assumption: σ^2 is known; the sampling procedure used is independent random sampling. The population is normally distributed or else the sample size is large enough to use a normal approximation for the distribution of $\overline{X}$.

2. State H_0 and H_1. For example,

$$H_0: \mu = \mu_0$$
$$H_1: \mu \neq \mu_0$$

3. Selection of significance level: We select the level of significance α to be, say, 0.05. (The choice of α is rather subjective and depends on assessing the consequences of falsely rejecting H_0. The usual choices are 0.05, 0.01, or 0.001.)

4. Selection of test statistic: To select the test statistic, recall that by the central limit theorem, when H_0 is true, the distribution of $Z = (\overline{X} - \mu_0)/(\sigma/\sqrt{n})$ is normally (or approximately normally) distributed with mean 0 and variance 1; hence the values of $\overline{X}$ will tend to be around μ_0 when H_0 is true, since that is where the hump in the density is. Hence, the observed value of Z should be close to zero. On the other hand, if H_1 is true, then the values of $\overline{X}$ will not necessarily be close to μ_0 (but rather, will be close to whatever the true mean actually is) so that the values of $(\overline{X} - \mu_0)/(\sigma/\sqrt{n}) = Z$ will tend to be around the tail of the normal distribution with mean 0 and variance 1. Thus using $(\overline{X} - \mu_0)/(\sigma/\sqrt{n})$ as a test statistic seems like a reasonable choice. It behaves differently under H_0 and H_1, and its value should give us a clue as to which hypothesis is true. Using $Z = (\overline{X} - \mu_0)/(\sigma/\sqrt{n})$ as a test statistic, we should reject H_0 for either extreme positive or negative values of Z, since large positive values seem to imply a mean much bigger than μ_0, and small negative values seem to imply a mean much smaller than μ_0.

5. Selection of the critical region: Since we want to fix the probability of a type 1 error at α, $P(\text{reject } H_0 \mid H_0 \text{ true}) = \alpha$, we select the critical region that is shaded in Figure 7.1. Since we know $P(-z_{\alpha/2} < Z \leq z_{\alpha/2}) = 1 - \alpha$, we

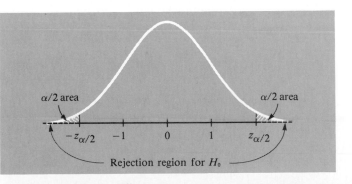

Figure 7 ■ 1
Choosing a rejection region.

choose to reject H_0 if and only if the observed value z of the random variable Z is greater than $z_{\alpha/2}$ or less than $-z_{\alpha/2}$. Then $P(\text{reject } H_0 \mid H_0 \text{ true}) = \alpha$, the total shaded area; and our critical region consists of all numbers less than $-z_{\alpha/2}$ and greater than $z_{\alpha/2}$. Here $z_{\alpha/2}$ is the upper $\alpha/2$ centile of the normal distribution with mean 0 and variance 1.

Since we reject H_0 if Z takes a value on *either* of the two shaded tails in Figure 7.1, we refer to this type of test as a two-tailed test.

Summary of Steps 1 through 5
Two-tailed test of the mean when σ^2 is known, independent random sampling is used, and $\overline{X}$ is at least approximately normal. At level of significance α reject H_0 if the observed value

$$z = \frac{\overline{x} - \mu_0}{\sigma/\sqrt{n}}$$

is greater than $z_{\alpha/2}$ or less than $-z_{\alpha/2}$.

Steps 6 through 8 of our eight-step procedure are then followed. The following example illustrates all eight steps.

Example 7 ■ 1   When a certain machine is adjusted correctly, it turns out parts with a mean length of 5 in. and a standard deviation of $\sigma = 0.6$ in. A sample of nine parts from this machine is tested and the sample mean is found to be $\overline{x} = 5.4$ in. Can we conclude that the machine is out of adjustment at the $\alpha = 0.05$ level of significance? We assume the standard deviation $\sigma = 0.6$ is the same whether or not the machine is in adjustment. We resolve this question by the hypothesis-testing method.

1. We assume that the sample mean $\overline{X}$ has at least approximately a normal distribution, σ^2 is known and the sampling used is independent random sampling.
2. $H_0: \mu = 5$ (the machine is in adjustment)
   $H_1: \mu \neq 5$.
3. $\alpha = 0.05$.
4. The test statistic used is $Z = (\overline{X} - 5)/(0.6/\sqrt{9})$, which has at least approximately a normal distribution with mean 0 and variance 1.
5. From Appendix A.1 we find $P(-1.96 < Z \leq 1.96) = 0.95$ so the rejection region consists of all values less than -1.96 and greater than 1.96 (see Fig. 7.2). Note that there is probability $\alpha/2 = 0.025$ on each tail, and this is how we know to take $z_{0.025} = 1.96$ as the cutoff value for the rejection region.
6. $\overline{x} = 5.4$ so that $z = (5.4 - 5)/(0.6/\sqrt{9}) = +0.4/0.2 = 2$.

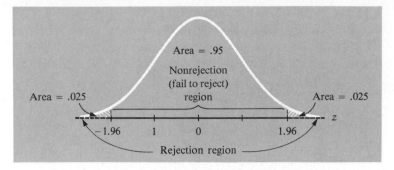

Figure 7 ▪ 2

7. Since $2 > 1.96$, the sample test statistic falls in the upper rejection region. Therefore, we reject H_0 and conclude at level of significance $\alpha = 0.05$ that $\mu \neq 5$ and that the average length of parts is significantly different from 5.
8. In view of the results in step 7, we are 95% confident that the machine is out of adjustment. We report to the production manager that repair is necessary.

Let us now turn to the situation of directional hypothesis tests about the mean μ. In this case, departures of μ from a fixed number μ_0 are of interest only if the departure is in one direction. For example, an advertiser is only concerned if the sales revenue *decreases* from the current level following adverse publicity, or a drug manufacturer is only concerned if the number of side effects from his drug *increases* among habitual users as compared with the number of side effects in the general population. When hypothesis testing in such cases we wish the hypothesis and the rejection region to reflect this concern with avoiding only one extreme. For example, if we wish to test whether a population has a mean μ that is greater than or equal to some given constant μ_0, we write

$$H_0: \mu \geq \mu_0$$
$$H_1: \mu < \mu_0.$$

In order to test this hypothesis at the level of significance α, we must find a critical region. The test statistic is again $Z = (\bar{X} - \mu_0)/(\sigma/\sqrt{n})$. The observed value $\bar{x}$ approximates μ; an $\bar{x}$ as large or larger than μ_0, (i.e., z positive) tends to support H_0, whereas $\bar{x}$ much less than μ_0 (z very negative) seems to be implausible or inconsistent with $\mu \geq \mu_0$, and hence tends to support H_1. We thus reject H_0 if the observed value $(\bar{x} - \mu_0)/(\sigma/\sqrt{n}) = z$ is too small a negative number. In order to ensure that the probability of a type 1 error is α, we reject H_0 if and only if we obtain a value z less than $-z_\alpha$. That is, we reject H_0 if the observed z is on the far left tail (see Fig. 7.3). For this reason directional hypothesis tests are called one-tailed tests. Notice that the "tail" used for the rejection region, $z < -z_\alpha$, has the same inequality sign as the one specified in H_1, namely, $<$.

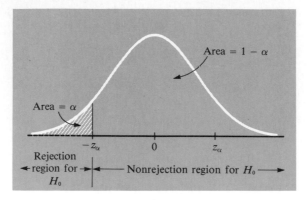

Figure 7 ▪ 3

Example 7 ▪ 2   A rope manufacturer advertised that its brand has a breaking strength of at least 1600 lb with a known standard deviation σ of 316.23 lb. A consumer agency wishing to test the claim sampled 10 ropes and found $\bar{x} = 1450$ lb. Should they sue the rope manufacturer for false advertising, or could this low value for $\bar{x}$ be due to chance? To answer this question, we use the hypothesis-testing method previously described. Since according to law the burden of proof is on the person who sues, we only advise suing if the sample evidence is overwhelming that the claim is false. Thus we give the manufacturer the benefit of the doubt and take the claim as H_0. We test

$$H_0 : \mu \geq 1600$$

$$H_1 : \mu < 1600$$

and we set $\alpha = 0.01$. If H_0 is true, we shall only wrongly prosecute 1% of the time. We shall reject H_0 (and sue the company) if and only if $\bar{x}$ is much smaller than 1600. Specifically, let $(\bar{x} - 1600)/(316.23/\sqrt{10}) = z$, and reject H_0 if we obtain an observed value z such that

$$z < -z_{0.01} = -2.326.$$

The picture given in Figure 7.3 is the same and the far left tail is the rejection region since the alternative hypothesis is "<." We calculate

$$z = (1450 - 1600)/316.23/\sqrt{10} = -1.50$$

which is > -2.326 so it is not in the rejection region; hence we fail to reject H_0. We do not advise suing.

It may also happen that we want to test if μ is smaller than μ_0; that is, we may want the directional hypothesis test

$$H_0: \mu \le \mu_0$$
$$H_1: \mu > \mu_0$$

The critical region is now the right tail since only values of $\bar{x}$ much larger than μ_0 (i.e., larger positive values of the z ratio) tend to cast doubt on the validity of H_0. See Figure 7.4 for the appropriate picture. At the level of significance α we reject H_0 if

$$z = \frac{\bar{x} - \mu_0}{\sigma/\sqrt{n}} > z_\alpha$$

Notice again that the inequality defining the rejection region (namely, >) is the same as that defining the alternative hypothesis.

Example 7 ▪ 3   A certain cat food producer wishes to keep the mean bacteria count for a batch of canned food below 4 parts per million (ppm) in order to sell the batch. It is known that the standard deviation of the bacteria count is 1 ppm. If a sample of 25 cans from a certain batch yields $\bar{x} = 4.43$ ppm, should he recall the batch of cans from the market or could the mean really be below 4 ppm for the entire batch and the 4.43 ppm be due to chance? Since the recall of canned food is costly, he will only recall the batch if $\bar{x}$ is significantly above the permissible 4 ppm level. We assume that the bacteria count per can follows a normal distribution. We test

$$H_0: \mu \le 4$$
$$H_1: \mu > 4$$

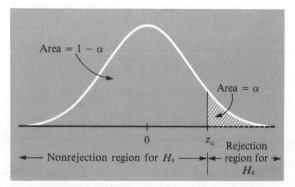

Figure 7 ▪ 4

and recall the batch if we reject H_0; that is, we recall the batch if the evidence makes it clear that $\mu > 4$. We select a level of significance $\alpha = 0.05$, because we can afford to "falsely" recall about 5% of the batches. The test statistic used is $(\overline{X} - 4)/(1/\sqrt{25}) = Z$. The rejection region consists of all values $z \geq z_{0.05} = 1.645$. We calculate $z = (4.43 - 4)/1/5 = 2.15 > 1.645$, so we reject H_0. The sample mean 4.43 is significantly larger (at the 5% level of significance) than we would expect just due to sampling variation if H_0 were true. A recall should be ordered.

Determination of the Probability of a Type 2 Error: Power of a Test (optional)

We wish to test

$$H_0: \mu = 70$$

$$H_1: \mu = 80$$

where μ is the mean of a normal population with $\sigma = 6$. Suppose $\alpha = 0.1$ and the sample size is $n = 9$. Since H_1 hypothesizes a value of μ greater than $\mu_0 = 70$, we use the right tail of the normal distribution: Figure 7.5 shows the critical value equal to $70 + z_\alpha(\sigma/\sqrt{n}) = 70 + 1.28(\frac{6}{3}) = 72.56$. If the observed value $\overline{x}$ is less than 72.56, we do not reject H_0. If $\overline{x}$ is greater than 72.56, we reject H_0 in favor of H_1. If H_0 is true, then the chance of the observed value $\overline{x}$ being greater than 72.56 is 0.1, the value of α. But what is the value of β, the probability of the type 2 error? To determine β, note that if H_1 is true, then the chance of the observed value $\overline{x}$ being less than 72.56 (which would result in incorrectly accepting H_0) is given by the area (see Fig. 7.5) under the normal curve with mean $\mu = 80$ and standard deviation $\sigma/\sqrt{n} = \frac{6}{3} = 2$ which is to the left of 72.56. To

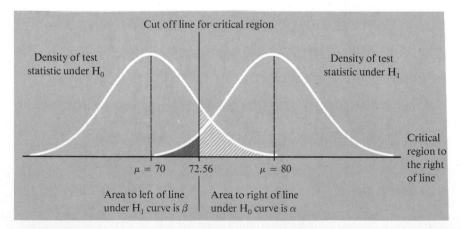

Figure 7 ▪ 5
The relationship between α (the size of the type 1 error) and β (the size of the type 2 error).

Cut off line for critical region

Density of test statistic under H_0

Density of test statistic under H_1

Critical region to the right of line

$\mu = 70$ 72.56 $\mu = 80$

Area to left of line under H_1 curve is β

Area to right of line under H_0 curve is α

find this area, we calculate

$$z = \frac{72.56 - 80}{\sigma/\sqrt{n}} = -\frac{7.44}{2} = -3.72$$

The area (found from Appendix A.1) to the left of -3.72 is 0.0001 which is the probability of the type 2 error in this case.

Notice that in example 7.3 we used as our critical region of size $\alpha = 0.1$, the right tail of the normal distribution from 72.56 on. This choice seems to be a commonsense one. But there are other critical regions of size $\alpha = 0.1$ we could have chosen. For example, if we had selected 69.12 to 70.88 as our critical region, the type 1 error would also be $\alpha = 0.1$. However, the value of β is 0.03, considerably higher than the type 2 error using the right tail as a critical region. In fact, it can be shown that for a given α the right tail for the hypothesis as given previously yields the smallest type 2 error of any critical region. This is the reason for the choice of the right tail as our critical region in this problem.

In this and subsequent chapters we give you rules for selecting the critical region for each test discussed. This critical region is *best* in the sense that, for any fixed α, it yields the smallest β. In much of the literature the value of $1 - \beta$ is called the *power* of the test. We always wish to choose the critical region with the largest power. The critical regions we lead you to use will always have largest power.

7 ■ 3   TESTING HYPOTHESES ABOUT THE MEAN OF A NORMAL DISTRIBUTION WHEN THE VARIANCE σ^2 IS UNKNOWN

In the previous section we saw how to make decisions or conclusions about the possible value of μ by hypothesis testing in the case where σ^2 is known. More frequently, we do not have any previous knowledge about the value of σ^2, and in fact, we only know the sample values $\bar{x}$ and s^2. When the population from which we are sampling is normally distributed we can still utilize the hypothesis-testing procedure introduced in the previous section, although of course we must modify our test statistic since σ^2 is unknown. It would be reasonable just to replace σ^2 by s^2 in the previous test statistic. Thus, to test hypotheses about the mean when σ^2 is unknown, we use the test statistic

Test statistic when σ^2 is unknown and population is normal:

$$T = \frac{\bar{X} - \mu}{S/\sqrt{n}}$$

which, as we saw in Chapter 6, has a t-distribution with $n - 1$ degrees of freedom. Now we proceed exactly as in the "σ^2 known" case, except of course σ^2 being unknown forces us to use the upper centile t-values instead of the upper centiles of the normal distribution. Other than this slight switch on which table we use to look up centile values, the procedure is just like before. We give some examples.

Employing the *t*-statistic

If we want to test if μ is some fixed number μ_0, then we want to test the hypothesis

$$H_0: \mu = \mu_0$$

$$H_1: \mu \neq \mu_0$$

at the level of significance α, and as before we use a two-tailed test. Namely, since $P\{-t_{\alpha/2} < T < t_{\alpha/2}\} = 1 - \alpha$, and T has a t-distribution with $n - 1$ degrees of freedom, we reject H_0 if the observed value t is greater than $t_{\alpha/2}$ or less than $-t_{\alpha/2}$ (see Fig. 7.6). Note the similarity between this hypothesis test and the one we used when σ^2 was known. Essentially they proceed the same way, only here we use the t-distribution, and there we used the normal distribution. The determination of the rejection region has the exact same logic, and the same picture as when σ^2 was known.

Example 7 ▪ 4   The Environmental Protection Agency (EPA) lists the average gas mileage on a certain model of car as 22.5 miles per gallon (mi/gal). A sample of 15 cars yields the mileages of 19, 20.3, 18, 26.1, 15, 19, 24, 20.5, 19.5, 17, 22.9, 24.5, 24.2, 22.9, and 20.9. Can we conclude at 0.05 level of significance that the average mileage μ is *not* 22.5 mi/gal? In this case we test $H_0: \mu = 22.5$ versus $H_1: \mu \neq 22.5$ since we can only conclude $\mu \neq 22.5$ if we reject $\mu = 22.5$. To perform the test, we first must assume the population of mileages form a normal distri-

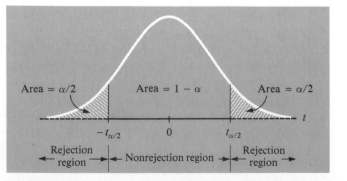

Figure 7 ▪ 6
t-distribution, showing rejection and nonrejection regions.

bution, and then calculate

$$\bar{x} = \frac{19 + 20.3 + \cdots + 20.9}{15} = 20.92$$

and

$$s^2 = \frac{(19 - 20.92)^2 + \cdots + (20.9 - 20.92)^2}{14} = 9.76$$

and

$$s = \sqrt{9.76} = 3.12$$

Since $\alpha = 0.05$ and the number of degrees of freedom is $15 - 1 = 14$, we reject H_0 if $t > t_{0.025} = 2.145$ or if t is less than $-t_{0.025} = -2.145$. We have

$$t = \frac{\bar{x} - 22.5}{3.12/\sqrt{15}} = \frac{20.92 - 22.5}{3.12/3.87} = -1.96$$

which is > -2.145 and not in the rejection region. Thus we fail to reject H_0 at $\alpha = 0.05$. That is, our data do not indicate that μ is significantly different from 22.5 mi/gal (at $\alpha = 0.05$). To rephrase our conclusion, the data we have do not detect that μ differs significantly from 22.5 mi/gal.

With modifications similar to the ones previously used in the two-tailed test (namely, using s instead of σ and t instead of z), we may also perform one-tailed or directional hypothesis tests about μ when σ^2 is unknown. For example, to see if μ is significantly greater than some fixed number μ_0, we test

$$H_0: \mu \leq \mu_0$$
$$H_1: \mu > \mu_0$$

at the level of significance α. We would reject H_0 if and only if $\bar{x}$ was significantly large. In other words, if

$$t = \frac{\bar{x} - \mu_0}{s/\sqrt{n}} > t_\alpha$$

the picture looks the same as before (see Fig. 7.7). Again, note that the rejection region is on the right tail ($t > t_\alpha$), which is consistent with the alternative

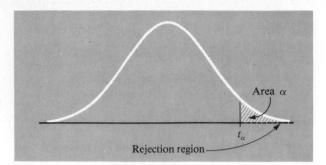

Figure 7 ▪ 7

hypothesis ($\mu > \mu_0$). To test

$$H_0: \mu \geq \mu_0$$
$$H_1: \mu < \mu_0$$

at the level of significance α, we reject H_0 (see Fig. 7.8) if

$$t = \frac{\bar{x} - \mu_0}{s/\sqrt{n}} < -t_\alpha$$

The left tail is now used for rejecting H_0 since the alternative hypothesis is $< \mu_0$.

Example 7 ▪ 5 Let us reconsider the data given in Example 7.4. If the EPA had claimed that the gas mileage is *at least* 22.5 mi/gal, we could test

$$H_0: \mu \geq 22.5$$
$$H_1: \mu < 22.5$$

At the level of significance $\alpha = 0.05$ we reject H_0 if

$$t = \frac{\bar{x} - 22.5}{s/\sqrt{15}} < -t_{0.05} = -1.761$$

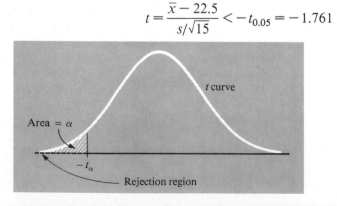

Figure 7 ▪ 8

where the degrees of freedom are $n - 1 = 14$, that is, df $= 14$. We calculate $t = (20.92 - 22.5)/(3.12/\sqrt{15}) = -1.96$, and so we reject H_0 in favor of H_1 at $\alpha = 0.05$. We conclude that the gas mileage is significantly less than 22.5 mi/gal. Note that our answer is different when we use a one-tailed test instead of a two-tailed test. This is because of the different rejection regions in the two tests. In the two-tailed test we split the 5% rejection region equally over the two tails, giving 2.5% on each tail. In this one-tailed directional hypothesis test we lumped all 5% on the same tail (see Fig. 7.9).

The *t*-Test for Small Samples; Summary of Tests

Note that in all our examples illustrating the *t*-test the sample size has been small, that is, less than 30. We use a small sample size in our examples because for larger samples the *t*-distribution is closely approximated by the normal. Hence we can utilize the normal for large sample sizes by using s^2 as though it were the true value of σ^2. For this reason it is often said that the *t*-statistic is used when you have small sample sizes, the normal is used when you have large sample sizes. But beware of the mistake found in some books that state you *always* use the *t*-statistic for small samples. This is not so! When σ is known you use the *z*-statistic even for small samples.

The following chart should help in testing hypotheses about the mean.

Assumption for 1a to 1c — population has normal distribution or the sample size *n* is large,		
a. $H_0: \mu = \mu_0$ $H_1: \mu \neq \mu_0$ **Two-tailed test**	**b.** $H_0: \mu \geq \mu_0$ $H_1: \mu < \mu_0$ **One-tailed test**	**c.** $H_0: \mu \leq \mu_0$ $H_1: \mu > \mu_0$ **One-tailed test**
1.a. σ^2 *known* *(example 7.1)* **Reject H_0 if** $z = \dfrac{\bar{x} - \mu_0}{\sigma/\sqrt{n}} > z_{\alpha/2}$ **or if** $z < -z_{\alpha/2}$	**1.b.** σ^2 *known* *(example 7.2)* **Reject H_0 if** $z = \dfrac{\bar{x} - \mu_0}{\sigma/\sqrt{n}} < -z_\alpha$	**1.c.** σ^2 *known* *(example 7.3)* **Reject H_0 if** $z = \dfrac{\bar{x} - \mu_0}{\sigma/\sqrt{n}} > z_\alpha$
2.a. σ^2 *unknown* *(example 7.4)* **Reject H_0 if** $t = \dfrac{\bar{x} - \mu_0}{s/\sqrt{n}} > t_{\alpha/2}$ **or** $t < -t_{\alpha/2}$ **with $n - 1$ degrees of freedom**	**2.b.** σ^2 *unknown* *(example 7.5)* **Reject H_0 if** $t = \dfrac{\bar{x} - \mu_0}{s/\sqrt{n}} < -t_\alpha$ **with $n - 1$ degrees of freedom**	**2.c.** σ^2 *unknown* **Reject H_0 if** $t = \dfrac{\bar{x} - \mu_0}{s/\sqrt{n}} > t_\alpha$ **with $n - 1$ degrees of freedom**
Assumption for 2a to 2c — population has a normal distribution.		

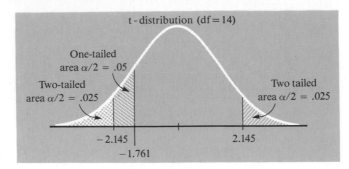

Figure 7 ▪ 9

7 ▪ 4   TESTING HYPOTHESES ABOUT THE POPULATION VARIANCE σ^2 OF A NORMAL DISTRIBUTION

In the quality control of manufactured items it is important to ascertain the value of the variance of the produced parts. For example, suppose nails are produced that are supposed to have an average length of 3 in. but half the nails produced are 1.5 in. long and half are 4.5 in. long. Such a situation is clearly unacceptable; the variation in the length of the nails is out of control even though the average is acceptable. As another example: For investors buying stock on margin, the variance of the price of the stock is important.

In trying to determine from the data whether or not the variation is acceptable, we would first establish the acceptable variance; let us say that for the manufacture of nails it is $\sigma^2 \le 0.25$ square inches (sq in.). We would then test the hypothesis

$$H_0: \sigma^2 \le 0.25$$

Since we are only concerned that the variation is too large, the alternative hypothesis is

$$H_1: \sigma^2 > 0.25.$$

The preceding is only meant to motivate the study of hypothesis testing concerning the variance. The illustration presents a one-tailed (directional test). However, both one-tailed and two-tailed (nondirectional) tests can be performed and in a manner similar to that used for testing hypotheses about the mean. The major difference is that we now use the test statistic $(n - 1)S^2/\sigma^2$. We observed in Chapter 6 that *when sampling from a normally distributed population,* the distribution of $(n - 1)S^2/\sigma^2$ is chi-square distributed with $n - 1$ degrees of freedom (n is the sample size) and the upper centile values are found in Appendix A.6. A level of significance α is chosen and, as before, the rejection region is determined from this and the centile values in Appendix A.6. The value of σ^2 to use in the test statistic is dictated by the null hypothesis.

Suppose we wish to test

$$H_0: \sigma^2 = \sigma_0^2$$

versus

$$H_1: \sigma^2 \neq \sigma_0^2$$

at the level of significance α. From Chapter 6 we know that $(n-1)S^2/\sigma_0^2$ has a chi-square distribution with $(n-1)$ degrees of freedom if H_0 is true. Thus we are to cut off the $\alpha/2$ area on both the right and left tails of the χ^2 distribution (which corresponds to rejecting H_0 for values of S^2 much larger or much smaller than the hypothesized σ_0^2) in order to find the rejection region (as shown in Figure 7.10). In summary, to test $H_0: \sigma^2 = \sigma_0^2$ the test statistic is $(n-1)S^2/\sigma_0^2$. We reject H_0 if the observed value of the test statistic is either greater than $\chi_{\alpha/2}^2$ or less than $\chi_{1-\alpha/2}^2$ as found in Appendix A.6. Let us give an example.

Example 7 ▪ 6   A teaching method that has been used by a certain school district yields a variance of 25 on a standardized test. A new method of teaching is used on a class of 31 students and the variance of their scores on the standardized test is found to be $s^2 = 13$. Is there a significant difference from the variance of the old method? We use $\alpha = 0.05$ as the level of significance, and assume the distribution of test scores follows a normal distribution. We test $H_0: \sigma^2 = 25$ versus $H_1: \sigma^2 \neq 25$. The number of degrees of freedom is df $= 31 - 1 = 30$. We reject H_0 if $(30)s^2/25 > \chi_{0.025}^2 = 46.98$ or if $(30)s^2/25 < \chi_{0.975}^2 = 16.79$ (see Fig. 7.11). Since $s^2 = 13$, the observed value of the test statistic is $(30)(13)/25 = 15.6$ which is less than 16.79. We may reject H_0 at $\alpha = 0.05$. We can thus state that the variance of the new method is significantly different from the old method at the level of significance $\alpha = 0.05$.

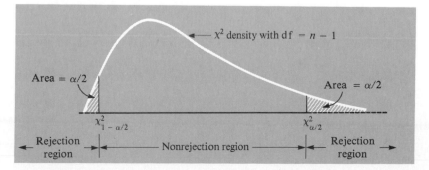

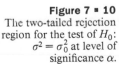

Figure 7 ▪ 10
The two-tailed rejection region for the test of H_0: $\sigma^2 = \sigma_0^2$ at level of significance α.

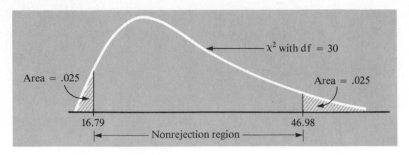

Figure 7 ▪ 11
The nonrejection and
rejection region.

One-tailed directional hypothesis tests may also be performed. To test the hypothesis $H_0: \sigma^2 \geq \sigma_0^2$ versus $H_1: \sigma^2 < \sigma_0^2$ at the level of significance α, we still use the test statistic $(n-1)S^2/\sigma_0^2$. As before, which tail of the density curve contains the rejection region is dictated by the inequality of the alternate hypothesis. If σ^2 is much less than σ_0^2, then we would expect $(n-1)S^2/\sigma_0^2$ to be relatively small, so we would reject H_0 if the observed value of the test statistic falls on the left tail; that is, if $(n-1)s^2/\sigma_0^2 < \chi_{1-\alpha}^2$. Note that since this is a one-tailed test, all α of the area goes onto the same one tail, and also the inequality defining the rejection region (namely, $<$) is the same as that defining H_1. (see Fig. 7.12).

Example 7 ▪ 7   A company wishes to purchase a new machine and has reduced the choices to two models. They both produce parts with the same mean length, and the variance in length of the parts produced by the first machine is known to be 18. The second machine costs more than the first so that the company will only buy the second machine if the variance of its parts is significantly less than the variance of the parts produced by the first machine. Assuming normally distributed lengths of parts, we test $H_0: \sigma^2 \geq 18$ versus $H_1: \sigma^2 < 18$ at the level of significance $\alpha = 0.01$. If we reject H_0, then there is at most a 1% chance that the sample variance observed could have come from a population with a population variance $\sigma^2 \geq 18$.

Figure 7 ▪ 12
The rejection region for
the one-tailed test H_0:
$\sigma^2 \geq \sigma_0^2$ with df $= 20$ and
$\alpha = 0.01$.

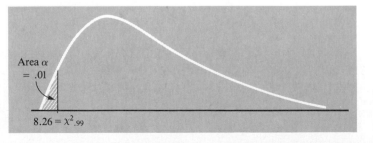

A sample of 21 parts from the second machine produces a sample variance $s^2 = 12$. At $\alpha = 0.01$ we reject H_0 if the test statistic $(n - 1)S^2/\sigma_0^2 = 20S^2/18$ has an observed value less than $\chi_{0.99}^2 = 8.26$, where $df = n - 1 = 20$ (see Fig. 7.12). Since $s^2 = 12$ we have the observed value of the test statistic $(21 - 1)12/18 = 13.33$, which is not in the rejection region, and we fail to reject H_0. Thus the sample variance is not sufficiently different from 18 to appear incompatible with H_0 (at $\alpha = 0.01$) and we would recommend buying the less expensive machine.

Often we wish to determine if the variance of a population is compatible with a given bound. In such cases we might wish to test $H_0: \sigma^2 \leq \sigma_0^2$ versus $H_1: \sigma^2 > \sigma_0^2$ at some given level of significance α. The logic we use now is the same as before, only the tail for the rejection region is changed to be consistent with the new H_1. The test statistic is still $(n - 1)S^2/\sigma_0^2$, and we now reject H_0 for an observed test statistic value that is too large, $(n - 1)s^2/\sigma_0^2 > \chi_\alpha^2$. The next sample is the prototype example of this sort.

Example 7 ▪ 8   A producer of chain links welds them together into chains. Since a chain is only as strong as its weakest link, the producer wishes to keep the variance of the breaking strength of the links below 30 square pounds (sq lb) of tension. A sample of 10 links yields the following breaking strengths: 1603, 1614, 1598, 1617, 1612, 1609, 1601, 1607, 1610, and 1599. Assuming the strengths are normally distributed, can we conclude that the variance is larger than 30? We use the level of significance $\alpha = 0.05$. The mean and variance of the sample are easily calculated.

$$\bar{x} = \frac{1603 + 1614 + \cdots + 1599}{10} = 1607$$

$$s^2 = \frac{(1603 - 1607)^2 + (1614 - 1607)^2 + \cdots + (1599 - 1607)^2}{9} = 42.7$$

Since we can conclude $\sigma^2 > 30$ is true only by rejecting the assertion $\sigma^2 \leq 30$ as being incompatible with the data, we test

$$H_0: \sigma^2 \leq 30$$

$$H_1: \sigma^2 > 30$$

At level of significance $\alpha = 0.05$, we reject H_0 if the observed value of the test statistic $(n - 1)S^2/\sigma_0^2 = 9S^2/30$ is greater than $\chi_\alpha^2 = \chi_{0.05}^2 = 16.919$, where $df = 9$. Since $s^2 = 42.7$, this observed ratio is $(9)(42.7)/30 = 12.81$ which is less than the critical value of 16.919, and so we fail to reject H_0. We cannot conclude that the strengths have a variance greater than 30.

We summarize the results of this section in the following chart.

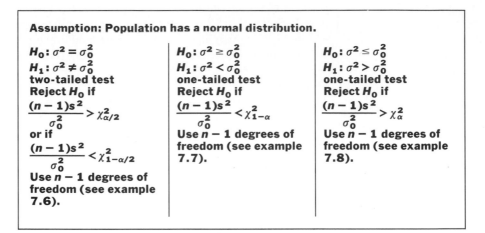

Assumption: Population has a normal distribution.

$H_0: \sigma^2 = \sigma_0^2$
$H_1: \sigma^2 \neq \sigma_0^2$
two-tailed test
Reject H_0 if
$$\frac{(n-1)s^2}{\sigma_0^2} > \chi_{\alpha/2}^2$$
or if
$$\frac{(n-1)s^2}{\sigma_0^2} < \chi_{1-\alpha/2}^2$$
Use $n-1$ degrees of freedom (see example 7.6).

$H_0: \sigma^2 \geq \sigma_0^2$
$H_1: \sigma^2 < \sigma_0^2$
one-tailed test
Reject H_0 if
$$\frac{(n-1)s^2}{\sigma_0^2} < \chi_{1-\alpha}^2$$
Use $n-1$ degrees of freedom (see example 7.7).

$H_0: \sigma^2 \leq \sigma_0^2$
$H_1: \sigma^2 > \sigma_0^2$
one-tailed test
Reject H_0 if
$$\frac{(n-1)s^2}{\sigma_0^2} > \chi_{\alpha}^2$$
Use $n-1$ degrees of freedom (see example 7.8).

7 ▪ 5 TEST OF HYPOTHESES CONCERNING PROPORTIONS

With the procedures used in the previous sections and when the sample size is large we are able to perform hypothesis tests about the proportion of a population that possesses a certain characteristic of interest. Indeed, if each member of the population is labeled with a one if it possesses the characteristic, and a zero otherwise, then the mean or expected value for the population is precisely p, the proportion of the population with the characteristic. If we use independent random sampling and if the sample size is large enough, then the central limit theorem holds and we can proceed as in Section 7.2. For example, suppose we wish to determine if a pair of dice is "loaded" in such a way that the sum 7 occurs a disproportionate number of times. If the dice are not loaded, then the probability of "getting a 7" on a throw of the dice is $\frac{6}{36} = \frac{1}{6}$. We let p denote the proportion of times 7 occurs, and we wish to test

$$H_0: p = \frac{1}{6}$$
$$H_1: p \neq \frac{1}{6}$$

We use the estimator $\hat{P}$, the proportion of times 7 occurs in n trials, to test the hypothesis H_0 versus H_1. We again assume independent random sampling. In Chapter 5 we learned that $\hat{P}$ is approximately normal with a mean of p and a variance of $p(1-p)/n$. Thus on standardizing we reject H_0 at the level of significance α if

$$z = \frac{\hat{p} - \frac{1}{6}}{\sqrt{(\frac{1}{6})(\frac{5}{6})/n}} > z_{\alpha/2} \qquad \text{or} \qquad \frac{\hat{p} - \frac{1}{6}}{\sqrt{(\frac{1}{6})(\frac{5}{6})/n}} < -z_{\alpha/2}$$

where $\hat{p}$ is the observed value of $\hat{P}$.

To perform the test, we roll the dice, calculate the value $\hat{p}$, place the result in the preceding expression, and reject H_0 if the observed value of the z ratio is either greater than $z_{\alpha/2}$ or less than $-z_{\alpha/2}$. Otherwise the data do not give a significant discrepancy from H_0, and we must fail to reject H_0.

Test of Proportions
When testing

$$H_0: p = p_0$$

$$H_1: p \neq p_0$$

at the level of significance α, we reject H_0 if

$$z = \frac{\hat{p} - p_0}{\sqrt{p_0(1 - p_0)/n}} > z_{\alpha/2}$$

or if

$$z = \frac{\hat{p} - p_0}{\sqrt{p_0(1 - p_0)/n}} < -z_{\alpha/2}$$

In a similar manner, we may perform one-tailed directional hypothesis tests about p as follows: To test

$$H_0: p \leq p_0$$

$$H_1: p > p_0$$

at the level of significance α, we reject H_0 if and only if

$$z = \frac{\hat{p} - p_0}{\sqrt{p_0(1 - p_0)/n}} > z_{\alpha}$$

To test

$$H_0: p \geq p_0$$
$$H_1: p < p_0$$

at the level of significance α, we reject H_0 if and only if

$$z = \frac{\hat{p} - p_0}{\sqrt{p_0(1 - p_0)/n}} < -z_\alpha$$

Example 7 ▪ 9   A pharmaceutical company claims that at most 1% of the users of a certain drug experience adverse reactions. To test its claim, an agency monitors 2500 patients taking the drug and observes adverse reactions in 64 cases. Can the agency accept the company's claim at a significance level of 0.05? We calculate* $25(1 - 2p)^2/p(1 - p) = 25(1 - 2(0.01))^2/(0.01)(0.99) = 2425.25$, so the sample size is large enough to use the normal approximation to test

$$H_0: p \leq 0.01$$
$$H_1: p > 0.01$$

at the level of significance $\alpha = 0.05$. We have $\hat{p} = 64/2500 = 0.0256$ so that

$$z = \frac{0.0256 - 0.01}{\sqrt{(0.01)(0.99)/2500}} = 7.8$$

This is much larger than the critical value for rejecting H_0, namely, $z > z_\alpha = z_{0.05} = 1.645$, so we reject H_0 and conclude there is significant evidence that $p > 0.01$.

An Improved Version of the Proportion Tests — The Continuity Correction (optional)

In Section 5.3 we discussed how to improve the approximation of the binomial by the normal distribution. This improvement involved a continuity correction. We now summarize the improved versions of those tests from those stated at the beginning of Section 7.5; these improved tests incorporate the continuity correction.

* See Section 5.4.

Tests about p by using correction factor:
Level of significance α.
Test: $H_0: p = p_0$ versus $H_1: p \neq p_0$
Reject H_0 if

$$z = \frac{\hat{p} - p_0 - 1/2n}{\sqrt{p_0(1 - p_0)/n}} > z_{\alpha/2}$$

or if

$$z = \frac{\hat{p} - p_0 + 1/2n}{\sqrt{p_0(1 - p_0)/n}} < -z_{\alpha/2}$$

Test: $H_0: p \geq p_0$ versus $H_1: p < p_0$
Reject H_0 if

$$z = \frac{\hat{p} - p_0 + 1/2n}{\sqrt{p_0(1 - p_0)/n}} < -z_{\alpha}$$

Test: $H_0: p \leq p_0$ versus $H_1: p > p_0$
Reject H_0 if

$$z = \frac{\hat{p} - p_0 - 1/2n}{\sqrt{p_0(1 - p_0)/n}} > z_{\alpha}$$

The eight-step procedure is used as outlined in Section 7.1 for any of the preceding tests.

Testing About the Number of Successes to Be Expected

In many important applications, the problem is phrased in terms of determining if the observed number of successes is greater or less than what would be expected from the theoretical binomial model. We can express our test statistic in these terms if we just take the formulas for testing about p using $\hat{p}$ (see the boxed formulas in Section 7.5), and multiply numerator and denominator by the sample size to obtain the test statistic

$$z = \frac{n\hat{p} - np_0}{\sqrt{np_0(1 - p_0)}}$$

$$= \frac{\text{observed number of successes} - \text{expected number of successes}}{\sqrt{np_0(1 - p_0)}}$$

Here np_0 is the expected number of successes in a binomial, $n\hat{p}$ is the observed number of successes in the actual data (since $\hat{p}$ = number of successes/n), and $\sqrt{np_0(1 - p_0)}$ is the standard deviation of the binomial. The test statistic z represents the number difference between the observed and expected, measured in number of standard deviations. Inference is performed exactly as in the previous sections by using the test statistic z and the normal centile values.

Application to the Law

In 1977, for the first time, the Supreme Court provided legal sanctity to the principles of hypothesis testing as discussed in this section. In *Castenda* vs. *Partida*** the U.S. Supreme Court considered the situation in a county in Texas wherein Spanish surnamed persons comprised 79.1% of the county's population. However, of 870 persons summoned to serve as grand jurors, only 39% were Spanish surnamed. The Supreme Court performed the test of hypothesis

$$H_0: p \geq 0.791$$

$$H_1: p < 0.791$$

(H_0 is the hypothesis that Spanish surnamed individuals are not discriminated against in being called to serve on grand juries where p is the proportion of Spanish surnamed individuals who should be summoned) and used the test statistic

$$z = \frac{n\hat{p} - np_0}{\sqrt{np_0(1 - p_0)}}$$

The Court found

$$z = \frac{(870)(0.39) - (870)(0.791)}{\sqrt{870(0.791)(1 - 0.791)}}$$

$$= -29$$

* 97 Supreme Court 1272 (1977).

which is significant at extremely small α, so small that the probability of such a large difference occurring by chance alone (i.e., the chance of a type 1 error) is about 1 divided by 10^{140}; this means that the probability of such a large difference is virtually zero. The Supreme Court relates this test as follows:

> If the jurors were drawn randomly from the general population, then the number of Mexican-Americans in the sample could be modeled by a binomial distribution. See Finklestein, The Application of Statistical Decision Theory to the Jury Discrimination Cases, 80 Harv. L. Rev. 338, 353–356 (1966). . . . Given that 79.1% of the population is Mexican-American, the expected number of Mexican-Americans among the 870 persons summoned to serve as grand jurors over the 11-year period is approximately 688. The observed number is 339. Of course in any given drawing some fluctuation from the expected number is predicted. The important point, however, is that the statistical model shows that the results of a random drawing are likely to fall in the vicinity of the expected value. . . . The measure of the predicted fluctuations from the expected value is the standard deviation, defined for the binomial distribution as the square root of the product of the total number in the sample (here 870) times the probability of selecting a Mexican-American (0.791) times the probability of selecting a non-Mexican-American (0.209). . . . *As a general rule for such large samples, if the difference between the expected value and the observed number is greater than two or three standard deviations, then the hypothesis that the jury drawing was random would be suspect to a social scientist.* (Emphasis supplied.)

After performing the test, the Court concluded:

> The 11-year data here reflected a difference between expected and observed number of Mexican-Americans of approximately *29 standard deviations*. A detailed calculation reveals that the likelihood that such a substantial departure from the expected value would occur by chance is less than 1 in 10^{140}.

Exact Test for *p* When *n* Is Small

One of the issues in a lawsuit* was the allegation that the defendant company discriminated against blacks in involuntary terminations. Blacks constituted 20% of the work force in question. Of 11 terminations, 8 were blacks. We wish to determine at the $\alpha = 0.05$ level if "too many" blacks were terminated. We assume that the terminations were mutually independent. If the terminations were made in a "color-blind" fashion, and if we regard each termination as a trial with outcomes "nonblack," "black," then the probability of black on any trial is $p = 0.2$, the proportion of blacks in the work force. We test the hypothesis

$$H_0: p \le 0.2 \quad \text{(no discrimination)}$$
$$H_1: p > 0.2$$

* *Booth* vs *National American Bank,* No. 78–92, USDC, ED LA.

a formulation that is favorable to the defendant since we have established a small type 1 error.

These assumptions enable us to use the "coin-tossing model." The sample size $n = 11$ is too small to use the normal approximation to the binomial. We must use the *exact* test. In this case we have a one-tailed test and use the upper tail. If $H_0: p = 0.2$ is true, then the probability of 11 blacks being terminated is 0 (to three places) (see Appendix A.8). The probability of 10, 9, and 8 each is 0 (again, to three places). The probability of 8, 9, 10, or 11 is the sum of the individual probabilities that is 0. From Appendix A.8 we see that (under the assumption that H_0 is true) the probability of 6 or more (6, 7, 8, 9, 10, or 11) is $0.010 + 0.002 + 0 = 0.012$. However, the probability of 5 or more is $0.039 + 0.012 = 0.051$. If we chose 5 or more as our critical region, then α would equal 0.051, which is larger than 0.05. To be conservative we want α to equal or be less than 0.05. Hence we choose our critical region to be 6 or more. Since 8 blacks were terminated, we reject H_0.

The judge in the lawsuit cited in the footnote, was skeptical of this type of analysis. He claimed the sample size was too small to come to a conclusion concerning the defendant's actions. Many people have a similar kind of fallacious intuition. It should be noted that small sample sizes favor H_0 (in this case the defendant) because the type 2 error (the chance of erroneously accepting H_0) is large, whereas the type 1 error is the same regardless of the sample size.

7 ▪ 6   WHAT CAN GO WRONG?

The Assumptions Underlying a Test May Not Be Valid

Blind use of statistical tests can lead to incorrect conclusions and decisions. The most common oversight in employing a test is failure to check the validity of those assumptions that underlie the test. For example, a test may require that the underlying population has a normal distribution. The investigator must be reasonably sure that the population has this distribution or one closely approximate to it before using the test.

As another example, investigators many times collect data without obtaining information on how the data were collected. If the data were collected using other than independent random sampling or a close approximation to it, and if the investigator used some of the tests described in this chapter, then the results *may* not be valid.* As another possibility, if the collected data were full of errors, then the results of the test might very well be misleading.

---

* As we shall see in studying sampling, random sampling is often done without replacement and it does not constitute independent random sampling. However, if the population is very large, then, even with sampling without replacement, it would be unlikely to obtain the same observation twice. In this case, sampling without replacement would "approximate" sampling with replacement.

A Statistically Significant Result May Not Be Significant or Important in a Nonstatistical Sense

Not only can blind use of statistical tests lead to incorrect decisions but so can the blind interpretation of such tests. Let us look at an example derived from an actual situation. A private hospital was accused by the Department of Health, Education, and Welfare of discriminating against racial minorities in admissions to the hospital. The hospital in question showed that only a certain part of the population is "available" to the hospital in the sense that they can financially afford the hospital. Of this available population, 12% are minorities. Thus the proportion of admissions that should consist of the racial minorities should be 0.12 *if admissions are random with respect to race;* that is, if each ill person is as likely to be admitted as any other ill person no matter the race of each person. Hence the proportion p of racial minorities in all admissions should be 0.12. We test

$$H_0: p \geq 0.12$$

$$H_1: p < 0.12$$

We compare $p_0 = 0.12$ to the observed value $\hat{p}$ by calculating z.

$$z = \frac{\hat{p} - p_0}{\sqrt{p_0(1 - p_0)/n}}$$

There were 10,000 admissions, so $n = 10{,}000$. The observed value $\hat{p} = 0.113$. Note that the difference between the *theoretical* value $p_0 = 0.12$ and the observed value $\hat{p} = 0.113$ is only 0.007 or 0.7%, yet the calculated value of z is -2.16 which indicates that the difference between p_0 and $\hat{p}$ is significant at $\alpha = 0.01$; hence we might affirm (as did a statistician in testimony on this matter) that the very serious charge against the hospital was supported by the statistical analysis *even though there is such a small difference between the theoretical and observed proportions of racial minorities.* The conclusion does not make sense—What has gone wrong?

To understand what has gone wrong, we observe that the object of a test is to detect a *difference* (in the statistical sense) between the theoretical and observed quantities *no matter how small and unimportant* that difference is. If the difference is very small, then it will probably not be detected using a small sample size; however, if a very large sample is employed, then a very small difference can be detected. Since there is usually *some* difference between a theoretical and observed quality, one can *usually obtain* a statistically significant result by increasing the sample size arbitrarily. Said another way, one can always detect trivial differences by using sufficiently large samples.

In short, a test is meant to detect differences; one may obtain a significant difference because a large sample is used, not because the difference is important.

7 ▪ 7   THE POISSON DISTRIBUTION FOR RARE EVENTS TESTS OF VERY SMALL PROBABILITIES (OPTIONAL)*

When n is small we can use the fact that $\hat{P} = Y/n$, where the distribution of Y is binomial to make confidence interval statements, hypothesis tests, and other inferences about the value of p. In the case where n is large but p is so small that n does not satisfy the rules of Section 5.4, then the normal distribution approximation does not yield acceptable results. Fortunately, the Poisson distribution (see Chaper 4) *does* yield an acceptable approximation to the distribution of Y. Since the probability p is very small, we often refer to the Poisson distribution as the *law of rare events*. We now review the Poisson distribution emphasizing its characteristics that are important here.

A random variable Y that counts the number of occurrences of a certain type of event is called a Poisson random variable with expected number of occurrences equal to λ if the probability of k occurrences is $P(Y = k) = e^{-\lambda}\lambda^k/k!$ for $k = 0, 1, 2, \ldots$. The symbol e represents a constant (approximately 2.718). If p is the proportion of the population of a certain type and if we sample n times from the population, then the expected number of occurrences of this type in the sample is $\lambda = np$. If n is very large and p is very small, then $n\hat{P} = Y$ is approximately Poisson distributed with $\lambda = np$.

The value of $P(Y = k)$ for several different choices of λ can be obtained with the aid of Appendix A.4.

Example 7 ▪ 10   An insurance company has determined that 0.1% of the homeowners of a particular city filed claims for the total loss of their home due to flooding during the past year. If the company holds 4000 policies for this city, what is the probability that they must pay off six claims in the next year? Here $p = 0.001$ so $np = (4000)(0.001) = 4$. Let Y be the number of policies paid off. Then $P(Y = 6) = 0.1042$.

Let us see how to apply the Poisson distribution to obtain hypothesis tests

---

* The reader should study Section 4.14 before reading this section.

about p when p is small and n is large. Suppose we wish to test

$$H_0: p = p_0$$
$$H_1: p \neq p_0$$

If H_0 is true, then $Y =$ the number of occurrences, has approximately a Poisson distribution, with $\lambda = np_0$. At the level of significance α we reject H_0 if $P(Y \geq n\hat{p}) < \alpha/2$ or if $P(Y \leq n\hat{p}) < \alpha/2$, where the preceding probabilities are found with the help of Appendix A.4. To implement the test, we first calculate $n\hat{p}$, the observed number of occurrences, and $\lambda = np_0$, the expected number of occurrences. We sum the probabilities greater than or equal to $n\hat{p}$ and sum the probabilities less than or equal to $n\hat{p}$. If either of these sums is smaller than $\alpha/2$, we reject H_0.

Example 7 ■ 11   A certain disease is known to occur in 0.057% of the population. A vaccine against this disease is developed and given to 5263 individuals. Of those inoculated, one contracted the disease. Can we conclude that being inoculated changes the proportion of people who will contract the disease? At $\alpha = 0.1$ we test

$$H_0: p = 0.00057$$
$$H_1: p \neq 0.00057$$

If H_0 is true, the number in the sample who contract the disease is approximately Poisson where $\lambda = (5263)(0.00057) = 3$. Since we observed 1 in our sample, we calculate $P(Y \leq 1) = P(Y = 0) + P(Y = 1) = 0.0498 + 0.1495 = 0.1993$. Since $\alpha = 0.1$, we have $0.1993 > \alpha/2 = 0.05$, hence we fail to reject H_0. We cannot conclude the vaccine changes the proportion of people who will contract the disease. Note that in practice we need only calculate the probabilities on one of the two tails, namely, the tail on which the observed value was located.

Using similar techniques to those just outlined, we may perform one-tailed directional hypothesis tests about very small p. To test

$$H_0: p \geq p_0$$
$$H_1: p < p_0$$

at level of significance α, we reject H_0 if $P(Y \leq n\hat{p}) < \alpha$, where Y has a Poisson distribution with $\lambda = np_0$. To test

$$H_0: p \leq p_0$$
$$H_1: p > p_0$$

at level of significance α, we reject H_0 if $P(Y \geq n\hat{p}) < \alpha$. Notice again (as in our previous tests) that the inequality that must be checked for performing the hypothesis test is the same inequality sign as that given in the alternative hypothesis.

Example 7 ■ 12 A sociologist wishes to investigate the effect of high density housing on mental health. Of 120 individuals who are randomly selected from a high density area of a large city, 3 of them are found to exhibit a certain type of psychotic behavior. If the proportion of this type of psychosis in the general population is 0.5%, can one conclude that at the 5% level of significance high density living increases the proportion of psychotics? We test

$$H_0: p \leq 0.005$$

$$H_1: p > 0.005$$

at level of significance $\alpha = 0.05$. Here $\lambda = (120)(0.005) = 0.6$. If Y is Poisson distributed, then we reject H_0 if $P(Y \geq 3) < 0.05$. We calculate $P(Y \geq 3) = P(Y = 3) + P(Y = 4) + P(Y = 5) + P(Y = 6) + \cdots + = 0.0198 + 0.0030 + 0.0004 + 0.0000 = 0.0232 < 0.05$, so we reject H_0 and conclude that high density living does increase this type of psychosis.

Summarizing, we have

Test of hypothesis about p if n is very large and p is very small. Level of significance α.

To test $H_0: p = p_0$ versus $H_1: p \neq p_0$
Let Y be Poisson with $\lambda = np_0$.
Reject H_0 if $P(Y \geq n\hat{p}) < \alpha/2$ or   if $P(Y \leq n\hat{p}) < \alpha/2$.

To test $H_0: p \geq p_0$ versus $H_1: p < p_0$
Let Y be Poisson with $\lambda = np_0$.
Reject H_0 if $P(Y \leq n\hat{p}) < \alpha$.

To test $H_0: p \leq p_0$ versus $H_1: p > p_0$.
Let Y be Poisson with $\lambda = np_0$.
Reject H_0 if $P(Y \geq n\hat{p}) < \alpha$.

Summary

In many investigations and studies the investigator is called on to make a decision about a parameter of a population. When this decision is based on a sample it is sometimes possible to use the methods of statistical hypothesis testing to aid in making the decision. In this chapter we have studied several tests of hypotheses. The first few tests concerned hypotheses about the mean of the population.

If H_0 represents the hypothesized value of the true mean, we learned to test several hypotheses about the mean *when σ^2 is known.* These were

a. $H_0: \mu = \mu_0$

   $H_1: \mu \neq \mu_0$

where H_0 is called the null hypothesis and H_1 is called the alternative hypothesis; μ_0 is the value of μ we are testing.

b. $H_0: \mu \geq \mu_0$

   $H_1: \mu < \mu_0$

c. $H_0: \mu \leq \mu_0$

   $H_1: \mu > \mu_0$

Since the decision to reject or not to reject H_0 must be based on the sample, a statistic (the value that depends on the sample) must be used to test H_0 versus H_1. In the cases (a) to (c) the statistic chosen was $z = (\bar{x} - \mu_0)/\sigma/\sqrt{n}$. We learned that if $\bar{x}$ is "near" the value μ_0 (i.e., z near to 0), we do not reject H_0; if $\bar{x}$ is "far" from the value μ_0 (i.e., z far from 0), we reject H_0. But what do near and far mean? To answer this question, we first note that z may take on a value far from 0 even when H_0 is true; that is, because of the random nature of the sample mean there is a chance that the observed value $\bar{x}$ will be far from μ_0 when H_0 is true. If $\bar{x}$ is far enough away from μ_0 (or z from 0), we may decide to reject H_0. The *set of values* of z for which we would reject H_0 is called the *critical region* or *rejection region.* If the observed value of z is one of the values in the critical region, we reject H_0. However, we emphasize again that because of the random nature of the sample mean, there is a chance that z will fall in the critical region even *when H_0 is true.* If z falls in the critical region when H_0 is true, we will incorrectly reject H_0; this error is called the type 1 error. We choose the critical region so that the chance of making a type 1 error is a fixed value α. If z happens by chance to fall in the "fail to reject" region when in fact H_1 is true, we have a type 2 error.

In the case of test (a), the critical region is (assuming independent random sampling)

$$z < -z_{\alpha/2} \quad \text{or} \quad z > z_{\alpha/2}$$

where $z_{\alpha/2}$ is the value such that

$$P(Z \geq z_{\alpha/2}) = \frac{\alpha}{2}$$

and Z is the random variable having the standard normal distribution. Since both regions at either end of the distribution are part of the critical region, the test is called a two-tailed test.

The critical region of test (b) alone is $z < -z_{\alpha}$, whereas that of test (c) is $z > z_{\alpha}$. These are called one-tailed tests.

The probability α is called the *significance level* of the tests.

Hypothesis testing concerning the mean was discussed when the variance σ^2 is unknown. In this case we used

$$t = \frac{\bar{x} - \mu_0}{s/\sqrt{n}}$$

where s is the sample standard deviation. The critical region of the test

$$H_0: \mu = \mu_0$$
$$H_1: \mu \neq \mu_0$$

is

$$t < -t_{\alpha/2} \quad \text{or} \quad t > t_{\alpha/2}$$

where $t_{\alpha/2}$ is the upper $(\alpha/2)$ percentile of the t-distribution with $n - 1$ degrees of freedom and n is the sample size.

The critical regions associated with the alternatives $H_1: \mu < \mu_0$ and $H_1: \mu > \mu_0$ are $t < t_{\alpha}$ and $t > t_{\alpha}$, respectively.

Another important group of tests is that associated with testing the hypothesis that the variance of a population is a given value σ_0^2. These tests are summarized in the table at the end of Section 7.4.

For all tests discussed in this chapter we assume that independent random sampling is used.

Tests concerning proportions and "rare events" are examined in Sections 7.5 and 7.7, respectively.

Of importance is the matter of how to choose the null hypothesis H_0. This hypothesis is selected as the one that is more serious to reject when true than to accept when false. We select H_0 as the one for which we wish to have a small type 1 error. If we cannot decide which hypothesis should have a small type 1 error, then we select as H_0 the hypothesis that represents the status quo, or the hypothesis of "no effect."

Key Words

Define each of the following terms:

independent random
    sampling
type 1 error
type 2 error
test statistic

null hypothesis
critical region
one-tail test
two-tail test
test of proportions

References*

North, D., and Miller, R.: *Economics of Public Issues.* New York, Harper and Row, 1980. Chapter 10 presents an excellent discussion of the type 1 and 2 errors and their pertinence to health and economics.

Tanur, J. M., et al.: *Statistics: Guide to the Unknown.* San Francisco, Holden-Day, 1978. This text includes 10 papers on applications of the subject matter discussed in this chapter. The applications include application to law, cloud seeding, drug screening, economics, health, and more.

Exercises

Section 7.1

1. Suppose a 195-lb man wants to purchase rope for mountain climbing. He would like to know if the mean breaking strength μ of the rope is greater than 200 lb. He wishes to test $H_0: \mu \le 200$ versus $H_1: \mu > 200$.
   a. What are the possible consequences of his making a type 1 error? a type 2 error?
   b. Which is the more critical error?
   c. What would you personally consider to be an acceptable size for α, the type 1 error if you were the mountain climber?

2. Consider a simple experiment in which there are only four possible outcomes (say, the flipping of two coins, or the number of successes in three repeated trials of some quality control problem). The outcomes are labeled 0, 1, 2, 3, and under H_0 their probabilities are $\frac{1}{8}, \frac{1}{8}, \frac{1}{2}, \frac{1}{4}$, respectively; under H_1 their probabilities are $\frac{3}{8}, \frac{1}{8}, \frac{3}{8}, \frac{1}{8}$. Let A represent the two outcomes $\{0, 1\}$, B the outcomes $\{1, 2\}$.
   a. Consider a test statistic that rejects H_0 if the outcome is in the set A. Find α, the

* See also the sources noted in problems 26, 31, 34, 35, and 45.

probability of the type 1 error, and β, the probability of the type 2 error for this test statistic.

b. Repeat (a) for the test statistic that rejects H_0 if the outcome is in the set B.

c. If you had to choose which of H_0 or H_1 was true using the rejection region A or B, which rejection region would you prefer? Why?

d. Use your answer to (c) to explain why just stating that you want a type 1 error of a certain amount is not generally sufficient to delineate what is a "good" rejection region. (*Note:* In general, we try to select the rejection region of a given size α that has β as small as possible.)

3. Explain why people say that you can gain information in a hypothesis test only when you reject H_0. (In this sense, hypothesis testing is a probabilistic version of "proof by contradiction" which you have seen in mathematics. Assume H_0 is true, and if the data lead you to a contradiction or a very improbable statement, then H_0 must have been false.)

4. State the null hypothesis (H_0) and the alternative hypothesis in each of the following situations.

a. You want to test whether the mean reading comprehension score (as measured by the SAT test) of high school seniors changed between 5 years ago and now. Take the mean 5 years ago as known.

b. You want to test whether women have significantly lower salaries in a particular job than do men. (Take the men's salary as known.)

c. The average trout caught in a particular river is 21.2 in long. A group of expert fishermen catch 20 trout that average 27.6 in. Can you conclude that the experts catch longer fish than does the general public?

d. The average length of stay in a particular

hotel before renovation of the building was 2 days. Test if the renovation has significantly increased the length stayed.

e. You wish to test whether the average age of night students is the same as that of day students at a particular university. Day students are known to have an average age of 20.2 years.

f. The average age of burglary suspects in Austin is 17.2 years. Is the average age significantly higher in New Orleans?

g. A certain drug is known to have side effects in 10% of the cases. A new drug is tested. Can we conclude that the mean percentage p of people with side effects to the new drug is significantly less than 10%?

5. A drug manufacturer is considering advertising a drug as being 90% effective in curing a certain ailment but is uncertain as to the true effectiveness. To perform a hypothesis test to see if the effectiveness is at least 90%, the manufacturer must keep in mind the false advertising laws. Answer (a) and (b) of problem 1 for this situation, taking the position of the manufacturer.

6. A federal agency wishes to test whether the drug company in problem 5 should be sued for false advertising. Such prosecution is expensive; however, they do not wish to let too many false advertisers get away with it. Perform (a) and (b) of problem 1, taking the agency's point of view.

Section 7.2

7. Using the following data, test each given hypothesis.

n	$\bar{x}$	σ^2	H_0	H_1	α
16	11	9	$\mu = 10$	$\mu \neq 10$	0.01
36	12	9	$\mu \geq 13$	$\mu < 13$	0.05
144	126.2	576	$\mu \leq 125$	$\mu > 125$	0.02
100	23	25	$\mu = 24$	$\mu \neq 24$	0.05

8. A rope manufacturer makes ropes whose breaking strengths are known to be normally distributed with a variance in strength of $\sigma^2 = 144$. A sample of nine ropes yields the following breaking strengths (in pounds): 193.43, 180.76, 185.81, 192.54, 181.01, 181.03, 185.12, 173.66, 215.08. Using these data, test the null hypothesis that the mean breaking strength is at least 200 lb. Use the eight-step procedure given in Section 7.2. Use $\alpha = 0.05$ and then $\alpha = 0.01$.

9. a. A dog food manufacturer produces a puppy chow that is supposed to cause a rapid increase in weight. Puppies fed this diet increase an average of 5 oz per week with a standard deviation of $\sigma = 2$ oz on this diet. A researcher employed by this company claims to have a slightly more expensive formula that will cause a 7-oz per week weight gain while maintaining the same variance of $\sigma^2 = 4$. To check the claim, the company feeds 11 randomly selected puppies the new feed and records the following weight gains (in ounces): 2.05, 3.30, 5.42, 7.53, 3.86, 7.49, 6.84, 7.40, 4.82, 2.60, 5.56. Test if the company should switch over to the new feed (i.e., is μ significantly larger than 5?). Use $\alpha = 0.05$ and $\alpha = 0.01$.

   b. (Optional) For each α in part a, calculate β assuming H_1: $\mu = 7$ oz/wk. Repeat the calculation of β for each of the following values for the alternative value of μ (in ounces): 6, 7.5, 8, 8.5, 9. Draw a curve of β versus the alternative values of μ. What information does this curve contain?

10. A certain street in New Orleans is paved to withhold a maximum 6.5 ton load. Twelve trucks using this street were stopped and weighed, yielding $\bar{x} = 6.7$ tons. If the standard deviation of trucks using this route is known to be 200 lb (0.1 ton), can we conclude that the average weight of all trucks using this street is larger than 6.5 tons? Test

H_0: $\mu = 6.5$ versus H_1: $\mu \geq 6.5$ tons at $\alpha = 0.05$ and $\alpha = 0.01$.

11. A psychologist developed an intensive course that was supposed to increase a student's score on the Stanford Binet IQ test. The average on this test is normally distributed with a mean equal to 100 and a standard deviation of 10. After giving the course to 36 randomly selected individuals, the researchers found that $\bar{x} = 112$. Test H_0: $\mu = 100$ against H_1: $\mu \neq 100$ at $\alpha = 0.05$ and $\alpha = 0.01$. Would you conclude that the mean IQ has changed since the course?

12. A bleach bottler sells bottles marked "contains 12 oz." The machine that fills the bottles is adjusted so that the standard deviation in the number of ounces dispensed is equal to 0.8 oz. A random sample of 16 bottles yielded an average measured volume of $\bar{x} = 11.6$ oz. Test H_0: $\mu = 12$ against the alternative H_1: $\mu \neq 12$. Use $\alpha = 0.05$ and $\alpha = 0.01$.

13. A sociologist wished to investigate whether conformity needs changed from generation to generation, so a test was designed that measures an individual's nonconformity rating. The larger the score on the test, the more nonconforming the individual. In 1956 the average score was 153. To see if a change occurred, the test was given to 25 randomly selected individuals in 1976 and the mean score was found to be $\bar{x} = 167$. Assuming $\sigma = 10$, test H_0: $\mu = 153$ versus H_1: $\mu \neq 153$ at $\alpha = 0.05$ and at $\alpha = 0.01$. State the experimental conclusions.

14. A paint manufacturer has a research and development department that is involved in product development for a new improved latex paint. The former version of this paint covered 1400 sq ft of wallboard per gallon with a standard deviation of 70 sq ft. The potential new formulation is tried with 16 gal and it is found that on the average the new paint covers 1425 sq ft per gallon. Perform a

statistical test to determine if the new formulation is different in coverage from the old formulation (use $\alpha = 0.05$).

15. *Statistical Abstracts of the United States, 1971*, states that the average marriage in the United States lasts 84 months. One particular town in Texas took a sample of 100 families and found the average marriage lasted 104 months. Assume that the standard deviation found in the town is the same as that of the whole United States and is $\sigma = 120$ months. Can this town be justified in concluding that their rural life-style increases the stability of marriages (use $\alpha = 0.01$)?

16. A researcher in pharmacokinetics and drug formulation wishes to design a tablet to carry a certain drug. A very slow dissolving rate is desired for uniform dispersal of the drug. The old formulation of the tablet had a 90% dissolution time of 15.7 min with a standard deviation of 3 min. A test of nine of the new formulation tablets had a 90% dissolution time of 17.2 min. Assume $\sigma = 3$ min is known, and test to see if the new formulation takes significantly longer to dissolve than did the old formulation (use $\alpha = 0.01$).

17. The response time of emergency personnel (fire, police, emergency medical service, and so on) is a critical factor in the decision on where to locate stations and also of which areas can easily be annexed by a city when expanding. The city claims that a potential annexing area is within an average 3-min response time. They will admit to a standard deviation of 1.5 min on their estimate. A neighborhood association fighting annexation does a test of response time on four police calls and finds it takes $\bar{x} = 4.1$ min. Can they claim their statistics contradict the city's claim (use $\alpha = 0.05$)?

18. Are humans taller now than in the past? Better nutrition, better information on vitamins, and better medical care might be reasons to assume that the answer is yes. In the past the average adult man stood about 5 ft 10 in. (70 in.). A sample of 25 men found that $\bar{x} = 6$ ft ($\bar{x} = 72$ in). Assume $\sigma = 1.5$ in. Are men significantly taller now? (Use $\alpha = 0.05$)

19. An elderly professor uses the same tests year in and year out. There is an attempt, however, to keep close control on the tests, and all of them are collected after each exam period. The data on all first exams over many years were collected and the mean score μ of the population was found to be 75, with $\sigma = 12$. This year the class of 36 students have an average of $\bar{x} = 80$ points. Should it be concluded that the exam leaked out (and should the professor go to all the trouble of making out new tests for the rest of the course?). Use $\alpha = 0.05$.

Section 7.3

20. Using the following data, test each given hypothesis.

n	$\bar{x}$	s^2	H_0	H_1	α
9	106	324	$\mu = 100$	$\mu \neq 100$	0.02
25	242	225	$\mu \geq 250$	$\mu < 250$	0.01
31	182	640	$\mu \leq 175$	$\mu > 175$	0.05
15	113	135	$\mu = 120$	$\mu \neq 120$	0.05

21. Test the hypothesis in problem 8 without assuming σ^2 known.

22. Without the unrealistic assumption that σ^2 is known in problem 9a, test the hypotheses given in problem 9a.

23. Test the hypothesis in problem 10 assuming $s = 276$ lb and σ is unknown.

24. Perform the analysis on problem 11 assuming $s^2 = 121$ and σ^2 is unknown.

25. Repeat problem 12 assuming $s^2 = 0.16$ and σ^2 is unknown.

26. In a study of dentifrices (toothpastes) published by Frankl and Alman, a group of 208 children were treated with a dentifrice over a

3-year period. They found this group had 19.98 new cavities per child over this period with a standard deviation of $s = 10.61$. Assuming that $\mu = 22.39$ is the "usual" cavity rate, test the hypothesis that the treatment significantly decreased the number of cavities (use $\alpha = 0.01$). (S. N. Frankl and J. E. Alman: Report of a three year clinical trial comparing a toothpaste containing sodium monofluorophosphate with two marked products. *Journal of Oral Therapeutics and Pharmacology, 4*:443–449, 1968.)

27. An aluminum extruding machine when set to the proper temperature and using the correct alloying formula averages 20 min/1000 ft. A sample of nine 1000-ft lengths yielded the following times: 18.7, 17.3, 19.8, 20.1, 27.2, 21.7, 18.1, 23.2, 24.1. Test the hypothesis that the settings are all correct (use $\alpha = 0.1$).

28. Production planning for a certain manufactured product unit necessitates first determining what the future demand for the product will be in the forthcoming year. From a sample of 25 clients it was found that the average demand would be 43.7 units, with an observed standard deviation of 10.3 units. Test the hypothesis that the average demand per customer next year will be at least 47 units (use $\alpha = 0.05$).

29. The new fad diet going around these days is the Dr. Slimright ice cream diet. According to Dr. Slimright's book, the patient is allowed to eat all the vanilla ice cream desired, but must avoid meats and vegetables (and of course nuts and cherries). He makes a plausible argument for the diet in his book and is very interesting on the radio and TV talk shows. To test his diet, a sample of 50 people is divided into two groups of 25 each. Each person is paired with another from the other group according to age, sex, and weight. One group goes on the Dr. Slimright diet, and the other group goes on a Weight Watchers diet.

At the end of 1 week, the individuals are weighed and the difference in weight for each person is determined. The diets are compared by subtracting the Weight Watchers dieters' gain or loss within each pair from the Slimright dieters' gain or loss. Using the 25 numbers so obtained, it is found that $\bar{x} = -3.4$ and $s = 1.0$. Test the hypothesis that the two diets are the same ($\mu = 0$) at $\alpha = 0.01$.

30. Return to problem 18 and assume that the variance σ^2 is unknown; however, the 25 men measured have $s = 1.75$ in. Test the hypothesis in problem 18 now.

31. In October 1972 the U.S. Congress passed a law known as the Bennett Amendment to the Social Security bill. This mandated a cessation of government-funded medical payments (Medicare, and so on) if the patient was not released from the hospital after a standard time (specified for each diagnosis). Physicians can appeal these standards. Since the Bennett Amendment is intended to improve medical care, it allows extensions of time of stay for exceptional cases while simultaneously reducing time and costs associated with other more normal cases.

This has been a controversial amendment, however, and the Reagan administration was considering plans to scrap it. In a paper by Churchhill, Cooper, and Govindarajan, a statistical analysis of the effect of the Bennett Amendment is given. Comparing an Ottawa hospital and a comparable Boston hospital gave the following results for acute appendicitis: $n = 130$ patients treated in Boston, April 1975–March 1977, mean length of stay 4.733 days, standard deviation 1.748 days. For Ottawa (which is not covered by the law) the mean stay for the same diagnosis during the same period was 5.691 days. Test if the law significantly reduced the length of stay by testing if the mean for the Boston hospital is at least 5.691 days. (Use $\alpha = .001$)

(N. C. Churchhill, W. W. Cooper, and V. Govindarajan: Effects of audits on the behavior of medical professionals under the Bennett Amendment. *Auditing—A Journal of Practice and Theory,* 1(2):Winter, 1982, pp. 69–90.)

32. A Coulter counter is a device for recording the red blood count in a laboratory analysis. A particular patient has readings (in million cells per cubic millimeter) of 4.3, 4.7, 4.6, 4.8, 4.5, 4.8, 4.9, 4.9, and 5.0. Test the hypothesis that his count is actually 4.9 or higher. Use $\alpha = 0.1$.

33. McWendy's is test marketing a new low-cost dessert for possible replacement of apple turnovers in the standard menu of its 500 hamburger stands. Five typical stands in a large metroplitan area have temporarily incorporated the dessert into their menu, eliminating apple turnovers. McWendy's will decide to make the change only if the daily demand in the five stands for the new dessert exceeds their historical mean of 1000 daily requests for apple turnovers. The test ran for 21 days and the results of the test gave a sample mean of 1020 with a sample standard deviation of 30. If McWendy's requires a 0.01 significance level, should they continue to offer apple turnovers on their menus, or should they replace them with the new dessert?

34. There are a lot of factors that predispose a bee to sting. For example, a person wearing dark clothes is more likely to be stung than a person in light clothes, and someone moving quickly and jerkily runs a higher risk than a slow and smoothly moving individual. A factor that is particularly important to beekeepers is to determine if bees more readily sting an object that has already been stung by another bee.

In an article by J. B. Free, the experimenter dangled eight cotton balls wrapped in muslin up and down in front of the entrance to a beehive. Four of the balls had just been exposed to a swarm of angry bees and were full of stingers. After a length of time, the balls were withdrawn and the number of new stingers in the balls was counted. The following data were obtained:

Trial	x Stings already present	y Fresh cotton balls	d = x − y
1	27	33	−6
2	9	9	0
3	33	21	12
4	33	15	18
5	4	6	−2
6	21	16	5
7	20	19	1
8	33	15	18
9	70	10	60
Total	250	144	106

$\Sigma d_i^2 = 4458.$

If bees do not have a predisposition to sting, then μ_D, the expected difference, should be zero. Test this hypothesis at $\alpha = 0.05$. (What is the alternative hypothesis?) (Free, J. B.: The stimuli releasing the stinging responses of honeybees. *Animal Behavior,* 9:193–196, 1961.)

35. To hunt flying insects, bats emit high frequency sounds and then listen for their echoes. Until an insect is located, these pulses are emitted at intervals of from 50–100 milliseconds (ms). When the insect is detected, the pulse-to-pulse interval suddenly decreases to about 10 ms so the bat can pinpoint the prey's position. To determine how far apart the bat and insect are when the bat first senses the insect is there, the experimenters videotaped and sound recorded 11 catches. By examining the film frames, the following distances were determined:

Catch No.	Distance (cm)	Catch No.	Distance (cm)
1	62	7	27
2	52	8	42
3	68	9	83
4	23	10	56
5	34	11	40
6	45		

Suppose the investigators have conjectured that the bat pinpoints the prey at a 68-cm distance or sooner. Use $\alpha = 0.05$ and test the null hypothesis that the distance is greater than 68 cm. (Taken from Griffin, et al.: *Animal Behavior, 8*:141–154, 1960.)

Section 7.4

36. Using the following data, test the following hypotheses:

n	s^2	H_0	H_1	α
16	10	$\sigma^2 = 4$	$\sigma^2 \neq 4$	0.01
25	7	$\sigma^2 \geq 10$	$\sigma < 10$	0.05
9	40	$\sigma^2 \leq 20$	$\sigma > 20$	0.10
35	9.3	$\sigma^2 = 15$	$\sigma^2 \neq 15$	0.01

37. To control the amount of bleach dispensed by an automatic machine, one must control the variance of the output. If the variance in the amount dispensed is too large, many of the bottles would be over- or underfilled. In a sample of 16 bottles it was found that the sample variance was $s^2 = 1$. Test that hypothesis $H_0: \sigma^2 \leq \frac{1}{2}$ versus the alternative H_1: $\sigma^2 > \frac{1}{2}$ at $\alpha = 0.05$.

38. A manufacturer of thermometers claims that the standard deviation in readings of his thermometers is no more than 0.5°C. A set of 16 thermometers is tested, with the resulting standard deviation of $s = 0.7$°C. Test the hypothesis H_0: $\sigma \leq 0.5$°C. versus H_1: $\sigma > 0.5$°C. at $\alpha = 0.01$.

39. An elementary school teacher tries to group a reading class into subgroups of approximately the same average reading ability and equal variance in ability. The nine children in the high reading group yield a variation in reading test scores of $s^2 = 8.5$. Test H_0: $\sigma^2 \leq 5$ versus H_1: $\sigma^2 > 5$ at $\alpha = 0.1$.

40. A fish food producer would like to compare the average protein content of his product to that of the known average protein content of his competitor's fish food. He wishes to assume that the σ^2 is known since he feels that the variation in protein content from can to can is equal to 3 g. A sample of nine cans yields the following protein contents (in grams): 114, 116, 117, 112, 116, 114, 110, 119, 115. Test the hypothesis H_0: $\sigma^2 = 3$ against the alternative H_1: $\sigma^2 \neq 3$. Use $\alpha = 0.05$.

41. One way to check the quality control of machined parts is to see if the variance of their part length is too large. A large variance in part length indicates possible poor quality control and may result in expensive unscheduled repair of the parts constructed. A machinist of automotive clutch parts found that the variance s^2 of 41 parts was $s^2 = 230,000$ sq mml. Test the hypotheses H_0: $\sigma^2 \leq 133,225$ sq mml. versus H_1: $\sigma^2 > 133,225$ sq mml. at $\alpha = 0.05$.

42. Return to the discussion of the Bennett Amendment (problem 31). Churchhill, Cooper, and Govindarajan considered the effect of the amendment not only on the length of stay in the hospital but also on the variance. For Ottawa for the period listed in problem 31 they found the standard deviation was 4.292 days. Use the data in problem 31 to test if the amendment significantly reduced the variance in length of stay as well (use $\alpha = 0.01$).

43. A group of 41 bricklayers were given the Kluge Dexterity Test. They averaged answer-

ing correctly to 89% of the questions, with a standard deviation of 3%. Assume that the test scores are normally distributed. Test the hypothesis that $\sigma = 5\%$ against the alternative $\sigma \neq 5\%$ at the significance level $\alpha = 0.05$.

Section 7.5

44. Using the following data, test each given hypothesis:

n	Number in sample with characteristic	H_0	H_1	α
60	15	$p = 0.2$	$p \neq 0.2$	0.10
120	96	$p \leq 0.75$	$p > 0.75$	0.05
250	200	$p \geq 0.82$	$p < 0.82$	0.01
300	175	$p \leq 0.55$	$p > 0.55$	0.025
500	210	$p = 0.35$	$p \neq 0.35$	0.02

45. Many studies have sought to characterize the nightmare sufferer. The general stereotype we get is of someone with high anxiety, low ego strength, feelings of inadequacy, and poor health. The purpose of the article referenced below was to ascertain if the proportion of men having nightmares often is the same as the proportion of women having nightmares often. Each of 160 men was interviewed and was asked if he had nightmares often or seldom. The following data were obtained: 55 men responded often. Take the proportion of women who responded often to be 30.6%. (From Hersen, M.: Personality characteristics of nightmare sufferers. *Journal of Nervous and Mental Diseases,* 153:29–31, 1971.)
   a. Define the parameters in this problem and state the appropriate null and alternative hypotheses.
   b. Test the hypothesis you gave in (a) and state your conclusions. (Use $\alpha = 0.05$.)

46. It is known that approximately 54.9% of the people whose family income is under

$10,000 smoke regularly. A sample of 240 people whose income is over $20,000 yields 43.9% smokers. To see if there is a relationship between income and smoking, test the hypothesis H_0: $p = 0.549$ versus H_1: $p \neq 0.549$, where p is the proportion of smokers in the over $20,000 income bracket. Use $\alpha = 0.01$.

47. A drug manufacturer is considering advertising its drug as being at least 90% effective in curing a certain ailment. A sample of 250 patients is given the drug and 210 patients recover. Should they advertise this effectiveness?

48. Consider the situation in problem 47 again. Should the Food and Drug Administration issue a cease and desist order on the advertising based on the data given in problem 47?

49. A sociological study is made to ascertain if a person's sex is related to his or her attitude toward a certain issue. In a poll of 100 people who were in favor of the issue, it was found that 59 respondents were males. Can we conclude that sex is related to a person's response? (*Hint:* Test if $p = 0.5$, where p is the proportion of respondents in favor of the issue who are males; use $\alpha = 0.01$)

50. In order to determine if jury selection in New Orleans during 1960 was nonrandom with respect to color, we first determine via census information that 34.8% of the jury population was black. Of the 150 jurors selected that year, there were 15 blacks. Can we conclude that the selection was nonrandom with respect to color (in the sense that blacks were discriminated against) at $\alpha = 0.0002$? (*Hint:* Test H_0: $p \geq 0.348$)

51. A manufacturing firm wants to guarantee its equipment for 90 days and so it tests 250 pieces of equipment for 90 days and finds 14 that wear out. The firm claims that it will not have to pay off any more than 5% of the time. Is this consistent with the data? (*Hint:* Test H_0: $p \geq 0.05$ at $\alpha = 0.2$)

52. An economist wants to find if the proportion of small businesses in a certain industry has increased over the last 20 years due to the changing economic climate. One thousand companies were selected in this industry and 312 of them were small companies. If 20 years before there were 30% small companies in this industry, can the economist feel confident that the proportion in question has increased? Use $\alpha = 0.1$.

53. In 1978 a questionnaire designed to determine the reaction to the American people to President Carter's energy policy was given to 250 people. Of these, 112 responded in favor of all aspects of the plan.

    Is it possible that a majority of Americans were actually in favor of all aspects of the Carter plan? (Use $\alpha = 0.05$.)

54. A doctrine of English law that is essential to the administration of justice is that a person be judged by a jury of his or her peers.

    a. *Swain* vs. *Alabama:* Swain, a black man, was convicted by an all white jury in Talladega County, Alabama, of raping a white woman. There seemed to be a pattern of excluding a portion of the population from service on juries since no black "within the memory of persons now living has ever served on any petit jury in any civil or criminal case tried in Talladega County, Alabama." Swain was condemned to die. His case was appealed to the U.S. Supreme Court on the grounds that the doctrine of "trial by peers" had been violated.

    The Supreme Court in 1965 denied the appeal because the panel from which the jury was selected (100 people) contained 8 blacks. The presence of 8 blacks on the panel (not on the jury) convinced the Supreme Court that "the overall percentage disparity has been small and reflects no studied attempt to include or exclude a specified number of Negroes."

    At that time in Alabama, 26% of eligible jurors were black. If 100 people were chosen at random, what is the chance that 8 or fewer would be black? Is the difference between the expected number of blacks and the observed number greater than four standard deviations?

    b. *Barksdale* vs. *Henderson:* Barksdale, a black man, was indicted for raping a white woman by a grand jury in New Orleans in September 1962. The grand jury venire (list) is drawn from a larger list (called the general venire) that was composed in 1962 from the city directory. The State of Louisiana claimed that on the 1962 general venire there were 269 blacks and 1538 whites, or a total of 1807. (The actual number of blacks had been disputed, but we use the figure provided by the state since of all the figures presented, it represents the largest number of blacks.)

    The percentage of eligible black jurors in the Orleans parish in 1962 was also in dispute; we use the *lowest* figure presented by the interested party, 32% that was offered by the state.

      i. Assume 1807 people are selected by independent random sampling from a population containing 32% blacks and the observed number of blacks in the sample of size 1807 is 269; test the hypothesis at $\alpha = 0.001$ that the chance p of a black being selected is 0.32, that is, $H_0: p \geq 0.32$. This is the hypothesis "no discrimination."

      ii. Is the number of standard deviations between the expected number of blacks and the actual number observed about 15? (If so, the chance of this difference if H_0 is true is about 1 in 10^{30}).

      iii. Barksdale was convicted and appealed to the federal court. Part of his appeal was based on the claim that blacks were systematically excluded from the general venire. Is there a basis for this claim?

iv. The following data show the figures representing the number of whites and blacks on the grand jury venires between March 1954 and September 1962. Again, assume the percentage of eligible black jurors in the population is 32%. Test the hypothesis that blacks are selected randomly for the grand jury venire against the hypothesis that attempts were made to exclude them.

Persons on Grand Jury Venires March 1954 to September 1962		
Total	White	Black
1550	1342	176

(Again, we use the figures supported by the state in the litigation.)

v. Part of Barkdale's appeal was based on the claim that blacks were systematically excluded from the grand jury venires. Is there evidence in part iv to support this claim?

The preceding description does not exhaust Barksdale's presentation of evidence. (See the first reference at the end of Chapter 10.) Barksdale's appeal was denied by a federal court in 1978. In 1979 a three-judge panel of the 5th circuit found for Barksdale and reversed the decision of the federal judge. The three-judge panel was reversed in 1980 by the entire 5th circuit by a *vote of 12 to 10*. Barksdale has been in jail since 1962.

55. A small town politician has decided that in order to be elected he shall tell the voters only what the majority wants to hear. There are 600 voters in the town and a sample of 150 of them yields the result that 78 are in favor of a particular issue and 72 are against it. Can he conclude that the majority of the town is in favor of the issue, or should he avoid answering questions about his stand on the issue (i.e., is it possible that $p \leq 0.5$?). Use $\alpha = 0.1$ and use a normal approximation.

Section 7.7

56. A certain medical operation has been shown to cause death in a 0.0008 proportion of the cases. A new technique is adopted and on the first 10,000 patients they found no fatalities. Can we conclude that this new technique is better? (Use $\alpha = 0.05$.)

57. A certain country has 0.5% of its populace classified as violently psychotic. A sample of 1000 people from a neighboring country has seven people classified as violently psychotic. Is the second country safer? (Use $\alpha = 0.1$.)

58. A certain drug in common usage causes an allergic reaction in 0.01% of the people who use it. A new drug is developed and is given to 50,000 people. Of these, 2 experience an allergic reaction. Is the new drug significantly safer? (Use $\alpha = 0.05$.)

59. A certain disease is very rare and in the past only 0.01% of the population has contracted the disease. An accidental chemical spill into the water supply of a town of 10,000 occurs, and within 1 year three cases of this disease are diagnosed in this town. Is this significantly more than we would expect? (Use $\alpha = 0.01$.)

60. The Centers for Disease Control in Atlanta, Georgia, stopped administration of the swine flu shots after a seemingly increased frequency of occurrence of the Guillain-Barré syndrome. Suppose we wished to determine if there actually was an increase. In a sample of 2000 people who were vaccinated with distilled water there were 5 who came down

with the Guillain-Barré syndrome, whereas of the 3000 persons inoculated with swine flu vaccine there were 10 who came down with the Guillain-Barré syndrome. Is the proportion of people who come down with the Guillain-Barré syndrome significantly higher for people inoculated with the swine flu vaccine? (Use $\alpha = 0.05$.) (These are hypothetical data.)

61. Some journals (e.g., *the Journal of Experimental Psychology*) will only publish an article if the results are "highly significant"; that is, if H_0 is the hypothesis that chance is responsible for the deviation between the data collected and the model postulated, then H_0 must be rejected at $\alpha \leq 0.01$. What are some of the drawbacks of this policy?

Additional problems

62. Hypergeometric distribution (review discussion of problem 37, Chapter 6.)

    Let us consider now the problem of hypothesis tests about p based on simple random sampling. Since we know that $Y = n\hat{P}$ has a hypergeometric distribution, we may use this fact to find a test. To test

    $$H_0: p = p_0$$
    $$H_1: p \neq p_0$$

    at the level of significance α, we assume Y has a hypergeometric distribution with $p_0 = M/N$, or equivalently, with $M = Np_0$. If $P[Y \geq n\hat{p}] \leq \alpha/2$ or if $P[Y \leq n\hat{p}] \leq \alpha/2$, we reject H_0.

Example 7 ▪ 13

An automobile manufacturer is expected to meet certain safety requirements. A consumer agency purchased 20 automobiles and tested 10 of them for compliance. Of these, one was found to be below standard. Let p denote the proportion of faulty automobiles purchased by the agency. Test

$$H_0: p = 0.25$$
$$H_1: p \neq 0.25$$

at $\alpha = 0.1$

We let Y be a hypergeometric random variable with $N = 20$ and $M = Np_0 = (20)(0.25) = 5$ and $n = 10$. We then calculate $P[Y \leq 1] = P[Y = 0] + P[Y = 1]$. We have

$$P[Y = 0] = \frac{\binom{5}{0}\binom{15}{10}}{\binom{20}{10}} = 0.0163$$

$$P[Y = 1] = \frac{\binom{5}{1}\binom{15}{9}}{\binom{20}{10}} = 0.135$$

Thus $P[Y \leq 1] = 0.0163 + 0.135 = 0.1513$. Since $0.1513 > 0.05$ we fail to reject H_0.

A similar method can be used to perform one-tailed directional hypotheses about the value of p. To test

$$H_0: p \geq p_0$$
$$H_1: p < p_0$$

at level of significance α, we reject H_0 if and only if $P[Y \leq n\hat{p}] \leq \alpha$, where Y has a hypergeometric distribution with $p_0 = M/N$, or equivalently $M = Np_0$. To test

$$H_0: p \leq p_0$$
$$H_1: p > p_0$$

we reject H_0 if $P[Y \geq n\hat{p}] < \alpha$.

a. A new experimental diagnostic test is administered to 20 patients. In a sample of 12 patients it was found that 2 registered positive. Let p be the proportion of positive tests among the 20 patients. Test

$$H_0: p \geq 0.15$$

$$H_1: p < 0.15$$

at $\alpha = 0.05$

(Answer: Let Y be a hypergeometric random variable with $N = 20$ and $M = (0.15)(20) = 12$. We calculate

$$P[Y \leq 2] = \frac{\binom{8}{0}\binom{12}{12}}{\binom{20}{12}} + \frac{\binom{8}{1}\binom{12}{11}}{\binom{20}{12}}$$

$$+ \frac{\binom{8}{2}\binom{12}{10}}{\binom{20}{12}} = 0.015 < 0.025$$

so we reject H_0 and conclude $p < 0.15$.)

b. An opinion poll of 200 people in a community of 2000 found that 91 people were in favor of a particular issue. Let p be the proportion in the community in favor.

Test

$$H_0: p \geq 0.5$$

$$H_1: p < 0.5$$

at level of significance 0.01. Since $N = 2000$ and $n = 200$ are large but n is relatively small with respect to N, we use the normal approximation.

Answer:

$$z = \frac{\hat{p} - \mu_{\hat{p}}}{\sigma_{\hat{p}}}$$

$$= \frac{\hat{p} - 0.5}{\sqrt{[(2000 - 200)/(2000 - 1)][(0.5)(0.5)/200]}}$$

$$= \frac{\hat{p} - 0.5}{0.0335}$$

We reject H_0 if and only if $z < -z_{0.99} = -2.326$. Since $\hat{p} = 91/200 = 0.455$, we have $z = -1.34$, so we fail to reject H_0.

c. With respect to the "Application to Ecology" (Section 4.14), test

H_0: proportion of marked beavers in the population of beavers = 0.2

H_1: this proportion < 0.2

if 10 marked beavers are found in 100 that are captured (use $\alpha = 0.05$).

63. Use the techniques of this chapter to come to some conclusions regarding the "blackmail case" described in problem 12 of Chapter 2.

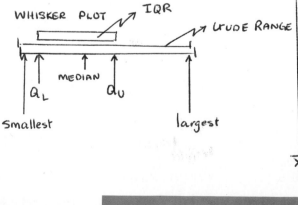

$f =$ frequency of a table 75 $\Big\}_{n=100}$

$P =$ relative freq $= .75$

$= f/n$

✹✹✹

SUMMATION NOTATION

$$\sum_{i=1}^{n} = \frac{n(n+1)}{2}$$

$$\bar{x} = \text{Mean} = \frac{\sum x}{n}$$

DEVIATION FROM MEAN

$$d_1 = x_1 - \bar{x}$$

$$d_2 = x_2 - \bar{x}$$

Variance (s^2)

$$= \frac{(\sum x_i)^2 - \frac{(\sum x_i)^2}{n}}{n-1}$$

CHAPTER · 8

Comparing Two Populations

*General impressions are never to be trusted. Unfortu-
nately when they are of long standing they become
fixed rules of life, and assume a prescriptive right not
to be questioned. Consequently those who are not
accustomed to original inquiry entertain a hatred and
a horror of statistics. They cannot endure the idea of
submitting their sacred impressions to cold-blooded
verification. But it is the triumph of scientific men to
rise superior to such superstitions, to desire tests by
which the value of beliefs may be ascertained and to
feel sufficiently masters of themselves to discard
contemptuously whatever may be found untrue.*

Francis Galton

Does a weight diet work; that is, will the dieters weigh
less after using the diet than before using it? Are
Scandinavians really taller than North Americans? Does
one brand of milk contain a more uniform distribution of
cream throughout than another brand? These are questions
that can be studied by using the techniques described in this
chapter. The first two questions compare the means of two
populations, whereas the third compares the variances of
two populations. This chapter is primarily concerned with
these topics.

8 ▪ 1   CONFIDENCE INTERVALS AND HYPOTHESIS TESTS COMPARING THE MEANS OF TWO POPULATIONS; VARIANCES KNOWN

Frequently the purpose of a statistical investigation is to compare two populations with respect to some parameter values. For example, in comparing the possible monetary returns for two different investment methods, we might choose the method that has a higher mean return. If a manufacturer buys two machines for producing parts of a certain type, he might then wish to compare the variances of the lengths of the parts produced by the two machines to determine which is more consistent. A medical doctor might compare the proportions of people cured by each of two different drugs before prescribing one of them.

We now assume two normal populations represented by the independent random variables X and Y. We assume X has a mean μ_X and variance σ_X^2, whereas Y has a mean μ_Y and variance of σ_Y^2. We wish to compare μ_X and μ_Y. Suppose we sample n_1 times from the first population and n_2 times from the second population, and the sampling technique used is *independent random sampling*. Our estimator of μ_X is $\overline{X}$ and of μ_Y is $\overline{Y}$. We know that $\overline{X}$ has a normal distribution with mean μ_X and variance σ_X^2/n_1, whereas $\overline{Y}$ has a normal distribution with mean μ_Y and variance σ_Y^2/n_2 (see Chapter 5). In general, the sum and difference of two independent normal random variables again have a normal distribution. Note also that

$$E(\overline{X} - \overline{Y}) = \mu_X - \mu_Y$$

and since $\overline{X}$ and $\overline{Y}$ are independent,

$$\text{Var}(\overline{X} - \overline{Y}) = \text{Var}(\overline{X}) + \text{Var}(\overline{Y}) = \frac{\sigma_X^2}{n_1} + \frac{\sigma_Y^2}{n_2}$$

Thus $\overline{X} - \overline{Y}$ also has a normal distribution with mean $\mu_X - \mu_Y$ and variance $\sigma_X^2/n_1 + \sigma_Y^2/n_2$. We use this crucial fact to find confidence intervals and perform hypothesis tests with respect to the difference $\mu_X - \mu_Y$.

Confidence Intervals for the Difference of Two Means

We now proceed to find the $(100)(1 - \alpha)\%$ confidence interval for $\mu_X - \mu_Y$. We note that since $\mu_X - \mu_Y$ is the mean and $\sigma_X^2/n_1 + \sigma_Y^2/n_2$ is the variance of $\overline{X} - \overline{Y}$, the standardized variable

$$Z = \frac{(\overline{X} - \overline{Y}) - (\mu_X - \mu_Y)}{\sqrt{\sigma_X^2/n_1 + \sigma_Y^2/n_2}}$$

has a normal distribution with a mean 0 and a variance 1. It follows immediately from our work on confidence intervals in Chapter 6 that

> **The $100(1 - \alpha)$% confidence interval for the difference $\mu_X - \mu_Y$ between means is from**
>
> $$\bar{x} - \bar{y} - z_{\alpha/2}\sqrt{\sigma_X^2/n_1 + \sigma_Y^2/n_2} \quad \text{to} \quad \bar{x} - \bar{y} + z_{\alpha/2}\sqrt{\sigma_X^2/n_1 + \sigma_Y^2/n_2} \qquad \text{(8.1)}$$

Example 8 ▪ 1 A company manufactures nuts and bolts on different machines. For the two pieces to fit together it is necessary that the difference of the diameters be within a certain range. If we sample 15 bolts from the population of bolts with known variance $\sigma_X^2 = 0.0025$, we obtain the sample mean diameter $\bar{x} = 0.52$. When we sample $n_2 = 17$ nuts from the population of nuts with known variance $\sigma_Y^2 = 0.003$, we obtain $\bar{y} = 0.49$. Find a 99% confidence interval for the mean difference in diameters.

Let μ_X be the mean diameter of the bolts and μ_Y the mean diameter of the nuts. Since $1 - \alpha = 0.99$, we have $z_{\alpha/2} = z_{0.005} = 2.576$. We assume the diameters of both the nuts and bolts form a normally distributed population of values. Then by formula (8.1) the 99% confidence interval is from

$$(0.52 - 0.49) - 2.576\sqrt{\frac{0.0025}{15} + \frac{0.003}{17}} \quad \text{to}$$

$$(0.52 - 0.49) + 2.576\sqrt{\frac{0.0025}{15} + \frac{0.003}{17}}$$

That is, on calculating the square root, we get from

$$0.03 - 2.576(0.0185) \quad \text{to} \quad 0.03 + 2.576(0.0185)$$

or finally, from

$$-0.018 \quad \text{to} \quad 0.078$$

This is the tolerance between the two machines, an estimate of how much difference in diameter we expect between a randomly selected nut and bolt.

Hypothesis Testing

We may also use the fact that

$$Z = \frac{(\bar{X} - \bar{Y}) - (\mu_X - \mu_Y)}{\sqrt{\sigma_X^2/n_1 + \sigma_Y^2/n_2}} \qquad \text{(8.2)}$$

has a normal distribution with mean 0 and variance 1 to develop tests of hypotheses about the difference $\mu_X - \mu_Y$. We test at the level of significance α.

$$H_0 : \mu_X - \mu_Y = c$$
$$H_1 : \mu_X - \mu_Y \neq c$$

where c is a known number representing the difference that the researcher expects to observe between the two means. Notice that when H_0 is true $\mu_X - \mu_Y = c$, and so we may replace $\mu_X - \mu_Y$ in equation (8.2) with c. With this substitution we reject H_0 if the observed value z of Z is more extreme than what the researcher would think is reasonable; that is, if z is greater than $z_{\alpha/2}$ or less than $-z_{\alpha/2}$, where

$$z = \frac{(\bar{x} - \bar{y}) - c}{\sqrt{\sigma_X^2/n_1 + \sigma_Y^2/n_2}}.$$

and the level α is chosen by the researcher.

Similarly, in the one-tailed tests the researcher might want to determine if an observed deviation in one direction between two sample means is significantly smaller than what he expected. If he expected a deviation of c, he would want to test

$$H_0 : \mu_X \geq \mu_Y + c$$
$$H_1 : \mu_X < \mu_Y + c$$

If the observed value of $\bar{x}$ is much smaller than the observed value of $\bar{y} + c$, then the researcher will tend to believe that H_1 is true. Thus the rejection region for H_0 is

$$z = \frac{(\bar{x} - \bar{y}) - c}{\sqrt{\sigma_X^2/n_1 + \sigma_Y^2/n_2}} \qquad < -z_\alpha$$

The same logic applies when testing

$$H_0 : \mu_X \leq \mu_Y + c$$
$$H_1 : \mu_X > \mu_Y + c$$

only now we use the right tail for the rejection region; reject H_0 if $z > z_\alpha$.

Summarizing, we have

To test hypotheses about the difference of two normal population means; variances known; level of significance α
Use:

$$z = \frac{(\bar{x} - \bar{y}) - c}{\sqrt{\sigma_X^2/n_1 + \sigma_Y^2/n_2}}$$

To test

$$H_0 : \mu_X = \mu_Y + c \qquad versus \qquad H_1 : \mu_X \neq \mu_Y + c$$

Reject H_0 if $z > z_{\alpha/2}$ or if $z < -z_{\alpha/2}$

To test

$$H_0 : \mu_X \leq \mu_Y + c \qquad versus \qquad H_1 : \mu_X > \mu_Y + c$$

Reject H_0 if $z > z_\alpha$.

To test

$$H_0 : \mu_X \geq \mu_Y + c \qquad versus \qquad H_1 : \mu_X < \mu_Y + c$$

Reject H_0 if $z < -z_\alpha$.

By taking $c = 0$ in these three tests, we can test $\mu_X = \mu_Y$, or $\mu_X \leq \mu_Y$ or $\mu_X \geq \mu_Y$, respectively.

Example 8 ▪ 2   A psychologist gives an intelligence test to ninth and tenth grade classes. It is known that the variance in scores of the tenth graders is $\sigma_X^2 = 10$ and of the ninth graders is $\sigma_Y^2 = 12$. It is commonly assumed that tenth grade students average at least 10 points higher than ninth grade students. The psychologist is suspicious of this common assumption and wants to show that it is not true. Thus he gives the test to 25 tenth graders and 15 ninth graders. It is found that $\bar{x} = 67$ for the tenth graders, and $\bar{y} = 59$ for the ninth graders. To determine if the data are consistent with the preceding assumption, we test

$$H_0 : \mu_X \geq \mu_Y + 10$$
$$H_1 : \mu_X < \mu_Y + 10$$

at the level of significance of, say, $\alpha = 0.02$. Using the fact that $\mu_X - \mu_Y = c = 10$ in equation (8.2), we would reject H_0 if

$$z = \frac{\bar{x} - \bar{y} - 10}{\sqrt{\dfrac{10}{25} + \dfrac{12}{15}}} \qquad < -z_{0.02} = -2.054$$

We calculate

$$z = \frac{67 - 59 - 10}{1.095} = -1.83 > -2.054$$

so we fail to reject H_0.

Thus there is insufficient evidence to conclude that $\mu_X < \mu_Y + 10$ at $\alpha = 0.02$, and the common assumption is not refuted.

Example 8 ▪ 3 An electronic component manufacturing plant provides microprocessing chips for a major computer builder. The plant manager suspects that there is a difference between the number of parts produced by the day and night shifts at the plant. Since labor negotiations are coming up soon, the manager wishes to determine if the difference is statistically significant. If the two shifts do not produce the same average number of parts, the manager will try to negotiate a differential in pay between the shifts. It is therefore requested that the production quantity be recorded for both shifts on a randomly selected set of days during the month. For the day shift it is known that $\sigma_X^2 = 20$ and for the night shift $\sigma_Y^2 = 50$. A sample of 14 days output from the day shift yields $\bar{x} = 576$ parts produced and a sample of 21 days output from the night shift yields $\bar{y} = 570$ parts produced. To determine if the difference is significant, we test

$$H_0 : \mu_X = \mu_Y$$
$$H_1 : \mu_X \neq \mu_Y$$

at $\alpha = 0.01$. We reject H_0 if

$$z = \frac{\bar{x} - \bar{y}}{\sqrt{\dfrac{20}{14} + \dfrac{50}{21}}} > z_{\alpha/2} = z_{0.005} = 2.576 \quad \text{or if} \quad z < -2.576$$

We calculate

$$z = \frac{6}{\sqrt{3.81}} = 3.07 > 2.576$$

so we reject H_0 and conclude that there is a significant difference between the output of the two shifts at $\alpha = 0.01$ and evidence that the day shift produces more.

8 ▪ 2   HYPOTHESIS TESTS COMPARING THE VARIANCES OF TWO POPULATIONS; *F*-DISTRIBUTION

It often becomes necessary to compare the variances of two populations. For example, one would use different techniques when teaching students with widely varying abilities than when teaching students with comparable abilities. As another example, insurance companies involved in casualty insurance (e.g., automotive insurance) might like to cross-classify their insured drivers so that in each driver class the risk as measured by the variance in losses is approximately equal. In financial security analysis the riskiness of an investment is often assumed to be measured by the variance of the monetary return produced by the investment. The security analyst selecting stocks for her portfolio must therefore compare variances. We show one way to compare such variances in the case of normal independent populations; that is, when the random variables X and Y are independent and normally distributed.

The *F*-Distribution

When comparing the variances of two normal independent populations, we do the natural thing—compute the sample variance from each population and then compare these values. Let S_X^2 and S_Y^2 represent the sample variance obtained from independent random sampling from the first and second populations, respectively. Let σ_X^2 and σ_Y^2 represent the population variances of the first and second populations, respectively, and n_1 and n_2 the size of the samples from the first and second populations, respectively. Then the random variable

$$F = \frac{(S_X^2/\sigma_X^2)}{(S_Y^2/\sigma_Y^2)}$$

should take on values near 1 since S_X^2 should be near σ_X^2 and S_Y^2 near σ_Y^2. If we could find the distribution of this ratio F, then we could use this ratio to test hypotheses about two variances.

Since the distribution of the numerator and denominator of F depends on their respective number of degrees of freedom, the preceding F-ratio depends on two parameters, $df(N)$ and $df(D)$, the degrees of freedom involved in the numerator and the denominator, respectively.

Summarizing, we have

The ratio

$$F = \frac{S_X^2/\sigma_X^2}{S_Y^2/\sigma_Y^2}$$

is called the *F*-ratio. The distribution is denoted *F*[df(*N*), df(*D*)] and depends on two parameters, the degrees of freedom in the numerator $df(N) = n_1 - 1$ and the degrees of freedom in the denominator $df(D) = n_2 - 1$, where n_1 and n_2 are the sample sizes from the *X* and *Y* populations respectively.

Since forming a table of $F[df(N), df(D)]$ for all values of $df(N)$ and $df(D)$ would require too many pages, we have only tabulated certain frequently used upper centile values.

The upper αth centile value for $F[df(N), df(D)]$ is denoted by $F_\alpha[df(N), df(D)]$, or just F_α for short.

For example, we see from Appendix A.7 that $F_{0.005}(15, 3) = 43.1$, $F_{0.995}(50, 10) = 0.334$, and $F_{0.025}(10, 40) = 2.39$. Figure 8.1 shows how the density for F changes as $df(N)$ and $df(D)$ change.

Hypothesis Tests

We may apply the F-ratio to obtain hypothesis tests about the relationship between σ_X^2 and σ_Y^2. For example, to test

$$H_0 : c\sigma_X^2 = \sigma_Y^2$$

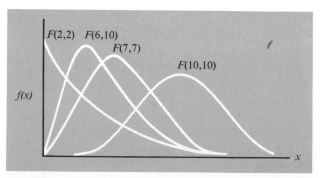

Figure 8 ▪ 1
The $F[df(N), df(D)]$ distribution for various choices of $df(N)$, $df(D)$.

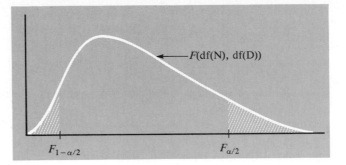

Figure 8 ■ 2
Two-tailed rejection region. Each tail cuts off areas $\alpha/2$, so the entire area cut off is α. Note the asymmetry.

versus

$$H_1 : c\sigma_X^2 \neq \sigma_Y^2$$

where c is a constant, we use the fact that $\sigma_Y^2/\sigma_X^2 = c$ if H_0 is true to obtain

---

Test statistic for comparing variances from normal populations:

$$F = \frac{(S_X^2/\sigma_X^2)}{(S_Y^2/\sigma_Y^2)} = \frac{cS_X^2}{S_Y^2} \qquad (8.3)$$

---

which has an $F(n_1 - 1, n_2 - 1)$ distribution where n_1 is the size of the sample from population X and n_2 is the size of the sample from population Y. Let s_X^2 be the observed value of S_X^2 and s_Y^2 be the observed value of S_Y^2. We reject H_0 at the level of significance α if and only if cs_X^2/s_Y^2 takes on a value greater than $F_{\alpha/2}(n_1 - 1, n_2 - 1)$ or if cs_X^2/s_Y^2 takes on a value less than $F_{1-\alpha/2}(n_1 - 1, n_2 - 1)$. In particular, to test

$$H_0 : \sigma_X^2 = \sigma_Y^2$$
$$H_1 : \sigma_X^2 \neq \sigma_Y^2,$$

we would reject H_0 if

$$\frac{s_X^2}{s_Y^2} > F_{\alpha/2}(n_1 - 1, n_2 - 1) \quad \text{or if} \quad \frac{s_X^2}{s_Y^2} < F_{1-\alpha/2}(n_1 - 1, n_2 - 1)$$

Graphically, this is illustrated in Figure 8.2.

Example 8 ■ 4   A new teaching method is claimed to cut the variance of scores in half as compared to the old teaching method. Test this hypothesis at $\alpha = 0.01$. Let σ_X^2

denote the variance of scores due to the old method. We wish to test

$$H_0 : \frac{1}{2} \sigma_X^2 = \sigma_Y^2 \quad \text{versus} \quad H_1 : \frac{1}{2} \sigma_X^2 \neq \sigma_Y^2$$

A class of 31 students is taught by the old method and the sample variance $s_X^2 = 410$ was obtained. A different class of 41 students was taught by the new method and $s_Y^2 = 163$ was obtained. Can we conclude that the variance was not cut in half? The right-hand cut-off value of $F_{0.005}(30, 40) = 2.40$ and the left-hand cut-off value of $F_{0.995}(30, 40) = 0.396$. We calculate

$$(0.5)s_X^2/s_Y^2 = (0.5)410/163 = 1.26.$$

Since this calculated value is not less than 0.396 nor greater than 2.40, we fail to reject H_0. We conclude that the data are not inconsistent with $\sigma_Y^2 = \frac{1}{2}\sigma_X^2$. That is, we do not have sufficient evidence to refute this claim at $\alpha = 0.01$.

We may also perform directional (one-tailed) hypothesis test about the ratio of the two population variances. To test

$$H_0 : c\sigma_X^2 \leq \sigma_Y^2 \quad \text{versus} \quad H_1 : c\sigma_X^2 > \sigma_Y^2$$

at the level of significance α, we reject H_0 if $F = cs_X^2/s_Y^2$ takes on a value greater than $F_\alpha(n_1 - 1, n_2 - 1)$.

To test

$$H_0 : c\sigma_X^2 \geq \sigma_Y^2 \quad \text{versus} \quad H_1 : c\sigma_X^2 < \sigma_Y^2$$

at level of significance α, we reject H_0 if $F = cs_X^2/s_Y^2$ takes on a value less than $F_{1-\alpha}(n_1 - 1, n_2 - 1)$. Notice that as before the alternative hypothesis inequality sign determines whether the rejection region is the right or left tail.

Example 8 ▪ 5   A manufacturer of links for making chains advertised that his links have a smaller variance in strength than do his competitor's links. A sample of 51 of his links have a variance of $s_X^2 = 308$, whereas a sample of 61 of the competitor's links have a variance in strength of $s_Y^2 = 256$. Can his competitor conclude that, in fact, the reverse is true and sue for false advertising? To resolve this issue using $\alpha = 0.05$, let σ_X^2 denote the variance in strength from the manufacturer and let σ_Y^2 be the variance in strength from the competitor. Since we are taking the competitor's point of view, we give the benefit of the doubt to the manufacturer. Thus we only advise suing if σ_X^2 is much larger than σ_Y^2. Here $c = 1$ and we test

$$H_0 : \sigma_X^2 \leq \sigma_Y^2 \quad \text{versus} \quad H_1 : \sigma_X^2 > \sigma_Y^2$$

at the level of significance $\alpha = 0.05$. We reject H_0 if

$$\frac{S_X^2}{S_Y^2} > F_{0.05}(50, 60) = 1.56^*$$

Since F takes on the value here of $308/256 = 1.20 < 1.56$, we fail to reject H_0 and consequently should not advise suing.

8 ▪ 3 CONFIDENCE INTERVALS AND HYPOTHESIS TESTS ABOUT THE DIFFERENCES OF MEANS OF TWO POPULATIONS

Often we wish to compare the means of two populations when the population variances are also unknown. We assume at the start that the two populations have the same variance and derive subsequently a sample statistic that can be used to find confidence intervals and hypothesis tests.

We assume that X and Y represent independent normal populations. Recall that if $X_1, X_2, \ldots, X_{n_1}$ is an independent random sample from population X and $Y_1, Y_2, \ldots, Y_{n_2}$ is an independent random sample from population X, then the distribution of $\overline{X} - \overline{Y}$ is normal with a mean $\mu_X - \mu_Y$ and a variance $\sigma_X^2/n_1 + \sigma_Y^2/n_2$. If we assume the two populations have the same variance $\sigma_X^2 = \sigma_Y^2 = \sigma^2$, then $\overline{X} - \overline{Y}$ has a variance of $\sigma^2(1/n_1 + 1/n_2)$. Since σ^2 is unknown, we must utilize the sample to estimate it. We know that

$$S_X^2 = \frac{1}{n_1 - 1} \sum_{i=1}^{n_1} (X_i - \overline{X})^2$$

is an estimator of $\sigma_X^2 = \sigma^2$, and also

$$S_Y^2 = \frac{1}{n_2 - 1} \sum_{i=1}^{n_2} (Y_i - \overline{Y})^2$$

is an estimator of $\sigma_Y^2 = \sigma^2$. Since S_X^2 and S_Y^2 are both estimators of σ^2, by

---

* $F_{0.05}(50, 60)$ does not appear in Appendix A.7. Interpolation is not needed to reach a decision in this problem, however. Use the fact that $F_{0.05}(50, \infty)$ is less than $F_{0.05}(50, 60)$ and $F_{0.05}(50, \infty) = 1.35$, which is larger than the observed value $F = 1.20$.

"pooling" the values of S_X^2 and S_Y^2 together, we would get an estimator of σ^2 which is better than either S_X^2 or S_Y^2 individually. We define the pooled sample variances S_p^2 by

$$S_p^2 = \frac{\sum_{i=1}^{n_1}(X_i - \bar{X})^2 + \sum_{i=1}^{n_2}(Y_i - \bar{Y})^2}{n_1 + n_2 - 2}$$

$$= \frac{(n_1 - 1)S_X^2 + (n_2 - 1)S_Y^2}{n_1 + n_2 - 2} \tag{8.4}$$

The idea here is that S_p^2 is calculated by using $n_1 + n_2$ observations, and hence should be a more accurate estimate of σ^2 than is either S_X^2 or S_Y^2 alone.

The number of degrees of freedom in the distribution of S_X^2 is $n_1 - 1$, and of S_Y^2 is $n_2 - 1$. When we pool the variances S_X^2 and S_Y^2, we add the degrees of freedom so that S_p^2 is calculated with $(n_1 - 1) + (n_2 - 1) = n_1 + n_2 - 2$ degrees of freedom. It is worth noting that the pooled variance estimator S_p^2 is just a weighted average of the two individual estimators S_X^2 and S_Y^2; that is,

$$S_p^2 = \left(\frac{n_1 - 1}{n_1 + n_2 - 2}\right)S_X^2 + \left(\frac{n_2 - 1}{n_1 + n_2 - 2}\right)S_Y^2$$

The weight that each individual estimator has is just the proportion of the total degrees of freedom $n_1 + n_2 - 2$ that is contributed by that estimator. This is so, since if one of the two sample sizes, say, n_1, is very large, then the corresponding estimator, S_X^2 in this case, is expected to be much closer to the true common value σ^2 and should have a much bigger influence in the estimation of σ^2. For example, if there are $n_1 = 201$ samples of X and $n_2 = 101$ samples of Y, then the samples of X constitute two thirds of the total free choices (degrees of freedom) in estimating σ^2. Accordingly, we give S_X^2, two thirds of the influence in the estimation, $S_p^2 = \frac{2}{3}S_X^2 + \frac{1}{3}S_Y^2$.

Using this pooled variance S_p^2 we obtain

When sampling n_1 times from X and n_2 times from Y, where the normally distributed populations have $\sigma_X^2 = \sigma_Y^2$, the test statistic

$$T = \frac{\bar{X} - \bar{Y} - (\mu_X - \mu_Y)}{\sqrt{S_p^2(1/n_1 + 1/n_2)}}$$

has a t-distribution with $n_1 + n_2 - 2$ degrees of freedom.

Confidence Intervals for $\mu_X - \mu_Y$ Variances Unknown but Equal

Using the fact that

$$T = \frac{\overline{X} - \overline{Y} - (\mu_X - \mu_Y)}{\sqrt{S_p^2(1/n_1 + 1/n_2)}} \tag{8.5}$$

has a t-distribution with $n_1 + n_2 - 2$ degrees of freedom, we know that

$$P(-t_{\alpha/2} < T < t_{\alpha/2}) = 1 - \alpha$$

where $t_{\alpha/2}$ is the cut-off point on the t-distribution with the $\alpha/2$ area to the right. From this expression, we find that the limits of $100(1 - \alpha)\%$ confidence interval for $\mu_X - \mu_Y$ are from

$$(\overline{x} - \overline{y}) - t_{\alpha/2} \sqrt{s_p^2 \left(\frac{1}{n_1} + \frac{1}{n_2}\right)} \quad \text{to} \quad (\overline{x} - \overline{y}) + t_{\alpha/2} \sqrt{s_p^2 \left(\frac{1}{n_1} + \frac{1}{n_2}\right)} \tag{8.6}$$

where $\overline{x}, \overline{y}, s_p^2$ are the observed values of $\overline{X}, \overline{Y}, S_p^2$, respectively. Notice that this is just the form we employed for calculating the confidence interval by using the t-distribution before. The only things that have changed are the degrees of freedom (now $n_1 + n_2 - 2$) and the estimate of the variance involved (now $s_p^2(1/n_1 + 1/n_2)$).

Example 8 ▪ 6 To determine the height difference between boys and girls of a certain age, a sample of 10 boys and 16 girls was taken. For the boys it was found that $\overline{x} = 57$ inches and $s_X^2 = 16$. For the girls, $\overline{y} = 60$ and $s_Y^2 = 24$. Find a 99% confidence interval for $\mu_X - \mu_Y$, assuming the populations are normal with variances that are equal.

Pooling the sample variances, we find

$$s_p^2 = \frac{9s_X^2 + 15s_Y^2}{24} = 21$$

From Appendix A.5 with $10 + 16 - 2 = 24$ degrees of freedom we have $t_{0.005} = 2.797$. The 99% confidence interval is then given by using equation (8.6); and is from

$$(57 - 60) - 2.797 \sqrt{21 \left(\frac{1}{10} + \frac{1}{16}\right)} \quad \text{to}$$

$$(57 - 60) + 2.797 \sqrt{21 \left(\frac{1}{10} + \frac{1}{16}\right)}$$

or on performing the calculations, we have for $\mu_X - \mu_Y$ the 99% confidence interval from

$$-8.17 \quad \text{to} \quad 2.17$$

Hypothesis Tests: Variances Unknown but Equal

Once it is known that

$$T = \frac{\overline{X} - \overline{Y} - (\mu_X - \mu_Y)}{\sqrt{S_p^2(1/n_1 + 1/n_2)}}$$

has a t-distribution with $n_1 + n_2 - 2$ degrees of freedom, we may develop hypothesis tests. For example, to test

$$H_0 : \mu_X - \mu_Y = c$$
$$H_1 : \mu_X - \mu_Y \neq c$$

at the level of significance α, we reject H_0 if and only if

$$t = \frac{\overline{x} - \overline{y} - c}{\sqrt{s_p^2(1/n_1 + 1/n_2)}} > t_{\alpha/2} \quad \text{or} \quad < -t_{\alpha/2} \quad \text{with df} = n_1 + n_2 - 2$$

To test the one-sided (directional) hypothesis

$$H_0 : \mu_X - \mu_Y \leq c \quad \text{versus} \quad H_1 : \mu_X - \mu_Y > c$$

at the level of significance α, we put all the permitted area on one tail and reject H_0 if $t = (\overline{x} - \overline{y} - c)/\sqrt{s_p^2(1/n_1 + 1/n_2)}$ takes on a value greater than t_α with df = $n_1 + n_2 - 2$. Finally, to test

$$H_0 : \mu_X - \mu_Y \geq c \quad \text{versus} \quad H_1 : \mu_X - \mu_Y < c$$

at the level of significance α, we reject H_0 if $t = (\overline{x} - \overline{y} - c)/\sqrt{s_p^2(1/n_1 + 1/n_2)}$ takes on a value less than $-t_\alpha$ with df = $n_1 + n_2 - 2$. Again, the pertinent tail to choose for the rejection region is determined by the alternative hypothesis. These t-tests are just like those done in the previous chapters.

Example 8 ▪ 7 A cigarette manufacturer measured the nicotine content of his cigarettes. From a sample of 61 cigarettes, he found $\overline{x} = 24$ mg and $s_X^2 = 9.63$. An independent laboratory tested 61 cigarettes of this brand and found $\overline{y} = 25$ mg and $s_Y^2 = 11.5$. Can we conclude that the mean nicotine content of the two populations from which the samples were taken is not the same? That is, is there a significant difference between the two results. To resolve this issue, first assume the populations are independently normally distributed with equal variances. We use $c = 0$ and test

$$H_0 : \mu_X = \mu_Y \quad \text{versus} \quad H_1 : \mu_X \neq \mu_Y$$

at the level of significance $\alpha = 0.05$.
   Then we calculate

$$s_p^2 = \frac{60s_X^2 + 60s_Y^2}{120} = 10.565$$

We reject H_0 if $t = (\bar{x} - \bar{y})/\sqrt{s_p^2/61 + s_p^2/61}$ takes on a value greater than $t_{0.025} = 1.98$ or less than $-t_{0.025} = -1.98$, where the number of degrees of freedom in the t-variable is $61 + 61 - 2 = 120$.
Since

$$t = \frac{24 - 25}{\sqrt{10.565/61 + 10.565/61}} = -1.70 > -1.98,$$

we fail to reject H_0. We conclude that the observed difference in nicotine content between the manufacturer and the independent laboratory is not statistically significant at $\alpha = 0.05$.

Example 8 ■ 8   It is claimed that a certain vitamin and mineral treatment will appreciably reduce the number of cavities a person will incur. To examine this claim the researchers gave this treatment to 31 children while another 31 received a placebo. After 6 months it was found that the treatment group averaged $\bar{x} = 5.9$ cavities with $s_X^2 = 2.56$ while the second group averaged $\bar{y} = 6.1$ cavities with $s_Y^2 = 2.25$. Is the observed difference significantly different from what we would expect if the treatment was not effective?
   We test

$$H_0 : \mu_X \geq \mu_Y$$
$$H_1 : \mu_X < \mu_Y$$

at $\alpha = 0.01$. If we assume independent normally distributed populations with a common variance, then our best estimate of the common variance is

$$s_p^2 = \frac{30s_X^2 + 30s_Y^2}{60} = 2.405$$

We reject H_0 (and conclude the treatment is effective) if $t = (\bar{x} - \bar{y})/\sqrt{s_p^2/31 + s_p^2/31}$ takes on a value less than $-t_{0.01} = -2.39$, where the number of degrees of freedom is $31 + 31 - 2 = 60$.

Since

$$t = \frac{5.9 - 6.1}{\sqrt{2.405/31 + 2.405/31}} = -0.508$$

we fail to reject H_0.

Example 8 ■ 9   A new filtering device is installed in a hospital's ventilation and air-conditioning system. This device is claimed to reduce the airborne bacteria count. A record of the bacteria count 1 week prior to installation and 1 week after installation yielded

Old device (X) *(Number of colonies per cubic foot of air)*	*New device (Y)* *(Number of colonies per cubic foot of air)*
$x_1 = 10.1$	$y_1 = 12.8$
$x_2 = 11.6$	$y_2 = 8.2$
$x_3 = 12.1$	$y_3 = 11.6$
$x_4 = 9.1$	$y_4 = 14.1$
$x_5 = 10.3$	$y_5 = 15.9$
$x_6 = 15.3$	$y_6 = 9.0$
$x_7 = 13.0$	$y_7 = 14.5$

We wish to test if the new filtering device actually reduces the airborne bacteria. We test $H_0 : \mu_X \leq \mu_Y$ versus $H_1 : \mu_X > \mu_Y$ at the level of significance $\alpha = 0.10$. We calculate $\bar{x} = 11.6$ and $\bar{y} = 12.3$ and $s_X^2 = 4.35$ and $s_Y^2 = 8.2$ and find

$$s_p^2 = \frac{6(4.35) + 6(8.2)}{12} = 6.3$$

We assume that each population is normally distributed with equal variances. We reject H_0 if $t = (\bar{x} - \bar{y})/\sqrt{6.3/7 + 6.3/7}$ takes on a value greater than $t_{0.1} = 1.356$, where the degree of freedom for the t-distribution is 12. Since $t = -0.7/1.34 = -0.52$, we fail to reject H_0. We cannot conclude that the new device significantly reduces the airborne bacteria. We would advise against purchasing the new device.

Population Variances Not Assumed Equal

If the variances of the two populations under investigation are assumed equal, then the pooled variance can be used to obtain an estimate for this common variance; however, if the variances of the two populations are not equal, you could be seriously misled by using the pooled variances in the t-statistic described in the first two sections. If you test the equality of the means at level α, say, of 0.05, and use the pooled variance when the population variances are not equal, your chances of a type 1 error may be very different from 0.05.

In this case we must modify the preceding t-statistic. The $100(1 - \alpha)\%$ confidence interval on $\mu_X - \mu_Y$ now goes from

$$\bar{x} - \bar{y} - t_{\alpha/2}\sqrt{s_X^2/n_1 + s_Y^2/n_2} \quad \text{to} \quad \bar{x} - \bar{y} + t_{\alpha/2}\sqrt{s_X^2/n_1 + s_Y^2/n_2}$$

where s_X^2, s_Y^2, n_1, and n_2 are defined as the sample variances and sample sizes of the populations and where the degrees of freedom of the t-distribution are approximately

$$\text{df} = \frac{[(s_X^2/n_1) + (s_Y^2/n_2)]^2}{\dfrac{(s_X^2/n_1)^2}{n_1} + \dfrac{(s_Y^2/n_2)^2}{n_2}}$$

To test $H_0 : \mu_X = \mu_Y$ versus a one- or two-tailed alternative, we use

$$t = \frac{\bar{x} - \bar{y}}{\sqrt{\dfrac{s_X^2}{n_1} + \dfrac{s_Y^2}{n_2}}}$$

where the df is as previously given. If the df is not an integer, then the critical value is found by interpolation in the t-table.

8 ▪ 4   CONFIDENCE INTERVALS AND TESTS OF HYPOTHESES ABOUT THE DIFFERENCE OF TWO MEANS FROM RELATED POPULATIONS

Under certain circumstances it is possible to greatly improve the accuracy of an experimental analysis by appropriately pairing together related samples rather than by using two independently drawn samples. In some cases (e.g., when a measurement is taken on the same sample unit before and after some treatment) it may be impossible to consider the two measurements that are under consideration as independent. In such instances we pair together samples from the two populations to obtain n *pairs* of observations. We might, for example, pair measurements taken under similar experimental circumstances, or on individuals who have the same IQ score. This technique is used often in medical studies to reduce the influence of variables that are not under active investigation. The pairs $(X_1, Y_1), \ldots, (X_n, Y_n)$ are formed; let $D_i = X_i - Y_i$ represent the difference for each pair. Testing $\mu_X = \mu_Y$ is equivalent to testing the mean difference $\mu_D = 0$, where $\mu_D = \mu_X - \mu_Y$.

Let $\overline{D} = \sum_{i=1}^{n} D_i/n$; recall that the expected value of $\overline{D}$ is

$$\mu_{\overline{D}} = E(\overline{X} - \overline{Y}) = \mu_X - \mu_Y = \mu_D$$

We assume the differences $D_1, D_2, \ldots, D_n$ are independent; hence, the variance of $\overline{D}$ can be calculated as before: Variance $(\overline{D}) = \sigma_D^2/n$. The estimator of σ_D^2 is

$$S_D^2 = \sum_{i=1}^{n} \frac{(D_i - \overline{D})^2}{(n-1)}$$

We assume we are sampling from a normally distributed population of D's. Then, as in Chapters 6 and 7, we see that

$$T = \frac{\overline{D} - \mu_D}{S_D/\sqrt{n}} \tag{8.7}$$

has a t-distribution with $n - 1$ degrees of freedom. Essentially, by pairing the data we have reduced the problem from a two-population problem to an analysis of a single population consisting of the differences. We already know how to handle the one-population situation by using the techniques of Chapters 6 and 7.

Confidence Intervals

The computation of confidence intervals and hypothesis tests is now carried out as was done in Chapters 6 and 7 with the single population of D_i's. Thus the limits of the $100(1 - \alpha)\%$ confidence interval for $\mu_D = \mu_X - \mu_Y$ are based on the t-distribution with $n - 1$ degrees of freedom. The interval is from

$$\overline{d} - t_{\alpha/2}\, s_D/\sqrt{n} \quad \text{to} \quad \overline{d} + t_{\alpha/2} s_D/\sqrt{n} \tag{8.8}$$

where $\overline{d}$ and s_D are the observed values of $\overline{D}$ and S_D, respectively. We have already worked some with these differences in the previous chapters, for example, problems 29 and 34 of Chapter 7.

Example 8 ▪ 10 Two hybrids of wheat were planted in various locations. To eliminate extraneous factors, the farmer divided each acre in half. Hybrid A was planted on one half and hybrid B was planted on the other half. The yields are as follows:

Acre	1	2	3	4	5	6	7	8	9
(x) Type A	6.1	5.9	7.0	7.2	6.9	7.0	7.5	6.0	6.5
(y) Type B	5.8	5.6	6.1	6.8	6.9	6.7	7.0	5.7	6.3

Find a 95% confidence interval for the difference between the mean yields of the two types of hybrid.

Assuming normally distributed yields, we calculate the observed differences $d = x - y$: $d_1 = 0.3$, $d_2 = 0.3$, $d_3 = 0.9$, $d_4 = 0.4$, $d_5 = 1.0$, $d_6 = 0.3$, $d_7 = 0.5$, $d_8 = 0.3$, and $d_9 = 0.2$. We find $\bar{d} = 0.47$, $s_D^2 = 0.0825$, and $s_D = 0.2872$. From Appendix A.5 we find that, with 8 degrees of freedom, $t_{0.025} = 2.306$. Using these in expression (8.8), we find the 95% confidence interval for $\mu_X - \mu_Y$ is from

$$0.47 - 2.306 \frac{(0.2872)}{\sqrt{9}} \quad \text{to} \quad 0.47 + 2.306 \frac{(0.2872)}{\sqrt{9}}$$

or from 0.25 to 0.69.

Hypothesis Testing

Using equation (8.7) also allows us to perform hypothesis tests about the mean difference $\mu_X - \mu_Y$ for two dependent or related normal populations. For example, if we are to test

$$H_0 : \mu_X = \mu_Y \quad \text{versus} \quad H_1 : \mu_X \neq \mu_Y$$

at the level of significance α, then we reject H_0 if the observed value of

$$T = \frac{\bar{X} - \bar{Y}}{S_D/\sqrt{n}} = \frac{\bar{D}}{S_D/\sqrt{n}}$$

takes on a value greater than $t_{\alpha/2}$ or less than $-t_{\alpha/2}$ with $n - 1$ degrees of freedom. This is just Chapter 7 revisited, using the population of differences as the underlying population of interest.

Example 8 ▪ 11 To test if diet influences learning ability, a sociologist chooses 61 children and gives them half of a standardized test. Each child is then fed a nutritious breakfast for a period of 3 weeks and retested using the second half of the test. The difference between their score prior to the nutritious breakfasts and their score after the breakfasts is taken as a measure of the influence of the diet. Because the same children are used in both parts of the study, the before and after test measurements are not independent. Suppose X is the score on the first half (before) and Y is the score on the second half (after) for each child. We want to test

$$H_0 : \mu_X = \mu_Y \quad \text{versus} \quad H_1 : \mu_X \neq \mu_Y$$

at level of significance $\alpha = 0.05$. We reject H_0 if $t = (\bar{x} - \bar{y})/\sqrt{s_D^2/61}$ takes on a value that is greater than $t_{0.025} = 2.000$ or less than $-t_{0.025} = -2.000$; here the number of degrees of freedom is $n - 1 = 60$. Suppose the sociologist finds that $\bar{x} = 83$, $\bar{y} = 85$, and $s_D^2 = 16.76$. Then $t = -2/0.524 = -3.82$, and we reject H_0; that is, the nutritious breakfast makes a significant difference at $\alpha = 0.05$.

We may also perform *one-tailed* (directional) hypothesis tests about the difference between two means of related measurements. To test

$$H_0 : \mu_X \geq \mu_Y$$
$$H_1 : \mu_X < \mu_Y$$

at the level of significance α, we reject H_0 if $t = (\bar{x} - \bar{y})/(s_D/\sqrt{n})$ takes on a value less than $-t_\alpha$ with $n - 1$ degrees of freedom. To test

$$H_0 : \mu_X \leq \mu_Y$$
$$H_1 : \mu_X > \mu_Y$$

at the level of significance α, we reject H_0 if $t = (\bar{x} - \bar{y})/(s_D/\sqrt{n})$ takes on a value greater than t_α. Again, the tail for the rejection region is determined by the inequality sign in H_1.

Example 8 ▪ 12   Suppose the sociologist in example 8.11 wished only to determine if nutritious breakfasts significantly *increased* a child's learning ability. We would then test

$$H_0 : \mu_X \geq \mu_Y$$
$$H_1 : \mu_X < \mu_Y$$

say, at the level of significance $\alpha = 0.01$. We then would reject H_0 if $t = (\bar{x} - \bar{y})/(s_D/\sqrt{61})$ takes on a value less than $-t_{0.01} = -2.39$, with 60 degrees of freedom. Since $t = -3.82$, we would still reject H_0 and conclude that the increase in learning ability is significant at the 1% level of significance.

We should also stress that pairing observations together from two populations increases the accuracy of the experimental analysis only provided the pairing is done correctly — that is, if the variation *between the different pairs* is larger than the variation *within the individual pairs.* For example, suppose we wish to study if the sex of students affects their grade point averages at graduation. We might suspect that their score on an entrance examination is a more significant indicator of their future performance. For this reason, we would pair men and women who achieve the same entrance test score. The variation in grade point average between pairs (those with different entrance scores) should be larger than that within the pairs (those who achieved the same entrance score). If we formed pairs based on some other variable, such as height, the pairing may actually be counterproductive. The variable you use to pair is very important.

Caution: Since the t-test based on pairing the observations has $n - 1$ degrees of freedom, whereas the t-test based on using two independent populations has $2n - 2$ degrees of freedom, if we pair observations unnecessarily, we lose information (lose degrees of freedom) and so lose accuracy in the sense that our type 2 error increases. This is because the effective sample size is smaller, and so there is a larger variance. Thus pairing the data may reduce the effect of exogenous variables, but not for free. The essential sample size as measured by the degrees of freedom also decreases. Hence pairing is not an omnibus solution to dependence; it only helps if the reduction in between group variance overcomes the loss in effective sample size.

8 ▪ 5   COMPARING TWO PROPORTIONS

In some type of experiments the data are presented as the proportion of the population exhibiting a certain characteristic, for example, the proportion of patients responding favorably to a drug therapy or the proportion of ethnic voters in a certain district. In this section we investigate methods for comparing two independent populations to determine which has a higher proportion of individuals with the given characteristic. Let $\hat{P}_1$ denote the sample proportion from the first population and $\hat{P}_2$ the sample proportion from the second population. The corresponding population proportions are denoted by p_1 and p_2. As seen in Chapter 6, sample proportions are really particular cases of sample means. For this reason the results of section 8.1 may be applied directly to this case provided the sample sizes n_1 and n_2 are sufficiently large. As in Chapter 6, we find the variance of $\hat{P}$ is $\sigma_{\hat{P}}^2 = p(1 - p)/n$ so that $\hat{P}_1 - \hat{P}_2$ has approximately a normal distribution with mean $p_1 - p_2$ and variance equal to $p_1(1 - p_1)/n_1 + p_2(1 - p_2)/n_2$. In most cases, of course, the values of p_1 and p_2 are unknown and must be estimated by the sample estimates $\hat{p}_1$ and $\hat{p}_2$, respectively.

Confidence Intervals

We may translate equation (8.1) directly to obtain the limits on the $100(1 - \alpha)\%$ confidence interval for $p_1 - p_2$. These are given by

---

The approximate $100(1 - \alpha)\%$ confidence interval for $p_1 - p_2$ is from

$$(\hat{p}_1 - \hat{p}_2) - z_{\alpha/2}\sqrt{\frac{\hat{p}_1(1 - \hat{p}_1)}{n_1} + \frac{\hat{p}_2(1 - \hat{p}_2)}{n_2}} \quad \text{to}$$

$$(\hat{p}_1 - \hat{p}_2) + z_{\alpha/2}\sqrt{\frac{\hat{p}_1(1 - \hat{p}_1)}{n_1} + \frac{\hat{p}_2(1 - \hat{p}_2)}{n_2}} \quad (8.9)$$

Example 8 ▪ 13   A theologian wished to see how the proportion of Americans of a specified religious group who were in favor of liberalizing church doctrines compared with the proportion of Western Europeans of this same religious group who were in favor of liberalizing church doctrines. A sample of 400 Americans yielded 331 in favor, whereas a sample of 250 Europeans yielded 153 in favor. Find a 95% confidence interval for the difference in the proportion of Americans and the proportion of Europeans favoring liberalization.

We first calculate

$$\hat{p}_1 = \frac{331}{400} = 0.828 \quad \text{and} \quad \hat{p}_2 = \frac{153}{250} = 0.612$$

from Appendix A.1 we find $z_{0.025} = 1.96$. Using these values in equation (8.9) yields the 95% confidence for $p_1 - p_2$ from

$$(0.828 - 0.612) - 1.96\sqrt{\frac{(0.828)(0.172)}{400} + \frac{(0.612)(0.388)}{250}} \quad \text{to}$$

$$(0.828 - 0.612) + 1.96\sqrt{\frac{(0.828)(0.172)}{400} + \frac{(0.612)(0.388)}{250}}$$

and so we have confidence 0.95 that $p_1 - p_2$ is somewhere in the interval from 0.145 to 0.287; that is, the theologian can be 95% certain that the proportion of Americans in favor of liberalization is from 14.5% to 28.7% higher than the proportion of their European counterparts.

Hypothesis Testing

By utilizing the fact that for large values of n_1 and n_2, $\hat{P}_1 - \hat{P}_2$ has a distribution that is approximately normal, we may construct hypothesis tests about the difference $p_1 - p_2$. The hypothesis tests we derive here are analogous to the hypothesis tests about the difference of two means developed in Section 8.1 (indeed, as noted before, $\hat{p}_1$ and $\hat{p}_2$ can be considered as observed sample means from the two populations). If we wish to test

$$H_0 : p_1 - p_2 = c$$

$$H_1 : p_1 - p_2 \neq c$$

where c is a constant (possibly zero), we reject H_0 if $\hat{p}_1 - \hat{p}_2 - c$ is too large or too small. Specifically, at the level of significance α we reject H_0 if

$$z = \frac{\hat{p}_1 - \hat{p}_2 - c}{\sqrt{\dfrac{\hat{p}_1(1 - \hat{p}_1)}{n_1} + \dfrac{\hat{p}_2(1 - \hat{p}_2)}{n_2}}} > z_{\alpha/2} \quad \text{or} \quad z < -z_{\alpha/2}$$

Example 8 ▪ 14   An investigator samples 100 men and finds 3 who are color-blind. Among a group of 75 women, 1 is color-blind. A colleague of the investigator hypothesizes that men and women have almost the same prevalence of colorblindness —that the difference in this prevalence between the sexes is 1%. Do the investigator's data depart significantly from the hypothesis? To solve this, let p_1 be the proportion of men who are color-blind and let p_2 denote the proportion of women who are color-blind. We wish to test

$$H_0 : p_1 - p_2 = 0.01$$

$$H_1 : p_1 - p_2 \neq 0.01$$

Let us use the level of significance $\alpha = 0.1$. We reject H_0 if $z > z_{0.05} = 1.645$ or if $z < -1.645$. Since $\hat{p}_1 = \frac{3}{100} = 0.03$ and $\hat{p}_2 = \frac{1}{75} = 0.0133$, we have

$$z = \frac{0.03 - 0.0133 - 0.01}{\sqrt{\dfrac{(0.03)(0.97)}{100} + \dfrac{(0.0133)(0.9867)}{75}}} = \frac{0.0067}{0.0047} = 0.310$$

Since $z = 0.310 < 1.645$, we cannot reject H_0. The difference is not significantly disparate from 1% at the level $\alpha = 0.05$. The colleague may be correct.

(When testing $H_0 : p_1 = p_2$ (which is the special case of $c = 0$), we can obtain an improvement by pooling $\hat{P}_1$ and $\hat{P}_2$, just as we improved our estimate of σ^2 when $\sigma_X^2 = \sigma_Y^2$ by pooling S_X^2 and S_Y^2. The pooled proportion estimator is $\hat{P}_p =$

$(n_1 \hat{P}_1 + n_2 \hat{P}_2)/(n_1 + n_2)$. The corresponding test statistic becomes

$$Z = \frac{\hat{P}_1 - \hat{P}_2}{\sqrt{\hat{P}_p(1 - \hat{P}_p)\left(\dfrac{1}{n_1} + \dfrac{1}{n_2}\right)}}$$

and we then proceed with our test as before.)

To perform one-tailed (directional) hypothesis tests comparing two population proportions, we may use the fact that $\hat{p}_1 - \hat{p}_2$ has approximately a normal distribution with mean $p_1 - p_2$ and a variance

$$\frac{p_1(1 - p_1)}{n_1} + \frac{p_2(1 - p_2)}{n_2}.$$

Suppose we wish to test

$$H_0 : p_1 \geq p_2 + c$$
$$H_1 : p_1 < p_2 + c$$

If H_1 is the true state of nature, then $p_1 - p_2 - c$ should be a negative number. Thus at level of significance α we would reject H_0 if

$$z = \frac{\hat{p}_1 - \hat{p}_2 - c}{\sqrt{\dfrac{\hat{p}_1(1 - \hat{p}_1)}{n_1} + \dfrac{\hat{p}_2(1 - \hat{p}_2)}{n_2}}} < -z_a$$

Example ■ 8.15   A criminal psychologist claims to have developed a prison rehabilitation program that can reduce the proportion of recidivists (repeat offenders) by more than 30%. To prove this claim, the psychologist convinced state officials to put 45 inmates through the program, and it was found that within 2 years after their release 28 to them were rearrested. Of the 85 other prisoners who underwent conventional rehabilitation therapy and who were released in the same time period, there were 28 who were rearrested. To analyze the psychologist's claim in light of the evidence, we let p_2 be the proportion of repeat offenders when using the conventional therapy method and p_1 denote the proportion of repeat offenders when using the new rehabilitation therapy. We are investigating the claim that the program works, so this must be the alternative hypothesis; that is, we wish to test

$$H_0 : p_1 \geq p_2 - 0.3 \quad \text{versus} \quad H_1 : p_1 < p_2 - 0.3$$

We use $\alpha = 0.05$ and reject H_0 if $z < -z_{0.05} = -1.645$. We calculate $\hat{p}_1 = \frac{28}{45} = 0.622$ and $\hat{p}_2 = \frac{72}{85} = 0.847$ and find that

$$z = \frac{0.622 - 0.847 - (-0.3)}{\sqrt{\dfrac{(0.847)(0.153)}{85} + \dfrac{(0.622)(0.378)}{45}}} = \frac{+0.075}{\sqrt{0.00675}} = +0.913$$

Since $+0.913 > -1.645$, we fail to reject H_0; that is, the criminal psychologist cannot prove the claim (the data are not inconsistent with a smaller than 30% improvement).

In a similar manner, we may test the hypothesis

$$H_0 : p_1 \leq p_2 + c \quad \text{versus} \quad H_1 : p_1 > p_2 + c$$

at the level of significance α. We reject H_0 if

$$z = \frac{\hat{p}_1 - \hat{p}_2 - c}{\sqrt{\hat{p}_1(1 - \hat{p}_1)/n_1 + \hat{p}_2(1 - \hat{p}_2)/n_2}} > z_\alpha.$$

Example 8 ▪ 16   Two drugs have possible use in treating a certain disease. The manufacturer of the second drug claims that it is superior to the first drug. In a test of 30 patients treated with the first drug, there were 18 favorable reactions, whereas in a test of 45 patients with the second drug, there were 23 favorable reactions. Can we reject the second manufacturer's claim? Let p_1 and p_2 denote the proportion of patients who respond favorably to drug number 1 and 2, respectively. We wish to test

$$H_0 : p_1 \leq p_2 \quad \text{versus} \quad H_1 : p_1 > p_2$$

We use $\alpha = 0.01$ and reject H_0 if $z > z_{0.01} = 2.326$. We calculate $\hat{p}_1 = \frac{18}{30} = 0.6$, $\hat{p}_2 = \frac{23}{45} = 0.511$ and thus

$$z = \frac{0.6 - 0.511}{\sqrt{\dfrac{(0.6)(0.4)}{30} + \dfrac{(0.511)(0.489)}{45}}} = \frac{0.089}{\sqrt{0.0136}} = 0.764$$

Since $z = 0.764 < 2.326$, we fail to reject H_0; that is, the second manufacturer could be telling the truth, we do not have sufficient evidence to contradict his claim.*

---

* Since $c = 0$, we could have used the pooled estimator $\hat{P}_p$. For large sample sizes, the results using this estimator are close to those using the procedure of this example.

8 ▪ 6 COMPARING TWO POISSON POPULATIONS (CONTRASTING TWO RARE EVENTS) (OPTIONAL)

As we saw in Chapters 4 and 7, the Poisson distribution is a model for finding the probability of observing rare events. Specifically, when we are dealing with an event that occurs with very small probability, and if we observe a large number of trials, then the number of times this rare event happens has a Poisson distribution where the parameter λ is the expected number of occurrences. We are really approximating a binomial with p very small and n very large by a Poisson distribution. We take

$$\lambda = (\text{number of trials}) \cdot (\text{probability of rare event}) = np$$

For example, suppose we are studying a certain type of congenital birth defect that only occurs with probability 0.0001 (one in 10,000). If there are 30,000 births in a particular area, then the number of babies born with this defect might reasonably be assumed to follow a Poisson distribution with $\lambda = (30000)(0.0001) = 3$. The probability of observing zero defects is 0.0498, one defect is 0.1494, and at most one defect is $0.0498 + 0.1494 = 0.1992$.

In many studies involving medicine, anthropology, psychology, genetics, or astronomy we have two populations that have Poisson, or approximately Poisson, distributions and we wish to compare them. For example, we may wish to see if a certain vaccine reduces the incidence of a rare disease or if catatonic schizophrenia is more prevalent in one culture than in another or if a certain rare astronomical phenomenon is more likely in one galaxy than in another. In all these examples we have two populations that have Poisson distributions.

To derive a hypothesis test for comparing two independent Poisson populations, let us suppose we sample n_1 times from the first population and let X denote the number of occurrences we observe in this sample. Similarly, we sample n_2 times from the second population and let Y denote the number of occurrences in this sample. Assume X has a Poisson distribution with mean $\lambda_1 = n_1 p_1$ and Y has a Poisson distribution with mean $\lambda_2 = n_2 p_2$. We wish to compare p_1 and p_2. Now, in general, it is true that if X has a Poisson distribution with mean λ_1 and Y has a Poisson distribution with mean λ_2, then $X + Y$ has a Poisson distribution with mean $\lambda_1 + \lambda_2$ (i.e., the expected number for the sum is the sum of the expected numbers). Using the definition of conditional probability, we may then calculate

$$P(X = k \mid X + Y = n) = \frac{P\{(X = k) \text{ and } (X + Y = n)\}}{P(X + Y = n)}$$

$$= \frac{P\{(X = k) \text{ and } (Y = n - k)\}}{P(X + Y = n)}$$

but since X and Y are independent, this equals

$$\frac{P(X=k)P(Y=n-k)}{P(X+Y=n)} \tag{8.10}$$

Now X has a Poisson distribution, so $P(X=k)=e^{-\lambda_1}(\lambda_1)^k/k!$ and Y has a Poisson distribution, so

$$P(Y=n-k)=\frac{e^{-\lambda_2}(\lambda_2)^{n-k}}{(n-k)!}$$

and $X+Y$ has a Poisson distribution so that

$$P(X+Y=n)=\frac{e^{-(\lambda_1+\lambda_2)}(\lambda_1+\lambda_2)^n}{n!}$$

Substituting these results in expression (8.10), we have

$$P(X=k\,|\,X+Y=n)=\frac{n!}{k!(n-k)!}\left(\frac{\lambda_1}{\lambda_1+\lambda_2}\right)^k\left(\frac{\lambda_2}{\lambda_1+\lambda_2}\right)^{n-k}$$

This formula tells us how likely it is in a total group of n occurrences that exactly k of them can be attributed to X. We now use this result for testing hypotheses about p_1 and p_2.

Suppose we wish to test

$$H_0:p_1=p_2$$
$$H_1:p_1\neq p_2$$

where p_1 and p_2 are very small, but n_1 and n_2 are large. If H_0 is true and the Poisson distribution is used, then $\lambda_1=n_1p$ and $\lambda_2=n_2p$, where p is the common value, $p=p_1=p_2$. We thus find

---

When $p_1=p_2$, we have

$$P[X=k\,|\,X+Y=n]=\binom{n}{k}\left(\frac{n_1}{n_1+n_2}\right)^k\left(\frac{n_2}{n_1+n_2}\right)^{n-k}$$

so that

$$P[X\leq k\,|\,X+Y=n]=\sum_{j=0}^{k}\binom{n}{j}\left(\frac{n_1}{n_1+n_2}\right)^j\left(\frac{n_2}{n_1+n_2}\right)^{n-j}$$

and

$$P[X\geq k\,|\,X+Y=n]=\sum_{j=k}^{n}\binom{n}{j}\left(\frac{n_1}{n_1+n_2}\right)^j\left(\frac{n_2}{n_1+n_2}\right)^{n-j}$$

Notice, these are the binomial probabilities!

Summarizing; in order to test

$$H_0 : p_1 = p_2$$
$$H_1 : p_1 \neq p_2$$

at the level of significance α, we let k denote the number of occurrences in the first sample and n denote the total number of occurrences in both samples combined. We reject H_0 if

$$P(X \leq k \,|\, X + Y = n) < \alpha/2 \quad \text{or if} \quad P(X \geq k \,|\, X + Y = n) < \alpha/2$$

This probability is the random chance of observing something *as extreme or more extreme than what we observed in fact.* If this probability is too small, then the credibility of the hypothesis $p_1 = p_2$ is in doubt. Intuitively, this says that if the first sample accounts for too few or too many of the total number of occurrences, then we reject H_0. The philosophy of this test is the same as our "proof by contradiction" interpretation of hypothesis testing. If an event has too low a probability under H_0, then reject H_0.

Example 8 ▪ 17   A large factory is considering switching the brand of light bulbs they use to a less expensive brand. They first wish to determine if the proportion of bulbs of brand 1 that burn out per week is the same as the proportion of bulbs of brand 2 that burn out per week. Let p_1 and p_2 denote these proportions, respectively. To test if $p_1 = p_2$, the company installs 240 bulbs of brand 1 and 360 bulbs of brand 2. They observe that one of the brand 1 bulbs burns out and 9 of the brand 2 bulbs burn out. Test if $p_1 = p_2$ at the level of significance $\alpha = 0.05$. Here we have $n = 1 + 9 = 10$ total failures, and $k = 1$ of these from brand 1. Then the chance under H_0 (i.e., under the assumption that the disparity is due to random variation) of observing something as extreme or more extreme than the observed 1 to 9 split of the 10 bulbs that burned out is

$$P(X \leq 1 \,|\, X + Y = 10) = \binom{10}{0}\left(\frac{240}{600}\right)^0 \left(\frac{360}{600}\right)^{10} + \binom{10}{1}\left(\frac{240}{600}\right)^1 \left(\frac{360}{600}\right)^9$$

$$= (0.6)^{10} + 10(0.4)(0.6)^9 = 0.0060 + 0.0403 = 0.0463$$

If the "equal probability" hypotheses is true, there is only a 0.0463 chance of observing what we actually did observe.

Since $\alpha/2 = 0.025 < 0.0463$, we fail to reject H_0, that is, the proportion of bulbs failing in the first week could be the same for the two brands. The factory should buy the less expensive brand. (Note at $\alpha = 0.1$ we would have rejected H_0.)

We may also utilize the preceding method to perform one-tailed (directional) hypothesis tests about p_1 and p_2 in the rare event situation. To test

$$H_0 : p_1 \geq p_2 \quad \text{versus} \quad H_1 : p_1 < p_2$$

at level of significance α, we would reject H_0 if the observed split of the n total occurrences is too unlikely under H_0, that is, if $P(X \leq k \mid X + Y = n) < \alpha$. Again, X is the number of occurrences in the first sample and n is the total number of occurrences. Similarly, to test

$$H_0 : p_1 \leq p_2 \quad \text{versus} \quad H_1 : p_1 > p_2$$

at level of significance α, we reject H_0 if $P(X \geq k \mid X + Y = n) < \alpha$. Again, we note that the inequality sign involved in the statistical test is in agreement with the sign of the alternative hypotheses.

Example 8 ▪ 18
A manufacturer has a machine that produces transistors. In a group of 15,000 transistors produced by the machine there are 11 defectives. A serviceman does maintenance work on the machine and among the next 15,000 transistors produced there are only 8 defectives. Can we conclude that the maintenance work decreased the proportion of defectives? We let p_1 be the proportion of defectives produced by the machine prior to maintenance and p_2 the proportion of defectives after maintenance. We test

$$H_0 : p_1 \leq p_2$$
$$H_1 : p_1 > p_2$$

at level of significance $\alpha = 0.05$. We have $n = 11 + 8 = 19$, $k = 11$, and $n_1/(n_1 + n_2) = 0.5$. Thus we reject H_0 if $P(X \geq 11 \mid X + Y = 19) < 0.05$. We calculate

$$P(X \geq 11 \mid X + Y = 19) = \binom{19}{11}(0.5)^{19} + \binom{19}{12}(0.5)^{19}$$

$$+ \cdots + \binom{19}{19}(0.5)^{19}$$

$$= 0.1442 + 0.0961 + \cdots + 0.0000019$$

$$= 0.324 > 0.05,$$

so we fail to reject H_0, that is, we cannot conclude that the maintenance reduced the proportion of defectives produced. Notice that in this problem, since $n_1 = n_2$ (so $n_1/(n_1 + n_2) = 0.5$) the preceding calculation is identical to the calculation of the chance that 19 flips of a coin will produce 11 or more heads.

8 ▪ 7 COMMENT: CONTROL AND STUDY GROUPS; THE INTERPLAY OF DATA COLLECTION AND ANALYSIS

Let us say that we have developed a vaccine, such as the Salk vaccine, to aid in preventing the onset of polio for, say, 1 year. The vaccine must be tested. A sample of the population called the *study* group is selected and given the vaccine. Another sample, called the *control* group is selected and does not receive the vaccine. At the end of 1 year the average number in each group who became ill with the disease is determined and tested. Note, H_0 is the hypothesis of "no effect" and so we test

$H_0 : \mu_s \geq \mu_c$ $\mu_s =$ theoretical mean number in the study group who become ill in 1 year

$H_1 : \mu_s \leq \mu_c$ $\mu_c =$ theoretical mean number in the control group who become ill in 1 year

The matter seems simple enough, but only because everything involved with the choice of the study and control group has been omitted. Suppose the study group consisted of adults between 30 and 45 years of age and the control group of children between 2 and 5. The selection of such groups would be a blunder, negating our test since polio is a childhood disease. Clearly, the study and control groups must be matched in some manner. One way is deliberately to try to find one child in the study group who matches one child in the control group. The matching might be done on the basis of age, sex, economic standing, and so on. Matching is often too difficult or expensive to do. Another experimental design often used instead is to select a sample from the pertinent population and *randomly* to assign members of the sample to the study and control groups. (Randomization will be discussed in more detail in Chapter 11.)

As can be seen from the preceding discussion, there are major problems in the selection of the study and control groups. (We discuss this matter again in Chapter 10.) But there are also difficulties in the *administration* of the experiment. It often happens that being given "medication" has the psychological effect of making one "feel better." To offset this well-documented effect, both the study group and control group are told they are being given medication, but the control group is actually given a placebo, a pill that appears to be medication but is not. Such studies are called single-blind studies in that the participants do not know who is getting the medication. However, it has also been discovered that sometimes there is an effect if the doctor who is administering the drug knows who is and is not receiving a placebo. The doctor, perhaps unconsciously, has a different attitude toward the control and study groups and may communicate this attitude to members of the groups. A study in which the doctor and participants both do not know who is getting the medication and placebo is called a double-blind study.

These are a few basic problem areas found in many scientific investigations. There are numerous others. Take, for example, a study conducted by the

Division of Human Reproduction of the World Health Organization in three countries (Egypt, Thailand, and Phillipines, 1976 to 1978) to determine the acceptability of side effects of contraceptive methods. Side effects such as "weight gain" might be unacceptable in some societies while perfectly acceptable or even desirable in others. To provide uniform and comparable measures between societies, the experimenters developed a questionnaire as an instrument for determining "acceptability." To minimize translation problems, they devised a precoded or categorical questionnaire (discussed in Chapter 15). The development of the questionnaire was no simple matter and even after it was finished (according to the notes of one of the investigators), "Serious doubts were raised as to whether the right questions were being asked, whether some issues had been left out that should have been included or whether too much had been included."

Such issues of data collection must be resolved if the analyses we perform, such as the t-test, are to have any value in advancing our knowledge. We simply cannot put the burden for valid results on the analysis. The design of the experiment itself is crucial in every experiment. This is a matter we pursue in subsequent chapters.

8 ■ 8 COMMENTS ON COMPUTER APPLICATIONS

We have discussed methods for obtaining confidence limits and for performing hypothesis tests about population means when we have either one or two populations. Most of our examples and exercises contain small data sets. When the variances are unknown but the sample sizes are large, we can use the normal (z) table instead of the t-table to find the required upper centile values; in this case we need no longer worry about the value of df. However, this simplification is often offset by the considerable increase in computation required to calculate the sample means and variances. Even for a sample size of about 100, the calculations become tedious and prone to error. This observation applies also to obtaining confidence intervals and testing hypotheses for the variance.

We have been rescued from these arduous tasks by the computer. The software is now available, even for the small science-oriented microcomputer, to perform these calculations quickly and easily. The standard software packages previously mentioned, SPSS, MINITAB, and so on, also contain software features that enable the automatic computation of confidence intervals and of those statistics required for testing hypotheses about the mean and variance.

Summary

In this chapter we learned how to find confidence limits on the difference between the means of two populations, both for the case when the variances are known, and also when the variances are unknown. In addition, tests of hypotheses were discussed for the difference of the means of two independent populations for the instances when the variances are known and unknown.

In Section 8.4 confidence intervals and tests of hypothesis were developed about the difference of the means from two related populations. Additional techniques were developed, assuming two independent populations, for placing confidence limits on the difference of two proportions and testing hypotheses about these differences. Such tests were also devised for the special case where the proportions are very small but the number of observations from the two populations is large.

Hypothesis tests about the variances of two independent populations were also developed; the F-test was introduced in this discussion.

Key Words

z-test for the difference
    between two means for
    two populations
t-test for the difference
    between two means for
    two populations
F-distribution
z-test for the difference
    between two proportions
    for two populations

test for the difference
    between two means from
    two associated popula-
    tions
pooled estimate of the
    variance
comparison of very small
    probabilities

References

See the sources cited in problems 6, 13, and 24.

Exercises

Section 8.1

1. For the following data, assume independent normal distributions. In each case
    a. Give a $100(1 - \alpha)\%$ confidence interval for $\mu_X - \mu_Y$.
    b. Perform the given hypothesis test.

n_1	n_2	σ_X^2	σ_Y^2	$\bar{x}$	$\bar{y}$	α	H_0	H_1
16	16	4	12	10.5	12.5	0.05	$\mu_X = \mu_Y$	$\mu_X \neq \mu_Y$
16	18	4	36	33	36	0.01	$\mu_X \geq \mu_Y$	$\mu_X < \mu_Y$
10	2	5	17	16	13.5	0.05	$\mu_X \leq \mu_Y$	$\mu_X > \mu_Y$
8	20	2	5	17	20.5	0.02	$\mu_X - \mu_Y = 3$	$\mu_X - \mu_Y \neq 3$
108	72	12	280	20	24.25	0.05	$\mu_X \geq \mu_Y + 1$	$\mu_X < \mu_Y + 1$

c. Use the data of Table 2.7 to test the hy-

pothesis that the populations of the two groups have equal systolic blood pressures. Assume that the standard deviation of the pressure for the 21 to 44 age group is 12.5 and for the other age group is 23.2. Use $\alpha = 0.05$.

2. Two different manufacturers make plastic for producing flexible black plastic plumbing pipe. Manufacturer A charges 33¢ per kilogram and has a ductility variance of 40. Manufacturer B sells plastic for 35¢ per kilogram and has a ductility variance of 32. Manufacturer B justifies the higher price by asserting

that the mean ductility measurement of B's plastic is three units higher than that of Manufacturer A. A customer chooses five samples of Manufacturer A's plastic and obtains an average ductility measurement of $\bar{x} = 38$. From three samples of Manufacturer B's plastic, the customer obtains an average ductility measurement of $\bar{y} = 46$. Does this information disprove Manufacturer B's claim at $\alpha = 0.01$? Also find a 98% confidence interval for $\mu_X - \mu_Y$.

3. An educational counselor wishes to determine if the grade point average of mathematics majors at a certain university is higher than that of history majors at the same university. It has been determined that the variance in grade point for history majors is $\sigma_X^2 = 2.5$ and for math majors is $\sigma_Y^2 = 1.5$. A sample of 24 history majors and 30 math majors yields $\bar{x} = 2.52$ and $\bar{y} = 3.1$, respectively. Can we conclude that the mean grade point average for math majors is higher than that for history majors? Use $\alpha = 0.05$. Find a 90% confidence interval for $\mu_X - \mu_Y$.

4. In portfolio management in finance one wishes to choose those investments that yield the largest return on the money invested. The riskiness of the investment is often measured by the standard deviation of the return variable, which in financial circles is called the volatility of the investment. An investment manager is considering two common stocks for possible inclusion in a portfolio. Both are known to have the same volatility measure $\sigma = 0.049$. The returns on the two stocks are estimated from past data. For stock A, 16 weeks worth of data give the average return of $\bar{x} = 0.12$, whereas for stock B, 25 weeks worth of data gives $\bar{y} = 0.10$. Can we conclude that the expected return μ_X on stock A is significantly larger than that on stock B? (Use $\alpha = 0.05$.)

5. Two different instructors are teaching the same statistics class and give common exams. The first instructor's students have a test variance known to be $\sigma_X^2 = 400$ and the second instructor's students have a test variance of $\sigma_Y^2 = 625$. The 22 students taught statistics by instructor 1 have a mean score on the final of $\bar{x} = 72$, whereas the 36 students taught statistics by instructor 2 have a mean score on the final of $\bar{y} = 79$. At $\alpha = 0.1$ can we conclude that the students taught by instructor 2 perform better than those taught by instructor 1?

6. German white wines are considered among the finest quality wines in the world, a fact that is surprising considering the geographical handicaps of the German vineyards. In a paper by Dyer and Ansher, an analytical profile of various German white wines was given. The wine regions are divided into the four areas of the German rivers: the Rhine, the Moselle, the Main, and the Neckar. For a sample of 10 Moselle wines, and 14 Rhine wines, they found the following:

Analyses of Moselle and Rhine Wines, 1970

	Total acids	*(g/100 ml)* Solids	Ash	*% v/v* Ethanol	Methanol
Moselle wines ($n = 10$ wines)					
Avg $\bar{x}$	0.61	3.29	0.218	8.31	0.005
SDs$_X$	0.17	0.40	0.027	1.28	0.001
Rhine wines ($n = 14$ wines)					
Avg $\bar{y}$	0.61	3.25	0.231	9.17	0.006
SDs$_Y$	0.13	0.41	0.040	0.96	0.002

(mg/100 ml)

	Ethyl acetate	n-Propyl alcohol	i-Butyl alcohol	i-Amyl alcohol
Moselle wines ($n = 10$ wines)				
$\bar{x}$	2.92	3.94	3.62	14.19
s_X	1.36	3.54	0.97	1.75
Rhine wines ($n = 14$ wines)				
$\bar{y}$	4.39	3.39	5.32	15.58
s_Y	2.11	1.56	2.24	4.42

ppm

	Na	K	Mg	Ca	Fe	Cu	Zn
Moselle wines ($n = 10$ wines)							
$\bar{x}$	21.6	769	78	61	0.31	0.19	0.4
s_X	10.4	114	11	14	0.11	0.13	0.3
Rhine wines ($n = 14$ wines)							
$\bar{y}$	16.5	901	83	56	0.31	0.31	0.3
s_Y	11.1	98	21	14	0.08	0.19	0.3

a. Using these data and assuming $\sigma_X = s_X$ and $\sigma_Y = s_Y$, find a 90% confidence interval for the expected difference in the ash content of Moselle and Rhine wines.

b. Test the hypothesis that Rhine wines have a higher copper content (Cu) than Moselle wines. Use $\alpha = 0.1$ and assume $\sigma_X = s_X$ and $\sigma_Y = s_Y$.

(Dyer, R. H., and A. F. Ansher: Analysis of German wines by gas chromatography and atomic absorption. *Journal of Food Science, 42*(2):534, 1977. Copyright Inst. Food Technologists)

Section 8.2

7. Test each of the following hypotheses at the given level of significance.

α	n_1	n_2	s_X^2	s_Y^2	H_0	H_1
0.01	16	11	25	110	$\sigma_X^2 = \sigma_Y^2$	$\sigma_X^2 \neq \sigma_Y^2$
0.05	21	121	64	34	$\sigma_Y^2 \geq \sigma_X^2$	$\sigma_Y^2 < \sigma_X^2$
0.1	31	21	4	7	$\sigma_Y^2 \leq \sigma_X^2$	$\sigma_Y^2 > \sigma_X^2$
0.02	61	101	121	145	$\sigma_Y^2 = 2\sigma_X^2$	$\sigma_Y^2 \neq 2\sigma_X^2$
0.05	11	21	144	160	$\sigma_Y^2 \geq 3\sigma_X^2$	$\sigma_Y^2 < 3\sigma_X^2$

8. Two suppliers of plastic for toys are competing for the contract with a major toy manufacturer. The toy manufacturer would like to choose the supplier whose product is most consistent or if there is no significant difference in their consistency, the least expensive supplier. The toy manufacturer takes 11 samples from Supplier X, makes plastic chips, and tests them for quality, rating each chip from 0 to 25. Similarly, 11 samples are taken from Supplier Y, chips made, and rated from 0 to 25. Consistency is determined by the size of the variance in rating. A large variance means poor consistency. Upon calculating the variance of the samples taken from the two suppliers, one finds that $s_X^2 = 25$ and $s_Y^2 = 12$. At $\alpha = 0.05$ should the toy manufacturer conclude that there is no significant difference between the two suppliers' consistency?

9. The consistency of the taste in a wine is of importance to buyers. The smaller the variance in concentration of the components in the wine is, the more consistent it is. In the limiting case of zero variance, the concentrations of the components are identical, and so the wines are also identical. By testing whether the variance is smaller, we are testing for consistency in the wine. Using the data from problem 6, test whether Rhine wines are more consistent than Moselle wines on potassium (K) concentration. (Use $\alpha = 0.05$.)

10. Repeat problem 9 by using the *i*-Amyl alcohol component.

11. The riskiness of an investment opportunity is often measured by the variance of the return distribution. For example, an investment in gold stock has high variance in returns, whereas a AAA rated bond has very low variance in returns. Consequently, gold is considered the riskier of the two investments. A security analyst considering which of two stocks to buy must weigh the risk of the investment against the potential return. A sample of 16 weeks past data on two stocks has led the analyst to believe stock A will have a higher expected return than stock B. Now we must examine the riskiness of the two investments. For stock A we have $n_1 = 21$ and $s_1^2 =$

0.01; whereas for stock B we have $n_2 = 21$ and $s_2^2 = 0.0025$. Test if Stock A is significantly riskier than stock B at $\alpha = 0.01$.

12. A bakery is considering new recipes for making dough for donuts. One potentially useful recipe can be mixed much faster and less expensively than the recipe currently being used; however, there is some concern that the variance in the shelf life of the donuts produced by this new recipe will be too great. Accordingly, a sample of 21 packages of donuts made with the old recipe and 21 packages made with the new recipe is taken and the variances in shelf life determined. It is found that $s_1^2 = 17.2$ sq days for the old and $s_2^2 = 23.7$ sq days for the new recipe. Can they conclude that the new recipe produces a larger variance in shelf life? (Use $\alpha = 0.05$.)

13. In a paper by Fraenkel and Blewett a comparison was made of the weights of pupae developed in 70% and 20% relative humidity. The larvae studied were put on the same diets, so that weight differences between the two groups could be attributed to the humidity factor. The following data were obtained:

70% RH			20% RH		
22.5	26.5	25.5	19.0	21.3	21.2
24.2	23.5	31.0	18.0	18.3	20.6
24.5	25.5	24.0	14.5	18.0	17.1
29.2	27.0	27.8	18.5	18.5	19.5
27.5	23.5	23.5	19.0	18.8	
24.5	23.5		19.5	17.9	

Test the hypothesis that the variances of the two groups are equal. (Use $\alpha = 0.1$.)
(Fraenkel, G., and Blewett, M.: The utilization of metabolic water in insects. Reproduced by permission by *Bulletin of Entomological Research,* 35:1944–1945. Publ. at Comm. Inst. of Entomology, London, U.K.)

Section 8.3

14. Calculate the pooled variance s_p^2 for the following data:

a. $n_1 = 12,\quad n_2 = 25,\quad s_X^2 = 164,\quad s_Y^2 = 275$

b. $n_1 = 30,\quad n_2 = 17,\quad \sum_{i=1}^{30} (x_i - \bar{x})^2 = 360,$

$\sum_{i=1}^{17} (y_i - \bar{y})^2 = 90$

c. $n_1 = 6,\quad n_2 = 12,\quad \sum_{i=1}^{6} x_i^2 = 1096,$

$\sum_{i=1}^{6} x_i = 24,\quad \sum_{i=1}^{12} y_i^2 = 1207,\quad \sum_{i=1}^{12} y_i = 108$

15. (Optional) Calculate the pooled variance s_p^2 for the following components of the white wine as given in problem 6: a. total acids, b. ash, c. ethanol, d. ethyl acetate, e. *i*-Butyl alcohol, f. potassium (K)

16. A hospital administrator wishes to determine if the average length of time spent in the hospital by persons with medical insurance is longer than the average length of time spent by persons without medical insurance. The administrator suspects this might be the case since most medical insurance pays a large percentage of in-hospital expenses, and hence the patient would be more inclined to remain hospitalized until fully recovered even if the patient could go home. To test these suspicions, the administrator gathered the following data on the length of stay of patients in the hospital for the same surgical operation:

With insurance	Without insurance
$n_1 = 31$	$n_2 = 13$
$\bar{x} = 23$	$\bar{y} = 24.5$
$s_X^2 = 1.5$	$s_Y^2 = 2.0$

a. Test if the two variances are equal. (Use $\alpha = 0.1$.)

b. Find a 90% confidence interval for $\mu_X - \mu_Y$ (assume equality of variances).

c. Test the administrator's suspicion using the level of significance $\alpha = 0.05$. State your conclusions.

17. A marine biologist suspects that the various manufacturing and chemical plants in the northeastern United States are dumping chemicals in the water that eventually show up in the cells of fish caught off the coast of Nantucket. To test this hypothesis, the biologist catches 50 cod close to the Atlantic Coast and another 50 cod far out to sea. On analyzing the tissues of the fish for the concentration of a certain heavy metal, the biologist obtained the following data.

Cod close to shore	Cod away from shore
$n_1 = 50$	$n_2 = 50$
$\bar{x} = 32.1$	$\bar{y} = 30.9$
$s_X^2 = 0.3$	$s_Y^2 = 0.21$

a. Calculate s_p^2.

b. Assume the two population variances are equal and test whether the fish caught close to shore have a significantly higher concentration of the heavy metal in their tissue (use $\alpha = 0.1$).

18. A criminologist wished to determine if chronic long-time use of alcohol actually has an inhibiting effect on the reflex reaction rate when sober. Twenty-one chronic alcohol users and 31 nondrinkers of the same age, social class, and so on were selected and their reaction rates tested when sober. The following data were obtained:

Drinkers	$\bar{x} = 4.8$	$s_X^2 = 2.1$
Nondrinkers	$\bar{y} = 3.6$	$s_Y^2 = 4.1$

Test if the chronic drinkers have a significantly slower reaction rate than the non-drinkers at $\alpha = 0.05$.

19. Returning to the German white wine data given in problem 6, use the pooled variance and test if Moselle wines contain significantly less ash than Rhine wines. (Use $\alpha = 0.1$.)

20. (Optional) Repeat the analysis of problem 15, only this time concerning the percentage ethanol in the two wines, and find a 95% confidence interval for the difference in ethanol percentage in the two wines.

21. Using the data of problem 13, test the hypothesis that higher humidity results in higher pupae weight. (Use $\alpha = 0.05$.)

Section 8.4

22. The long-range environmental effects of crude oil spills are not yet fully understood. Consider the following fictional data. Measurements were made on the amount of dissolved oxygen in various rivers both upstream and downstream from an oil spill. Since different rivers have intrinsically different amounts of oxygen in them, the data were considered as paired.

Spill number	Oxygen upstream (ppm)	Oxygen downstream (ppm)
1	48	47
2	20	20
3	16	14
4	50	51
5	32	30

a. Find a 90% confidence interval for the mean difference in oxygen concentration above and below the oil spill.

b. Test if the spills significantly decrease the oxygen concentration below the spill. (Use $\alpha = 0.1$.)

23. A psychologist gives a group of subjects a questionnaire on their views on violence. The subjects are then shown a violent film and afterward the subjects are given a second questionnaire similar to the first. We wish to ascertain whether viewing violence changes a person's attitude toward violence. For 10 subjects the following scores were obtained:

Subject number	First questionnaire	Second questionnaire
1	28	29
2	31	29
3	29	32
4	21	23
5	20	24
6	30	30
7	28	31
8	23	25
9	22	21
10	24	26

State and test the appropriate hypothesis at $\alpha = 0.1$ and $\alpha = 0.01$.

24. Smith and Roe consider two different procedures for determining the amount of amylase in body fluids. The original method and the new method are compared on the basis of the amount of amylase found in the sample. The following data are given in units per millimeter:

Subject	Original	New	Subject	Original	New
1	38	46	8	46	58
2	48	57	9	69	85
3	58	73	10	59	74
4	53	60	11	81	96
5	75	86	12	44	55
6	58	67	13	56	71
7	59	65	14	50	63
			15	60	74

Test if the new method finds significantly more amylase than the old method (use $\alpha = 0.1$.) (Smith, B. W., and Roe, J. H.: A micro-modification of the Smith and Roe method for the determination of amylase in body fluids. *Journal of Biological Chemistry, 227*: pp. 357–362, 1957. By permission of copyright owner Amer. Soc. of Biological Chemists, Inc.)

25. Once it was considered usual for a new home buyer to spend approximately 25% of his or her income on housing. In 1981, with high inflation and interest rates, the national average was running at about 33% and residents of some cities such as San Diego, California, averaged paying 42% of their annual income on housing (*Wall Street Journal,* February 10, 1982). Consider a hypothetical survey of 36 home buyers in two cities; one city in Texas, and one in California. For the 36 buyers in Texas the percentage of monthly income spent on housing averaged 30% with $s^2 = 25$, whereas for the 36 buyers in California, this percentage averaged 39% with $s^2 = 30$. Can we conclude that the buyers in the Texas city pay a smaller proportion of their monthly income for housing than do the buyers in the California city? Use $\alpha = 0.05$, and assume the variance in percent spent is the same for both.

Section 8.5

26. A manufacturer has two machines producing identical parts and naturally wishes to give the same guarantee to all parts produced. It is desirable therefore to test if the proportion of defective parts produced by machine 1 is the same as the proportion produced by machine 2. Of 100 parts tested from machine 1, 5 defectives were found and from machine 2, of 150 parts tested, 12 defectives were found. At $\alpha = 0.05$, can we conclude that the proportion of defectives produced by the machines are not equal? Find a 95% confidence interval for the difference in the proportions of defectives produced.

27. A bank auditor is attempting to determine why a certain large bank folded. She samples a large number of loans from the failed bank and from a successful bank in the same city. She then compares the observed proportion of bad loans issued by the insolvent bank with the observed proportion of bad loans issued by the successful bank. The following is found:

Insolvent bank

number of loans in the sample — 1523
number of bad loans — 76

Solvent comparable bank

number of loans in the sample — 2120
number of bad loans — 64

At $\alpha = 0.05$, can the auditor claim that a bad loan policy existed for the insolvent bank; that is, is the proportion of bad loans issued by the insolvent bank larger than the proportion issued by the solvent bank? Find a 95% confidence interval for the difference between the banks.

28. According to Mr. Fred Turner, Chairman of the Board of McDonald's hamburgers, 70% of all customers arriving at McDonald's after the breakfast hour order French fries, and the French fries sales total over a billion dollars a year (*Wall Street Journal,* February 8, 1982). Needless to say, the company is very particular about the potatoes it uses (preferably Russet Burbank potatoes). One problem in expanding to the European market is that Russet Burbank potatoes are not grown in Europe, and the Common Market has rules against importing potatoes from outside. Consider a hypothetical situation in which two stores' sales are monitored to see if there is really a difference in sales when using Rus-

set Burbank versus another variety of potatoes. Potentially millions of dollars could rest on the answer. Suppose Store 1 sells fries made the usual way, and Store 2 sells fries made with the new variety. Of 100 people who came into Store 1, 73 buy fries; whereas of 100 people who come into Store 2 only 64 buy fries. Can we conclude that use of the new variety of potatoes significantly reduces the percentage of customers who buy French fries? (Use $\alpha = 0.01$.)

29. An exterminator is trying to determine which of two pesticides is more effective. To do this, the exterminator tests each pesticide on 150 cockroaches and finds that the more expensive pesticide has an 82% kill proportion and the less expensive pesticide has a 70% kill rate. Does this information imply that the more expensive pesticide is significantly more effective? (Use $\alpha = 0.1$.) Find an 80% confidence interval for the difference between the two proportions of cockroaches killed by the pesticides.

Section 8.6

30. A hospital administrator has two doctors on the staff who perform similar operations. Doctor A has performed 500 operations and had 3 fatalities, whereas Doctor B performed 750 operations and had 2 fatalities. Can the administrator conclude that the probabilities of a fatality for the doctors are unequal? (Use $\alpha = 0.1$.)

31. A certain rare disease occurs in 1% of the 200 people who are not vaccinated. Of the 200 persons who are vaccinated, only one contracted the disease. Does this show the vaccine is effective? If your answer is no, how would you change the experiment in order to get more definitive results?

32. A current controversy in public policy for the insurance industry is the extent to which de-

mographic characteristics should be allowed to be used in rate making. For example, is it acceptable to use sex, age, race, or economic status as rate-making variables, or should rates be set in a manner that is strictly neutral to these variables? A plausible argument in favor of using certain of these variables might be that they are significantly related to the loss experience of the insurance company. Hence the group responsible for the loss should pay for the increased loss experience through higher premiums. (Otherwise the other groups are paying more than their fair share.) Consider two classes of insurance policyholders, A and B. There are 1050 policyholders in Class A and 2500 policyholders in Class B. Class A records show two accidents during the year; and Class B records show three accidents during the year. Can we conclude that the proportion of accidents in Class A is larger than the proportion for Class B? (Use $\alpha = 0.1$.)

VARIANCE $= \Delta^2 = \dfrac{\sum (x_i - \bar{x})^2}{n-1} = \dfrac{\sum\limits_{i=1}^{n} x_i^2 - \dfrac{(\sum x_i)^2}{n}}{(n-1)}$

To Compute $\bar{y} : \Delta_y^2$ from $\bar{x} : s^2$

$\bar{y} = a\bar{x} + b$

$s_y^2 = a^2 \Delta_x^2$

* * * * * * *

NORMAL SAMPLE DISTIN'

$\mu_{\bar{x}} = \mu$

$\sigma_{\bar{x}} = \dfrac{\sigma}{\sqrt{n}}$

THE Z statistic $= \dfrac{x - \mu}{\dfrac{\sigma}{\sqrt{n}}}$ or S or z where; $\mu_{\bar{x}} = \mu$ $\sigma_{\bar{x}} = \dfrac{\sigma}{\sqrt{n}}$

chi squared   Let $s^2 =$ sample variance

$x^2 = \dfrac{(n-1)s^2}{\sigma^2}$

$x = $ distn $= (n-1)$ df

$\Delta^2 = \dfrac{\sum\limits_{i-1}^{n} (x_i - \bar{x})^2}{n-1}$

$Z = \dfrac{\sqrt{n}\,(\bar{x} - \mu)}{\sigma}$

* * * *

Sampling Theorm

$P = 1 - \alpha$

$P = .95$ $\alpha = .05$ $\alpha/2 = .025$

$P = .99$ $\alpha = .01$ $\alpha/2 = .005$

then $Z_{\alpha/2} = Z_{.025} = 1.96$

$= Z_{.005} = 2.576$

THEREFORE

$P[-1.96 \le Z \le 1.96] = .95$

$P[-2.576 \le Z \le 2.576] = .99$

$P\left[\bar{x} - Z_{\alpha/2}\dfrac{\sigma}{\sqrt{n}} \le \mu \le \bar{x} + Z_{\alpha/2}\dfrac{\sigma}{\sqrt{n}} \right] = P = 1 - \alpha$

$P \times 100\%$ chance of capturing μ.

ESTIMATE   AVE PULSE RATE   ERROR BOUNND $= 2$ beats/min $\varepsilon(1 - \alpha) = .99$ $\alpha = .01$. How many subjects do we need.

$n = \dfrac{Z_{\alpha/2}^2 \; \delta^2}{B^2} = \dfrac{(Z_{.005}^2)(\delta^2)}{B^2}$

if $\alpha = .01$ TAKE PRELIM STUDY where $\Delta = \delta$

if $\Delta = 15.7$ b/min   intitial sample is 20

// $n \simeq \dfrac{(2.576)^2 (15.17)^2}{2^2} \longrightarrow \delta^2$ $(\cdot B^2)$

$Z_{\alpha/2}^2$

$\dfrac{\bar{x} - \mu}{\dfrac{\sigma}{\sqrt{n}}}$

Measuring Blood Pressure

Variability is known for any age group

$\sigma = 10$

We want to estimate μ = mean BP for individual Aged 35 yrs. (unknown)

take sample $n = 200$ people

$\bar{X}$ = mean BP for sample = 141.4 ⟶ Point estimate of μ.

Confidence limits for μ

$$\bar{X} - Z_{\alpha/2} \frac{\sigma}{\sqrt{n}} \quad \text{to} \quad \bar{X} + Z_{\alpha/2} \frac{\sigma}{\sqrt{n}}$$

$n = 200$ $\sigma = 10$ $\bar{X} = 141.4$ $Z_{\alpha/2}$

$\alpha = .05$ $Z_{\alpha/2} = 1.96$

$\alpha = .01$ $Z_{\alpha/2} = 2.576$

95%

FOR $\alpha = .05$

$$\bar{X} - Z_{\alpha/2} \frac{\sigma}{\sqrt{n}} \quad \text{to} \quad \bar{X} + Z_{\alpha/2} \frac{\sigma}{\sqrt{n}}$$

$$141.4 - (1.96)\left(\frac{10}{\sqrt{200}}\right) \quad \text{to} \quad 141.4 + (1.96)\left(\frac{10}{\sqrt{200}}\right)$$

$$\underline{140.01} \quad \text{to} \quad 142.79 \quad = 95\% \text{ Con Int}$$

99%

For $\alpha = .01$

$$141.4 - (2.576)\left(\frac{10}{\sqrt{200}}\right) \quad \text{to} \quad 141.4 + 2.576\left(\frac{10}{\sqrt{200}}\right)$$

$$\underline{140.58} \quad \text{to} \quad 143.22 \quad = 99\% \text{ Confid Int.}$$

* * *

Estimation by Confidence Int

$$\bar{X} \pm Z_{\alpha/2} \frac{\sigma}{\sqrt{n}} \quad (\sigma \text{ known})$$

$$\bar{X} \pm t_{\alpha/2} \frac{s}{\sqrt{n}} \quad \begin{array}{l}(\sigma \text{ unknown} \\ \text{estimated by} \\ s)\end{array}$$

$$\begin{array}{c} 1.96 \text{ or} \\ 2.576. \end{array}$$
$Z_{\alpha/2}$ table = $\alpha/2$ critical value for norm standard distn

$t_{\alpha/2}$ table = "   "   " t distn with $(n-1)$ degree freedom.

$\mu_X = P$ ⟶ population proportion

$\sigma_P = \sqrt{npq}$

$\hat{P} = \frac{x}{n}$ = # of successes

$\quad = \left(\frac{1}{n}\right)x + 0 = ax + b \quad \begin{array}{l} a = \frac{1}{n} \\ b = 0 \end{array}$

$\mu_{\hat{P}} = a\,\mu_X + b = \frac{1}{n}(np) + 0 = P$

$\sigma_{\hat{P}} = |a|\,\sigma_X = \left(\frac{1}{n}\right)\sqrt{npq} = \sqrt{\frac{npq}{n^2}} = \sqrt{\frac{pq}{n}}$

$$Z = \frac{\hat{P} - P}{\sigma_{\hat{P}}}$$

$$P\left[-Z_{\alpha/2} \leq Z \leq Z_{\alpha/2}\right] = 1 - \alpha$$

$$P\left[-Z_{\alpha/2} \leq \frac{\hat{P} - P}{\sigma_{\hat{P}}} \leq Z_{\alpha/2}\right] = 1 - \alpha$$

$$P\left[\hat{P} - Z_{\alpha/2}\,\sigma_{\hat{P}} \leq P \leq \hat{P} + Z_{\alpha/2}\,\sigma_{\hat{P}}\right] = 1 - \alpha$$

CHAPTER · 9

Prediction and Association: Regression and Correlation

By the prickings of my thumbs,
Something wicked this way comes.

Second Witch
Macbeth, Act IV, Sc.1.

$\mathbf{W}$e are constantly attempting in our own daily lives to determine the degree of association or the relationship between two (or more) variables. Will I be healthier if I buy my food at a natural food or health store? (Is my health affected by this food?) Do increases in the prime lending rate of banks increase the worth of the U.S. dollar abroad? How is the number of hours I study related (if at all) to the grade I receive? Scientific investigations are also often concerned with determining the relationship and association between variables. In this chapter we discuss some methods for modeling, predicting, and exploiting associations between variables.

Handwritten margin notes:

Example (NonSmoker MORTALITY RATE)

n = 1067 people

X = 117 deaths

$\hat{p} = \dfrac{X}{n} = \dfrac{117}{1067} = .1097$

10.97% estimate mortality rate.

95% Confidence Limit.

$Z_{\alpha/2} = Z_{.025} = 1.96$

$\sigma_{\hat{p}} \cong \sqrt{\dfrac{\hat{p}(1-\hat{p})}{n}}$

$= 0.0096$

then

$\hat{p} \pm Z_{\alpha/2}\,\sigma_{\hat{p}} = (.1097) \pm (1.96)(.0096)$

.0909 to .1284

A 95% Con Int conclude that a mortality rate will happen bet 9.09% & 12.84%

* * *

SAMPLE SIZE DETERMINATION

$B = Z_{\alpha/2}\,\dfrac{\sigma}{\sqrt{n}}$ Size of B determines the accuracy of the limit.

$n = \dfrac{Z_{\alpha/2}^2\,\sigma^2}{B^2}$

error bound.

PRELIM SAMPLE

$n \cong \dfrac{Z_{\alpha/2}^2\,\delta^2}{B^2}$

From earliest times the human race has been in the prediction business. Which side will win the next big battle? Are there more fish upstream than two leagues downstream, or is there a place somewhere between where they are more abundant? Will it rain tomorrow? These are the types of questions that are continually being asked. Not so long ago, various professions sprung up to answer these questions: Magicians, soothsayers, witches and warlocks, and ancient publishers of the *Farmer's Almanac,* all hung out their shingles as reliable prophesizers. Most of these professions are now defunct; they all have lost credibility because their predictions were either not reliably valid or were couched in such vague and ambiguous language (like those of Nostradamus or the Greek Delphi) that no one could understand what it was they were predicting.

Clearly, the first requirement of a prediction is that it be stated so that everyone can understand it. Secondly, some measure of how far "off" the prediction might be must be established. Most ancient soothsayers and their fans did not recognize the chance that a prediction could be wrong. By insisting on their infallibility and thereby not establishing some measure of fallibility, they were doomed to lose credibility completely as their predictions went awry.

To satisfy both of the preceding requirements, that of the unambiguous interpretation of a prediction and the development of a measure of its inaccuracy, we must first clearly define what is being predicted and also those variables associated with the prediction. Statisticians have taken such steps, and so some forms of prophesy have now become subfields of statistics.

To begin, what is being predicted? What we predict here is the observed value of some random variable. There are many practical applications that involve this type of prediction. As some examples: Insurance companies need to furnish predictions of their future claims when appealing for rate increases; medical investigators need to predict the effect on mortality of a suspected carcinogen in the work place; educators sometimes wish to predict the grade point averages of students going through graduate school. The collection of techniques used to develop such predictions is called "regression analyses." These techniques have been applied in numerous fields, even in the law, a field that might seem quite remote from statistical concerns. In the case of *Vyuanavich* vs. *Republic National Bank,* referenced at the end of this chapter, a bitter battle over charges of race and sex discrimination against the bank was waged, with experts on both sides using regression analyses to support their respective cases.

9 ▪ 1 FITTING A STRAIGHT LINE

As previously stated, we wish to predict the observed value of some random variable, and also to determine the pertinent variables associated with the random variable in question. Thus to help make the prediction of a student's grade point average, y, in graduate school, the educator may observe the student's grade point average, x, in undergraduate school. If the relationship between the variables x and y is known, then knowledge of the value of x will help predict the value of y. To illustrate, suppose it is known that, in general, students do approximately one fourth a grade point lower in graduate school than in undergraduate school. This relationship may be expressed as $y = x - 0.25$. An educator wishing to predict what a particular student's graduate grade point will be, may obtain information to help her guess by looking up the student's undergraduate grade point. Suppose, for example, it is $x = 3.85$. Using the known relationship between x and y, she would then predict $y = 3.85 - 0.25 = 3.60$ as a graduate grade point average for the student.

The preceding example assumes a deterministic relationship between x, the student's undergraduate grade point, and y, the student's graduate grade point; that is, it is assumed that once the student's undergraduate grade point is determined, the student's graduate grade point is determined. We know from experience that such a deterministic relationship does not ordinarily exist between the two variables; there is a relationship but it is not deterministic. For example, many students who have the same undergraduate grade point go on to graduate school but do not wind up with the same graduate grade point average. In fact, we can regard the set of grade point averages Y for a given undergraduate average x as a *random variable;* Y has a range of values having a probability distribution.

Let us take another example. Suppose we wish to investigate crop yield and know that the amount of fertilizer used affects this yield. We randomly select 25 acres and apply to them various amounts of fertilizer. After a suitable time has elapsed, we record the yields. To illustrate our point, consider the following results where x_i denotes the amount of fertilizer used on the ith acre and y_i denotes the yield on the ith acre.

x_i	50	50	50	70	70	70	90	90	90
y_i	32	27	35	38	47	45	45	47	40

x_i	110	110	110	130	130	130	150	150	150
y_i	59	56	48	58	56	64	63	58	65

x_i	170	170	170	190	190	190	210
y_i	64	68	75	68	77	91	78

To see if there is an evident visual relationship between x and y, we plot all the points (x_i, y_i) on a graph to obtain Figure 9.1. Such a graphical representation of the outcome is called a *scatter diagram.* By observing Figure 9.1, we can see that indeed there appears to be a relationship between x and the observed values y of

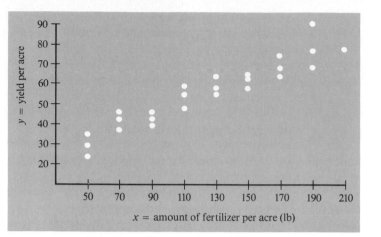

Figure 9 ▪ 1
Scatter diagram drawn
from observed data.

the random variable Y; and in fact, almost a straight line relationship. However, the points do not fall exactly on a straight line. This is due to various other factors such as soil condition, rainfall, and so on, which may vary from acre to acre in a random manner. Nonetheless, we can use scatter diagrams such as Figure 9.1 to give "seat of the pants" estimates of the yield corresponding to a given amount of fertilizer. For example, by fertilizing with $x = 160$ points of fertilizer, we would predict a yield y between 58 and 75 bushels. However, we would like to obtain better estimates than those provided by such a crude range. Let us consider this problem more deeply.

If we look carefully at the scatter diagram, Figure 9.1, we notice for each value of x several different observed values of Y. For example, the value $x = 50$ lb has three different observed Y values that correspond to it. This variation in the observed values of Y could be due, for instance, to either random errors in measurement of the variable or random variation in the nature of the soil from acre to acre treated by 50 lb of fertilizer. To take into account the variability of the Y scores for a single value of x, we assume the observed values corresponding to a particular x value are reflective of a sample from the entire population of possible values in the range of Y; and the apparent straight line nature of the relationship is modeled by assuming $Y = \gamma + \beta_1 x + E$, where E is a random variable representing the "random error" and γ and β_1 are constants.

We assume, on the average, the errors cancel each other out so that the average error is equal to zero; that is, μ_E, the expected value of E, $= 0$. Now suppose you wish to predict y, the observed value of Y, corresponding to the value x. Since there is a whole distribution of possible Y's corresponding to x, the best guess would be to predict the mean value of Y given the value x. We denote this mean value by $\mu_{Y|x}$. Using the preceding equation for Y in terms of x, we have the equation for this prediction.

$$\mu_{Y|x} = \gamma + \beta_1 x + \mu_E = \gamma + \beta_1 x; \qquad \text{since } \mu_E = 0 \qquad (9.1)$$

Equation 9.1 represents the equation of a straight line. The reader can confirm this fact by choosing a value for γ and one for β_1 and then plotting on graph paper the values of $\mu_{Y|x}$ for different values of x. Parameter γ is called the *intercept* of the line and represents the value of $\mu_{Y|0}$, the place where the line intersects the vertical $\mu_{Y|x}$ axis; the parameter β_1 is called the *slope* of the line and indicates the steepness of the line.*

If we know the precise value of γ and β_1 in equation (9.1), then we can estimate the value of Y corresponding to x by using $\mu_{Y|x}$ as our predicted value of Y. Equation (9.1) is called the *regression equation* of Y on x. This example illustrates that there are essentially three questions that arise in attempting to predict the value of a variable Y by this method.

---

1. **What variable x is related to Y that will help us gain knowledge about the value $\mu_{Y|x}$ that we expect Y to have?**
2. **Suppose we do find a variable x that we know is related to Y, can we *estimate* an equation for the value that we expect Y to have in terms of x?**
3. **Suppose we have an equation for Y in terms of x, then to predict the observed value of Y that corresponds to a value x, we merely plug the value x into the equation and obtain our estimate. How accurate is this estimate, and can we find confidence intervals for the estimate in order to gain some idea about the precision of our estimate?**

The purpose of this chapter is to investigate these three questions. Of particular interest are the last two questions that we shall investigate in detail for the case in which $\mu_{Y|x}$ is a linear (i.e., straight line) function of x.

We emphasize the two distinct purposes for obtaining the regression equation relating x and Y. Note that we take observations within a given interval of x values; this is called the observation interval. The observation interval in Figure 9.1 is from $x = 50$ to $x = 210$. The points at which observations are actually made within this interval, such as $x = 50$, $x = 70$, $x = 90$, and so on, are called the observation points. The first purpose of the regression equation is to show how much of the data are explained by their relationship with x within the observation interval. The second purpose is to *predict* the observations we might obtain at points other than the observation points. An important special case is that in which x is time. For example, we might take several observations on the price Y of a given stock during an observation interval of a week. First, the regression equation explains how much of the past data are explained through the relationship they express between Y and x. In this role the regression equation is normative in the sense of modeling the past relationship be-

---

* Many authors use the symbols α and β or a and b to represent the intercept and slope parameters. We shall not do this since we have already used α and β for the probability of type 1 and type 2 errors in hypothesis testing; we do not use a and b since we want to use Greek letters for population parameters whenever it is reasonable to do so.

tween Y and x. Also, the regression equation *predicts* values y at points other than the observation points. Assuming this relationship continues into the future (a sometimes heroic assumption), we can then use the estimated regression line to predict the future with respect to observed values of Y. As we illustrate at the end of the chapter, it is a logical error to assume that just because a particular regression model fits past data very well, it will therefore predict future values as well. These two roles of the regression line are distinct (but of course related). Each role has its use in applying regression techniques to actual real-life situations.

9 ▪ 2 LINEAR REGRESSION AND THE BEST-FITTING STRAIGHT LINE

A scatter diagram such as Figure 9.1 indicates that there is some relationship between the variable x and the random variable Y so that knowing the value of x will aid us in predicting the value of Y. From here on in this chapter, we assume the functional relationship between x and Y that is suggested by the scatter diagram is a straight line $Y = \gamma + \beta_1 x + E$, where E is the random error with mean 0. We do this for two reasons: The straight line relationship between two variables is easy to express and in many practical situations the relationship between x and Y is either actually, or can be closely, approximated by a straight line. To paraphrase Albert Einstein, we should make the model as simple as possible, but no simpler. We call a straight line relationship a linear relationship between x and Y. In general, regression analysis involves predicting values of Y when our observed data $(x_1, y_1), (x_2, y_2), \ldots, (x_n, y_n)$ has the property that the values of x are chosen prior to the experiment and are not random.

It should also be noted that the relationship between x and Y need not always be a straight line. Indeed, as the scatter diagrams in Figure 9.2 show, any relationship is possible. Often the physics, economics, or biology of the particular application under investigation will dictate which form of the relationship is assumed for the particular study.

In the particular problem here, we assume $Y = \gamma + \beta_1 x + E$, where E has mean 0; so $\mu_{Y|x} = \gamma + \beta_1 x$. The process of predicting the value of Y for a given value x is accomplished by means of estimating the parameters γ and β_1 in the *regression* equation $\mu_{Y|x} = \gamma + \beta_1 x$. Since there are many possible ways to estimate the parameters γ and β_1, and consequently many possible estimates of the regression line, we must single out an estimate that gives a line that "best fits" the observed data on the scatter diagram. Let us suppose the prediction equation is $\hat{y}_x = a + bx$, where a is an estimate of γ, b is an estimate of β_1, and $\hat{y}_x$ is the predicted value of Y for the given value x. One method of determining how good the straight line $\hat{y}_x = a + bx$ fits the data is to measure the distance

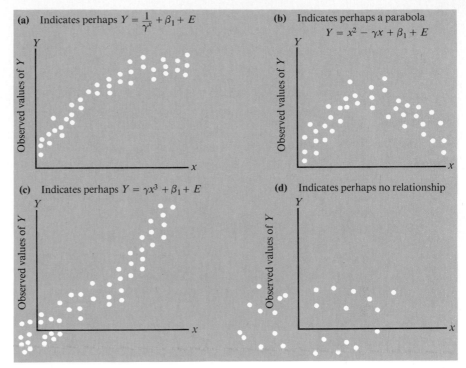

(a) Indicates perhaps $Y = \dfrac{1}{\gamma^x} + \beta_1 + E$

(b) Indicates perhaps a parabola
$Y = x^2 - \gamma x + \beta_1 + E$

(c) Indicates perhaps $Y = \gamma x^3 + \beta_1 + E$

(d) Indicates perhaps no relationship

Figure 9 ▪ 2
Various scatter diagrams.

between the *observed value* y_i corresponding to x_i and the predicted value $\hat{y}_{x_i}$ corresponding to x_i (see Fig. 9.3). The square of this difference $(y_i - \hat{y}_{x_i})^2$ measures how well the line $\hat{y}_x = a + bx$ predicts the value of Y corresponding to the observation points x_i. The value $e_i^2 = (y_i - \hat{y}_{x_i})^2$ is the square of the error of predicting y_i based on x_i. The sum of all these errors is $\sum\limits_{i=1}^{n} e_i^2$, which measures how well the line fits *all* the data $(x_1, y_1), \ \ldots, (x_n, y_n)$. The smaller the value of $\sum\limits_{i=1}^{n} e_i^2$ is, the better the line $\hat{y}_x = a + bx$ fits the data. Thus we have a criterion for what we mean by a "best-fitting straight line." A line $\hat{y}_x = a + bx$ is called a *best-fitting line* if $\sum\limits_{i=1}^{n} e_i^2 = \sum\limits_{i=1}^{n} (y_i - \hat{y}_{x_i})^2$ is as small as possible. In the terminology of the previous section, we pick estimates to give the best possible *normative model* of the observed data.

   The method we have used to determine a best-fitting line is called the *method of least squares*. The estimates a of γ and b of β_1 are called the least squares estimates of γ and β_1. The resulting line $\hat{y}_x = a + bx$ is called the *least squares line*. The estimates a and b are given by the following formulas:

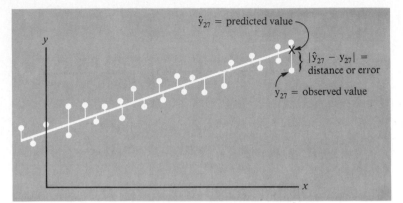

Figure 9 ▪ 3
The best-fitting straight
line minimizes the sum of
the squared distance
between the observed and
predicted values of Y.

$$b = \frac{\sum_{i=1}^{n} \{(x_i - \bar{x})(y_i - \bar{y})\}}{\sum_{i=1}^{n} (x_i - \bar{x})^2}$$

$$a = \bar{y} - b\bar{x}$$

(9.2)

The line $\hat{y}_x = a + bx$ is the best-fitting line in the sense that it best fits the past data by minimizing the sum of the squares of the errors $\Sigma\, e_i^2 = \sum_i (y_i - \hat{y}_{x_i})^2$.

In the *prediction* problems of the type referred to previously, we must go through five steps to solve the problem.

---

1. **Collect the data (x_1, y_1), . . . , (x_n, y_n).**
2. **Plot a scatter diagram of the observed values of Y versus x.**
3. **If the scatter diagram indicates a linear (straight line) relationship between x and Y, use equation (9.2) to obtain a best-fitting line for the data. (Otherwise use a more sophisticated equation as found in the last reference at the end of this chapter.)**
4. **(Optional) Superimpose the graph of the regression equation obtained in (3) on the scatter diagram obtained in (2).**
5. **Using the regression equation, predict the values of Y for desired values of x.**

Example 9 ▪ 1   An experiment was run to determine the effects of acidity on a certain chemical process. The experimenter varied the ph, x, of the solution and recorded the yield of the process.

Step 1: The results were $(x, y) = (1, 1.7), (1, 1.5), (3, 2), (3, 3), (3, 3.1), (4, 3.5)$.
Step 2: The scatter diagram is shown in Figure 9.4.

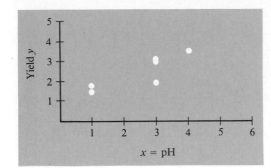

Figure 9 ▪ 4

Step 3: Since this could reasonably indicate a linear relationship, we use equation (9.2) to estimate the parameters of the regression line $\hat{y}_x = a + bx$:

x_i	y_i	$(x_i - \bar{x})$	$(x_i - \bar{x})^2$	$(y_i - \bar{y})$	$(x_i - \bar{x})(y_i - \bar{y})$
1	1.7	-1.5	2.25	-0.77	1.155
1	1.5	-1.5	2.25	-0.97	1.455
3	2	0.5	0.25	-0.47	-0.235
3	3	0.5	0.25	0.53	0.265
3	3.1	0.5	0.25	0.63	0.315
4	3.5	1.5	2.25	1.03	1.545
Total 15	14.8		7.50		4.5
$\bar{x} = 2.5$	$\bar{y} = 2.47$				

We have

$$b = \frac{\Sigma\,(x_i - \bar{x})(y_i - \bar{y})}{\Sigma\,(x_i - \bar{x})^2} = \frac{4.5}{7.50} = 0.6$$

$$a = \bar{y} - b\bar{x} = 2.47 - (0.6)(2.5) = 0.97$$

Thus the regression equation of Y on x is $\hat{y}_x = 0.97 + 0.6x$.

Step 4: We now superimpose the graph of $\hat{y}_x = 0.97 + 0.6x$ on the scatter diagram. To graph $\hat{y}_x = 0.97 + 0.6x$, we choose three points x and graph the $\hat{y}_x$ values.* (See Fig. 9.5.)

Step 5: If we wish to predict the yield for a ph of 4.5, we would predict $\hat{y}_{4.5} = 0.97 + 0.6(4.5) = 3.67$. Note that the point $(\bar{x}, \bar{y})$ is always on the line so that this point may always be chosen when graphing $\hat{y}_x$. This is a charactcristic property of linear regression and eases and computations involved in plotting the regression line.

It should be noted that when using a calculator it is perhaps more convenient to use computational formulas for a and b instead of equation (9.2). These may

---

* We need only choose two values of x, find the two corresponding values $\hat{y}_x$ and connect the points to draw the straight line; we choose three points of x as a check on our calculations. If the straight line goes through all three points, we are pretty sure no calculation errors have been made.

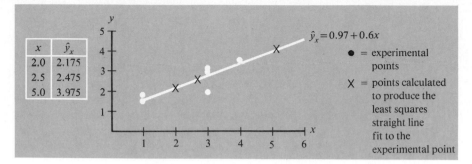

x	$\hat{y}_x$
2.0	2.175
2.5	2.475
5.0	3.975

$\hat{y}_x = 0.97 + 0.6x$

● = experimental points

X = points calculated to produce the least squares straight line fit to the experimental point

Figure 9 ▪ 5

be derived by multiplying out the squares in the formula for b to obtain

$$b = \frac{\sum\limits_{i=1}^{n} x_i y_i - (\sum x_i)(\sum y_i)/n}{\sum\limits_{i=1}^{n} x_i^2 - \left(\sum\limits_{i=1}^{n} x_i\right)^2 \Big/ n} = \frac{\text{sum of products} - (\text{product of sums})/n}{\text{sum of squares} - (\text{square of sum})/n}$$

$$a = \bar{y} - b\bar{x} \tag{9.3}$$

Example 9 ▪ 2   An educational psychologist wishes to determine the relationship between a student's IQ score and the student's grade point average in college. Fifty persons are selected at random and the pair (x, y) is recorded, where x is the person's IQ score on a Stanford-Binet IQ test and y is the student's grade point average. To predict the grade point average of a student with an IQ of 110, the psychologist examined the data from 50 students. The following data were obtained:

$$\sum_{i=1}^{50} x_i = 5250, \qquad \sum_{i=1}^{50} x_i^2 = 552,220, \qquad \sum_{i=1}^{50} y_i^2 = 409.2,$$

$$\sum_{i=1}^{50} y_i = 142.5, \qquad \sum_{i=1}^{50} x_i y_i = 15,011$$

Using these data, we may calculate

$$b = \frac{\sum xy - (\sum x_i)(\sum y_i)/n}{\sum x_i^2 - (\sum x_i)^2/n}$$

$$= \frac{15,011 - (5250)(142.5)/50}{552,220 - (5250)^2/50}$$

$$= \frac{15,011 - 14,962.5}{552,220 - 551,250} = \frac{48.5}{970} = 0.05$$

$$a = \bar{y} - b\bar{x} = \left(\frac{142.5}{50}\right) - (0.05)\left(\frac{5250}{50}\right) = -2.4$$

So the *estimated* regression of Y on x is

$$\hat{y}_x = -2.4 + 0.05x$$

To estimate the grade point of a person with an IQ of 110, we use the value

$$\hat{y}_{110} = -2.4 + 0.05(110) = 3.1$$

This example also illustrates that it is very dangerous to utilize the regression-line technique to predict Y values corresponding to x's that are far outside the interval sampled. In this example the relation between the IQ and grade point for college students seemed linear for x values between about 90 to 125. To predict that someone with an IQ of 135 would have a grade point $\hat{y}_{135} = 4.35$ is ridiculous in a system where 4 is the highest possible grade point. Similarly, it is doubtful if a person with an IQ of 75 could do as well as the predicted value of 1.35. These errors are due to attempts to predict by a straight line far outside the sampled x values. Within a particular range the preceding regression line fits well, but clearly fails outside the range.

9 ▪ 3   THE ISSUE OF ASSOCIATION: CORRELATION; THE SAMPLE CORRELATION COEFFICIENT

How Much Does the Straight Line Explain? How Well Does It Fit?

Figure 9.6 displays some data obtained on the variable y for some fixed values of x. Figure 9.7 shows the least squares line that was fitted to the data. This line passes through all the data points exactly. We can assert that the random variable Y is indeed associated with the variable x since the observed values y are so clearly related to x by the straight line.

The fitted line is said to "explain" all the data in the sense that the derived relationship $\hat{y}_x = x + 1$ between Y and x yields all the data points that are actually observed. The errors are all zero, the line fits perfectly, and all of the variability in the observed values of Y is accounted for once we fit the regression line.

Now consider the scatter diagram of Figure 9.8 and the fitted least squares straight line. The straight line is said to be a poor "explanation" of the data because the data points are so widely dispersed about the fitted line; moreover, the fitted relationship $\hat{y}_x = 2x + 3$ does not account for much of the variability that we observed in the Y data. In fact, most of the observed values of Y are far from the straight line. In this case the straight line explains little of the variation in the data, whereas that of Figure 9.7 explains all the variation in the values of y. We can see (Fig. 9.8) that Y is poorly *associated* with x.

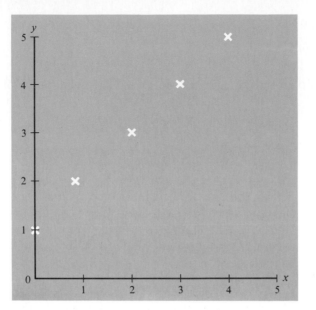

Figure 9 ▪ 6
Observed values y for
various values of x.

A useful descriptive statistic to have would be a measure of how well a fitted least squares straight line explains the data; that is, how tightly grouped around the regression line the y values fall; or, in other words, how closely Y is associated with x. We now develop such a measure.

From the preceding discussion we see that the measure should be closer to

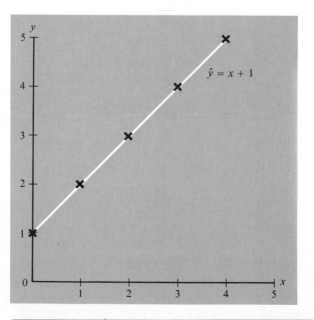

Figure 9 ▪ 7

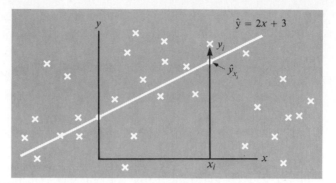

$\hat{y} = 2x + 3$

Figure 9 ∎ 8

zero the farther the data points are from the straight line. With reference to Figure 9.8, the square of the distance of the observed point y_i from the line is $(y_i - \hat{y}_{x_i})^2$. The sum of squares

$$\sum_i e_i^2 = \sum_i (y_i - \hat{y}_{x_i})^2$$

is called the *residual sum of squares.* This value is a measure of the deviation of the observed values y_i from the fitted line. The smaller it is, the more tightly clustered about the line are the data.

Recall that a basic measure of the variation of the observed values y_i among themselves is $\sum_i (y_i - \bar{y})^2$, where $\bar{y}$ is the average of the observed values. This sum is called the *total sum of squares.* If the residual sum of squares is the same as the total sum of squares, then the fitted straight line does not explain any of the variation in the data since the variation in the data is the same whether or not one uses the straight line. If the straight line explains any of the variation in the data, namely, the variation in the observed values y_i, then the variation of the values y_i about the straight line should be less than the variation of the values y_i among themselves; otherwise the straight line does not explain (reduce) the variation of the data from what it was before the line was obtained.

In short, when

$$\sum_i e_i^2 = \sum_i (y_i - \hat{y}_{x_i})^2 = \sum_i (y_i - \bar{y})^2$$

the straight line does not explain any of the data. In another form, when

$$\frac{\sum_i (y_i - \hat{y}_{x_i})^2}{\sum_i (y_i - \bar{y})^2} = 1$$

then the straight line does not explain any of the data; or more conveniently, when the following expression, denoted r^2, is zero we say that the straight line does not explain the data. This expression

$$r^2 = 1 - \frac{\sum_i (y_i - \hat{y}_{x_i})^2}{\sum_i (y_i - \bar{y})^2} = 1 - \frac{\text{sum of squares around the least squares line}}{\text{total sum of squares around } \bar{y}}$$

is called the *coefficient* of *determination.* It is the proportion of the total sum of squared deviations that is explained by the regression line. Note that if $\sum (y_i - \hat{y}_{x_i})^2$ is zero, then each value $(y_i - \hat{y}_{x_i})^2$ is zero; that is, each data point lies exactly on the straight line. In this case the straight line explains *all* the variation in the data and r^2 takes on the value of 1; that is 100% of the variation is explained by the line.

The value of 1 is the largest value r^2 can have as we see from the following identity.*

$$\sum_i (y_i - \bar{y})^2 = \sum_i (\hat{y}_{x_i} - \bar{y})^2 + \sum_i (y_i - \hat{y}_{x_i})^2$$

From this expression, the residual sum of squares must be less than or equal to the total sum of squares $\sum_i (y_i - \bar{y})^2$; hence r^2 must be less than or equal to 1.

The square root of r^2 is called "the *sample* coefficient of correlation between the set of X's and Y" and is denoted br r; we emphasize that r is the observed value of a random variable since it is a function of the observed values y of the random variable Y. Hence the observed value r can vary from sample to sample. Summarizing, we have

---

The sample correlation coefficient is

$$r = \pm \sqrt{1 - \frac{\sum_i (y_i - \hat{y}_{x_i})^2}{\sum_i (y_i - \bar{y})^2}}$$

The + sign is used if the slope of the straight line is positive and the − sign if the slope is negative. The coefficient of determination r^2 is the proportion of the total sum of squared deviations explained by the line.

It is usually much easier to calculate r from the following formula that also

---

* To show that this is an identity, expand both sides. It becomes clear after doing this that the only remaining term is the sum of the terms $(\hat{y}_{x_i} - \bar{y})(y_i - \hat{y}_{x_i}) = (a + bx_i - \bar{y})(y_i - a - bx_i)$. Without loss of generality, let $\sum x_i = 0$. Then use the fact that $a = \bar{y}$ and $b = \sum x_i y_i / \sum x_i$ to show that the sum of these terms is zero.

automatically yields the correct sign for r:

$$r = \frac{\sum\limits_{i=1}^{n} x_i y_i - (\sum x_i)(\sum y_i)/n}{\sqrt{(\sum x_i^2 - (\sum x_i)^2/n)(\sum y_i^2 - (\sum y_i)^2/n)}} \tag{9.4}$$

where n is the sample size. Note that the numerator of r is exactly the same as the numerator of the regression-line slope estimate b. Thus $r = 0$ if and only if the best-fitting line is horizontal, $y = \bar{y}$.

Example 9 ▪ 3   Using the data of example 9.2, calculate the correlation coefficient between the IQ of a student and the student's grade point. We have

$$r = \frac{(15011) - (5250)(142.5)/50}{\sqrt{(552220 - (5250)^2/50)(409.2 - (142.5)^2/50)}}$$

$$= \frac{48.5}{\sqrt{(970)(3.075)}} = \frac{48.5}{\sqrt{2982.75}} = 0.888$$

Thus $r^2 = 0.789$, so the knowledge of IQ accounts for 78.9% of the total observed sum of deviations of the grade point average.

The sample correlation coefficient r is always between -1 and $+1$. A correlation of $r = +1$ indicates a perfect straight-line relationship between x and y with large x's corresponding to large y's. A correlation of $r = -1$ indicates a perfect straight-line relationship between x and y with large x's corresponding to small y's. A correlation of $r = 0$ indicates that x and y are not linearly related.

The Correlation Coefficient as a Measure of Association Between Two Random Variables X and Y

The regression problem we have discussed is the one where a random variable is related to values x through the formula

$$Y = \gamma + \beta_1 x + E; \qquad \mu_E = 0$$

where the x values are not randomly selected values. In this case the expected or mean value of Y varies with x.

$$\mu_{Y|x} = \gamma + \beta_1 x \tag{9.5}$$

However, two random variables Y and X may also be linearly related as follows:

$$Y = \gamma + \beta_1 X \tag{9.6}$$

Hence the average of Y is now linearly related to the *average* of X, $E(Y) = \gamma + \beta_1 E(X)$. The sample correlation coefficient r now measures how well the model (9.6) explains the data. A correlation of $r = +1$ indicates a perfect straight-line relationship between the observed values of X and Y with large observed values of X corresponding to large observed values of Y. A correlation of $r = -1$ indicates a perfect straight-line relationship between the observed values of X and Y with large observed values of X corresponding to small observed values of Y. A correlation of $r = 0$ indicates that X and Y may not be linearly related.

If X and Y are independent random variables, we should obtain a value of r close to zero.

Interpretation of the Sample Correlation Coefficient: A Caution

Perhaps the most popularly misinterpreted statistic is the sample correlation coefficient. It is very tempting to conclude, for instance, from a correlation coefficient of 0.9 between high blood pressure and heart attack that a high blood pressure measure will "cause" a heart attack. Possibly, the two variables are unrelated to each other but their high correlation is due to the fact that they are related to an unknown third variable. Thus, in the previous example, it is possible that the high value of the correlation coefficient between blood pressure and heart attack is due to a third variable such as stress, anxiety, high cholesterol, or hardening of the arteries. It could be, for example, that heavy smoking causes both high blood pressure and heart attacks. Therefore a high frequency of heart attacks would be associated with high blood pressure, and the sample correlation coefficient would perhaps be high. In general, one should be quite careful about interpreting the sample correlation coefficient as evidence of a causal relationship between the two variables. The correlation coefficient between the cost of living index and my daughter's age is very close to 1, but increasing cost of living has not caused my daughter to age, or vice versa.

9 ■ 4 INTERPRETATION OF ρ, THE POPULATION CORRELATION COEFFICIENT

Testing the Independence of Two Random Variables: The Correlation Coefficient, ρ, of Two Random Variables

The sample correlation coefficient r is also an estimate of the true correlation ρ between two random variables X and Y. The true correlation coefficient is found from a formula similar to that of r but it uses all the values in the ranges of

X and Y. The true correlation coefficient measures the actual straight-line association of two random variables, whereas r only estimates this association through its estimation of ρ. If $\rho = 0$, then the two random variables are not related by the model $Y = \gamma + \beta_1 X$. If $\rho = 1$ or $\rho = -1$, then they are exactly so related in the sense that if the value x is an observed value of X, then the corresponding observed value of Y will be $\gamma + \beta_1 x$.

We have previously discussed another measure of association between two random variables X and Y; namely, the *independence* of X and Y. Recall that X is independent of Y if the probability that $X = x$ (for every x in the range of X) is unchanged given that a value y (no matter which one it is) of Y has occurred. For example, let $X = 0$ if heads occurs on a toss of a coin and $X = 1$ otherwise, whereas Y is the volume of eggs sold in Chicago today. The probability of any value of X is unchanged no matter what we are told about the occurrence of the possible values of Y. The random variables X and Y are *independent*.

What is interesting is that r, which we have used as a measure of association (because it estimates ρ), can also be used for testing whether or not two random variables are independent. If X and Y are independent, then Y should not depend on X in any way. If we know that X and Y are independent, then knowledge about Y does not depend on X; that is, the regression line $\hat{y}_x = a + bx$ should be horizontal. This line is horizontal when $b = 0$, in which case $\hat{y}_x = a$. Since the numerators of r and b are the same, if b is equal to or close to zero, then r must be equal to or close to zero. In short, if X and Y are independent, then r should be near zero.

Now what can we conclude if r is near 0? Can we conclude that X and Y are independent? Not always. We can do so under special conditions, though. The special conditions that are especially useful to us are: X and Y each is normally distributed and the joint probability* of X and Y has a population density histogram of the type shown in Figure 9.9, which is called a *joint normal distribution*. Under these conditions, $\rho = 0$ if and only if X and Y are independent. Under these conditions, then, we may use the value of r (which estimates ρ) to aid us in testing the hypothesis

$H_0 : X$ and Y are independent versus $H_1 : X$ and Y are not independent

We reject H_0 if the *sample* correlation coefficient r is significantly different from zero. We now describe the test.

Assuming that X and Y are joint normally distributed, to test

$$H_0 : \rho = 0 \qquad (X \text{ and } Y \text{ are independent})$$

$$H_1 : \rho \neq 0 \qquad (X \text{ and } Y \text{ are not independent})$$

Figure 9 ■ 9
Histogram or density for a two-dimensional normal distribution.

---

* Recall that the joint probability of X and Y is the probability that X lies between two values and Y lies between two values: for example, $P(u < X \leq v \text{ and } s < Y \leq t)$.

we use the statistic

$$t = r \sqrt{\frac{n-2}{1-r^2}} \tag{9.6}$$

which is the observed value of the random variable T which has a t-distribution with $n-2$ degrees of freedom. We reject H_0 at the level of significance α if the observed value t is greater than $t_{\alpha/2}$ or less than $-t_{\alpha/2}$; df $= n - 2$.

Example 9 ▪ 4   In an article in *Scientific American,* a psychologist wished to determine the manifestations of anxiety. He measured a patient's heart rate variability standard score and the patient's standardized anxiety score on a test. He obtained the following data which we now summarize.

Anxiety score	Heart rate variability	Anxiety score	Heart rate variability
−1.8	−1.55	−1.1	0.9
−1.75	−1.01	−1.05	−0.65
−1.3	−1.25	−1.05	−0.3
−1.3	+0.25	−1.05	−1.75
−1.25	−0.50	0.2	−0.45
−1.25	−1.80	0.3	1.05
−1.15	0.45	0.4	0.35
−1.00	1.05	0.4	0.90
−0.75	−0.5	0.6	0.5
−0.6	−0.7	0.7	0.2
−0.55	0.2	0.9	1.85
−0.3	0.35	0.95	1.8
−0.25	0.6	1.1	1.1
−0.25	0.9	1.1	−0.25
−1.17	0.3	1.3	−0.4
−1.1	−1.8	1.5	−1.2
−1.1	−1.75	1.6	1.8
−1.1	−0.65	1.75	−1.25
		1.9	1.7

Let X denote the anxiety score and Y denote the heart rate variability. As before, x_i is an observed value of X and y_i is an observed value of Y. Then we calculate

$$\Sigma \, x_i y_i = 20.6, \qquad \Sigma \, x_i = -7.52, \qquad \Sigma \, x_i^2 = 44.7,$$

$$\Sigma \, y_i = -1.51, \qquad \Sigma \, y_i^2 = 42.9, \qquad n = 37$$

so that

$$r = \frac{(20.6) - (-7.52)(-1.51)/37}{\sqrt{\left(44.7 - \frac{(-7.52)^2}{37}\right)\left(42.9 - \frac{(-1.51)^2}{37}\right)}} = 0.472$$

At $\alpha = 0.05$, we use the observed value t and reject H_0 if $t > t_{0.025} = 2.03$ or if $t < -t_{0.025}$ with $37 - 2 = 35$ degrees of freedom. Since

$$t = 0.472 \sqrt{\frac{35}{1 - (0.472)^2}} = 3.17 > 2.03$$

we reject H_0 and conclude that the anxiety score and heart rate variability are not independent. There appears to be support for assuming a relationship exists between the two variables.*

Another word of caution regarding the test for the independence of two variables—a correlation coefficient of zero does not, in general, imply that the two variables are independent. We are only certain of this for the case when X and Y are joint normally distributed. Otherwise it is quite possible for a definite relationship to exist between X and Y as shown in Figure 9.10f and still have $r = 0$. We emphasize again that the sample correlation coefficient only measures the strength of the *linear* relationship between two variables and a zero sample correlation coefficient implies only the absence of a linear relationship between the two variables. The preceding hypothesis test can be used to show that two variables are not independent; however, as always, when we fail to reject H_0 we cannot necessarily assume H_0 is true. Thus failure to reject independence does not imply X and Y are actually independent.

Other Tests Concerning ρ

The previous section treats the problem of testing for the independence of two normally distributed random variables; that is, testing

$$H_0 : \rho = 0$$

$$H_1 : \rho \neq 0$$

The test statistic t described in the previous section can also be used to test this null hypothesis versus $H : \rho > 0$ or $H_1 : \rho < 0$ using the appropriate one-tailed version of the preceding t-test.

* Data condensed from "The Nature and Measurement of Anxiety" by Raymond B. Cantell, copyright *Scientific American,* March 1963. All rights reserved.

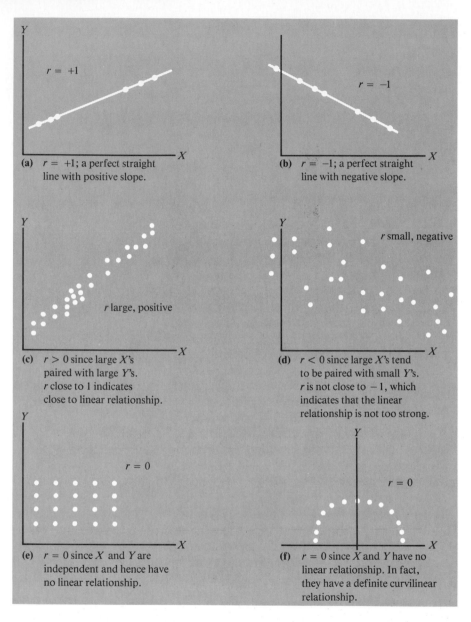

(a) $r = +1$; a perfect straight line with positive slope.

(b) $r = -1$; a perfect straight line with negative slope.

(c) $r > 0$ since large X's paired with large Y's. r close to 1 indicates close to linear relationship.

(d) $r < 0$ since large X's tend to be paired with small Y's. r is not close to -1, which indicates that the linear relationship is not too strong.

(e) $r = 0$ since X and Y are independent and hence have no linear relationship.

(f) $r = 0$ since X and Y have no linear relationship. In fact, they have a definite curvilinear relationship.

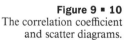

Figure 9 ■ 10
The correlation coefficient and scatter diagrams.

Another technique, however, must be used to test the hypothesis

$$H_0 : \rho = \rho_0 \quad \text{versus} \quad H_1 : \rho \neq \rho_0 \quad \text{when} \quad \rho_0 \neq 0$$

(or versus the appropriate "one-tailed" alternative). In this case we use the statistic

$$z = [F(r) - F(\rho_0)] \sqrt{n-3}; \qquad n = \text{sample size}$$

where $F(r)$ and $F(\rho_0)$ are found from Appendix A.9.

The observed value z comes from a random variable with approximately a normal distribution with $\mu = 0$ and $\sigma = 1$. Hence we can use the z-test to test $H_0 : \rho = \rho_0$, when ρ_0 is not zero.

Example 9 ▪ 4 continued

Using the data of example 9.4, let us test

$$H_0 : \rho = 0.2 \quad \text{versus} \quad H_1 : \rho > 0.2$$

with $\alpha = 0.05$. In this case $\rho_0 = 0.2$ and the estimate $r = 0.472$, so looking up the values in Appendix A.9 we find $F(0.472) = 0.51$ and $F(0.2) = 0.203$.

Thus we calculate the observed value

$$z = (0.51 - 0.203) \sqrt{34} = 1.79$$

Comparing z to $z_\alpha = 1.65$, we reject H_0.

9 ▪ 5   INFERENCE CONCERNING THE BEST-FITTING STRAIGHT LINE

We have only superficially considered the issue of "prediction." Until now we have been concerned primarily with the issue of *association,* the matter of measuring the degree to which two variables are related. Even so, we can now briefly reflect on the methods previously introduced and begin to understand why modern statistical methods of prediction are superior, if more modest, than those of the ancient soothsayers and magicians. First, we explicitly display our data in quantified form. If we are to believe Shakespeare and other authors, the ancient practitioners did not do likewise, their data often consisting of "eye of newt and toe of frog, wool of bat and tongue of dog, adder's fork and blind worm's sting, lizard leg and howlet wing," and so on.* Among the many differences between our methods and theirs is that we have *quantified* our data, establishing, on the basis of past experience, a model in *quantified terms* relating the variables used in prediction and establishing a measure of "association" between them. We are ready to admit the failure of our model relating two variables if the measure of association indicates no such relationship between the two exists. On the contrary, an ancient practitioner (and modern followers

* (*Macbeth*, Act IV, Scene 1.)

of superstition) would and could never admit such failure. For example, what would the "second witch" in *Macbeth* have said after declaring "by the prickings of my thumbs, something wicked this way comes" if "something wicked" had not come? The foul air would have been filled with loud but lame excuses. But if there were a measure of the association between wickedness and the number or intensity of thumb prickings, and if that association were found to be low, then the witch might very well have avoided relying on thumb prickings as a predictor of wickedness.

Modesty, then, is a characteristic of our methods. Once we realize the possibility that we can make mistakes, the need becomes apparent to have as part of our methodology a measure for evaluating this possibility. These considerations lead us naturally to the topics of the following sections.

Recall that in the study of linear regression we assume that the relationship between x and Y is given by

$$Y = \gamma + \beta_1 x + E \qquad (9.7a)$$

where E is the random error term. This equation can be rewritten as

$$E = Y - (\gamma + \beta_1 x) \qquad (9.7b)$$

We again assume that the mean value of E for any value x is equal to zero. At this point we also assume that E has a normal distribution with variance σ_e^2, which is the same no matter the value of x.* In summary,

> Assumption: The error E is normally distributed with mean 0 and variance σ_e^2 for all x. In this case Y is normally distributed with mean $\gamma + \beta_1 x$ and variance σ_e^2.

Under this assumption we subsequently find confidence limit estimates on the parameters γ and β_1 of the regression line as well as on the predictions we make using this line. These estimates will give us some insight into the trustworthiness of our predictions.

To begin, note from equation (9.7a) that the only source of variation in Y is the variation in the random error E. Suppose the variation in E were zero; that

---

* We use the notation σ_e^2 rather than σ_E^2 to conform with standard notation.

is, suppose $\sigma_e^2 = 0$. Then our prediction would be perfect, since, given γ, β_1, and any value of x, our predicted value must be the true value $\gamma + \beta_1 x$ and there would be no uncertainty in this prediction. However, the larger σ_e^2 becomes, the more our predicted value may depart from $\gamma + \beta_1 x$. To develop some confidence limit estimates, we must first estimate σ_e^2, the variation of E. The expression $E = Y - (\gamma + \beta_1 x)$ gives us some insight into what the estimate of σ_e^2 should be. E is the difference between Y and the true regression line. An estimate of this difference at x_i should be $e_i = y_i - (a + bx_i)$, where y_i is an observed value of Y at x_i. The square of this difference, e_i^2, gives us a measure of how far the observed value y_i deviates from the least squares regression line. If we average all these squared deviations, we obtain an estimate s_e^2 of the observed data points from the line. Hence

$$s_e^2 = \frac{1}{n-2} \sum_{i=1}^{n} e_i^2$$

is an estimate* of σ_e^2.

The square root of s_e^2, denoted s_e, is called the standard error of the estimated regression line. A simpler computational formula for the standard error, s_e, is

---

Computation formula for the standard error of the regression line:

$$s_e = \sqrt{\frac{(n-1)(s_Y^2 - b^2 s_x^2)}{n-2}} \tag{9.8}$$

where s_Y^2 and s_x^2 are the observed variances calculated for observed Y and the set of x values, respectively.

Example 9 ▪ 5   Consider the data taken from the *Scientific American* article referred to in example 9.4. We have

$$\Sigma x_i = -7.53, \quad \Sigma x_i^2 = 44.7, \quad \Sigma x_i y_i = 20.6,$$
$$\Sigma y_i = -1.51, \quad \Sigma y_i^2 = 42.9$$

* We divide $\sum_{i=1}^{n} e_i^2$ by $n-2$ instead of n. To understand why this is so, note that $\sum_{i=1}^{n} e_i^2 = \sum_{i=1}^{n} [y_i - (a + bx_i)]^2$, where a, b estimate γ and β_1, respectively. Recall that every time we estimate a parameter we lose a degree of freedom. In this case we estimate two parameters; hence there are really only $n-2$ free choices for the deviations that we use to define s_e^2.

Thus

$$s_x^2 = \frac{44.7 - (-7.53)^2/37}{36} = 1.2$$

$$s_Y^2 = \frac{42.9 - (-1.51)^2/37}{36} = 1.19$$

$$b = \frac{20.6 - (-7.53)(-1.51)/37}{44.7 - (-7.53)^2/37} = 0.47;$$

$$a = -\frac{1.51}{37} - 0.47\left(\frac{-7.53}{37}\right) = 0.055$$

$$s_e^2 = \frac{36}{35}(1.19 - (0.47)^2(1.2)) = 0.95$$

$$\hat{y}_x = 0.055 + 0.47x.$$

Assuming the errors are normally distributed, we estimate that for a fixed value of anxiety score x, the variability in heart rate in people with this anxiety is normally distributed with mean $0.055 + 0.47x$ and variance 0.95. Thus patients with anxiety score of -1.3 should on the average have a negative heart rate variability of -0.556. We subsequently develop a confidence limit estimate on this prediction.

Confidence Intervals γ, β_1, and σ_e^2

Recall that a and b are estimates of γ and β_1, respectively. Hence a and b are observed values of random variables that we denote by A and B, respectively. In order to make inferences about the true values of γ and β_1, we need the estimated variances for A and B. These are respectively

$$s_A^2 = s_e^2\left(\frac{1}{n} + \frac{\bar{x}^2}{(n-1)s_x^2}\right) \tag{9.9}$$

$$s_B^2 = \frac{s_e^2}{(n-1)s_x^2} \tag{9.10}$$

For the numerical data from example 9.5, we have

$$s_A^2 = (0.95)\left(\frac{1}{37} + \frac{(-0.20)^2}{36(1.2)}\right) = 0.027$$

and

$$s_B^2 = \frac{0.95}{36(1.2)} = 0.022$$

If we assume that Y is normally distributed with mean $\gamma + \beta_1 x$ and variance σ_e^2, then $100(1 - \alpha)$% confidence interval for γ is from

$$a - s_A t_{\alpha/2} \quad \text{to} \quad a + s_A t_{\alpha/2} \tag{9.11}$$

and the $100(1 - \alpha)$% confidence interval for β_1 is from

$$b - s_B t_{\alpha/2} \quad \text{to} \quad b + s_B t_{\alpha/2} \tag{9.12}$$

The number of degrees of freedom in both cases in $n - 2$, since s_A^2 and s_B^2 both have $n - 2$ degrees of freedom.

Example 9 ▪ 6   In example 9.2 we estimated the linear relation between x, a person's IQ, and Y, the grade point average of students. This estimated relationship was expressed as

$$\hat{y}_x = -2.4 + 0.05x$$

The slope of the line is 0.05; this slope is the rate of increase in grade point average with IQ. We see that for every point increase in IQ there is an estimated 0.05 increase in grade point average. To evaluate how good this estimate of 0.05 is, we find the 99% confidence interval for β_1 that is from $b - s_B t_{0.005}$ to $b + s_B t_{0.005}$, where $b = 0.05$ and using Appendix A-1 because df $= 48$, $t_{0.005}$ is about 2.58. We have

$$s_x^2 = \frac{552220 - (5250)^2/50}{49} = 19.8$$

$$s_Y^2 = \frac{409.2 - (142.5)^2/50}{49} = 0.0628$$

$$s_e^2 = \frac{n-1}{n-2}(s_Y^2 - b^2 s_x^2) = \frac{49}{48}(0.0628 - (0.05)^2(19.8) = 0.0136$$

$$s_B^2 = \frac{0.0136}{49(19.8)} = 0.00001; \; s_B = 0.0037$$

The 99% confidence interval for β_1 is from

$$0.05 - (0.0037)(2.58) \quad \text{to} \quad 0.05 + (0.0037)(2.58)$$

that is, from 0.04 to 0.06.

Thus we see that with "99% confidence" the rate of increase of grade point average with IQ is between 0.04 and 0.06.

As an exercise we can also find 99% confidence limits for the intercept γ. This interval is from $a - s_A t_{0.005}$ to $a + s_A t_{0.005}$, where $a = -2.4$. Again, using the data of example 9.2, we obtain

$$s_A^2 = (0.0136)\left(\frac{1}{50} + \frac{(105)^2}{49(19.8)}\right) = 0.155; \qquad s_A = 0.394$$

Thus the 99% confidence interval for γ is from $-2.4 - 0.394(2.58)$ to $-2.4 + 0.394(2.58)$, that is, from -3.417 to -1.383.

The estimate of the intercept γ, and the determination of its confidence limits, is often of crucial importance in many problems. See problem 28 for an example of this kind.

We may also find confidence limits for the error variance σ_e^2. In fact, these limits have essentially been derived in Chapter 7; the $100(1 - \alpha)\%$ confidence limit for σ_e^2 is from

$$\frac{(n-2)s_e^2}{\chi_{\alpha/2}^2} \quad \text{to} \quad \frac{(n-2)s_e^2}{\chi_{1-\alpha/2}^2}$$

where Appendix A.6 is used to find the values in the preceding denominator using $n - 2$ degrees of freedom.

Confidence Intervals for $\mu_{Y|x}$, the Expected Y Value for Fixed x

As we have seen, $\mu_{Y|x}$ is the center of the population of Y values that correspond to x. If we wish to predict the value of Y for a given value x, we use the estimate $\hat{y}_x$ of $\mu_{Y|x}$. The estimate $\hat{y}_x$ is unbiased, that is, $E(\hat{Y}_x) = \mu$, where $\hat{Y}_x$ is the estimator that takes on the possible values $\hat{y}_x$. In order to calculate confidence intervals for the mean $\mu_{y|x}$, we must determine exactly how much our estimate $\hat{y}_x$ varies from sample to sample; that is, we must find the sampling distribution of $\hat{Y}_x$. To do this, we need to estimate the variance of $\hat{Y}_x$. The variance of $\hat{Y}_x$ should depend on the variability of A and B since both appear in the formula $\hat{Y}_x = A + Bx$. This variance should be smaller the closer x is to $\bar{x}$. This is in fact the case, and the estimate of the variance of $\hat{Y}_x$ is given by

$$s_{\hat{Y}_x}^2 = s_A^2 + (x^2 - 2x\bar{x})s_B^2 = s_e^2\left(\frac{1}{n} + \frac{(x-\bar{x})^2}{(n-1)s_x^2}\right)$$

The confidence interval for $\mu_{Y|x}$ is given by

$100(1-\alpha)$% confidence interval for the mean value $\mu_{Y|x}$ of Y for fixed x,

from $\hat{y}_x - s_{\hat{Y}_x}t_{\alpha/2}$ to $\hat{y}_x + s_{\hat{Y}_x}t_{\alpha/2}$ **(9.13)**

with df $= n - 2$.

Example 9 ▪ 7   Suppose the educational psychologist in example 9.2 encounters a group of students, all of whom have a 110 IQ. It is desired to estimate the average grade point average for the group with a 95% confidence interval. To construct the 95% confidence interval for $\mu_{Y|110}$, we note that since

$$s_{\hat{Y}_{110}}^2 = 0.0136\left\{\frac{1}{50} + \frac{(110-105)^2}{970}\right\} = 0.00062, \quad \text{and}$$

$$t_{0.025} = 2.02, \quad \text{with df} = 48 \quad \text{and} \quad \hat{y}_{110} = 3.1$$

the 95% confidence interval for the mean grade point $\mu_{Y|110}$ is from

$$3.1 - 2.02\sqrt{0.00062} \quad \text{to} \quad 3.1 + 2.02\sqrt{0.00062}$$

or from

$$3.05 \quad \text{to} \quad 3.15$$

Figure 9.11 shows a graph of the confidence interval (9.13) that forms a band on $\mu_{Y|x}$; it also shows the line $\hat{y}_x$. Note that the band is smallest when x is equal to $\bar{x}$ and that the band widens dramatically as x moves farther away from $\bar{x}$. In most situations the widening of the band becomes most severe as we move outside the observation interval. This result provides another reason for caution in making predictions outside this interval.

Prediction Intervals for a Single Observation

In example 9.7 we found a confidence interval for the mean $\mu_{Y|x}$ of all Y scores corresponding to a specific x value. It frequently happens that we wish to find a

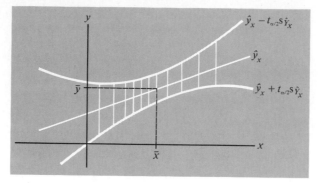

Figure 9 ▪ 11
Confidence interval for $\mu_{Y|x}$. Note the band is smallest when x is equal to $\bar{x}$.

confidence or prediction interval for a single *observation* of Y, rather than for the mean value of all such Y's. Let Y_x be the Y value corresponding to a given value x. We have previously noted that Y_x has a normal distribution with mean $\mu_{Y|x} = \gamma + \beta_1 x$ and a variance σ_e^2. Keep in mind that the observed value of Y_x does not come from the regression line. It is some future observed value. Now the estimator of the regression line $\mu_{Y|x}$ is $\hat{Y}_x = A + Bx$. Since $\hat{Y}_x$ is normal with mean $\gamma + \beta_1 x$ and variance $\sigma_e^2[1/n + (x - \bar{x})^2/\Sigma\,(x_i - \bar{x})^2]$, and since Y_x and $\hat{Y}_x$ are independent random variables, we know that $Y_x - \hat{Y}_x$ is also normally distributed with a mean 0 and a variance $\sigma_e^2[(1 + 1/n) + (x - \bar{x})^2/\Sigma\,(x_i - \bar{x})^2]$, which is just the sum of the variances of Y_x and $\hat{Y}_x$. Thus,

$$\frac{Y_x - \hat{Y}_x}{\sqrt{S_e^2\left[1 + \dfrac{1}{n} + \dfrac{(x - \bar{x})^2}{\Sigma\,(x_i - \bar{x})^2}\right]}}$$

has a t-distribution with $n - 2$ degrees of freedom. Thus with $\mathrm{df} = n - 2$ the $100(1 - \alpha)\%$ prediction interval for Y_x is from

$$\hat{y}_x - t_{\alpha/2}\sqrt{s_e^2\left[1 + \frac{1}{n} + \frac{(x - \bar{x})^2}{\Sigma\,(x_i - \bar{x})^2}\right]} \quad \text{to}$$

$$\hat{y}_x + t_{\alpha/2}\sqrt{s_e^2\left[1 + \frac{1}{n} + \frac{(x - \bar{x})^2}{\Sigma\,(x_i - \bar{x})^2}\right]}$$

Note that we use the term "prediction interval" when estimating a future observation of Y_x, but use the term "confidence interval" when estimating a parameter like $\mu_{Y|x}$.

Example 9 ▪ 8   Suppose the educational psychologist in example 9.2 wished to find a prediction interval for the grade point of a single student who has an IQ of 120. Since this is a single observation, the 99% prediction interval for her grade point is determined as follows (see example 9.6 for the value of s_e^2 and s_x^2):

$$t_{0.005} = 2.58 \qquad df = 48$$

$$\frac{(x - \bar{x})^2}{\Sigma (x_i - \bar{x})^2} = \frac{(120 - 105)^2}{(n - 1)s_x^2} = \frac{(120 - 105)^2}{49(19.8)} = 0.23$$

$$\hat{y}_{120} = -2.4 + 0.05(120) = 3.6$$

so the prediction interval is from

$$3.6 - 2.58 \sqrt{0.0136 \left(1 + \frac{1}{50} + 0.23\right)} \quad \text{to}$$

$$3.6 + 2.58 \sqrt{0.0136 \left(1 + \frac{1}{50} + 0.23\right)}$$

or from

$$3.26 \quad \text{to} \quad 3.94$$

Remarks on Methodology

A review of the previous sections reveals that we have accomplished what we set out to do: (1) established a measure r^2 of "goodness of fit" of our model to the data and (2) through confidence intervals and prediction intervals, we have measures of the "trustworthiness" of our predictions.

Our methods have been restricted to straight-line relationships between quantitative variables. Much of the vast literature on regression methods deals with using other than straight-line regressions. A few references are given at the end of this chapter. However, this methodology would not satisfy Macbeth's witch (see the quote opening this chapter) who could rightfully claim that most of her predictions involve *qualities* (such as wickedness) rather than *quantities*. Recently, considerable work has been done in studying regression problems involving qualitative data. A reference is given in the next section.

One weakness in the methods described in this chapter is that there is only one independent variable x in the basic model $Y = \gamma + \beta_1 x + E$. In most problems found in practice Y depends on many variables; in these cases the model is said to be multivariate. There is also a vast literature in this area with some references given at the end of this chapter.

9 ■ 6 REMARKS ON COMPUTER SOFTWARE

Since regression analysis is used so often, and since the computations for large data sets are so extensive, every statistical software package previously mentioned in this text, MINITAB, SPSS, SAS, and Bio-Med, all have extensive software that perform multivariate linear and nonlinear regressions. Moreover, the latter three packages also enable such regressions to be performed using qualitative data. The Bio-Med package has a program for time prediction called the "Cox regression" which enables regressions to be performed on censored data; this is time data, such as "length of life," in which not all participants have reached the end of life so that we do not know how long they will live. This regression program is of great practical use in many biomedical investigations.

Summary

This chapter has discussed some techniques for ascertaining the relationship or correlation between two variables. The first model studied was

$$Y = \gamma + \beta_1 x + E$$

where γ and β_1 are fixed but unknown constants, x is a selected observation point, and E is a random variable with mean $\mu_E = 0$. Hence the mean value for Y corresponding to the value x is

$$\mu_{Y|x} = \gamma + \beta_1 x$$

One problem discussed was how to obtain best estimates of γ and β_1 from the data. The estimates a and b respectively that minimize the sum of the squared distances between the observed values of Y and the predicted values of Y were chosen. The line $a + bx$ is called the least squares estimate and is best in the sense that the sum of squared errors is smallest.

A measure of association between two random variables, X and Y, called the correlation coefficient ρ was introduced. It is estimated by the sample correlation coefficient r. If X and Y are joint normally distributed, then $\rho = 0$ if and only if X and Y are independent. This fact permits us to test for the independence of X and Y by testing

$$H_0 : \rho = 0$$
$$H_1 : \rho \neq 0$$

Other tests of hypothesis concerning ρ were also developed.

The sample correlation coefficient r also gives us some information regarding how well an estimated line $a + bx$ fits the data. If $r = +1$ or $r = -1$, the straight line fits the data perfectly; if $r = 0$, one can conclude that the relationship between $\mu_{Y|x}$ and x is not a linear relationship. This use of r enables us to interpret r and ρ (since r is an estimate of ρ): r measures the degree of the linear relationship between $\mu_{Y|x}$ and x; ρ measures the degree of the linear relationship between two random variables. The coefficient of determination r^2 is interpreted as the proportion of the total variation of Y values that has been explained by the regression relationship.

Formulas for the confidence limit estimates on γ, β_1, and $\mu_{Y|x}$ were also obtained as well as prediction intervals for a particular predicted value of Y.

Key Words

Define the following terms:

sample correlation coefficient
true correlation coefficient
coefficient of determination
least squares fit
$\mu_{Y|x}$

linear relationship
prediction interval
scatter diagram
straight line

References*

Excellent discussions of the application of regression analysis to the law can be found in
Baldus, D. C., and J. W. Cole: *Statistical Proof of Discrimination.* New York, McGraw-Hill, 1980.
Fischer, A.: Multiple regression in legal proceedings. *Columbia Law Review, 80*:702–708, 1980.

The reference to the legal case *Vuyanavich* vs. *Republic National Bank of Dallas* referred to at the beginning of this chapter is CA-3-6982-G, U.S. District Court for the Northern District of Texas, Dallas Division. This action was the subject of three prior published opinions: 409 F. Supp. 1083 (1976); 78 FRD 352 (1978); 82 FRD 420 (1979), all ND Texas.

An application of regression to a physical problem can be found in
Taur, J. M., et al.: *Statistics: A Guide to the Unknown.* San Francisco, Holden-Day, 1978, pp. 421–423.

An important study in which the correlation between heart disease and other factors is the central issue can be found in
Dawber, T., et al.: Coffee and cardiovascular disease: Observation from the Framingham Study. *New England Journal of Medicine, 291*:871–874, 1974.

A discussion of multivariate techniques can be found in
Kleinbaum, D. G., and L. L. Kupper: *Applied Regression Analysis and Other Multivariable Methods.* Boston, Duxbury Press, 1978.

* See also the sources cited in problems 2, 14, 15, 20, 21, 27.

Exercises

Section 9.1

1. The following is a list of the grades in a statistics class prior to the final exam, and then on the final exam. Plot a scatter diagram for these data. Would you say that it indicates a straight-line relationship between variables? Would you say the grade prior to the final is a good indicator of the final exam grade?

Student	Observed percentage prior to final (x)	Final exam observed grade (y)
1	59	35
2	94	91
3	87	93
4	80	76
5	89	89
6	93	65
7	96	92
8	99	92
9	98	93
10	75	80
11	93	96
12	96	91
13	87	83
14	69	44
15	88	97
16	86	88
17	73	55
18	88	82
19	85	75
20	72	51
21	93	86
22	71	54
23	87	86
24	96	87
25	83	86

2. According to the *Dairy Record* (70:23, 1970), the monthly production of cream cottage cheese (x) and ice cream (Y) for 1969 was

Observed x (in millions of pounds)	Observed y (in millions of gallons)
70.9	54.9
72.4	54.1
81.6	61.6
79.8	64.3
85.2	68.5
77.6	75.7
80.7	80.1
75.5	77.0
73.9	69.4
77.6	63.1
69.8	49.9
69.3	54.1

   a. Without examining the data, would you expect that the relationship between x and Y would be positive or negative. Explain.
   b. Plot a scatter diagram for the data.

3. A questionnaire is designed to detect a tendency toward paranoid schizophrenia in patients who were admitted to a state mental hospital. Seven patients are given the questionnaire and their scores recorded. These patients are then interviewed individually by the staff psychologists and psychiatrists, with each evaluator rating the patient's tendency toward schizophrenia on a scale of 1 to 25. The average for each patient is then recorded.

Patient	Questionnaire score (x)	Average psychiatric rating (y)
1	1	7.5
2	6	12.5
3	8	16.5
4	3	6
5	5	15
6	9	20
7	4	11.5

a. Plot a scatter diagram for these data.
b. Estimate a straight line through the scatter diagram in (a)
4. When selling colored plastic to a toy manufacturer, a certain chemical company must try to match exactly the standard color chip presented by the toy manufacturer. It is the job of the quality control matcher to ascertain how well the produced plastic matches the standard in such respects as color, opacity, and tensile strength. A rating of 0 indicates a perfect match and a rating of 8 indicates a very poor match. The price per kilogram that can be charged by the plastic company depends on the quality of the plastic. The company collected the following data:

Quality rating (x)	Observed selling price per kilo (y)
1	0.33
1	0.36
2	0.36
2	0.30
3	0.26
3	0.30
4	0.24
4	0.30
5	0.21
5	0.25
6	0.23
6	0.20
7	0.16
8	0.15

a. Plot a scatter diagram for these data.
b. Estimate by sight the price at which you think a kilogram of plastic with a zero rating should sell.
5. Fire ants pose a particular problem to farmers in the South and Southwest. In the past they have been controlled through the use of the pesticide Mirex. Unfortunately, Mirex degrades into Kepone, a highly toxic, long-lasting chemical poison that attacks the nervous system of humans as well as of ants. As a substitute pesticide, a certain chemical company develops a compound named Qukil. Data collected on the percentage kill of fire ants for several different levels of Qukil follow:

Kilograms per acre of Qukil (x)	Observed percent of fire ants killed (y)
2	57, 50, 48
6	58, 62, 66
10	70, 72, 68
15	100, 100

Plot a scatter diagram for these data.
6. A chemical reaction is monitored for several different levels of catalyst present. The speed of reaction is then recorded.

Amount of catalyst present (x grams)	Speed of reaction (y seconds)
1	38, 40
2	37, 39
3	32, 34
4	26, 27
5	21, 20

Plot a scatter diagram for the amount of catalyst versus the speed of reaction.

Section 9.2

7. For the following data, calculate $\bar{x}$, $\bar{y}$, a, and b. Plot the scatter diagram and the estimated regression line $\hat{y}_x$. State the predicted value of Y for $x = 0$.

a.

x	−1	−1	1	2	3
y	1, 2	1.5, 2.3	1.9, 2.5	2, 2.8	3, 3.1

b.

x	1	2	3	4	5
y	3	2.5	1	0	−1.5

8. Using the data of problem 1, calculate the regression or prediction line $\hat{y}_x$, and sketch its graph. What final exam score would you predict for a student who had a 85% average prior to the final?

$$(\Sigma\, x = 2137, \quad \Sigma\, y = 1967,$$

$$\Sigma\, xy = 171{,}917, \quad \Sigma\, x^2 = 185{,}283,$$

$$\Sigma\, y^2 = 162{,}177)$$

9. Using the data of problem 2, calculate the regression line for predicting Y based on x. How many million gallons of ice cream would you predict as being produced in a month that has 80 million pounds of cream cottage cheese produced? (Treat the observed values of x as given.)

10. Using the questionnaire data given in problem 3, calculate the regression or prediction line $\hat{y}_x$ and sketch the graph. If a patient had a score of 7 on the psychiatric questionnaire, what would you predict his average psychiatric rating to be? How about a score of 2?

11. What price per kilogram would you predict the plastic in problem 4 could be sold for if it had a quality control rating of 2? Find and plot the regression line $\hat{y}_x$.

12. Utilize the data in problem 5 to derive the regression line $\hat{y}_x$. Predict the percent of fire ants that will be killed if one uses 8 kilograms per acre of Qukil.

$$(\Sigma\, x = 84, \quad \Sigma\, y = 751,$$

$$\Sigma\, xy = 6526, \quad \Sigma\, x^2 = 870,$$

$$\Sigma\, y^2 = 54{,}325)$$

13. Estimate the speed of the chemical reaction in problem 6 by using the regression equation

$\hat{y}_x$ for 2.5 g of catalyst. Sketch the regression equation.

14. An investigation on weight gain in baby turkeys as a function of the amount of the chemical biotin fed was conducted by Dobson. He found the following:

Biotin level	Average weight at 3 weeks (in grams)
0	206
25	222
50	225
75	296
100	324
125	375
150	363
175	408
200	407
225	434

a. Plot a scatter plot of these data.
b. Find the regression equation for predicting weight based on the biotin level.
c. Estimate the weight for turkeys given 130 g of biotin by a single number.

(Dobson, D.: Biotin requirements of turkey poults. *Poultry Science, 49*:1970, pp. 546–553.)

15. The *Life Insurance Fact Book,* published by the Institute of Life Insurance in 1967, gives the total sales of life insurance for each year 1951 to 1962. In billions of dollars we have

Year	Sales	Year	Sales
1951	27.5	1957	66.7
1952	31.5	1958	66.8
1953	36.2	1959	70.8
1954	45.4	1960	74.4
1955	48.4	1961	79.0
1956	55.3	1962	79.6

a. Plot a scatter diagram for predicting sales based on years.
b. Find the equation for the regression of sales on years.
c. Predict the volume of sales in 1963.

Sections 9.3 and 9.4

16. For the data in problem 1 calculate the sample correlation coefficient and the coefficient of determination of the two scores. Test if the two scores are independent. (Use $\alpha = 0.05$.) (See problem 8 for the necessary sums.)

17. Calculate the correlation coefficient between the production of cream cottage cheese and ice cream production by using the data in problem 2. Test if the two production levels are independent by using $\alpha = 0.05$.

18. Calculate the regression line $\hat{x}_y$ of the questionnaire score on the psychiatric rating by using the data in problem 3. What score would you predict for a new patient whom the psychologists rated 15? What is the correlation coefficient between the questionnaire score and the psychologists' rating? Test if the two evaluations are independent by using the value of r. (To calculate $\hat{x}_y$, just reverse x and y in all the formulas.) Use $\alpha = 0.01$.

19. Find the correlation coefficient between the amount of Qukil pesticide used per acre and the percent of fire ants killed by using the data from problem 5 (note that $\Sigma x_i = 84$, $\Sigma y_i = 751$, $\Sigma x_i y_i = 6526$, $\Sigma x_i^2 = 870$, and $\Sigma y_i^2 = 54{,}325$). Test the hypothesis that the correlation coefficient is 0.6 at $\alpha = 0.05$.

20. Light beers are now a very high selling item. The following table lists some beers and their calories and carbohydrate content:

	Calories	Carbohydrates
Pearl Light	68	1.5 grams
Olympia Gold	70	2.0
Pabst Light	70	3.1
Schlitz Light	96	5.0
Miller Lite	96	2.8
A-B Light	97	5.5
Coors Light	110	5.2
Michelob Light	134	not known.

Austin American Statesman, December 23, 1978.

a. What is the correlation between calories and carbohydrates for these beers?
b. Test if calories and carbohydrates are independent.
c. Determine the regression line for predicting carbohydrates based on calories, and predict the carbohydrate concentration of Michelob Light.

Section 9.5

21. An experiment performed by Ulric Neisser was designed to determine the speed at which people scan material in search of a critical item. In one example he randomly placed a letter K among lines consisting of four letters. The subject was to search down the list until he found the line containing the letter K. The following data were recorded:

Line containing critical item (x)	Time in seconds (y)
9	8
11	8
15	7
17	9
19	14
21	13
24	14
25	16
29	17
32	19
34	21
36	21
39	26
41	27

a. Calculate the regression line $\hat{y}_x$, and give a confidence interval for the expected time it would take a subject to spot a critical item on line 18.
b. Calculate the correlation coefficient r.
c. Calculate s_e^2, s_Y^2, s_x^2, s_A^2, and s_B^2.
d. Calculate a 90% confidence interval for γ

and β_1, the intercept and slope parameters.

e. Give a 95% confidence interval estimate for the mean number of seconds ($\mu_{Y|18}$) it would take to find a critical item located on the eighteenth line.

f. Give a 95% confidence interval estimate for the number of seconds it would take for a person to find a critical item located on the eighteenth line in a single trial.

(Neisser, U.: Visual search. *Scientific American,* June 1964, pp. 94–102. All rights reserved.)

22. Using the data of problem 1 (and problem 8)
    a. Find s_e^2, s_Y^2, s_x^2, s_A^2 and s_B^2.
    b. Give a 90% confidence interval for $\mu_{Y|85}$, the mean final exam score for students with an 85% prior to the final.
    c. Give a 90% prediction interval for the final exam score of a single student who had an 85% prior to the final.

23. Use the data from problems 2 and 9 to find a 90% confidence interval for the average amount of ice cream produced when 80 million pounds of cream cottage cheese are produced.

24. Using the data of problem 3,
    a. Calculate s_e^2, s_Y^2, s_x^2, s_A^2, and s_B^2.
    b. Give a 95% confidence interval for γ and β_1, the intercept and slope parameters.
    c. Give a 99% confidence interval for the mean psychiatric rating $\mu_{Y|7}$ for patients scoring 7 on the questionnaire.
    d. Calculate a 99% prediction interval for the psychiatric rating of a single patient who scores 7 on the questionnaire.

25. Using the data on biotin level versus weight gain of baby turkeys given in problem 14, give a 90% confidence interval for the weight of a single turkey with a biotin level of 130.

26. Using the life insurance sales data from problem 15, give a 90% confidence interval for the predicted sales in 1963.

27. The internal stress characteristics of various metals under various temperature conditions is of extreme importance in aircraft and spacecraft safety considerations. One quantity related to stress in ferromagnetic materials is the magnetomechanical damping peak height. This was studied by Roberts and Barrand, who obtained the following data:

x: Temperature (centigrade)	y: Magnetomech. damp. peak height (mm.)
−92	0.375
−50	0.331
−40	0.300
−3	0.251
23	0.224
40	0.216
52	0.204
78	0.146
125	0.081
162	0.056
176	0.035

a. Find the linear prediction equation $\hat{y}_x$.
b. Test at $\alpha = 0.05$ if the slope is different from zero. (This can be done by determining if zero is in the 95% confidence interval for β_1.)
c. Give a 90% confidence interval for the estimated or predicted mean value of Y for $x = 30°C$.
d. Give a 90% prediction interval for the predicted value of Y for a single observation at $x = 30°$.

(Roberts, J. T. A., and Barrand, P.: Anisotropy effects in the magnetochemical damping of nickel-copper alloys. *Journal of the Physics D: Applied Physics,* 3:9, 1970, pp. 1340–1345. Copyright Institute of Physics.)

28. In financial analysis one often desires to predict the return to be expected from a particular stock or asset. One method frequently used is the so-called capital asset pricing model (CAPM) developed by Sharpe, Lintner, and Mossin. Essentially this as-

sumes that the return on an asset has a linear regression relationship with the return on the market portfolio (which is usually taken to be the Dow-Jones industrials, or the listed stocks on the New York Stock Exchange). Consider the following data on an asset and the Dow-Jones industrials for a 12-month period:

Month	Asset A annualized return	D-J industrials annualized return
1	0.12	0.10
2	0.14	0.12
3	0.10	0.11
4	0.11	0.09
5	0.13	0.11
6	0.10	0.08
7	0.10	0.10
8	0.13	0.12
9	0.14	0.13
10	0.12	0.09
11	0.12	0.10
12	0.15	0.11

a. Calculate the regression line for predicting asset A's returns based on the Dow-Jones averages.

b. Find a 90% confidence interval for the slope coefficient.

c. Find a 90% confidence interval for the intercept coefficient.

29. In carcinogenic testing a frequently used procedure is to give an experimental animal a very high dosage of a suspected carcinogen, and then record the time until a tumor occurs. Regression techniques are then used to predict the time until a tumor occurs at more moderate doses. This is the method used to determine so-called safe levels or dosages of potentially harmful chemical agents. The permissible level of lead in the work place was determined in just this way. Let x denote the natural logarithm of the dosage given (in grams) and Y denote the natural logarithm of the time until tumor occurs (in days) for an experimental group of mice.

x	y
2.30	3.56
2.48	3.10
2.71	2.30
2.83	2.90
3.04	2.10
3.58	1.98
3.69	1.10

a. Find the regression line $\hat{y}_x$.

b. Find a 90% prediction interval estimate for the survival time of a single mouse given a dosage of 7.39 g ($x = 2$).

ESTIMATING DIFFERENCE

$\mu_P = \mu_X - \mu_Y$

$\sigma_P = \sqrt{\sigma_X^2 + \sigma_Y^2}$

$x \; ; \; y$ normal ; independent.

* * *

$\hat{P}_1, \; \hat{P}_2 =$ 2 populations (normal) being estimated.

$\mu_{\hat{P}_1} = P_1 \qquad\qquad \mu_{\hat{P}_2} = P_2$

$\sigma_{\hat{P}_1} = \sqrt{\dfrac{P_1(1-P_1)}{n_1}} \qquad \sigma_{\hat{P}_2} = \sqrt{\dfrac{P_2(1-P_2)}{n_2}}$

difference $= \hat{P}_1 - \hat{P}_2$

$\mu_d = \mu_{\hat{P}_1} - \hat{P}_2 = P_1 - P_2$

$d = \sqrt{\sigma_{\hat{P}_1}^2 + \sigma_{\hat{P}_2}^2} = \sqrt{\dfrac{P_1(1-P_1)}{n_1} + \dfrac{P_2(1-P_2)}{n_2}}$

* **

Mortality in Smokers
age (60-70)

DATA	Pipe $(\hat{P}_1)$	NonSmokers $(\hat{P}_2)$
n	$402 \; n_1$	$1067 \; n_2$
# deaths	$54 \, (x_1)$	$117 \, (x_2)$

$\dfrac{\hat{P}_1 - x_1}{n_1} =$ estimate mortality rate for pipe smokers.

$= \dfrac{54 \, (x_1)}{402 \, (n_1)} = .1343$

$\hat{P}_2 = \dfrac{117 \, (x_2)}{1067 \, (n_2)} = .1097$

Difference $P_1 - P_2$

$\hat{P}_1 - \hat{P}_2 = .0247$

ESTIMATE INC. MORT   2.47% due to PIPE

95% confidence $Z_{.025} = 1.96$

$(\hat{P}_1 - \hat{P}_2) \pm (1.96) \sqrt{\dfrac{\hat{P}_1(1-\hat{P}_1)}{n_1} + \dfrac{\hat{P}_2(1-\hat{P}_2)}{n_2}}$

$.0247 \pm .0382$

$= -.0135$ to $+ .0629$

95% confidence limit for $\uparrow$ in mortality due to pipe = 1.35% to 6.29%

SAMPLE SIZE FOR DIFFERENCES.

Confidence limit

$\bar{x} - \bar{y} \pm Z_{\alpha/2} \sqrt{\dfrac{\sigma_x^2}{n} + \dfrac{\sigma_v^2}{m}}$

$\overset{OR}{=} \bar{x}_1 - \bar{x}_2 \pm Z_{\alpha/2} \sqrt{\dfrac{\sigma_1^2}{n_2} + \dfrac{\sigma_2^2}{m_2}}$

EST. 2 PROPORTIONS

$\hat{P}_1 - \hat{P}_2 \pm Z_{\alpha/2} \sqrt{\dfrac{P_1(1-P_1)}{n_1} + \dfrac{P_2(1-P_2)}{n_2}}$

Means
$\bar{x}_1 - \bar{x}_2 \pm B$

Proportion
$\hat{P}_1 - \hat{P}_2 \pm B$

$B = Z_{\alpha/2} \sqrt{\dfrac{\sigma_1^2}{n_1} + \dfrac{\sigma_2^2}{m}}$

$\subset$ DIFFERENCE IN MEANS

$B = Z_{\alpha/2} \sqrt{\dfrac{P_1(1-P_1)}{n_1} + \dfrac{P_2(1-P_2)}{n_2}}$

$\subset$ DIFF IN PROPORTIONS

WE CAN REPLACE

σ_1^2 by $P_1(1-P_1)$

σ_2^2 " $P_2(1-P_2)$

* * *

$\dfrac{B_0}{Z_{\alpha/2}} = \sqrt{\dfrac{\sigma_1^2}{n_1} + \dfrac{\sigma_2^2}{n_2}}$

IF $n_1 ; n_2$ are unknown ASSUME THEY ARE EQUAL.

$\therefore \dfrac{Z_{\alpha/2}^2}{B_0^2} = (\sigma_1^2 + \sigma_2^2)$

OR $N = n_1 + n_2$ (minimized)

$\therefore n_1 = \dfrac{Z_{\alpha/2}^2}{B_0^2} (\sigma_1^2 + \sigma_1 \times \sigma_2)$

$n_2 = \dfrac{Z_{\alpha/2}^2}{B_0^2} (\sigma_2^2 + \sigma_1 \times \sigma_2$

CHAPTER · 10

Evaluation of Categorical Data; Contingency Tables

I know the Kings of England, and I quote the fights historical
From Marathon to Waterloo, in order categorical.

Sir William S. Gilbert
The Pirates of Penzance, Act I

Until now we have been mainly studying techniques for analyzing quantitative data. Often our data do not consist of observations of quantities, such as blood pressure or height, but of attributes, such as sex or color, in which the information falls into categories (male, female; red, blue, orange). In this chapter we study analytical tools that are applicable to data that can be divided into categories.

In many statistical applications the data are not quantitative, such as height or weight, but are *qualitative* or categorical in nature. For example, to determine if there is racial discrimination in the selection of a jury in a trial, the relevant information is how many observations (selections for the jury) fall into the category "black" and how many in the category "white." If we are studying genetics, we might be interested in the number of offspring that exhibit certain characteristics such as "eye color." In "acceptance sampling" from an industrial production line we are interested in the proportion of objects produced on the line that falls into the category "defective" and the proportion that falls into the category "nondefective." In gathering data in the social sciences, we can use a very common instrument, the categorical response questionnaire—a questionnaire that places respondents in various categories such as male, female, lower, upper, middle class, and so on.

10 ▪ 1 GOODNESS OF FIT

One important class of problems associated with categorical data is the determination of how well a particular probability model fits the data. Up to now we have taken the model as given. Now we can check it through the methods that are described in this chapter. Suppose we have n observations where each observation falls into one of k categories. We have a *theoretical model* that enables us to determine how many observations we "expect" to find in each category; that is, from the model we are able to compute the expected number of observations in each category. If the *actual number* of observations in each category is close to the *expected number,* then we say that the "theoretical model fits the data." Thus from the data we are able to test whether or not a theoretical model is acceptable as a description of the state of nature. For example, suppose we have a model for the responses to a political question where the permitted answers are "yes," "no," and "no opinion." The model postulates that the probability of yes is 0.3, of no is 0.5, and no opinion is 0.2. We conduct 100 interviews on randomly selected people, and the expected number of yes answers is $0.3 \times 100 = 30$, of no answers is 50, and of no opinion is 20. Suppose we observe in the 100 interviews 0 yes, 0 no, and 100 no opinion answers. We must conclude that the theoretical model previously described does not fit the data.

To begin to develop a test of theoretical models, let us again suppose that we have n observations and k categories. We may display the data along with the theoretical distribution as follows:

Category	Observed frequency	Theoretical probability	Expected frequency
1	o_1	p_1	$e_1 = np_1$
2	o_2	p_2	$e_2 = np_2$
.	.	.	.
.	.	.	.
.	.	.	.
k	o_k	p_k	$e_k = np_k$
Total	n	1	n

We let o_i denote the observed frequency in category i; that is, o_i is the observed actual number of objects that fall into category i. The theoretical probabilities p_i are given to us by the model. If there are n total objects and p_i is the probability of falling into category i, then we expect np_i objects to fall into category i. This expected frequency is denoted by e_i, and we wish to determine how well the observed frequencies agree with the expected frequencies. The expression

$$\sum_i^k (o_i - e_i)^2$$

measures how close the observed frequencies are to the expected frequencies. However, merely summing the squared deviations of the observed frequency from the expected frequency does not adequately describe how well the observed and expected frequencies match. Being off by 1 in 5 is much different from being off by 1 in 1000. By dividing by the number of observations expected in each cell, we can obtain a proportional error for each category. It is the sum of these proportional errors that yields a good measure of how close the observed frequencies are to the expected frequencies. The formula for this measure is

$$x^2 = \sum_{i=1}^k \frac{(o_i - e_i)^2}{e_i}$$

Note that the observed frequency o_i is the number of observations in our sample that fall into category i. If we took another sample, this number might very well change. Thus o_i is the observed value in the range of a random variable that we denote O_i. Correspondingly, x^2 is the observed value of the random variable X^2.

$$X^2 = \sum_{i=1}^k \frac{(O_i - e_i)^2}{e_i}$$

This statistic is called the Pearson chi-squared statistic.

We see that x^2, the observed value of X^2, equals zero when the observed frequencies *exactly* match the expected frequencies; and the worse the match, the larger the value of this sum.

If n, the total number of observations, is very large, and the observations are independent, then as the name implies, the distribution of X^2 is approximately the χ^2-distribution with $k - 1$ degrees of freedom, where k is the number of categories. Thus the critical values of X^2 can be found in Appendix A.6 for n large. The rule of thumb we use is that n is large enough when the expected number e_i is each category i is at least 5.

Example 10 ▪ 1   A marketing analyst claims that 20% of the customers will be attracted to packaging type 1, 30% to type 2, 15% to type 3, 25% to type 4, and 10% to packaging type 5. A survey of 200 customers' attitudes yielded the following responses:

Category	Frequency observed
Attracted to type 1	48
" to type 2	56
" to type 3	34
" to type 4	48
" to type 5	14

From these data we form the table

Category	o_i	p_i	e_i	$(o_i - e_i)^2$	$\dfrac{(o_i - e_i)^2}{e_i}$
1	48	0.2	40	64	1.6
2	56	0.3	60	16	0.267
3	34	0.15	30	16	0.533
4	48	0.25	50	4	0.08
5	14	0.10	20	36	1.8
Total	200	1	200		$x^2 = 4.280$

Thus we find $x^2 = 4.28$ is a measure of how close the observed frequencies are to the hypothesized frequencies in the fourth column. If we test the hypothesis that the observations come from the stated theoretical distribution, then we know that X^2 has a χ^2-distribution with $5 - 1 = 4$ degrees of freedom. We reject this hypothesis if x^2 is too large and thereby conclude that the stated distribution poorly fits the data. In this example, at $\alpha = 0.05$ level of significance, we reject the hypothesis if $x^2 \geq \chi^2_{0.05} = 9.49$. Since $x^2 = 4.28 < 9.49$, we fail to reject the hypothesis and hence conclude that there is insufficient evidence to refute the analyst's theoretical distribution.

As remarked earlier, the distribution of the random variable X^2 is approximately χ^2 with $k - 1$ degrees of freedom provided n is sufficiently large. When the degrees of freedom are equal to 1, we can improve the approximation by using the formula

$$X^2 = \frac{(|O_1 - e_1| - \frac{1}{2})^2}{e_1} + \frac{(|O_2 - e_2| - \frac{1}{2})^2}{e_2}$$

which is approximately χ^2 distributed with 1 degree of freedom. The subtraction of $\frac{1}{2}$ in the numerator is called a continuity correction. We previously encountered this when approximating the binomial by the normal distribution in Section 5.3.

Example 10 ▪ 2   In his original experiments in genetics Mendel hypothesized that there are two genes that determine inheritance properties. He tested sweet peas for long-stem offspring versus short-stem offspring. When mating two second-generation plants possessing a dominant factor (long-stemmed) and a recessive factor (short-stemmed), his model predicted that the number of long-stemmed offspring to short-stemmed offspring should be in the ratio of 3 to 1. That is, the proportion of long-stemmed offspring should be three fourths and of short-stemmed should be one fourth. He performed the experiment on 1064 plants and obtained the following data:

Category	o_i	e_i	$(\|o_i - e_i\| - \frac{1}{2})^2$
Long stem	787	798	110.25
Short stem	277	266	110.25
	1064	1064	

therefore

$$x^2 = \frac{110.25}{798} + \frac{110.25}{266} = 0.138 + 0.415 = 0.553$$

For comparison we observe from Appendix A.6 that $\chi^2_{0.1} = 2.71$. Thus even at a high level of significance we would fail to reject Mendel's hypothesized 3 to 1 ratio.

Example 10 ▪ 3   Suppose we suspect that a machine is producing a constant proportion 0.1 of defective parts. We observe 150 parts and obtain

Category	Observed frequency o_i	Theoretical probability	Expected frequency e_i
Defective	7	0.1	15
Good	143	0.9	135
	150	1.0	150

To test the hypothesis that the proportion of defectives is 0.1 at $\alpha = 0.01$ level of significance, we use

$$x^2 = \frac{(|7 - 15| - \frac{1}{2})^2}{15} + \frac{(|143 - 135| - \frac{1}{2})^2}{135}$$

$$= 3.75 + 0.4166 = 4.17$$

We reject the theoretical model if $x^2 \geq \chi_{0.05}^2 = 3.84$. Since $x^2 = 4.17 > 3.84$, we reject the proposed model and conclude that 0.1 is not the appropriate proportion of defectives.

Example 10 ▪ 4   A certain manufacturer produces parts that sell in boxes of four parts. The number of defective parts per box is believed to follow a binomial distribution with the probability of a defective equal to 0.4. To test this theoretical model, one selects 200 boxes and records the number of defectives per box.

No. of defectives	No. of boxes with this no. of defectives	Theoretical probabilities	Expected no. of boxes
0	30	0.1296	25.92
1	72	0.3456	69.12
2	73	0.3456	69.12
3	20	0.1536	30.72
4	5	0.0256	5.12
	200	1.00	200

The observed value x^2 of X^2 is equal to

$$\frac{(30 - 25.92)^2}{25.92} + \frac{(72 - 69.12)^2}{69.12} + \frac{(73 - 69.12)^2}{69.12} + \frac{(20 - 30.72)^2}{30.72}$$

$$+ \frac{(5 - 5.12)^2}{5.12} = 4.72$$

We know that X^2 has a χ^2-distribution with $5 - 1 = 4$ degrees of freedom, so we reject the theoretical model at $\alpha = 0.01$ if the value of $x^2 \geq \chi^2_{0.01} = 13.28$ with df $= 4$. Since $x^2 = 4.72 < 13.28$, we fail to reject the theoretical model.

In general, whenever we are testing a "goodness of fit" of a theoretical model, our null hypothesis H_0 is always "H_0: The proposed theoretical model fits," and we use a one-tailed test. Only if the x^2 statistic is *larger* than the cut-off value do we reject H_0.

10 ▪ 2   AN IMPORTANT EXAMPLE: TESTING FOR NORMALITY

In discussing tests we have constantly stated that *if* the population under consideration *has* a normal distribution, then certain tests of hypotheses about means and variances can be performed and certain confidence intervals may be found. The question that naturally arises is how do we determine if a distribution is different from a normal distribution; that is, if the assumption of normality is unjustified? The methods introduced in this section allow us to detect deviations from normality.

Example 10 ▪ 5 A research analyst for a tire manufacturer wishes to find a 95% confidence interval for the life expectancy in miles of his tires. To do this, the analyst must first check if the distribution of wear life deviates significantly from a normal distribution. It is decided to test 400 tires. The results are as follows:

	Interval (miles)	Midpoint m_i	No. observed o_i
1	Below 23,000	22,500	33
2	23,000–24,000	23,500	31
3	24,000–25,000	24,500	66
4	25,000–26,000	25,500	100
5	26,000–27,000	26,500	84
6	27,000–28,000	27,500	57
7	28,000 and up	28,500	29

Now, to see if these data deviate from a normal distribution, we must estimate both the mean μ and the variance σ^2 of the distribution. The best estimate of μ is $\bar{x}_g$ and the best estimate of σ^2 is s_g^2.

$$\bar{x}_g = \frac{(22,500)(33) + (23,500)(31) + (24,500)(66) + \cdots + (28,500)(29)}{400}$$

$$= 25,645$$

$$s_g^2 = \frac{(22,500 - 25,645)^2(33) + \cdots + (28,500 - 25,645)^2(29)}{399}$$

$$= 2,635,563.9$$

Thus the analyst wishes to see if the data fit a normal distribution with mean 25,645 and variance 2,635,563.6. To calculate the theoretical probabilities for each interval, first standardize each endpoint. Thus the endpoints become

$$\frac{23,000 - 25.645}{1623.45} = -1.63, \quad \frac{24,000 - 25,645}{1623.45} = -1.01, \text{ and so on.}$$

Then the probability for each category is obtained using Appendix A.1.

$$p_1 = P(Z \le -1.63) = 0.0516,$$

$$p_2 = P(-1.63 < Z \le 1.01) = 0.1047, \text{ and so on}$$

The following table is developed:

Interval	Observed frequencies o_i	Standardized interval	Theoretical probabilities p_i	Expected frequencies $e_i = 400p_i$
Below 23,000	33	Below -1.63	0.0516	20.64
23,000–24,000	31	-1.63 to -1.01	0.1047	41.88
24,000–25,000	66	-1.01 to -0.397	0.190	76
25,000–26,000	100	-0.397 to 0.219	0.240	96
26,000–27,000	84	0.219 to 0.835	0.212	84.8
27,000–28,000	57	0.835 to 1.45	0.128	51.2
28,000 and up	29	1.45 and up	0.0737	29.48
Total	400		1	400

Now the chi-squared test may be performed as before,

$$x^2 = \sum_i \frac{(o_i - e_i)^2}{e_i} = \frac{(33 - 20.64)^2}{20.64} + \frac{(31 - 41.88)^2}{41.88} + \cdots + \frac{(29 - 29.48)^2}{29.48}$$

$$= 12.38$$

Before determining the significance level of this test, we must find the number of degrees of freedom involved. There are seven categories (intervals) involved so the analysis starts with $7 - 1 = 6$ degrees of freedom. Note that both μ and σ^2 were *estimated* by using the grouped data formulas. In general, every time a population parameter is estimated we lose a degree of freedom; hence there are $6 - 2 = 4$ degrees of freedom. For comparison $\chi^2_{0.025} = 11.14$ and $\chi^2_{0.01} = 13.24$. If we test H_0: The population is normal, at $\alpha = 0.01$ the data are not sufficient to reject, but if the test is performed at $\alpha = 0.025$, we reject H_0.

We emphasize that the choice of intervals in testing for normality is quite arbitrary. We require only that each interval have an *expected* frequency of at least 5.

Technical Note:

A technical consideration worth mentioning at this point concerns the applicability of the χ^2-distribution approximation when μ and σ^2 are both estimated from the data. If the estimates $\bar{x}$ and s^2 are calculated by using the grouped data formulas and the intervals used in testing, then the number of degrees of freedom is $(k-1) - 2 = k - 3$. This is what we did in the preceding example. If μ and σ^2 are known, then the number of degrees of freedom is $k - 1$. In the case where $\bar{x}$ and s^2 are calculated by using the entire ungrouped data set, the number of degrees of freedom is somewhere between $k - 3$ and $k - 1$, and closer to $k - 1$ as n gets larger. Thus the easiest solution is to advise calculating $\bar{x}$ and s^2 from the data interval grouping as given, and then use $df = k - 3$. If $\bar{x}$ and s^2 are calculated by using the entire data set, then we advise looking up the αth centile under both $df = k - 1$ and $df = k - 3$ to get a range of critical values. The χ^2_α value with $df = k - 3$ will be conservative. See the DeGroot reference at the end of the chapter for more details on this problem.

As a final word of caution, in this section, as in all hypothesis tests, the failure to reject H_0 does not imply that we should accept H_0 as true. If we test normality and fail to reject it, the most that can be said is that the data are not inconsistent (at the level α) with normality.

10 ▪ 3   TWO VARIABLES OF CLASSIFICATION: CONTINGENCY TABLES; TEST OF INDEPENDENCE

We now consider the important special case in which each observation is placed into one of several classifications. For example, suppose we wish to determine whether sex has any effect on opinion with respect to a certain political issue. We sample the opinions of 400 people and obtain the following table:

Table 10 ▪ 1   Observed Frequencies

	Favor	Against	No opinion	Total
Men	125	60	10	195
Women	95	90	20	205
Total	220	150	30	400

We have two variables or categories here, sex and opinion.

Each of the cells in Table 10.1 is considered as a classification for the goodness-of-fit test introduced in Section 10.1. We must only find the theoretical probabilities and expected frequencies in order to apply the χ^2-test. Let p equal the proportion of men in the population and $(1 - p)$ the proportion of women. Let π_1 denote the proportion of the population in favor of the issue and π_2 the proportion against the issue, leaving $1 - \pi_1 - \pi_2$ with no opinion. Recall that two events A and B are independent if and only if $P(A \text{ and } B) = P(A)P(B)$. Using this fact, we can calculate the theoretical probability for each cell assuming independence of sex and response. For example, the probability of selecting a person who is both a man and in favor of the issue is $p\pi_1$. Thus of the 400 people in the study, we expect $(400)p\pi_1$ to be men in favor, $(400)p\pi_2$ to be men against, and so on. We can write this in the following table:

Expected Frequencies			
Favor	*Against*	*No opinion*	*Total*
Men $(400)p\pi_1$	$(400)p\pi_2$	$(400)p(1 - \pi_1 - \pi_2)$	$(400)p$
Women $(400)(1 - p)\pi_1$	$(400)(1 - p)\pi_2$	$(400)(1 - p)(1 - \pi_1 - \pi_2)$	$(400)(1 - p)$
Total $(400)\pi_1$	$(400)\pi_2$	$(400)(1 - \pi_1 - \pi_2)$	400

Because we do not actually know the population probabilities p, π_1 and π_2 we must estimate them by the sample probabilities 195/400, 220/400, and 150/440, respectively. The expected frequencies are then calculated by substituting these values in the preceding table.

Expected Frequencies			
	Favor	*Against*	*No opinion*	*Total*
Men	107.25	73.125	14.625	195
Women	112.75	76.875	15.375	205
Total	220	150	30	400

We can now proceed as in the previous section by considering each cell in the preceding boxes a separate category. Then we obtain

$$x^2 = \sum_{\text{all cells}} \frac{(o_i - e_i)^2}{e_i}$$

$$= \frac{(125 - 107.25)^2}{107.25} + \frac{(95 - 112.75)^2}{112.75} + \frac{(60 - 73.125)^2}{73.125} + \frac{(90 - 76.875)^2}{76.875}$$

$$+ \frac{(10 - 14.625)^2}{14.625} + \frac{(20 - 15.375)^2}{15.375} = 13.18$$

To calculate the number of degrees of freedom, we first started with six cells, hence $6 - 1 = 5$ degrees of freedom. We were forced to estimate p, π_1, and π_2 and hence we lose 3 degrees of freedom. Thus X^2 has a χ^2-distribution with $5 - 3 = 2$ degrees of freedom.

If we wish to test the hypothesis of independence at the level of significance $\alpha = 0.025$, then the critical region is $x^2 \geq \chi^2_{0.025} = 7.38$, with df $= 2$. Since $x^2 = 13.18 > 7.38$, we reject the hypothesis of independence.

We may utilize the previous example as a guide for determining a test of independence of the two variables (categories) of classification in the general situation. Suppose we have r classifications of category I and c classifications of category II and each observation must fall into exactly one classification of category I and one classification of category II. We let O_{ij} denote the random variable with range of the observed frequencies in cell (i, j), the cell with category I classification i and category II classification j. We write R_i for the total number of observations in row i, C_j for the total number of observations in column j, and n for the grand total number of observations.

Observed Frequencies

		Category II				Total
		A_1	A_2	$\cdots$	A_c	
	a_1	O_{11}	O_{12}	$\cdots$	O_{1c}	R_1
Category I	a_2	O_{21}	O_{22}	$\cdots$	O_{2c}	R_2
	.	.	.		.	.
	.	.	.		.	.
	.	.	.		.	.
	a_r	O_{r1}	O_{r2}	$\cdots$	O_{rc}	R_r
Total		C_1	C_2	$\cdots$	C_c	n

This type of table is called a "$r \times c$ contingency table."

If categories I and II are independent, and if R_1/n of the observations fall in row 1, we would expect this same proportion R_1/n of the C_1 observations in column 1 to fall into cell $(1, 1)$, R_1/n of the C_2 observations in column 2 to fall into cell $(1, 2)$, and so on; that is, we expect $C_1(R_1/n)$ in cell $(1, 1)$, $C_2(R_1/n)$ in cell $(1, 2)$, . . . , and, in general, we take the expected frequency in cell (i, j) as $e_{ij} = C_j(R_i/n) = R_i C_j/n$. Summarizing, we have

Expected Frequencies						
			Category II			*Total*
		A_1	A_2	$\cdots$	A_c	
Category I	a_1	$R_1 C_1/n$	$R_1 C_2/n$	$\cdots$	$R_1 C_c/n$	R_1
	a_2	$R_2 C_1/n$	$R_2 C_2/n$	$\cdots$	$R_2 C_c/n$	R_2
	.	.	.		.	.
	.	.	.		.	.
	.	.	.		.	.
	a_r	$R_r C_1/n$	$R_r C_2/n$		$R_r C_c/n$	R_r
Total		C_1	C_2	$\cdots$	C_c	n

We form the chi-squared statistic as before

$$X^2 = \sum_{j=1}^{c} \sum_{i=1}^{r} \frac{(O_{ij} - e_{ij})^2}{e_{ij}} = \frac{(O_{11} - R_1 C_1/n)^2}{R_1 C_1/n} + \cdots + \frac{(O_{rc} - R_r C_c/n)^2}{R_r C_c/n}$$

that has $(r - 1)(c - 1)$ degrees of freedom.* To test the hypothesis that the two types of classifications are independent, we form the X^2-statistic and reject the hypothesis at the level of significance α if $x^2 \geq \chi_\alpha^2$ with df $= (r - 1)(c - 1)$.

Example 10 ■ 6   A sociologist wishes to determine if watching violence on television leads to aggressive behavior. One hundred children are chosen to watch violence-prone shows and 100 to watch nonviolent programs. After 2 weeks of observing their behavior, each child is classified as either aggressive or nonaggressive. The following data were obtained:

	Television Show Type		
Behavior type	Violent	Nonviolent	Total
Aggressive	63	29	92
Nonaggressive	37	71	108
Total	100	100	200

This is a 2 × 2 contingency table.

* To see this, we note that there are rc cells so that if all the theoretical cell probabilities were known there would be $rc - 1$ degrees of freedom. However, we had to estimate $r - 1$ of the row probabilities by R_i/n and $c - 1$ of the column probabilities by C_j/n in order to find $e_{ij} = n(R_i/n)(C_j/n) = R_i C_j/n$. Thus the number of degrees of freedom is $rc - 1 - (r - 1) - (c - 1) = (r - 1)(c - 1)$.

At $\alpha = 0.005$, test the null hypothesis that behavior is independent of viewing habits.

To perform this test, we find the expected frequencies in each cell as before; that is, for cell (1, 1) we expect $(100)(92/200) = 46$, and so on.

Television Type: Expected Frequencies			
Behavior type	Violent	Nonviolent	Total
Aggressive	46	46	92
Nonaggressive	54	54	108
Total	100	100	200

Here we have $(2-1)(2-1) = 1$ degree of freedom, so we use the continuity correction in calculating x^2; that is, we subtract $\frac{1}{2}$ from each of the two terms before we square it.

$$x^2 = \frac{(|63-46|-\frac{1}{2})^2}{46} + \frac{(|37-54|-\frac{1}{2})^2}{54} + \frac{(|29-46|-\frac{1}{2})^2}{46}$$

$$+ \frac{(|71-54|-\frac{1}{2})^2}{54}$$

$$= 21.92$$

Since $x^2 = 21.91 > \chi^2_{0.005} = 7.88$, we reject the hypothesis of independence. The investigator must conclude that there is significant evidence of a dependence between TV viewing and behavior.

Because 2×2 contingency tables occur so often it is useful to have a computational formula for x^2 in this case. Suppose we have the table

		Total
a	b	$a+b$
c	d	$c+d$
Total $a+c$	$b+d$	n

Where a, b, c, and d are the observed frequencies. Then it may be shown that

$$x^2 = \frac{n(|ad-bc|-n/2)^2}{(a+b)(c+d)(a+c)(b+d)} \qquad (10.1)$$

which is used to test independence in a 2×2 table.

10 ■ 4   ASSUMPTIONS UNDERLYING THE χ^2-TEST FOR CATEGORICAL DATA; A CAUTION

To correctly apply the χ^2-test we must be sure that the underlying assumptions are met. First of all, we must be certain that each of the observations is independent of all the other observations. This implies, in particular, that we should proceed with caution before using the χ^2-test when dealing with repeated measurements on the same subjects. We must convince ourselves in this case that the independence assumption is valid. Another assumption is that every observation can be classified into one and only one cell. Hence, with two variables of classification, each observation must fall into one and only one cell.

Care must be exercised to avoid designing classifications that permit an observation to fall into more than one cell. Consider the following situation. A study was conducted to determine whether decisions in the U.S. Supreme Court favor U.S. citizens against foreign citizens. One might be tempted to set up a 2 × 2 contingency table as follows:

		Decision	
		For U.S. citizen	*Against U.S. citizen*
Citizenship	*United States*	*	
	Foreign	**	

The problem here is that if a particular court action is classified, it is done once for the U.S. citizen and once for the foreign citizen. For example, if the court decides in favor of the U.S. citizen, then it is simultaneously entered into the table once as a U.S. citizen winning (*), and once as a foreigner losing (**). But these are the same trial. The entries are not independent and a particular case does not fall into one and only one category.

This is an actual example of a situation considered by M. E. Gates.* This problem was solved by redefining the classifications to exclude double counting. Instead, each case was classified by the citizenship of the plaintiff, for example,

		Decision		
		For U.S.	*Against U.S.*	
Citizen status of plaintiff	*United States*	51	90	141
	Foreign	34	67	101
		85	157	242

* Gates, M. E.: Priority patterns in the application of international law and municipal law norms in the courts of the Federal Republic of Germany and the USA, PhD thesis, 1977, Tulane University (Call No. at Tulane Library RARE-LD5434 t915, 1977gc).

For this example we find from equation (10.1) of the previous section

$$x^2 = \frac{242\left(\left|51 \cdot 67 - 90 \cdot 34\right| - \dfrac{242}{2}\right)^2}{141 \cdot 101 \cdot 85 \cdot 157} = 0.0709$$

At a level of significance $\alpha = 0.1$ we would fail to reject the null hypothesis of independence since, with df $= 1$, we have $\chi^2_{0.1} = 2.71 > 0.0709$. In fact, this example illustrates another point, namely, the value $x^2 = 0.0709$ is almost too small to believe. By this we mean that (using the left tail of the χ^2-distribution now) we only expect to see a value this small in about 21% of the cases in which independence really does hold. Summarizing, a significantly large observed value of the random variable X^2 makes us suspect that the hypothesis of independence is not true, but a significantly small observed value of X^2 should also arouse our suspicions that the results are almost too good to be true, and maybe we should investigate why.

Perhaps the most important assumption made in order to perform the chi-squared test is that the sample size n be sufficiently large. As already mentioned, the distribution of X^2 is only approximately chi-squared; the approximation is better, the larger the sample size is. The question that immediately arises is, "How large must n be in order to use the χ^2-test?" In this book we use the following (rather conservative) convention that n is sufficiently large provided that the *expected* frequencies are all greater than or equal to 5.

10 ▪ 5   TESTS OF ASSOCIATION FOR CATEGORICAL DATA (OPTIONAL)

Up to now we have been using the χ^2-distribution to test for independence of the attributes of classification when we are dealing with categorical data. We say that two attributes are associated if they are not independent. Provided that the requirements for the χ^2-distribution are satisfied (as outlined in the previous section), we can ascertain at a given level of significance whether the attributes are independent or associated. Unfortunately, the χ^2-test for independence is very sensitive to sample size, and for n sufficiently large, will detect even the slightest departure from independence. Since it can be argued that in reality no two events are actually completely unrelated, quite often we can show significant association merely by taking extremely large sample sizes. In such a case it becomes important to measure the *strength* of the association. We wish to determine how much knowledge of one attribute is gained by knowing the other attribute. Put another way, although the two attributes appear to have significant statistical dependence, do they have any practical exploitable dependence?

One measure of this strength of association (developed by Goodman and Kruskal) is called the index of predictive association. Essentially, it measures the proportional decrease in error in guessing the category of one attribute when the other attribute is known. It tells us how exploitable (in the sense of guessing the right category) the dependent random variable's information is.

For example, suppose for some population the following category probabilities are known:

| | Attribute A | | | |
Attribute B	A_1	A_2	A_3	Total
B_1	0.08	0.07	0.10	0.25
B_2	0.07	0.03	0.05	0.15
B_3	0.05	0.40	0.15	0.60
Total	0.20	0.50	0.30	1.00

For example, the probability of choosing an individual from the population exhibiting both attributes A_2 and B_3 is 0.40.

Now suppose you were asked to predict which A attribute an individual will exhibit and you had no knowledge of what B attribute the individual had. Naturally you would choose A_2, that is the A attribute that has the highest probability of occurring; that is, $P(A_2) = \text{Max}_{j=1,2,3} P(A_j)$. The probability of error when attribute B is not used for prediction is $1 - P(A_2) = 1 - 0.5 = 0.5$. On the other hand, if attribute B is used and you wish to predict which A attribute occurs, we have by the rules of conditional probability,

$P[\text{error when attribute } B \text{ is used}] = P[\text{error and attribute } B \text{ known to be } B_1]$

$\quad + P[\text{error and attribute } B \text{ known to be } B_2]$

$\quad + P[\text{error and attribute } B \text{ known to be } B_3]$

$\quad = P[\text{error} \mid B_1]P(B_1) + P[\text{error} \mid B_2)P[B_2] + P[\text{error} \mid B_3]P[B_3]$

However, if B_i is known to occur, then our best guess for the A attribute is to pick the A attribute in the ith row that has the highest probability; that is, we choose the A_j that maximizes $P[A_j \text{ and } B_i]$. For example, given B_1 we would choose A_3. Then

$$P[\text{error} \mid B_1] = 1 - P(A_3 \mid B_1) = 1 - \frac{P(A_3 \text{ and } B_1)}{P(B_1)}$$

$$= 1 - \frac{0.1}{0.25}$$

Hence

$P[\text{error when attribute } B \text{ is used}]$

$$= \left(1 - \frac{0.1}{0.25}\right)(0.25) + \left(1 - \frac{0.07}{0.15}\right)(0.15) + \left(1 - \frac{0.4}{0.6}\right)(0.6) = 0.43$$

The decrease in error of guessing by using B is

P[error when B attribute is not used] $-$ P[error when B attribute is used]

$$= 0.5 - 0.43 = 0.07$$

This says that using B has decreased our error in guessing A by 7%. This percentage represents the average decrease in the error over the three specific attributes B_1, B_2, B_3. The *proportional* reduction in error is called the *index of predictive association* for A and is denoted by λ_A,

$$\lambda_A = \frac{P(\text{error when } B \text{ not used}) - P(\text{error when } B \text{ used})}{P(\text{error when } B \text{ not used})}$$

$$= \frac{0.07}{0.5} = 0.14 \tag{10.2}$$

Hence 14% of the error is eliminated on the average by using the B attribute of the individual. In general, for c classifications of type A and r classifications of type B, we define λ_A for predicting A as in expression (10.2.) Here $\text{Max}_{1 \le i \le c} P(A_i)$ is the largest attribute A probability, so the attribute with the largest probability (column total) is our best guess if B is not used, and the error is

$$P(\text{error when } B \text{ not used}) = 1 - \underset{1 \le i \le c}{\text{Max}} P(A_i)$$

Similarly, when B is used, we find the best guess for each B (the biggest probability within each row). The error is

$$P(\text{error when } B \text{ used}) = \sum_{j=1}^{r} \left\{ 1 - \underset{1 \le i \le c}{\text{Max}} P(A_i \mid B_j) P(B_j) \right\}$$

If we have a chart with frequency f_{ij} for cell (i, j) such as

		A_1	A_2	A_3	$\cdots$	A_c	Total
				Attribute A			
	B_1	f_{11}	f_{12}	f_{13}	$\cdots$	f_{1c}	R_1
	B_2	f_{21}	f_{22}	f_{23}	$\cdots$	f_{2c}	R_2
Attribute B	.	.	.	.		.	.
	.	.	.	.	$\cdots$	.	.
	.	.	.	.		.	.
	B_r	f_{r1}	f_{r2}	f_{r3}	$\cdots$	f_{rc}	R_r
	Total	C_1	C_2	C_3	$\cdots$	C_c	n

Then we may approximate λ_A by

$$l_A = \frac{\sum_i \left(\underset{j}{\text{Max }} f_{ij}\right) - \underset{j}{\text{Max }} C_j}{n - \underset{j}{\text{Max }} C_j} = \frac{(\text{Max in row 1}) + \cdots + (\text{Max in row } r) - \text{Max column total}}{\text{Sample size} - \text{Max column total}}$$

Example 10 ▪ 7   Let us return to the problem of determining the effect of watching violence on television on aggressive behavior that was discussed in example 10.6. We wish to find the percentage decrease in our error in guessing whether a person watches violent programs by knowing whether he or she is aggressive or nonaggressive. From the table in example 10.6 we have

$$\underset{j}{\text{Max }} f_{1j} = 63; \qquad \underset{j}{\text{Max }} f_{2j} = 71$$

$$\underset{j}{\text{Max }} C_j = 100; \qquad n = 200$$

so the observed value is

$$l_A = \frac{63 + 71 - 100}{200 - 100} = 0.34$$

Thus if we must guess if a child watches violent television programs, the knowledge of his or her aggressiveness classification allows us to reduce by 34% our error in guessing.

It should be noted that if the two attributes are indeed independent, then $\lambda_A = 0$, whereas if the prediction of A is completely determined by knowledge of B, then $\lambda_A = 1$. The value $\lambda_A = 0$ does not, however, imply that the two attributes are independent, but rather, that knowledge of B does not improve our ability to guess attribute A; that is, the dependence between A and B cannot be exploited for guessing A.

When we are interested in predicting B based on A, we would switch the rows and columns in the preceding analysis, now putting the B attributes as columns and the A attributes as rows. The resulting measure λ_B (or the estimate l_B) is now calculated by using the same formula as before. Note, however, that $\lambda_A \neq \lambda_B$ in general.

There are many measures of association for categorical data, and the index of predictive association is only one such measure. For an interesting account of various measures see the references at the end of the chapter.

10 ▪ 6   COMMENT: INTERPLAY OF DATA COLLECTION AND ANALYSIS

In the late 1940's and early 1950's Doctors Douglas McAlpine and Nigel Compston performed an experiment to investigate whether physical trauma

was a strong contributing factor to the onset of multiple sclerosis.* They selected 250 multiple sclerosis patients (study group) and 250 people as a control
group. The control group was selected from patients who were hospitalized in
two specific hospitals. The control group was matched to the study group but
the publication by McAlpine and Compston does not clearly describe how. At
any rate, the two groups were queried as to whether or not they had suffered
trauma within a 3-month period prior to the onset of their illnesses. (A study
that uses past data as part of its analysis is called a *retrospective* study.) The
authors tested the hypothesis H_0: Trauma and multiple sclerosis are *not* associated. The results were as follows:

	Suffered trauma	*Suffered no trauma*	*Total*
Study group	36	214	250
Control group	13	229	242
$x^2 = 10.2$ (significant at $\alpha < 0.005$)			

Eight of the control patients had been hospitalized for trauma and were not
included in the calculation. Although the investigators do not explain why they
did not include these eight patients, one can conjecture that they eliminated
these trauma patients because their status was ambiguous. Since they suffered
trauma they should have been included with the 13 other cases that suffered
trauma: yet if they had had no trauma in the 3-month period before they had
been hospitalized for trauma, should they not be included with the 229? It
appears this uncertainty led to their elimination. However, eliminating them
produces a biased sample. If a person has been hospitalized for trauma, then
there is virtually no chance he or she had been hospitalized for another, different trauma within the preceding 3-month period, for if they had such a terrible
accident then they would be either still in the hospital (and so excluded from the
study) or at home convalescing and so unable to get into another serious
accident. In other words, all people who suffered a serious accident that did not
result in multiple sclerosis within 3 months were essentially eliminated from the
study.

The investigators blundered by selecting hospitalized patients as their control
group. By proceeding as they did, they essentially excluded all hospitalized
trauma cases that did not result in multiple sclerosis and so they biased the data
against the null hypothesis H_0. Thus the 8 trauma study cases must be included
with the 13 in the preceding table, yielding 21. With this change, $x^2 = 3.88$,
which is significant at $\alpha = 0.049$, just barely significant.

The investigators in their paper repeatedly emphasized that multiple sclerosis patients are much more likely to recall trauma (or to exaggerate it) in

* They published their results in the *Quarterly Journal of Medicine*, New Series (England),
21(82):135–167, April 1952.

relation to the onset of their disease than are other types of patients. If only one member of the 229 in the control group failed to recall a pertinent trauma, the value of $x^2 = 3.29$, which is not significant at $\alpha = 0.05$.

The preceding case history illustrates that analysis is not enough. The design of the experiment is important to the validity of the ultimate conclusions.

Summary

We have discussed how to test the fit of a model to data by placing the observations into categories and comparing the number (frequency) of observations in each category with the number (frequency) predicted by the model. The X^2-statistic (Section 10.1) is used in this test; it is approximately χ^2-distributed with degrees of freedom df $= k - 1$, where k is the number of categories. An important application of the statistic X^2 is to test data for departures from normality.

It turns out that the χ^2-distribution is also important in testing whether two categorical variables are independent of one another. The statistic X^2 (as developed in Section 10.2 for contingency tables) is approximately χ^2-distributed, now with degrees of freedom df $= (r - 1)(c - 1)$, where r is the number of rows and c is the number of columns of the contingency table.

In Section 10.5 we developed an approach to evaluating the degree of the association between two variables by using conditional probabilities. This measure represents the proportional decrease in guessing one attribute based on whether or not the other attribute is used.

Key Terms

association
categorical data
$(r \times c)$ contingency table
independence of attributes
cross-classified data

Pearson chi-squared
  statistic
assessing departures from
  normality
goodness of fit
tests for independence

References

An excellent explanation of the statistic

$$X^2 = \sum_{j=1}^{k} \frac{(O_i - e_i)^2}{e_i}$$

(see Section 10.1)

and its application to the law (particularly to the Barksdale case discussed in problem 54, Chapter 7) is given in

Finkelstein M. O.: The application of statistical decision theory to the jury discrimination cases. *Harvard Law Review, 80:*365–371, 1960.

There are several good (but slightly more advanced) books concerning the analysis of categorical data. In particular, they consider modeling and testing hypothesis by using categorical data for examining more complex models than just the goodness of fit and independence of characteristics we have considered here. We have just seen the tip of a large iceberg. The interested reader is directed to the following books:

Haberman, S. J.: Introductory topics of *Analysis of Qualitative Data,* Vol. 1. New York, Academic Press, 1978.

Reynolds, H. T.: *The Analysis of Cross Classifications.* New York, Free Press/Macmillan, 1977.

Reynolds, H. T.: *Analysis of Nominal Data.* Beverly Hills and London, Sage Publications, 1977.

These books also contain discussions concerning measures of association and present several measures of association other than the index of predictive association that we presented in this chapter.

Various measures of assessing goodness of fit are discussed in

DeGroot, M. H.: *Probability and Statistics.* Reading, MA, Addison-Wesley, 1975.

He shows (Section 9.2 of Chapter 9) that the X^2 statistic we have used in this chapter for testing normality should have $\bar{x}$ and s^2 calculated from the same interval grouping as that used for the test, or else the degrees of freedom must be slightly adjusted as we noted in the text.

The following references discuss some situations where one should not use the correction factor $\frac{1}{2}$ (called Yate's correction) used in examples 10.2, 10.3, and 10.6:

Conover, W. J.: Uses and abuses of Yate's correction. *Biometriks, 24*:1028, 1968.

Conover, W. J.: Some reasons for not using Yate's correction. *Journal of American Statistical Society, 69*:374–382, 1974.

Camilli, G., and K. D. Hopkins: Applicability of χ^2 to 2×2 contingency tables with small expected cell frequencies. *Psychological Bulletin, 85*:163–167; 1978.

Exercises

Section 10.1

1. Normally a certain species of water fowl lays four eggs and has a 10% chance of all four hatching, 36% three hatching, 41% of two hatching, 10% of one hatching, and 3% of none hatching. After a chemical spill occurred, ecologists monitored 200 nests and discovered that 12 nests had all eggs hatch, 57 had three hatch, 89 had two hatch, 30 had one hatch, and 12 had no eggs hatch.
   a. State and test the hypothesis that the chemical spill had no effect on the pro-

portions of eggs that hatch in the nests. Choose $\alpha = 0.05$.
   b. What is your test statistic value and your critical region?
   c. What is your conclusion concerning the effect of the chemical spill?

2. A certain utility company has been accused of discrimination in hiring practices. Of the 100 persons hired by this company this year, 70 were men and 30 were women. Of the applicants who qualified for these positions it

was found that 60% were men and 40% were women.

a. At $\alpha = 0.1$, can we conclude that there was a significant departure from the expected 60:40 hiring ratio?

b. How about at $\alpha = 0.01$?

c. Before concluding that discrimination took place, we must examine other possible reasons for a departure from the expected 60:40 hiring ratio. Give some plausible "other reasons" that might have accounted for this departure.

3. A questionnaire was handed out to 150 persons in a certain city supposedly at random. To check whether this was indeed the case, the person analyzing the questionnaire utilized the answers to divide the respondents into three categories: upper class, middle class, and lower class. Utilizing census figures, she determined that 10% of the city is classified as upper class, 60% as middle class, and 30% as lower class. Among the 150 persons polled, she finds 22 classified as upper class, 89 as middle class, and 39 as lower class. Should she conclude that the respondents were not randomly selected from the city at $\alpha = 0.025$?

4. In order to properly schedule and staff college preparation classes in a certain school district, the school board wishes to determine at the 0.1 level of significance if the college-bound high school students are uniformly spread throughout the three high schools in the district. (The high schools are each of the same size.) They collect the following data:

	High school			
	A	B	C	Total
Number of college-bound students	25	45	20	90

Is this consistent with uniform distribution of the college-bound students?

5. A political scientist studying a certain state's political parties hypothesized that Democrats would be more sympathetic to striker's demands, and hence there would be twice the probability of a strike under a Democratic governor as under a Republican governor. Over a period of years the following data were collected:

	Democratic governor	Republican governor	Total
Number of strikes	378	222	600

Do these data support the political scientist's assertion at $\alpha = 0.1$? (Assume that the two administrations were in office for the same number of years.)

6. The role of the human in system malfunctions is a particularly important consideration in many industrial malfunctions. For example, in nuclear power plants such human-caused problems can have devastating financial and environmental consequences. In a study by Robinson, Deutsch, and Rogers the causes of 213 vehicle malfunctions were classified as follows:

Primarily human cause	25
Primarily equipment cause	158
Combination human-equipment	30

How well does this fit a theoretical model that the probabilities are in the relationship $\frac{1}{6}$, $\frac{2}{3}$, and $\frac{1}{6}$, respectively, for the three categories. Use the chi-squared test with $\alpha = 0.05$. (Robinson, J. E.; Deutsch, W. E.; and Rogers, J. G.: The field maintenance interface between human engineering and maintainability engineering. *Human Factors, 12:*3, 1970, pp. 253–259. Copyright 1970 by the Human Factors Soc. Inc. Reproduced by permission.)

7. A study of the relationship between life stresses and illness is described in Uhlenhuth, Lipman, Balter, and Stern. Respondents were asked to note which of certain life events had occurred in the previous 18 months. If no relationship existed at all in the data collected between the time of stress and the time of illness, then one might expect the number of people reporting any given month as their last stress date to follow a uniform distribution. The following data were obtained:

Months before interview	Number of subjects	Percentage of subjects
1	15	10.2
2	11	7.5
3	14	9.5
4	17	11.6
5	5	3.4
6	11	7.5
7	10	6.8
8	4	2.7
9	8	5.4
10	10	6.8
11	7	4.8
12	9	6.1
13	11	7.5
14	3	2.0
15	6	4.1
16	1	0.7
17	1	0.7
18	4	2.7
Total	147	100.0

Test the hypothesis that the data are from a uniform distribution over the 18 categories given. (Use $\alpha = 0.01$.) (E. H. Uhlenhuth, et al.: Symptom intensity and life stress in the city. *Archives of General Psychiatry* (Chi-

cago), *31*:759–764, 1974. Copyright 1974 American Medical Assoc.)

8. A question on social class membership was asked in the 1975 General Social Survey of the National Opinion Research Center (University of Chicago). People were asked to classify themselves according to their social class and the following data resulted:

Response	Score	Number responding
Lower class	1	72
Working class	2	714
Middle class	3	655
Upper class	4	41
Total		1482

(Data taken from Haberman, S.: *Analysis of Qualitative Data.* New York, Academic Press, 1978. Reproduced by permission of National Opinion Research Center.) Test the hypotheses that the probabilities for the four classes are 0.05, 0.45, 0.45, and 0.05, respectively. (Use $\alpha = 0.05$.)

9. In "true-false" type exams we always assume that both types of answers are equally likely. In one scientific investigation by Metfessel and Sax a typical result was 61 true and 39 false. Using the continuity correction to the χ^2, test if these results are consistent with the equal likelihood of true and false answers. (Use $\alpha = 0.05$.) (Metfessel and Sax: Response set patterns in published instructors manuals in education and psychology. *California Journal of Education Research, 8*(5):195–197, 1957.)

10. Do bus accidents occur at random, or are some drivers accident prone whereas others are extremely safe? In a study of accidents among the bus drivers of the Ulster (Northern Ireland) Transport Authority during a 3-year period, 708 drivers were examined

and the number of accidents they had were recorded. If accidents were random, then each time an accident occurred, each of the 708 drivers was equally likely to have been involved. This theoretical model was used to calculate the expected number of drivers in each class of the following table. The technique is equivalent to taking each of the 1623 accidents and assigning a driver at random to that accident. The following data are given by Cresswell and Froggatt:

Number of accidents in 3-year period	Number of drivers	
	Observed	Expected
0	117	71.5
1	157	164.0
2	158	187.9
3	115	143.6
4	78	82.3
5	44	37.7
6	21	14.4
7 or more	18	6.6
Total	708	708.0

Test the hypothesis that the accidents occur at random (i.e., test the preceding model). Use $\alpha = 0.05$. (Cresswell and Froggatt: *The Causation of Bus Driver Accidents.* New York, Oxford University Press, 1963.)

Section 10.2

11. An investigation was undertaken by Finn, Jones, Tweedie, Hall, Dinsdale, and Bourdillon to determine uric acid levels among 267 normal males. The object of the study concerns the genetic mechanism for gout, a disease characterized by high uric acid levels. The data they obtained are given in the table that follows:

Serum uric acid (mg/100 ml)	Number of men
3.0–3.4	2
3.5–3.9	15
4.0–4.4	33
4.5–4.9	40
5.0–5.4	54
5.5–5.9	47
6.0–6.4	38
6.5–6.9	16
7.0–7.4	15
7.5–7.9	3
8.0–8.4	1
8.5–8.9	3
Total	267

Test ($\alpha = 0.05$) if these data are consistent with the normal distribution for serum uric acid level. (Finn, et al.: Frequency curve of uric acid in the general population. *Lancet,* 2:185, 1966.)

12. The following table of 100 blood pressure readings for males age 29 through 59 is reproduced from Chapter 2.

Interval	Frequency
90–99	5
100–109	8
110–119	22
120–129	27
130–139	17
140–149	9
150–159	5
160–169	5
170–179	2
Total	100

The mean was $\bar{x} = 126.37$ and the standard deviation was 17.97. Test the hypothesis that the data follow a normal distribution by using $\alpha = 0.05$.

Section 10.3

13. In a study concerning the use of propranolol in the treatment of myocardial infarction, each patient with myocardial infarction was randomly put into one of two groups. One group was treated with propranolol, and the other was not. The status of the 91 patients studied by Snow follows:

Status 28 days after admission	Propranolol-treated patients	Control patients
Dead	7	17
Alive	38	29
Total	45	46

Test if survival after 28 days is independent of treatment. (Use $\alpha = 0.05$.) (Snow: Effect of propranolol in myocardial infarction. *Lancet, 2:*551, 1965.)

14. In a study to determine if past performance of portfolio managers for mutual funds is any indication of their ability to perform well in the future, Brightman and Haslanger collected data on 52 managers over a 6-year period. They ranked each manager's performance during the 1971 to 1973 period according to their quartile, and again the same managers were ranked during the 1974 to 1976 period. The following table was given:

		1971–1973 Performance Quartile				
		1	2	3	4	Total
1974–1976	1	3	1	1	8	13
performance	2	2	3	5	3	13
quartile	3	4	3	5	1	13
	4	4	6	2	1	13
Total		13	13	13	13	52

Test the hypothesis that rank in 1974 to 1976 is independent of rank in 1971 to 1973 at $\alpha = 0.05$. (Brightman and Haslanger: Past investment performance: Seductive but deceptive. *Journal of Portfolio Management,* V.6:43–45, 1980. Copyright Institutional Investor Inc.)

15. In an attempt to determine if performance on ESP tests is affected by the person's belief in ESP, Ryzl scored people according to their relative ability on an ESP test. He obtained the following data:

Performance	Belief in ESP	
	Yes	No
Above average	24	1
Below average	28	10

Does this provide enough evidence to reject the independence of performance and attitude at $\alpha = 0.05$? (Ryzl, M.: Precognition scoring and attitude toward ESP. *Journal of Parapsychology, 32:*1–8, 1968.)

16. In the famous medical experiment concerning sterilization, Joseph Lister used carbolic acid to sterilize the operating room prior to performing an amputation. He obtained the following data:

Patient lived	Carbolic acid used		
	Yes	No	
Yes	34	19	43
No	6	16	22
	40	35	

Can we conclude that sterilization significantly reduces the chance of death at $\alpha = 0.01$.

17. The following table concerns the relationship

of thromboembolic disease of oral contraceptive users and blood group type. The following data are given in Jick, Slone, Westerholm, Inman, Vessey, Shapiro, Lewis, and Worcester.

Blood group	Women with thromboembolism	Healthy women
	Number	Number
A	32	51
B	8	19
AB	6	5
O	9	70
Total	55	145

Is there evidence to conclude that blood type is related to thromboembolic disease of oral contraceptive users? (Use $\alpha = 0.05$.) (Jick, et al.: Venous thromboembolic disease and ABO blood types. *Lancet, 1*:539, 1969.)

18. In a study of the relationship between socioeconomic status and cheating in school, Harshorne and May obtained the following data:

	Socioeconomic status			
Cheated?	Lower	Middle	Higher	Row totals
Yes	28	72	37	137
No	16	71	176	263
Total	44	143	213	400

Is there a significant relationship between cheating and socioeconomic status? (Use $\alpha = 0.05$.) (Harshorne, and May: *Studies in the Nature of Character. I. Studies in Deceit.* New York, Macmillan, 1928.)

19. The following table gives the assessment after 6 months for patients admitted to the hospital for pulmonary tuberculosis. Some patients were treated with streptomycin and bed rest, whereas others were treated with bed rest alone.

Radiological assessment	Patients treated with streptomycin and bed rest	Patients treated with bed rest alone
Considerable improvement	28	4
Moderate or slight improvement	10	13
No material change	2	3
Moderate or slight deterioration	5	12
Considerable deterioration	6	6
Deaths	4	14
Total	55	52

Does this provide sufficient evidence to conclude that treatment and assessment are associated? (Medical Research Council: Streptomycin treatment of pulmonary tuberculosis. *British Medical Journal, 2*:769, 1948.)

Section 10.5

20. In a study of 1405 students the participation in religious activities was compared with the answer to the question, "How happy has your home life been?" C. L. Stone found the following:

		Answer to "How happy has your home life been?"		
		Very happy	Fairly happy	Un-happy	Row Total
Participation in religious activities	Not at all	105	78	25	208
	Very little	257	149	30	436
	Somewhat	368	153	24	545
	Very much	151	52	13	216
	Column total	881	432	92	1405

a. Test if the answer is independent of the participation in religious activities.
b. Calculate an estimate of the index of predictive association for predicting religious participation based on their happiness answer.
c. Calculate an estimate of the index of predictive association for predicting happiness based on religious participation. (Stone, C. L.: Church participation and social adjustment of high school and college youth. *Rural Sociology Series on Youth, No. 12,* Bulletin 550, Dept. of Rural Sociology, Agricultural Research Center, Pullman: Washington State University, Pullman, WA: 1954.)

21. Using the data from problem 13, calculate the index of predictive association for pre-

dicting survival status based on treatment group classification.

22. Calculate an estimate of the predictive value (as measured by λ_A) of knowing an investment manager's previous relative ranking. Use the data from problem 14.

23. Using Lister's amputation data (problem 16) estimate the predictive association for determining survival status based on whether or not carbolic acid was used.

24. Using the data of problem 17, estimate λ_A, the index of predictive association for determining if an oral contraceptive user will have thromboembolic disease based on her blood type.

25. Using the data of problem 18, determine the extent to which knowledge of a student's socioeconomic status is useful for predicting whether he or she will cheat.

HYPOTHESIS TESTING

A. THE HYP. $H_0 = H_A =$ $\mu = \mu_o$ $p \leq .80$ where μ_o = hypothesized value for
$\mu \neq \mu_o$ $p \geq .80$ actual μ

B. THE TEST STAT Q = Acceptance Region = Accept H_0
C = Critical Region = Accept H_A

TYPE I ERROR - when you reject H_0 ! it is true
TYPE II ERROR - " " accept " " " " false

C. SIGNIFICANCE LEVEL
level of confidence to reject H_0 $(.95 - .99)$ $.95$ $.99$
$\alpha = .05$ or $.01$
$\alpha = .035$ 965%

D. THE Z TEST

$$Z = \frac{\bar{x} - \mu}{\sigma / \sqrt{n}}$$ $\bar{x}$ = hypothesized value for μ

may use α if $n > 30$
Acceptance Region = $-b \leq Z \leq b$
Critical Region = $Z \leq -b$ or $Z > +b$

if $n \leq 30$
$$t = \frac{\bar{x} - \mu_o}{s / \sqrt{n}} = (n-1)df$$
if σ unknown.
If $\alpha = .05$
$t = .025 = 2.776$
$n-1 = 4 df$
if $n = 5$

TO TEST PROPORTIONS

p = population proportion (unknown)
$H_0 = p = P_o$ H_a: $p \neq P_o$

$\hat{p} = \frac{x}{n}$

$Z = \frac{\hat{p} - P_o}{\sigma \hat{p}} = \frac{\hat{p} - P_o}{\sqrt{\frac{P_o(1-P_o)}{n}}}$

$\sigma \hat{p} = \sqrt{\frac{P(1-P)}{n}}$ where $P = P_o$ if H_0 = true

TESTING 2 PROPORTIONS
P_1 = pop. one
P_2 = pop. 2

TEST H_0: $P_1 = P_2$ H_A $P_1 \neq P_2$
$P_1 - P_2 = 0$ $P_1 - P_2 \neq 0$

P_1 P_2 sample pop = $\hat{P}_1 = \frac{x_1}{n_1}$
x_1 x_2
n_1 n_2 $\hat{P}_2 = \frac{x_2}{n_2}$

TESTING 2 POPS.

$n_1 = 200$ ♀ with $\quad x_1 = 7$

$n_2 = 200$ ♀ without $\quad x_2 = 3$

$\hat{P}_1 = \frac{7}{200} = .035$

$\hat{P}_2 = \frac{3}{200} = .015$

THE TEST STAT

$$Z = \frac{\hat{P}_1 - \hat{P}_2}{\sqrt{\hat{P}(1-\hat{P})\left(\frac{1}{n_1} + \frac{1}{n_2}\right)}}$$

WHERE

$$\hat{P} = \frac{x_1 + x_2}{n_1 + n_2} \simeq \sqrt{\hat{P}(1-\hat{P})\left(\frac{1}{n_1} + \frac{1}{n_2}\right)}$$

$$\boxed{\frac{x_1 + x_2}{n_1 + n_2}}$$

mean $= \mu_{\hat{P}_1} - \hat{P}_2 = P_1 - P_2$

std dev $= \sigma_{\hat{P}_1} - \hat{P}_2 = \sqrt{\frac{P_1(1-P_1)}{n_1} + \frac{P_2(1-P_2)}{n_2}}$

Acceptance Region

Accept H_0 if $\quad -Z\alpha_{/2} \leq Z \leq Z\alpha_{/2}$

Reject

Reject H_0 if $\quad Z \leq - Z\alpha_{/2}$

$\text{OR } Z \geq Z\alpha_{/2}$

EXAMPLE

$n_1 = 1067 \qquad n_2 \quad 54$

$x_1 \quad 117 \qquad x_2 \quad 402$

$P_1 = \frac{x}{n} = .1097 \quad P_2 \frac{x}{n} = .1343$

$\frac{x_1 + x_2}{n_1 + n_2} = \hat{P} = .1164$

$Z = $ diff bet the 2 pops

$$Z = \frac{P_1 - P_2}{\sqrt{\hat{P}(1-\hat{P})\left(\frac{1}{n_1} + \frac{1}{n_2}\right)}}$$

$$= \frac{.1097 - .1343}{\sqrt{(.1164)(.8836)\left(\frac{1}{1067} + \frac{1}{402}\right)}}$$

$Z = 1.131$

---

p-value prob that Z is more extreme than -1.131

$$P[Z > 1.131] + P[Z < -1.131]$$

$$.0968 + .0968$$

$$19.4\% = 19.4\%$$

TESTING DIF OF 2 MEANS

$\mu_1 = $ = first mean

$\mu_2 = $ 2nd "

$H_0 = \mu_1 = \mu_2 \qquad H_A = \mu_1 \neq \mu_2$

$\rightarrow \mu_1 < \mu_2$ OR $\mu_1 > \mu_2$

take sample size n_1^d compute $\bar{x}_1, \Delta_1^2$

then " " n_2^d " $\bar{x}_2 \colon \Delta_2^2$

$\Delta =$

Test STAT

$$Z = \frac{\bar{x}_1 - \bar{x}_2}{\sqrt{\frac{\sigma_1^2}{n_1} + \frac{\sigma_2^2}{n_2}}} \simeq \frac{\bar{x}_1 - \bar{x}_2}{\sqrt{\frac{\Delta_1^2}{n_1} + \frac{\Delta_2^2}{n_2}}} \quad \text{if } n_1 \colon n_2 > 30$$

Accept

$$-Z\alpha_{/2} \leq Z \leq Z\alpha_{/2}$$

2 tailed Crit Region

$$-Z\alpha_{/2} \geq Z \qquad Z > Z\alpha_{/2}$$

TO ESTIMATE $\mu_1 - \mu_2$

$$\bar{x}_1 - \bar{x}_2 \pm Z\alpha_{/2}\sqrt{\frac{\Delta_1^2}{n_1} + \frac{\Delta_2^2}{n_2}}$$

CHAPTER 11

Analysis of Variance

The unity of all science consists alone in its methods, not in its material.

Karl Pearson
The Grammar of Science, 1892

Frequently it is necessary to study several populations simultaneously. In Chapter 8 we considered examples in which we studied two populations and developed both confidence limit estimates on the difference between the means of two populations and techniques for testing hypotheses concerning the differences of two population means. Here we deal with a special class of problems encountered in studying more than two populations. Problems to be solved in this chapter concern how to decide whether observed differences among more than two sample means can be attributed to chance. For instance, we may wish to examine several populations, each of which follows a different dinner diet in order to determine if all the diets have the same effect on weight gain. As another example, we may want to determine if there is really a difference in the average mileage obtained by four different types of automobiles. We now develop methods that allow such determinations.

Handwritten notes:

t test for dif in means

2 normal pops

$n_1 \quad \bar{x}_1 =$ sample mean

$\quad \Delta_1 =$ sample st dev.

$n_2 \quad \bar{x}_2 \quad \Delta_2$

$$Z = \frac{\bar{x}_1 - \bar{x}_2 - (\mu_1 - \mu_2)}{\sqrt{\frac{\sigma_1^2}{n_1} + \frac{\sigma_2^2}{n_2}}}$$

If Assume $\sigma_1 = \sigma_2$

then
$$Z = \frac{(\bar{x}_1 - \bar{x}_2) - (\mu_1 - \mu_2)}{\sigma \sqrt{\frac{1}{n_1} + \frac{1}{n_2}}}$$

$\Delta_1 =$ estimate σ_1

$\Delta_2 = $ " σ_2

POOLED ESTIMATE

$$\Delta_P = \sqrt{\frac{(n_1 - 1)\Delta_1^2 + (n_2 - 1)\Delta_2^2}{n_1 + n_2 - 2}}$$

must know σ & n_1 & n_2 must be small.

$$STAT = t = \frac{\bar{x}_1 + \bar{x}_2 - (\mu_1 - \mu_2)}{\Delta_P \sqrt{\frac{1}{n_1} + \frac{1}{n_2}}}$$

$n_1 + n_2 + n_2 - 2 = df$

11 ■ 1   ONE-WAY ANALYSIS OF VARIANCE

Suppose a food manufacturer develops a product and wishes to ascertain its marketability in various packages; in particular, he wishes to determine if his product sells best in either bags, jars, cartons, or cans. He will test market the product in the various packages and note the number of packages of each type sold per week. He will then select a particular package as the one that he will finally use only if the data clearly indicate that this package outsells all the others; that is, he will accept the hypothesis that the average number of packages sold per week is the same for each package unless the evidence demonstrates otherwise. Put in another way, he will test the hypothesis:

$$H_0: \mu_1 = \mu_2 = \mu_3 = \mu_4$$

$$H_1: \text{not all are equal}$$

where

μ_1 = true mean number of bags sold per week

μ_2 = true mean number of jars sold per week

μ_3 = true mean number of cartons sold per week

μ_4 = true mean number of cans sold per week

If the manufacturer does not reject H_0, he will make his decision in favor of one of the packages on some basis other than marketability; for example, he may choose the package that is easiest to use or the least costly to produce.

We have not yet developed methods to test the preceding H_0. We have, so far, developed techniques only for comparing means pairwise. We now develop methods for testing hypotheses concerning the equality of means for three or more populations. To explain these methods, let us continue the example of the food manufacturer who wishes to test various types of packaging.

Suppose he picks stores at random from the stores that sell his product and puts each of the packaging types in the 4 selected stores. At the end of the week the sales in each of the 4 stores are recorded, yielding the data shown in Table 11.1.

Using these data, the manufacturer wishes to ascertain if the mean sales for all the types are equal. Thus he wants to test $H_0: \mu_1 = \mu_2 = \mu_3 = \mu_4$ versus H_1: not all the means are equal.

We use a methodology called analysis of variance (ANOVA) to test H_0 versus H_1. The idea behind the analysis of variance is that if H_0 is true, then all the sample means $\bar{x}_i$ estimate the same value, and hence should be close together; that is, the *variance* of the collection of the four sample means ($\bar{x}_1, \bar{x}_2, \bar{x}_3$, and

	Population 1 bags	Population 2 jars	Population 3 cartons	Population 4 cans	Grand totals
	21	12	21	18	
	19	19	25	19	
	23	17	15	16	
	12	10	17	20	
Total	$T_1 = 75$	$T_2 = 58$	$T_3 = 78$	$T_4 = 73$	$T = 284$
Mean	$\bar{x}_1 = 18.75$	$\bar{x}_2 = 14.5$	$\bar{x}_3 = 19.5$	$\bar{x}_4 = 18.25$	$\bar{\bar{x}} = 17.75$
	$s_1^2 = 22.9$	$s_2^2 = 17.7$	$s_3^2 = 19.67$	$s_4^2 = 2.9$	$s_p^2 = 15.8$

Table 11 ▪ 1   Data on Weekly Sales in Various Packages

$\bar{x}_4$) should not be large. On the other hand, if H_0 is not true, then at least one of the population means is different, and hence the variance should be large among the four sample means.

We assume independent random sampling and that *each population is normally distributed.* Let us also suppose that each of the populations (see Table 11.1) has the same variance σ^2. Note that each value s_1^2, s_2^2, s_3^2, and s_4^2 in Table 11.1 is an estimate of σ^2, since each population has a variance equal to σ^2. We may obtain a better estimate of σ^2 by averaging or pooling the values of s_1^2, s_2^2, s_3^2, and s_4^2 together just as we did in using the pooled variance from two populations for the *t*-test in Chapter 8 (see formula 8.4). Since the sample sizes are all equal to 4, we have

$$s_p^2 = \frac{3(s_1^2 + s_2^2 + s_3^2 + s_4^2)}{16 - 4} = 15.8$$

where 16 is the total sample size and 4 is subtracted from 16 because 4 degrees of freedom are lost in this calculation, one for each of the sample estimates s_1^2, s_2^2, s_3^2, and s_4^2. The sample variance s_p^2 is called the *pooled estimate* of the variance σ^2.

Now recall that the estimate $\bar{x}$ comes from a distribution of $\bar{X}$ with a variance equal to σ^2/n, where n is the sample size. Hence under H_0 the values $\bar{x}_1, \bar{x}_2, \bar{x}_3$, and $\bar{x}_4$ each come from the same distribution with variance $\sigma^2/4$ ($n = 4$ in this case). Moreover, if H_0 is true, each of these distributions has the same mean μ which is estimated by $\bar{\bar{x}}$, the average of *all* the data

$$\bar{\bar{x}} = \frac{\bar{x}_1 + \bar{x}_2 + \bar{x}_3 + \bar{x}_4}{4}$$

while $\Sigma_{i=1}^{4} (\bar{x}_i - \bar{\bar{x}})^2/(4 - 1)$ is an estimate of $\sigma^2/4$. If we multiply this last expression by 4, then the result, which we denote by s_B^2, is also an estimate of σ^2. That is,

$$s_B^2 = 4 \frac{\Sigma(\bar{x}_i - \bar{\bar{x}})^2}{4 - 1} = 4(4.96) = 19.8$$

is an estimate of σ^2.

If H_0 is true, then s_p^2 and s_B^2 should be approximately equal since both estimate σ^2. If H_0 is not true, s_B^2 should be much larger than s_p^2 since s_B^2 will incorporate not only the intrinsic variation σ^2 of the population but also the variation between the true means. In other words, if H_0 is true, the ratio s_B^2/s_p^2 should be near 1, whereas if H_0 is not true s_B^2/s_p^2 should be much greater than 1. Thus we should reject H_0 if s_B^2/s_p^2 is greater than some *critical value* that itself is greater than 1. But what is the value of the *critical value;* how do we find it?

To answer these questions, we must know that the random variable S_B^2/S_p^2 (where S_B^2 is the estimator yielding the estimate s_B^2 and S_p^2 is the estimator yielding s_p^2) has the F-distribution $F(3, 12)$. The numbers 3 and 12 in $F(3, 12)$ refer to the degrees of freedom for the numerator, S_B^2, and denominator, S_p^2, respectively. The degrees of freedom for the numerator is 3 since S_B^2 has 3 degrees of freedom (the number of populations $- 1$), whereas the degrees of freedom for the denominator is 12 since S_p^2 has 12 degrees of freedom (the total sample size $-$ the number of populations).

In our example, the food manufacturer would fail to reject H_0 at $\alpha = 0.05$ since the data in Table 11.1 yield*

$$F = s_B^2/s_p^2 = \frac{19.8}{15.8} = 1.25 < 3.49 = F_{0.05} \text{ with (3, 12) degrees of freedom†}$$

Thus the food manufacturer should use the most convenient packaging type since package type does not seem to significantly affect the marketability of his product.

We have discussed the method for testing the equality of the means of four

---

* The observed value s_B^2/s_p^2 is denoted by F. This notation is contrary to the convention we have established to indicate observed values by lower-case letters; however, since this notation using F is commonly used, we employ it here. As we shall see, capital letters are generally used to indicate observed quantities in the analysis of variance so do not be surprised when you encounter this notation. This has already been seen in Table 11.1 where T_i is an observed number (not a random variable) representing the total number of observations in column i.

† The value of $F_{0.05}$ with (3, 12) degrees of freedom is not in Appendix A.7. It can be obtained by noting that $F_{0.05}(3, 12) = 1/F_{0.95}(12, 3)$. However, the conclusion not to reject H_0 can also be arrived at by noting that $F_{0.05}$ with (3, 20) degrees of freedom is in Appendix A.7 and is equal to 3.10, which is also larger than $F = 1.25$. Since $F_{0.05}$ with (3, 12) degrees of freedom must be larger than $F_{0.05}$ with (3, 20) degrees of freedom, it follows at once that that we do not reject H_0.

populations. This method is applicable to testing the equality of means for any number, k, of populations, that is, for testing

$$H_0\colon \mu_1 = \mu_2 = \cdots = \mu_k$$

$$H_1\colon \text{not all the means are equal}$$

even when the number of observations from each population is not the same. We subsequently go into some detail concerning the notation used to describe the formulas employed for the general case since this notation is found in many textbooks and reports dealing with the analysis of variance.

One comment is appropriate at this point. The type of analysis of variance (ANOVA) previously discussed is called "one-way ANOVA"; one-way because, even though several populations are involved, there is only one *variable* involved. To understand this point, consider the example of the manufacturer who wishes to determine if different types of packages produce the same sales. The variable here is "packages"; it is a variable because there is not just one package but several of them. We shall subsequently consider analyses of variance involving two variables.

Now let us consider the one-way ANOVA where there are k populations and where the number of observations from each population is not necessarily the same. We call the numerator of s_B^2 the sum of squares *between* populations (SS_B) and the numerator of s_p^2 the sum of squares for *pooled* (or sum of squares within) populations (SS_W). If we are presented with the data as listed in Table 11.2, we use the following notation:

$$T = \sum_{j=1}^{k} T_j = \text{grand total}$$

$$\bar{\bar{x}} = T \Big/ \sum_{j=1}^{k} n_j = \text{overall mean}$$

Table 11 ▪ 2

	Observations				Total	Means	Variance
1	x_{11}	x_{12}	$\cdots$	x_{1n_1}	$T_1 = \sum_{i=1}^{n_1} x_{1i}$	$\bar{x}_1 = T_1/n_1$	s_1^2
2	x_{21}	x_{22}	$\cdots$	x_{2n_2}	$T_2 = \sum_{i=1}^{n_2} x_{2i}$	$\bar{x}_2 = T_2/n_2$	s_2^2
.	.			.	.	.	.
.	.			.	.	.	.
.	.			.	.	.	.
k	x_{k1}	x_{k2}	$\cdots$	x_{kn_k}	$T_k = \sum_{i=1}^{n_k} x_{ki}$	$\bar{x}_k = T_k/n_k$	s_k^2

Populations sampled

We have previously discussed two types of variance, namely, s_B^2 and s_p^2. We also have a third measure of total variation, s_T^2, the variation of the individual observations x_{ij} from the overall mean $\bar{\bar{x}}$;

$$s_T^2 = \frac{\sum_{i=1}^{k} \sum_{j=1}^{n_i} (x_{ij} - \bar{\bar{x}})^2}{n - 1}; \quad n = \sum_{i=1}^{k} n_i = \text{total sample size}$$

The numerator of s_T^2 is called the "total sum of squares," SS_T.

Let n_i equal the number of observations taken from the ith population, whereas $n = \Sigma\, n_i$ denotes the total number of observations. Again, s_B^2 is computed utilizing the means, $\bar{x}_1, \bar{x}_2, \ldots, \bar{x}_k$, and hence has $k - 1$ degrees of freedom. In this case we use the formula

$$s_B^2 = \frac{\sum_{i=1}^{k} n_i(\bar{x}_i - \bar{\bar{x}})^2}{k - 1}$$

for the variation between means since $n_i(\bar{x}_i - \bar{\bar{x}})^2$ is an estimate of σ^2 and hence the average

$$\frac{\sum_{i=1}^{k} n_i(\bar{x}_i - \bar{\bar{x}})^2}{k - 1}$$

is also an estimate of σ^2. Similarly, s_p^2 is the estimated pooled variance which is given by the formula

$$s_p^2 = \frac{\sum_{i=1}^{k} (n_i - 1)s_i^2}{n - k}$$

[This is consistent with our previous method of weighting each sample variance s_i^2 according to the proportion of the total degrees of freedom furnished by that sample, that is, by $(n_i - 1)/\Sigma\, n_i - k)$.]

We usually summarize our results in what is referred to as an ANOVA table. In Table 11.3 we present such a table with the computational formulas for SS_B, SS_W, and SS_T simplified.*

* The formulas were obtained by writing $\bar{x}_i = T_i/n_i$, $\bar{\bar{x}} = T/n$, and

$$s_i^2 = (\Sigma_j\, x_{ij}^2 - T_i^2)/[n_i/(n_i - 1)]$$

in the formulas for s_B^2 and s_p^2 and then simplifying.

Table 11 ▪ 3 One-way Analysis of Variance Table

	Sum of squares	Degrees of freedom	Mean square	F-ratio
Between population means	$\sum \dfrac{T_i^2}{n_i} - \dfrac{T^2}{n} = SS_B$	$k - 1$	$s_B^2 = \dfrac{SS_B}{k-1}$	$F = \dfrac{s_B^2}{s_p^2}$
Pooled or within populations	$\sum \sum x_{ij}^2 - \sum \dfrac{T_i^2}{n_i} = SS_W$	$n - k$	$s_p^2 = \dfrac{SS_W}{n-k}$	
Total	$\sum \sum x_{ij}^2 - \dfrac{T^2}{n} = SS_T$	$n - 1$		

The idea behind the partitioning of the sum of squares into SS_B and SS_W is given by Figure 11.1, which shows how one can explain the total amount of variation SS_T in the observed data. The figure shows that differences *between* populations are one source of variation and differences *within* populations are another.

This sort of separation of the components of variation was encountered before in Chapter 9, where the total variation in the Y data in a regression was partitioned into the sum of squares due to the regression line and the sum of squares around the regression line.

Example 11 ▪ 1   A dietician wishes to determine which factors in the diet lead to increased weight. Seven different diets are fed to mice yielding the following weights in grams:

Diet

Weight (in grams)	1	2	3	4	5	6	7
	110	90	104	108	114	127	108
	113	109	122	109	131	114	109
	108	98	121	118	111	119	118
	103	95	116	123	130	122	117
	119	115	188		134	132	124
			140		121		
			115				

$n_1 = 5$	$n_2 = 5$	$n_3 = 7$	$n_4 = 4$	$n_5 = 6$	$n_6 = 5$	$n_7 = 5$	$n = 37$
$T_1 = 553$	$T_2 = 507$	$T_3 = 906$	$T_4 = 458$	$T_5 = 741$	$T_6 = 614$	$T_7 = 576$	$T = 4355$
							$k = 7$

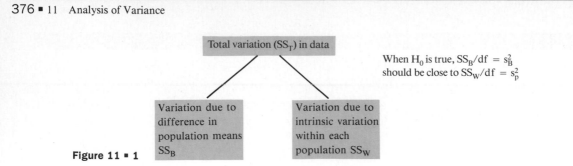

Figure 11 ▪ 1

We utilize the formulas in Table 11.3 to calculate the sum of squares (SS):

$$SS_B = \frac{(553)^2}{5} + \frac{(507)^2}{5} + \frac{(906)^2}{7} + \frac{(458)^2}{4} + \frac{(741)^2}{6} + \frac{(614)^2}{5}$$

$$+ \frac{(576)^2}{5} - \frac{(4355)^2}{37} = 2947.52 \quad \text{and}$$

$$SS_W = 110^2 + 113^2 + 108^2 + 103^2 + 119^2 + 90^2 + 109^2$$

$$+ \cdots + 124^2 - \left(\frac{(553)^2}{5} + \cdots + \frac{(576)^2}{5} \right)$$

$$= 521,805 - 515,542.79 = 6262.21$$

We put these values into the ANOVA table.

	Sum of squares	Degrees of freedom	Mean square	F-ratio
Between	2947.52	$7 - 1 = 6$	491.25	$F = 2.35$
Within	6262.21	$37 - 7 = 30$	208.74	

To ascertain if these diets do not have the same effect, we assume normality and test

$$H_0: \mu_1 = \mu_2 = \mu_3 = \mu_4 = \mu_5 = \mu_6 = \mu_7$$

$$H_1: \text{not all means are equal}$$

At the $\alpha = 0.05$ level of significance we reject H_0 if $F > F_{0.05} = 2.42$ with df $= (6, 30)$. Since we calculated $F = 2.35$ in the ANOVA table, we do not reject H_0 and conclude that the various diets do not have a significantly different effect on weight gain.

11 ▪ 2   TWO-WAY CLASSIFICATIONS WITH ONE OBSERVATION PER CELL

Often one can save time and money by performing essentially two experiments simultaneously. For example, suppose a farmer wishes to determine what the optimal amount of irrigation is for his fields, and also what the best type of fertilizer is for his fields. He has five brands of fertilizer to test and three levels of irrigation. By planting one acre of crops at each level of irrigation and with each fertilizer, he obtains the yields listed in Table 11.4.

Note that each acre is classified according to two attributes or variables — the type of fertilizer and the level of irrigation that were used. The type of ANOVA in which there are two variables is called a "two-way ANOVA." Since there are three possible levels of irrigation and five fertilizer types, we refer to Table 11.4 as a 3 by 5 (3 × 5) ANOVA table.

In this experiment there are at least two different tests that may be performed. $H_0^{(1)}$: There is no column effect (no difference among fertilizers); that is,

$$H_0^{(1)}: \quad \mu_A = \mu_B = \mu_C = \mu_D = \mu_E$$

and $H_0^{(2)}$: There is no row effect (no irrigation effect); that is,

$$H_0^{(2)}: \quad \mu_l = \mu_m = \mu_h$$

μ_l = mean yield for low irrigated fields

μ_m = mean yield for moderately irrigated fields

μ_h = mean yield for heavily irrigated fields

μ_A = mean yield for fields fertilized with product A

μ_B = mean yield for fields fertilized with product B

μ_C = mean yield for fields fertilized with product C

μ_D = mean yield for fields fertilized with product D

μ_E = mean yield for fields fertilized with product E

We assume again that each population is normal and has the same variance σ^2.

Table 11 ▪ 4

			Fertilizer type			
	A	B	C	D	E	Totals
Low	33.7	35.2	30.0	34.6	29.5	163
Moderate	37.2	39.1	32.1	37.1	32.8	178.3
Heavy	35.6	38.1	35.1	30.1	35.1	174
Totals	106.5	112.4	97.2	101.8	97.4	515.3

To develop our hypotheses tests $H_0^{(1)}$ and $H_0^{(2)}$, we use the same procedure as before, finding estimates for the common variance σ^2. To do this, we proceed as if there was no distinction among the rows to obtain the sum of squares for the column means, SS_C (this formula is the same as SS_B in Table 11.3 except that we use it here only for the column means):

$$SS_C = \frac{(106.5)^2}{3} + \frac{(112.4)^4}{3} + \frac{(97.2)^2}{3} + \frac{(101.8)^2}{3} + \frac{(97.4)^2}{3} - \frac{(515.3)^2}{15}$$

$$= 17,757.93 - 17,702.27 = 55.66$$

which has 4 degrees of freedom since there are five columns.

Similarly, we may calculate the sum of squares for the row means, SS_R (this formula is the same as SS_B in Table 11.3 except that we use it here only for the row means):

$$SS_R = \frac{(163)^2}{5} + \frac{(178.3)^2}{5} + \frac{(174)^2}{5} - \frac{(515.3)^2}{15} = 24.9$$

which has 2 degrees of freedom since there are three rows.

Again, as in the case of one-way analysis of variance, we may calculate the sum of squares for the entire table, SS_T (see Table 11.3 for this formula):

$$SS_T = 33.7^2 + 37.2^2 + 35.6^2 + 35.2^2 + 39.1^2 + 38.1^2 + 30.0^2$$

$$+ \cdots + 35.1^2 - \frac{(515)^2}{15} = 17826.45 - 17702.27 = 124.18$$

which has 14 degrees of freedom since it is calculated from all 15 cell values. Notice that we have not yet accounted for all the variation in the data, since

$$SS_T - SS_R - SS_C \neq 0$$

What is left over is called the *error* sum of squares SS_E, and is due to the intrinsic variability of the data, without regard to any possible row or column effects. We define SS_E as the error sum of squares,

$$SS_E = SS_T - SS_R - SS_C.$$

In this example $SS_E = 43.62$. Since it measures the intrinsic variation of the data, the variance $s_E^2 = SS_E/\mathrm{df}$ is also an estimate of σ^2, where df is the degrees of freedom for SS_E; in this case df $= 14 - 4 - 2 = 8$.

We may represent this partitioning of the total variation in the data by the diagram shown in Figure 11.2.

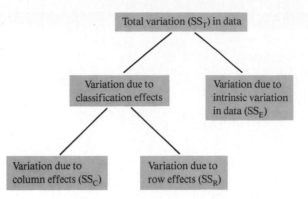

Figure 11 ▪ 2

> **If $H_0^{(1)}$ is true, $SS_C/df = s_C^2$ should be close to SS_E/df since they both estimate σ^2. If $H_0^{(2)}$ is true, $SS_R/df = s_R^2$ should be close to $SS_E/df = s_E^2$ since they both estimate σ^2.**

In comparing $s_E^2 = 43.6/8 = 5.45$ with $s_C^2 = SS_C/4 = 13.92$, we can utilize the F-test to test for column effects. The ratio s_C^2/s_E^2 should be near 1 if $H_0^{(1)}$ is true since both s_C^2 and s_E^2 are estimates of σ^2 if $H_0^{(1)}$ is true; whereas if $H_0^{(1)}$ is not true, this ratio should be much larger than 1. Thus to test

$$H_0^{(1)}: \text{no column effect, } \mu_A = \mu_B = \mu_C = \mu_D = \mu_E$$

versus

$$H_1^{(1)}: \text{not all column means are equal}$$

We form the F-ratio;

$$F_C = \frac{s_C^2}{s_E^2} = \frac{13.92}{5.45} = 2.55$$

and reject $H_0^{(1)}$ if this ratio is too large. At $\alpha = 0.05$ we would reject $H_0^{(1)}$ if $F_C > F_{0.05}$ with df $= (4, 8)$. Since $F_{0.05} = 3.84^*$ and $F_C = 2.55$, we fail to reject $H_0^{(1)}$ and conclude that there is no significant column effect; that is, fertilizer type does not significantly affect the mean yield.

In a similar manner, the farmer could test if irrigation level affects the yield by

---

* Note that $F_{0.05}$ with df $= (4, 8)$ is not listed in Appendix A.7. This value may be found by interpolation. However, interpolation is not really needed to come to the conclusion above since $F_{0.05}$ with df $= (4, 10)$ is listed in Appendix A.7 and is equal to 3.48. Since this value of 3.48 must be less than $F_{0.05}$ with df $= (4, 8)$ and since $3.48 > 2.55$ we conclude that we fail to reject H_0.

Table 11 ▪ 5

Classification I

Classification II

	C_1	C_2	C_3	C_4	$\cdots$	C_c	Total
R_1	x_{11}	x_{12}	x_{13}	x_{14}	$\cdots$	x_{1c}	T_{R_1} = total of row 1
R_2	x_{21}	x_{22}	x_{23}	x_{24}	$\cdots$	x_{2c}	T_{R_2} = total of row 2
.	.					.	.
.	.					.	.
.	.					.	.
R_r	x_{r1}	x_{r2}	x_{r3}	x_{r4}	$\cdots$	x_{rc}	T_{R_r} = total of row r
Total	T_{C_1} Total of column 1	T_{C_2} Total of column 2	T_{C_3} Total of column 3	T_{C_4} Total of column 4	$\cdots$	T_{C_c} Total of column C	$T = \Sigma\, T_{R_i}$ $= \Sigma\, T_{C_i}$ = grand total

comparing $s_E^2 = 5.45$ with $s_R^2 = SS_R/2 = 12.45$. The hypotheses are

$$H_0^{(2)}: \text{there is no row effect, } \mu_l = \mu_m = \mu_h$$

$$H_1^{(2)}: \text{not all row means are equal}$$

Using the F-ratio, $F_R = s_R^2/s_E^2$, we reject $H_0^{(2)}$ at $\alpha = 0.05$ if $F_R > F_{0.05}$ with df = (2, 8), since $F_{0.05} = 4.46$ and $F_R = 12.45/5.45 = 2.28$, we fail to reject $H_0^{(2)}$ and conclude that the irrigation level does not significantly affect the yield of this type of plant.

This example serves as a model for the general situation of tests for row and column effects when there is a single observation (as illustrated in Table 11.5) in

Table 11 ▪ 6 Two-way ANOVA Table: Two-way Classification with a Single Observation per Cell

	Sum of squares	Degrees of freedom	Mean square $= \dfrac{SS}{df}$	F-ratio
Column means	$SS_C = \Sigma\, \dfrac{T_{C_i}^2}{r} - \dfrac{T^2}{rc}$	$c - 1$	$s_C^2 = \dfrac{SS_C}{c-1}$	$F_C = \dfrac{s_C^2}{s_E^2}$
Row means	$SS_R = \Sigma\, \dfrac{T_{R_i}^2}{c} - \dfrac{T^2}{rc}$	$r - 1$	$s_R^2 = \dfrac{SS_R}{r-1}$	$F_R = \dfrac{s_R^2}{s_E^2}$
Error or within	$SS_E = SS_T - SS_C - SS_R$	$(r-1)(c-1)$	$s_E^2 = \dfrac{SS_E}{(r-1)(c-1)}$	
	$SS_T = \Sigma\,\Sigma\, x_{ij}^2 - \dfrac{T^2}{rc}$	$rc - 1$		

Table 11 ■ 7 Metal Durability Readings (4 × 5 Table)

	Temperature					
	−4°	0°	50°	100°	200°	Total
I	19.5	18.8	26.8	21.3	14.2	$T_{R_1} = 100.6$
II	11.6	15.1	15.0	18.1	13.0	$T_{R_2} = 72.8$
III	10.0	12.1	16.7	15.1	14.0	$T_{R_3} = 67.9$
IV	7.2	17.1	18.1	10.2	8.1	$T_{R_4} = 60.7$
Total	$T_{C_1} = 48.3$	$T_{C_2} = 63.1$	$T_{C_3} = 76.6$	$T_{C_4} = 64.7$	$T_{C_5} = 49.3$	$T = 302.0$

Alloy technique

each row-column classification; there are c classifications of variable I and r classifications of variable II. The preceding array of numbers is called a matrix, and each classification into a row and column is called a cell; for example, the number x_{ij} is the observation in cell (i, j) formed at the intersection of the ith row and the jth column. We treat here the special case of one observation per cell. We may calculate an ANOVA table by utilizing the formulas given in Table 11.6. The justification of these formulas is the same as for the previous example.

Example 11 ■ 2   A manufacturer of metal components for aircraft wing fabrication needs to determine the durability of his components under different temperatures and different alloying techniques. He performs durability tests on metal pieces under different temperatures and obtains the data given in Table 11.7.

Using the formulas given in Table 11.6, we may calculate

$$SS_C = \frac{(48.3)^2}{4} + \frac{(63.1)^2}{4} + \frac{(76.6)^2}{4} + \frac{(64.7)^2}{4} + \frac{(49.3)^2}{4} - \frac{(302)^2}{20}$$

$$= 4699.63 - 4560.2 = 139.43$$

$$SS_R = \frac{(100.6)^2}{5} + \frac{(72.8)^2}{5} + \frac{(67.9)^2}{5} + \frac{(60.7)^2}{5} - \frac{(302)^2}{20}$$

$$= 4743.02 - 4560.2 = 182.82$$

$$SS_T = (19.5)^2 + (11.6)^2 + \cdots + (8.1)^2 - \frac{(302)^2}{20}$$

$$= 4982.26 - 4560.2 = 422.06$$

$$SS_E = 422.06 - 182.82 - 139.43 = 99.81$$

We summarize these results in the following ANOVA table:

	SS	df	Mean squares $= \dfrac{SS}{df}$	F ratio
Column means (temp)	139.46	4	$s_C^2 = 139.43/4 = 34.86$	$F_C = 34.87/8.32 = 4.2$
Row means (alloying)	182.82	3	$s_R^2 = 182.82/3 = 60.94$	$F_R = 60.94/8.32 = 7.3$
Error or within	99.82	12	$s_E^2 = 99.81/12 = 8.32$	
Total	422.09	19		

To test

$H_0^{(1)}$ there is no effect of temperature on durability (i.e., no column effect)

$H_1^{(1)}$ there is a temperature effect

we reject $H_0^{(1)}$ if $F_C = s_C^2/s_E^2 > F_\alpha$ with degrees of freedom df = (4, 12). At the level of significance 0.01 we have $F_{0.01} = 5.41$ with degrees of freedom df = (4, 12) and since $F_C = 4.2 < 5$, we fail to reject $H_0^{(1)}$ and conclude that temperature (in the range tested) does not significantly affect the durability of the metal.

To ascertain if the alloying technique significantly affects the durability, we test

$H_0^{(2)}$: there is no row effect

$H_1^{(2)}$: there is a significant row effect

We reject $H_0^{(2)}$ at $\alpha = 0.05$ if $F_R = s_R^2/s_E^2 > F_{0.05} = 3.49$ with (3, 12) degrees of freedom. Since $F_R = 7.3 > 3.49$, we reject H_0 and conclude that the alloying technique does have a significant effect on the durability of the metal.

11 ▪ 3   SUMMARY OF THE CHARACTERISTICS OF THE ANOVA MODEL: LINEAR MODEL

The ANOVA tests require that the following assumptions be made:

1. There is independent random sampling.
2. The data come from normal populations.
3. The normal distribution of the data from each population (or category) has the same variance σ^2. (This assumption is often termed "homogeneity of variances.")

These assumptions provide a picture or *model* of those physical situations to which the ANOVA tests are applicable. But it is not the complete model for

ANOVA tests. The following discussion presents the remaining parts of the ANOVA model.

Let us return to the one-way ANOVA (Section 11.1). Suppose we have k populations: If X_i is a random variable from population i, then we assume

$$E(X_i) = \mu_i = \mu + \alpha_i \qquad (11.1)$$

that is, the mean μ_i of population i is equal to μ which is the overall population mean plus a mean α_i which is specific to population i. (Since μ is the overall mean, $\mu = (\mu_1 + \mu_2 + \cdots + \mu_k)/k$ where k is the number of populations). If each $\alpha_i = 0$, then the mean of each population is the same. Hence the hypothesis that all population means are equal

$$H_0: \mu_1 = \mu_2 = \cdots = \mu_k$$

is the same as the hypothesis

$$H_0: \alpha_1 = \alpha_2 = \cdots = \alpha_k = 0$$

Example 11 ▪ 3   Suppose there are four machines in a plant that produce what are supposed to be identical ball bearings. The overall mean is $\mu = 2$ cm. Each machine is supposed to produce ball bearings with a diameter having a population mean of 2 cm. If all machines are working as they should, the population diameter mean of ball bearings should equal the value of 2 cm. for each of the four machines. However, because of wear, $\mu_1 = 1.92$ cm., $\mu_2 = 2.02$ cm., $\mu_3 = 1.97$ cm., $\mu_4 = 2.09$ cm., where μ_i is the mean of the ball bearings produced by the ith machine. The overall mean $\mu = 2$ and $\alpha_1 = -0.08$ cm., $\alpha_2 = 0.02$ cm., $\alpha_3 = -0.03$ cm., $\alpha_4 = 0.09$ cm.

In the two-way classification discussed in Section 11.2, the main assumption (in addition to the usual assumptions of independent random sampling, normality, and homogeneity of variances) is that if X_{ij} is a random variable with the range of possible observations from the population of row category i and column category j, then

$$E(X_{ij}) = \mu + \alpha_i + \beta_j \qquad (11.2)$$

that is, the mean of the population of row i and column j is equal to an overall mean μ plus a mean α_i associated with row i and a mean β_j associated with column j. If $\alpha_i = 0$ for every row i, (i.e., there is no row effects), then the mean of each row i is the average of the means in each cell of that row; that is, the mean for each row is

$$\frac{1}{c} \sum_{j=1}^{c} E(X_{ij}) = \mu + \frac{1}{c}[\beta_1 + \beta_2 + \cdots + \beta_c]$$

where c is the number of columns. Hence each row has the same mean. Therefore the hypothesis

$$H_0^{(1)}: \text{row means all equal}$$

is equivalent to the hypothesis

$$H_0^{(1)}: \alpha_1 = \alpha_2 = \cdots = \alpha_r = 0$$

where r is the number of rows. Similarly, the hypothesis

$$H_0^{(2)}: \text{column means all equal}$$

is equivalent to the hypothesis

$$H_0^{(2)}: \beta_1 = \beta_2 = \cdots = \beta_c = 0$$

Example 11 ▪ 4   There are four machines and four operators. The operators rotate among the machines that produce ball bearings. The foreman notices that the population mean of the diameter of the ball bearings is different from machine to machine. Moreover, for each machine, this population mean changes when the operator changes. He assumes the mean (μ_{ij}) of machine i with operator j is equal to $\mu + \alpha_i + \beta_j$, where μ is the value of the mean diameter that should be produced (2 cm) if there were no machine or operator effects, α_i is the average effect of the wear of machine i, and β_j is the average effect on the average diameter of the technique of the operator j.

If X_{ij} is a random variable from the population row i and column j, then we sometimes find the following model useful:

$$E(X_{ij}) = \mu + \alpha_i + \beta_j + \delta_{ij} \qquad (11.3)$$

The term δ_{ij} is called the *interaction;* the inclusion of δ_{ij}, the interaction, suggests that there is possibly an effect associated with each *cell.* Thus this model postulates a synergystic effect of rows and columns. Certain combinations produce more or less than what is expected just due to the row and column effects.

Interaction will be studied in the next section where we discuss the testing of the additional hypothesis

$$H_0^{(3)}: \delta_{11} = \delta_{12} = \cdots = \delta_{1c} = \delta_{21} = \cdots = \delta_{2c} = \cdots = \delta_{rc} = 0$$

Assumptions of the type (11.1), (11.2), and (11.3) are called *linearity* assumptions since they state that there is a linear, a straight line relationship, between the mean of the individual observations and the other means. A model that contains this kind of assumption is called a linear model. Hence

---

The ANOVA model is
1. **a linear model, where**
2. **sampling is independent random sampling**
3. **each population is normal**
4. **all population variances are equal**

---

11 ▪ 4 TWO-WAY CLASSIFICATIONS WITH MULTIPLE OBSERVATIONS PER CELL: INTERACTION

In many cases one may suspect that the difference between the total variation SS_T and the sum of the treatment variations $SS_C + SS_R$ is not due entirely to intrinsic errors in the data, but might be caused at least in part by some sort of interaction between the two variables under consideration. For an illustration consider the situation in example 11.2 where a manufacturer was investigating the effects of temperature and alloying techniques on the durability of the metal produced. It could be that the manufacturer has designed one technique to perform well in low temperatures and another technique for good performance in high temperatures. In this case the variation in the different cells is due not only to the particular row and column to which the cell belongs but also in part to an interaction effect between the temperature and the alloying technique. An interaction is simply a lack of additivity of the row effects and column effects. Such interaction effects are quite common in medical studies. For example, the effect on mortality of diabetes and obesity is not additive.

To get an estimate of the interaction effect between two variables of classification we must be able to compute the variance of the observations within each cell, and hence if we suspect interaction effects are present, we must take repeated measurements in each cell. We discuss here the analysis needed when we have the same number of observations per cell.

Suppose, for example, the manufacturer of metal parts (see Ex. 11.2) suspects that alloying techniques and temperature interact to affect the durability of the metal. To estimate the interaction effect as well as the row and column effect he replicates (or repeats) the experiment three times and obtains the data in Table 11.8.

The observations that are at the same level of each variable constitute what is known as a cell. Thus the observations at $-4°C$ and technique A constitute cell $(1, 1)$, $0°C$ and technique A constitute cell $(1, 2)$, and so on.

Now we have the option of making several hypothesis tests. By considering each cell as a separate population, we may utilize the one-way ANOVA tech-

Table 11 ▪ 8 Metal Durability Readings Replicated Three Times (4 × 5 Table with 3 Replications)

		Temperatures (°C)				Row total
	−4	0	50	100	200	
Technique A	19.5	18.8	26.8	21.3	14.2	
	18.3	19.7	20.1	19.8	13.7	
	17.8	20.1	19.8	20.3	12.2	
A-cell totals	$T_{11} = 55.6$	$T_{12} = 58.6$	$T_{13} = 66.7$	$T_{14} = 61.4$	$T_{15} = 40.1$	$T_{R_1} = 282.4$
Technique B	11.6	15.1	15.0	18.1	13.0	
	16.3	18.1	19.3	16.6	12.2	
	12.4	11.3	18.1	19.2	17.0	
B-cell totals	$T_{21} = 40.3$	$T_{22} = 44.5$	$T_{23} = 52.4$	$T_{24} = 53.9$	$T_{25} = 42.2$	$T_{R_2} = 233.3$
Technique C	10.0	12.1	16.7	15.1	14.0	
	11.3	16.1	19.3	17.8	16.2	
	13.2	14.0	20.0	19.3	13.8	
C-cell totals	$T_{31} = 34.5$	$T_{32} = 42.2$	$T_{33} = 56.0$	$T_{34} = 52.2$	$T_{35} = 44$	$T_{R_3} = 228.9$
Technique D	7.2	17.1	18.1	10.2	8.7	
	10.4	19.3	19.3	12.3	10.3	
	9.2	16.1	20.1	15.0	12.0	
D-cell totals	$T_{41} = 26.8$	$T_{42} = 52.5$	$T_{43} = 57.5$	$T_{44} = 37.5$	$T_{45} = 31.0$	$T_{R_4} = 205.3$
Column total	$T_{C_1} = 157.2$	$T_{C_2} = 197.8$	$T_{C_3} = 232.6$	$T_{C_4} = 205$	$T_{C_5} = 157.3$	$T = 949.9$

Alloying technique (row label)

nique to test

$$H_0^{(1)}: \text{ all 20 cell means are equal}$$

$$H_1^{(1)}: \text{ not all the cell means are equal}$$

To do this, we proceed as in Section 11.1. Here $n_i = 3$ observations per cell and $n = 60$ observations total: We calculate the sum of squares between cells (SS_B) and the sum of squares within cells (SS_W) as before.

$$SS_B = \sum_{\text{all cells}} \frac{T_{ij}^2}{3} - \frac{T^2}{60} = \frac{(55.6)^2}{3} + \frac{(58.6)^2}{3} + \cdots + \frac{(31.0)^2}{3} - \frac{(949.9)^2}{60}$$

$$= 15767.98 - 15038.50 = 729.48$$

$$SS_W = (19.5)^2 + (18.3)^2 + (17.8)^2 + \cdots + (8.7)^2 + (10.3)^2 + (12.0)^2$$

$$- \sum_{\text{all cells}} \frac{T_{ij}^2}{3} = 15{,}928.99 - 15{,}767.98 = 161.01$$

Since we are considering 4 rows and 5 columns, we have $k = 20$ populations so SS_B has 19 degrees of freedom. Since we are pooling the 20 population variances, we see that SS_W has $20(3 - 1) = 40$ degrees of freedom. Note that $s_B = SS_B/19$, $s_p^2 = SS_W/40$, and so $F = s_B^2/s_p^2$. We reject $H_0^{(1)}$ if $F > F_\alpha$ with df $= (19, 40)$. For $\alpha = 0.05$, $F = (729/19)/(161/40) = 9.53 > F_{0.05}$, so we reject H_0 and conclude that all cell means are not equal. This result encourages us to further investigations. To understand why we must now look more deeply into the data, recall that one assumption underlying our model is that each cell mean is the sum of an overall mean, a mean row effect, a mean column effect, and an interaction: We write this assumption in symbols as follows:

$$\text{Average of the } i, j\text{th cell} = \mu_{ij} = \mu + \alpha_i + \beta_j + \delta_{ij}$$

μ = overall mean

α_i = mean of the ith row

β_j = mean of the jth row

δ_{ij} = mean of interaction in ith row and jth column

Since the μ_{ij} are not all equal, there must be some difference among the row means or column means or interactions. That is, there must be a row, column, or interaction effect. We must test to determine which of these effects exist. These tests are described subsequently. (It should be noted that when doing a sequence of successive hypothesis tests, the overall level of significance for the collection of tests is different from the individual levels used. We consider each test separately in the sequel).

Now there are several components that compose the variation between the populations (SS_B) and we can represent these as shown in Figure 11.3.

We can now calculate SS_R and SS_C as we did in Section 11.2, keeping in mind that each row total T_{R_i} is based on 15 observations and each column total

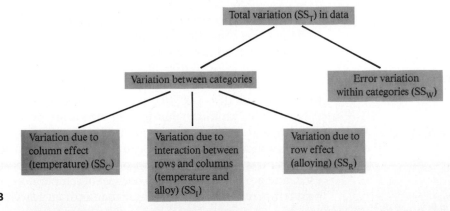

Figure 11 ▪ 3

T_{C_i} is based on 12 observations. We calculate the row sum of squares (SS_R):

$$SS_R = \sum \frac{T_{R_i}^2}{15} - \frac{T^2}{60}$$

$$= \frac{(282.4)^2}{15} + \frac{(233.3)^2}{15} + \frac{(228.9)^2}{15} + \frac{(205.3)^2}{15} - \frac{(949.9)^2}{60}$$

$$= 15{,}248.12 - 15{,}038.5 = 209.62$$

and $SS_C = 353.8$ (verify this result). The interaction sum of squares SS_I is defined by $SS_I = SS_B - SS_C - SS_R$. The following ANOVA table summarizes all the calculations:

	Sum of squares	df	Mean square	F-value
Column means	$SS_C = 353.8$	4	88.5	21.96
Row means	$SS_R = 209.62$	3	69.9	17.35
Interaction	$SS_I = SS_B - SS_C - SS_R = 166.1$	$4 \cdot 3 = 12$	13.8	3.42
Subtotal (or Between)	$SS_B = 729.48$	19	38.4	9.5
Within	$SS_W = 161.01$	40	4.03	
Total	$SS_T = 890.5$	$5 \cdot 4 \cdot 3 - 1 = 59$	15.1	

$$SS_T = SS_W + SS_B \text{ or } SS_T = \sum \sum x_{ij}^2 - \frac{T^2}{n}$$

n = total sample size

Each F-value shown is obtained by dividing the appropriate sum of squares (SS_C, SS_R, SS_I, SS_B) by SS_W. We test each hypothesis at $\alpha = 0.05$.

To test the hypothesis concerning the equality of column means, we find that 21.96 is larger than $F_{0.05}$ with df = (4, 40), and so we reject this hypothesis.

To test the hypothesis concerning equality of row means, we find that 17.35 is greater than $F_{0.05}$ with df = (3, 40), and so we reject this hypothesis.

To test the hypothesis that there is no interaction, we find that 3.42 is greater than $F_{0.05}$ with df = (12, 40), and so we reject this hypothesis as well.

No wonder we found that the cell means are not all equal: There is a row, column, and interaction effect! For this problem this means that a significant effect is noticed for the durability of the various alloys that is due to changes in temperatures, a significant effect on durability depending on which alloy was used, and a significant interactive effect of temperature and alloy together (like heat-tempered steel being stronger than ordinary steel).

SENSITIVITY OF ANOVA TO DEPARTURES FROM THE UNDERLYING ASSUMPTIONS

Assumptions that form part of the model for the ANOVA tests are that the samples are randomly chosen from normal populations having equal variances. However, investigation has shown that the results of the ANOVA tests are changed very little by *moderate* violations of these two assumptions. If you encounter a situation in practice in which these assumptions are violated, you may still be able to employ the ANOVA tests. It is best, in these situations, to refer to the literature listed in the references at the end of this chapter.

It turns out that the ANOVA tests are *robust.* Robust as used in statistics when referring to the qualities of a test means that the test works well even when the assumptions underlying the test are somewhat violated. Some of the references at the end of the chapter discuss "robustness" in general in statistics as well as with respect to ANOVA.

In some important special cases, when the normality assumption is violated, namely, when the data have a binomial or a Poisson distribution, the data can be transformed so that they are approximately normally distributed. There are many such references that address this subject. One should study these references carefully to make sure that the data in the situation being studied can be transformed to "normally distributed" data. Some such references are given at the end of this chapter.

INTERPLAY OF DATA COLLECTION AND ANALYSIS: DESIGN OF EXPERIMENTS

In Sections 8.7 and 10.6 we discussed the controlled experiment in which all factors in the experiment except the one under study are held fixed. For example, let us consider again the problem that opened this chapter. A food manufacturer wishes to ascertain the marketability of his product in various package types: bags, jars, cartons, or cans. However, there are a number of factors other than package type that can influence sales. Among these are the price of each package type, the location of each package-type in the store and on the shelves, the salesperson who sells each package type and the store itself, whether or not it is in a good commercial location. Since we are interested only in how package type influences sales, these other factors are called extraneous factors. In a controlled experiment the manufacturer would fix the store and price for all package types, place the package types in equally accessible locations in the store, use the same salesperson to sell all of them, and so forth.

The controlled experiment is often used in scientific experiments, but this methodology is sometimes difficult and expensive to implement. Moreover, we may not be able to generalize our conclusions outside the controls we have used.

For example, our conclusions about package type may not be valid for any store other than the one in which we did our experiment.

There are other procedures that we can employ to try to protect ourselves from extraneous factors. One such method is called *randomization.* We randomly select several stores from the population of stores that will be used to sell the product. We may stock each selected store with the same number of package types but assign shelf space and location randomly to the package types. The salespeople may be selected at random from the population of salespeople. The variations in the results caused by the extraneous factors can then be considered due to "chance." This type of experiment in which we randomly make particular choices of the extraneous factors is called a *completely randomized design.* Using this design, we find that the one-way ANOVA discussed in this chapter provides the model and the analysis for testing the null hypothesis that all package types are equally marketable.

The two-way ANOVA gives us another method of analysis, when used with an appropriate experimental design, to take into account the effects of extraneous factors. We have previously alluded to a "store effect" on sales of the package types. One store may sell better than the other because it is in a good commercial district. If we had used only one package type, say, jars, in this store we might have concluded that jars sold best of all package types. These better sales may have been due simply to the fact that jars were located in a store that sold *everything* better. We can explicitly take into account store effect by selecting a number of stores at random from those stores used to sell the product and then assign the same number of each package type to each selected store. In this case the experiment would be called a *randomized-block design.* The stores in the example are called the *blocks.* The two-way ANOVA can be used to test store effect.

One need not choose the stores at random. A number of stores could be deliberately selected to reflect the range of store types. For example, there may be many different categories of stores: large department stores, local markets, "mom and pop" stores, health food stores, and so on. The stores may differ, and so be categorized, by location: urban, rural, and so on. If an equal number of observations is made for every combination of categories in the variable of classification, the experiment is called a *factorial experiment.*

The two-way ANOVA can be used with both randomized-block designs and factorial experiments.

Latin Squares

When several variables of classification are included in an experiment and each variable is divided into a number of categories, then the required sample size may reach unrealistic bounds. If we had four teaching methods that we wanted to test and used eight schools with six different types of textbooks with five replications, we would need $4 \times 8 \times 6 \times 5 = 960$ students in the study. This sample size may be too large for our resources in time, money, and manpower. One type of design that permits us to reduce the size of the experiment while permitting us to test null hypotheses about variable effects is the Latin square. This design may be used when each classification variable has the same number of categories and there is no interaction effect. Consider the case where we wish to test the four package types, which we now denote as *A, B, C,* and *D*, in four different stores (I, II, III, and IV) and in four locations (1, 2, 3, and 4). The following matrix shows the construction of this design:

		Store type			
		I	*II*	*III*	*IV*
	1	*A*	*B*	*C*	*D*
	2	*D*	*A*	*B*	*C*
Location	*3*	*C*	*D*	*A*	*B*
	4	*B*	*C*	*D*	*A*

In this case not every package type is used in every store and location. Package type *A* is used in store I and location 1, in store II and location 2, and so on.

The preceding matrix is called a Latin square because each letter appears exactly once in each row and column. We can use this design to test null hypotheses concerning package effect, store effect, and location effect. We detail these tests (which are part of the ANOVA methodology) in the problem section. The point we make here is that, by the judicious choice of an experimental design, we can make practical the implementation of the experiment and also achieve valid results.

Summary

In this chapter we studied procedures for testing for differences among the means of two or more populations. The model (assumptions) underlying these procedures are

1. The model is a linear model.
2. Sampling is independent random sampling.
3. Populations are normally distributed.
4. Population variances are all equal to σ^2.

The following fact underlies the methodology used here: If means of subgroups differ greatly, the variance between the sample means is much larger than the variance within the samples. The test of the differences between the means is based on this fact. Hence we develop the F-statistic which is essentially equal to

$$\frac{\text{Estimate of } \sigma^2 \text{ based on variation among } \bar{x}\text{'s}}{\text{Estimate of } \sigma^2 \text{ based on the variation within the sample}}$$

If the population means of the subgroups are greatly different from one another, this ratio should be much larger than 1. The question is, "How much larger than one must the ratio be before we reject the null hypothesis that the means of all the subgroups are equal?" The answer is given by the F-test.

We studied three applications of the preceding methodology: The one-way analysis of variance (ANOVA), the two-way ANOVA with and without interaction. These analyses rest on a separation of the variance of all the observations into parts, each of which measures variability of some specific part of the observations, such as variation *within* each of the several subgroups (populations) and variation *between* populations. It is from this breakdown of the variance of the observations (all of which go to determining the numerator and denominator of F as previously described) that this methodology obtains its name, analysis of variance.

Key Terms

analysis of variance	column mean
interaction	row mean
linear model	within groups
mean square	sum of squares
replication	error sum of squares
robust	ANOVA table

References

There are many excellent texts that discuss analysis of variance in detail. Among them are:

Dixon, W. J., and Massey, F. J.: *Introduction to Statistical Analysis.* New York, McGraw-Hill Book Co., 1969.

Federer, W. T.: *Experimental Design: Theory and Application.* New York, Macmillan, 1963.

The following highly technical paper discusses the robustness of ANOVA:

Donaldson, T. S., Robustness of the F-test to errors of both kinds and the correlation between the numerator and the denominator of the F-ratio, *Journal of the American Statistical Association, 63*:660–676, 1968.

The following papers discuss the use of transformations for making nonnormal data normal or approximately normal and the difficulties in using such transformations:

Bartlett, M. S.: The use of transformations, *Biometrics, 3*:39–52, 1947.

Levine, D. W., and W. P. Dunlap: Power of the F-test with skewed data: Should one transform or not? *Psychological Bulletin, 92*:272–280, 1982.

Smith, J. E. K.: Data transformation in analysis of variance, *Journal of Verbal Learning and Verbal Behavior, 15*:339–346, 1976.

For a discussion of what to do when the hypothesis of "no effect" is rejected see

Bickel, P. J., and K. A. Doksum: *Mathematical Statistics.* San Francisco, Holden-Day, 1977, pp. 294–296.

Unfortunately, this is an advanced text and should not be approached by the beginner without guidance. A less forbidding exposition can be found in

Scheffé, H.: A method for judging all contrasts in the analysis of variance. *Biometrika, 40*:87–104, 1952.

Exercises

Section 11.1

1. Three groups of mice were injected with different types of tranquilizers and the number of seconds it took for them to fall asleep was recorded. The results were:

   Tranquilizer A 8, 7, 10, 11, 6, 9, 10

   Tranquilizer B 9, 12, 15, 13, 9, 10, 8, 10, 11

   Tranquilizer C 11, 10, 9, 8, 6, 12, 8, 9, 7

   a. Form a one-way ANOVA table for this data.

   b. Test the hypothesis (at $\alpha = 0.05$) that all three drugs give the same average time for putting the experimental animals to sleep.

2. The marketing research department for a manufacturer of microwave ovens conducted an experiment to determine the relationship between the price of their product and the perceived quality of their product. Each subject was asked to rate the quality of one particular microwave on a scale of 1 to 100. In each case the subject was told the price of the oven was in one of four price categories. The responses for the 60 subjects were:

Price category	Computational quantities	
250–350	$\sum_{i=1}^{15} x_{1i} = 1050$	$\sum_{i=1}^{15} x_{1i}^2 = 125,000$
350–450	$\sum_{i=1}^{15} x_{2i} = 1200$	$\sum_{i=1}^{15} x_{2i}^2 = 142,000$
450–550	$\sum_{i=1}^{15} x_{3i} = 1275$	$\sum_{i=1}^{15} x_{3i}^2 = 120,000$
550–650	$\sum_{i=1}^{15} x_{4i} = 1350$	$\sum_{i=1}^{15} x_{4i}^2 = 132,000$

   a. Construct an ANOVA table for these data.

   b. Using $\alpha = 0.1$, test the hypothesis that the perceived quality was the same for all price classifications.

3. A researcher in small business functions conducts an investigation to determine if there is a relationship between the annual profits and the length of time in business for small real estate firms. Firms of approximately the same size were classified into three groups of ten, each depending on how long they had

been in business. The profitability for each group was measured in multiples of $1000 and the following data were found:

Short time $\sum_{i=1}^{10} x_{1i} = 50$ $\Sigma x_{1i}^2 = 520$

Medium time $\sum_{i=1}^{10} x_{2i} = 30$ $\Sigma x_{2i}^2 = 400$

Long time $\sum_{i=1}^{10} x_{3i} = 70$ $\Sigma x_i^2 = 570$

a. Construct an ANOVA table for these data.
b. Test if the mean profitability is different for the three groups. (Use $\alpha = 0.1$.)

4. A medical researcher investigating the severity of side effects of various different treatments for a certain ailment has collected the following data using a severity scale 0 to 100:

Treatment	Number of subjects	Sum of scores	Sum of squared scores
A	10	500	31,500
B	12	744	52,100
C	10	710	53,500
D	15	720	41,200

a. Form an ANOVA table using these data.
b. Test the hypothesis that the average severity of side effects is the same for each treatment. (Use $\alpha = 0.05$.)

5. A leather manufacturer is considering which of five different tanning formulas to use for tanning skins. The criterion he will use to decide among the formulas is how easy the resulting tanned skin is to work with in producing leather products. Accordingly, each skin is rated on a scale of 0 to 20, depending on his perception of the "workability" of the skin. The following data were obtained:

Tanning formula

	1	2	3	4	5
Ratings	17	10	12	19	16
	12	8	18	15	8
	18	16	13	14	12
	16	12	9	9	10

a. Form an ANOVA table using these data.
b. Test the hypothesis that all five formulas give the same expected rating. (Use $\alpha = 0.05$.)

6. The following summary presents the results obtained from interviews with 50 people in each of four cities. Each person projected his yearly earnings 5 years hence. Are residents of each city on the average equally optimistic about the future?

City	Sample size	Sum of projections (in millions of dollars)	Sum of the squares of each (multiply each by 1 million)
A	50	2	103.2
B	50	1.13	161.1
C	50	9.6	121.8
D	50	1.2	111.4

Section 11.2

7. An instructor of a large class wishes to pass out three different tests to her class. She wants to check that all the tests are about the same in difficulty; four students are selected and given the three tests with the following results:

	Test A	Test B	Test C
Student 1	82.3	81.2	60.2
Student 2	71.5	70.5	59.8
Student 3	83.6	71.4	58.7
Student 4	90.1	90.9	87.6

At the significance level $\alpha = 0.05$, is there evidence that

a. The tests are not all the same?

b. The students are not all the same? (*Note:* $\Sigma x^2 = 70,236.5$; $\Sigma x = 907.8$)

8. The following data were obtained from an experiment concerned with the octane rating of five gasolines labeled *A, B, C, D,* and *E,* and four different methods for determining the ratings:

| Rating method | Gasolines | | | | |
	A	*B*	*C*	*D*	*E*
1	91.7	91.7	92.4	91.8	93.1
2	91.2	91.9	91.2	92.2	92.9
3	91.9	90.9	91.6	92.0	92.4
4	90.6	90.9	91.0	91.4	92.4
$\bar{x}$	91.35	91.35	91.55	91.85	92.7
s_i^2	0.34	0.28	0.38	0.12	0.13

a. Construct an ANOVA table for these data.

b. Test if there is a significant difference between the various methods used for determining octane ratings. (Use $\alpha = 0.05$.)

c. Test if there is a significant difference in octane ratings between the different gasolines. (Use $\alpha = 0.05$.)

9. A sociologist in studying family size by city size and region of a country found the following:

| | | Region | | | |
		North	South	West	East
	Village	7	6	8	7
City	*Town*	2	4	4	4
Size	*Metropolis*	4	6	5	3

Each entry is the size of one family. Suppose that the assumptions underlying ANOVA are satisfied; test the hypotheses that there is no difference among row means and that there is no difference among column means.

Section 11.4

10. A questionnaire was designed to rate the quality of life in several villages of a Latin American country. To ensure that the questionnaire is measuring the correct attributes, a group of experts is asked to rate each village on a 1 to 100 scale, and the questionnaire is administered, being scored 1 to 100 also. The results were

Area of villages	Expert rating	Questionnaire rating
Northern	67	61
	83	82
	79	75
Central	72	91
	87	83
	61	70
Southern	86	75
	87	86
	83	82

a. Calculate an ANOVA table for the preceding data.

b. The first question of interest is to determine if there is an interaction; that is, does one method of rating (i.e., expert opinion vs. questionnaire) tend to be better or worse than expected in certain geographical areas? Test H_0: no interaction at $\alpha = 0.05$.

c. Test H_0: no row effect at $\alpha = 0.05$; that is, does there appear to be geographical effect on quality of life?

d. Test H_0: no column effect at $\alpha = 0.05$; that is, does the rating technique matter?

11. Three brands of automobile tires were tested under two different road conditions. Three tires of each brand were used for each type of road condition. The following are the results of the computations on the observation of the life of the brands of tires under the road conditions:

Sums of squares

Tire	56.89
Road	20.33
Subtotal	78.67
Total	96.00

Using ANOVA, test the appropriate hypotheses concerning "tires," "road" and the interaction between roads and tires. (Use $\alpha = 0.05$.)

Section 11.6

12. The ANOVA for the "Latin Square" design involves the calculation of SS_c, SS_r, SS_t just as in the two-way ANOVA without interac-

tion but involves the calculation of an additional sum of squares called the treatment sum of squares, $SS(Tr)$. To illustrate, suppose we have the following Latin Square design:

Stores

		I	II	III
	1	194 A	730 B	1187 C
Location	2	758 C	311 A	589 B
	3	369 B	558 C	311 A

Package type (treatment)
A = bags
B = cartons
C = cans

The number shown in each cell is the number of the package type (treatment) sold for the month in the store location indicated by the cell. For example, there were 194 bags sold in store I at location 1. The treatment total for bags is denoted T_A and is equal to 816. For a $\ell \times \ell$ Latin square, the value of $SS(Tr) = (T_A^2 + T_B^2 + \cdots)/\ell - (T/\ell)^2$ where, as before, T is the sum of all the data values. The ANOVA table is

	Sum of squares	Degrees of freedom	Mean square	F
Column Means	SS_C	$\ell - 1$	$s_C^2 = \dfrac{SS_C}{\ell - 1}$	$F_c = \dfrac{s_C^2}{s_E^2}$
Row Means	SS_R	$\ell - 1$	$s_R^2 = \dfrac{SS_R}{\ell - 1}$	$F_R = \dfrac{s_R^2}{s_E^2}$
Treatments	$SS(Tr)$	$\ell - 1$	$s_{Tr}^2 = \dfrac{SS(Tr)}{\ell - 1}$	$F_{Tr} = \dfrac{s_{Tr}^2}{s_E^2}$
Error or within	$SS_E = SS_T - SS_C - SS_R - SS(Tr)$	$(\ell - 1)(\ell - 2)$	$s_E^2 = \dfrac{SS_E}{(\ell - 1)(\ell - 2)}$	
Total	SS_T	$\ell^2 - 1$		

(The formulas for SS_E, SS_R, SS_T are given in Table 11.6.)

a. For the data just given, show that $F_C = 1.84$, $F_R = 2.33$, $F_{Tr} = 8.69$.

b. Test whether there is a store effect; a location effect.

c. Is there evidence that the package types do not sell equally well?

CHAPTER 12

Nonparametric Statistics

In this sign shalt thou conquer.

Eusebius
Life of Constantine, I, 28

In Chapter 10 we studied some techniques that are applicable to analyzing qualitative data and some types of quantitative data. We learn additional techniques here through the study of nonparametric statistics. This is the field of study that develops statistics for hypothesis testing and confidence intervals when the exact form of the population distribution is not known. Generally, we test hypotheses in this field about population *distributions* rather than about population parameters, such as μ or σ^2, hence the name *nonparametric* statistics.

12 ▪ 1 INTRODUCTION

In order to use the tests of hypotheses discussed in previous chapters assumptions must be made about the form of the distribution of the population sampled. For example, when we tested the null hypotheses concerning means we assumed a normal distribution for the population. Using this assumption, we were able to apply the z-test or t-test, depending on whether or not we knew the value of the population variance σ^2, or whether, in the absence of a knowledge of σ^2, the sample size was large enough to permit the use of the z-test instead of the t-test.

The study of *nonparametric* statistics is aimed at finding test statistics with fewer assumptions about the form of the population distribution. Consequently, such tests can be applied to a wider variety of situations than the tests we have previously studied. Moreover, even when assumptions are made about the form of the population distributions so that standard tests like the z-test and t-test can be applied, often a nonparametric test is preferred because it is easier to apply than is the standard test. In this chapter we consider some of the more common nonparametric techniques.

Before beginning our study, let us review some matters we discussed earlier.

The Median of a Distribution

The median of a distribution is a measure of central tendency of the distribution. It is defined as the number m such that $P(X \geq m) \geq 0.5$ and $P(X \leq m) \geq 0.5$; that is, the median, m, is a number for which one half the probability distribution is around and on either side of m. When the random variable X has a continuous distribution, the median is $x_{0.5}$, the fiftieth percentile value exactly.

Example 12 ▪ 1   If a distribution is normal with mean μ, then the median is also μ since the normal distribution is symmetric about the mean μ.

This example makes an important point: If a distribution is symmetrical about its mean, then the mean and median are equal.

The Coin-Tossing Model — A Review of the Binomial

In Chapter 4 we studied the model of coin tossing, which consists of the following assumptions:

1. There are a fixed number n of trials.
2. The trials are mutually independent.
3. There are only two possible outcomes, say, "+" (plus) and "−" (minus) on each trial.
4. The probability of + is the same on each trial.

Under these assumptions, the probability of x plus signs in n trials is

$$\binom{n}{x} p^x (1-p)^{n-x}, \qquad x = 0, 1, \ldots, n \qquad (12.1)$$

where p is the probability of $+$ on any given trial and $(1-p)$ is the probability of $-$ on any given trial. Remember that the distribution (12.1) is called the binomial distribution. We shall see that it is very pertinent to tests about the median of a population.

12 ▪ 2 THE SIGN TEST

The preceding assumptions 1 to 4 also underly a group of nonparametric tests called the "sign test." We now study some of these tests.

One-Sample Sign Test

Suppose we wish to test the null hypothesis

$$H_0: \text{median of the population} \le m_0$$
$$H_1: \text{median of the population} > m_0$$

The assumption underlying this test is that the population distribution is continuous.

We take n independent observations and denote by a "$+$" any observation that falls above m_0 and by minus any observation that falls below m_0. If H_0 is true, then the chance of obtaining a plus on any trial is greater than or equal to $\frac{1}{2}$. To illustrate how this fact can be used to test H_0, consider the following example.

Example 12 ▪ 2 An experimental car is designed to travel 54 mi per gallon of gas. In 10 runs it travels the following distances on 1 gal:

54.6, 53.2, 58.0, 53.1, 54.2, 54.3, 53.9, 53.8, 52.0, 49.0

Since it is an experimental car, the null hypothesis is that in a majority of such runs the car will not get better than 54 mi per gallon; that is,

$$H_0: m \le 54$$
$$H_1: m > 54$$

In the following test, we assume that the distribution of miles per gallon is continuous.

Let $+$ denote an observed value that is greater than 54 and $-$ denote a value that is less than 54. In terms of the signs, the data obtained are

$$+ - + - + + - - - -$$

We reject H_0 if the number of $+$ signs is too large since this would indicate to the researcher that 54 was not the median and the median must be larger. But what is "too large"? To answer this question, let $\alpha = 0.011$. For 10 observations the chance of 9 or 10 $+$ signs for $p = \frac{1}{2}$ and $n = 10$ is 0.011 (see Appendix A.8). Hence to reject H_0 at $\alpha = 0.011$, we would need to observe 9 or 10 $+$ signs. Since we observe four $+$ signs, we cannot reject H_0 at $\alpha = 0.011$.* There is insufficient evidence, given the information at hand, to reject the hypothesis $m \leq 54$ as opposed to $m > 54$.

Notice the critical role played by the assumptions we made about the available information. If we *assumed additionally* that the number of miles per gallon was normally distributed, then due to the symmetry of the normal distribution, the mean and median are the same, and so we could have used a t-test about the mean, say, $H_0: \mu \leq 54$ versus $H_1: \mu > 54$. In general, if this extra information is available, it is preferable to use the t-test since this test uses the exact numerical values of the observations made in calculating the test statistic, whereas the sign test only uses the information on whether each observation is bigger or less than the hypothesized value $m_0 = 54$. On the other hand, if the distribution is very far from a normal distribution and has fat tails, then the assumptions for the t-test are not met and the sign test may be preferable. The sign test is also simpler to calculate and use, and hence is often used in applications like signal detection in sonar and radar where speed of calculation is important and precise measurements are not necessary.

The sign test can be used for both one-tailed and two-tailed tests. We now summarize the one sample sign test.

Sign Test
Assumptions: 1. Population has a continuous distribution.
**   2. The *n* observations are mutually independent.**
Procedure:
   For each observation, record a + if the observation is greater than m_0 and a − sign if the observation is less than m_0, where m_0 is specified by the null hypothesis. Exclude observations that are equal to m_0. The sample size is reduced by the number of these excluded observations.

---

* By stating H_0 as $H_0: m \leq 54$, the probability of the type 1 error we obtain is less than or equal to the chosen significance level α. In this sense, our type 1 error is conservative. To understand why this is so, note that if $p = 0.5$ and $m = 54$, then the probability of the type 1 error is exactly α. If H_0 is true but $m < 54$, and $p > 0.5$ then the probability of the type 1 error is less than α if the distribution of miles per gallon is continuous.

Let k_α denote the largest integer such that the sum of the binomial probabilities with n trials and $p = \frac{1}{2}$ is less than or equal to α. (Example: For $n = 10$, $\alpha = 0.011$, from Appendix A.8, $k_{0.011} = 1$.)

To test (at significant level α)

$$H_0: m = m_0 \quad \text{versus} \quad H_1: m \neq m_0$$

Reject H_0 if the observed number of $+$ signs is greater than or equal to $n - k_{\alpha/2}$ or less than or equal to $k_{\alpha/2}$.

To test (at significant level α)

$$H_0: m \geq m_0 \quad \text{versus} \quad H_1: m < m_0$$

Reject H_0 if the observed number of $+$ signs is less than or equal to k_α.

To test (at significant level α)

$$H_0: m \leq m_0 \quad \text{versus} \quad H_1: m > m_0$$

Reject H_0 if the observed number of $+$ signs is greater than or equal to $n - k_\alpha$.

In many applications we must use the sign test because the data are qualitative rather than quantitative.

Example 12 ▪ 3   Every month an *equal* number of men and women apply for jobs at Dark Energy Corporation. A $+$ sign indicates a woman is hired, a $-$ sign indicates a man is hired. (Note that in this case the data are qualitative.) In the first 3 months of the year there were 15 hirees in the following order

$$- - - + - - - - - - - + - - - -$$

Do these data indicate a potential bias against women? We test ($\alpha = 0.005$)

$$H_0: p \geq \frac{1}{2} \qquad \begin{array}{l}(p \text{ is the probability of a woman} \\ \text{being hired})\end{array}$$

$$H_1: p < \frac{1}{2}$$

Before presenting the solution to this problem, note that we are really just doing a test of proportions by using a binomial model, the same test discussed in Section 7.5. We are *not* testing for the median of some continuous population. Hence the previous assumption that the population is continuous does not apply here.

From Appendix A.8 we see that the chance of two or fewer plus signs ($p = 0.5$, $n = 15$) is 0.003. We cannot obtain a critical region with $\alpha = 0.005$ since there is no entry 0.005 in the table. We use the region with α closest *but smaller* than the *desired* α. Thus we use as our rejection region "2 or fewer + signs." We observe two + signs. Hence we reject H_0 at $\alpha = 0.005$. We would conclude that there is prima facie evidence of discrimination and should investigate further why this statistical discrepancy in the company occurred.

The preceding examples demonstrated one-tailed tests. The following example illustrates a *two-tailed* test.

Example 12 ▪ 4   A medical experiment is being conducted, and patients are randomly allocated to the treatment or control group. A + is noted if the patient is placed in treatment and a − is noted if the patient goes into the control group. The following sequence is obtained for 10 patients who are supposedly placed at random:

$$- - - + - - - - - -$$

Is the placement biased? If not, then the chance would be $p = 0.5$ of a patient being placed into either group.

Let $\alpha = 0.05$. We test

$$H_0: p = 0.5$$

$$H_1: p \neq 0.5$$

Using Appendix A.8 ($p = 0.5$, $n = 10$), we find the chance of 1 or 0 or 9 or 10 occurrences of a + to be $0.001 + 0.01 + 0.01 + 0.001 = 0.022$, whereas the chance of 2 or fewer or 8 or more + signs is $0.001 + 0.01 + 0.044 + 0.044 + 0.01 + 0.001 = 0.11$. Hence we take as the rejection region "one of the numbers 0, 1, 9, or 10 of + signs" since the resulting probability of type 1 error (0.022) is less than $\alpha = 0.05$. Since we observe only one + sign in the 10 experimental assignments, we reject H_0 at $\alpha = 0.05$. We conclude that there is a bias in placement that should be explained.

The Two-Sample Sign Test

In many investigations often comparisons are made of the results of two different experiments on the same material to determine if the results are the same.

Table 12 ▪ 1 Typing Errors of Two Typists						
	Number of errors made on page					
Typist	*1*	*2*	*3*	*4*	*5*	*6*
A	8	4	4	9	3	6
B	6	6	7	8	6	2
Sign of $A - B$	+	−	−	+	−	+

For example, suppose I am undecided as to which of two typists, A and B, to hire. I give each applicant the same six pages to type, each page consisting of a different kind of material requiring different typing skills; for example, one page is a list of numbers, another a set of tables, and so forth. I count the number of typing errors each has made on each page and display these results in Table 12.1.

I wish to test the hypothesis that the two typists are equally good. I look at the difference in the errors between A and B on each page; $d_i =$ number of errors on page i made by A minus the number of errors made by B on the same page. If d_i is positive, I assign a + sign to page i; otherwise I assign a − sign. For example, a + is assigned to pages 1, 4, and 6 in Table 12.1 and a − to 2, 3, and 5. If the typists are equally good, then the probability p of a + sign should equal the probability of a − sign, and p should equal $\frac{1}{2}$. Hence the hypothesis that the typists are equally good is equivalent to

$$H_0: p = 0.5$$

$$H_1: p \neq 0.5$$

The two-sample sign test is used when
1. **We have a sequence of paired random variables**
   $(X_i, Y_i);$ $i = 1, 2, \ldots, n$
2. **The differences $D_i = X_i - Y_i$ are mutually independent.**
   Let d_i represent an observed value of D_i. If $d_i = x_i - y_i$ is positive, we assign a + to the difference, a − if the difference d_i is negative, whereas the result is excluded if $d_i = 0$.
   The null hypothesis is $H_0: p = 0.5$, or $p \leq 0.5$, or $p \geq 0.5$, where p is the probability of a +.

Continuing the example of the typists illustrated in Table 12.1 we are testing

$$H_0: p = 0.5$$

$$H_1: p \neq 0.5$$

Suppose we wish to test H_0 at $\alpha = 0.05$. We observe from Appendix A.8 ($p = 0.5$, $n = 7$) that the probability of either $0 + $ signs or all $+$ signs is $0.08 + 0.08 = 0.016$, whereas the probability of one or fewer $+$ signs *or* six or seven $+$ signs is $0.008 + 0.055 + 0.055 + 0.008 = 0.126$. Hence we cannot obtain a critical region with $\alpha = 0.05$. We choose the critical region "no $+$ signs or all $+$ signs" and so have an $\alpha = 0.016$.

The data of Table 12.1 reveals three $+$ signs so we do not reject H_0; thus we cannot conclude that there is a significant difference between the two typists.

We could have tested H_0: $p \leq 0.5$ with the *one-sided* (one-tailed) alternative

$$H_1: p > 0.5$$

that is, with the alternative that A is a better typist than B. Using $\alpha = 0.008$, we reject H_0 if we find seven $+$ signs.

Using the data of Table 12.1, we do not reject H_0.

Example 12 ▪ 5 Two judges rate each of 12 contestants in a beauty contest on a scale of 0 to 10. A judge can give more than one contestant a particular rating. The results are shown in Table 12.2.

We wish to determine whether the two judges have significantly different "tastes" in beauty. We test

$$H_0: p = 0.5$$
$$H_1: p \neq 0.5$$

where p is the probability of a $+$ sign.

From Appendix A.8 the probability of 2 or fewer or 10 or more $+$ signs is $0.003 + 0.016 + 0.016 + 0.003 = 0.038$. Hence if $\alpha = 0.038$, we reject H_0. We conclude (at $\alpha = 0.038$) that there is a significant difference between the judges.

Comparison of the Sign Test and *t*-Test

In either the one-sample or two-sample case, if the data are quantitative and satisfy the other assumptions underlying the *t*-test, then one might prefer the

Table 12 ▪ 2 Rankings of Contestants by Two Judges

	1	2	3	4	5	6	7	8	9	10	11	12
Judge I	5	6	10	7	0	9	7	10	9	6	9	9
Judge II	4	1	7	5	8	5	5	6	8	10	5	4
Sign of difference I − II	+	+	+	+	−	+	+	+	+	−	+	+

t-test because it uses the numerical values of the observations, whereas the sign test loses such information. Generally, the more information a test uses, the smaller will be the probability of a type 2 error. However, if computational simplicity is of prime importance, then the sign test might be preferred. In this age of high-speed computers and sophisticated hand calculators, when there is a relatively large amount of data to reduce, time and costs may still be important enough to use a simple counting method in place of multiple calculations, even if each calculation is simple. The authors often find this to be so, particularly while working in the field and trying to obtain preliminary results before sending the data to the collection center. In our experiences in Third World countries, which do not have easily accessible computer centers, we have discovered that nonparametric methods are very useful. Simple methods are invaluable while trekking through deserts or living in remote mountain villages.

What Can Go Wrong?

One assumption underlying both the one- and two-sample tests is that the sequences of outcomes of $+$ and $-$ signs are mutually independent; that is, the sign outcome on any trial is independent of the outcomes on any other trial. Notice that this assumption is violated in example 12.6 that follows.

Example 12 ■ 6   Two quality control inspectors rank each of 10 products. No ranking can be given to more than 1 product. The rankings of the inspectors of 10 products are as follows:

	Product									
	A	*B*	*C*	*D*	*E*	*F*	*G*	*H*	*I*	*J*
Inspector I	1	4	6	2	3	9	8	7	5	10
Inspector II	4	3	7	5	2	10	9	8	6	1
Sign of difference I − II	−	+	−	−	+	−	−	−	−	+

We should like to test the hypothesis, the inspectors have the "same judgment" in quality. However, we cannot use the sign test because the signs are not independent of each other. To understand this, note that the last sign in the example *must be* $+$; since Inspector I has used up all his scores except 10, and Inspector II has used up all his scores except 1. Thus, given the previous scores, the final score of each inspector is fixed; that is, it is not independent of the previous scores. Here the final sign ($+$ in this case) is not independent of the previous signs.

The sign test cannot be used if the outcomes of any trials depend on the outcomes of other trials.

The lesson to be learned here is that one must be sure that the assumptions underlying any test are satisfied before using it. If the assumptions are not satisfied, one must ascertain if the departures from the assumptions affect, for practical purposes, the applicability of the test.

Large Samples

The coin-tossing or binomial model is applicable to both the sign tests previously described and to those that follow; that is, the number of + signs has a binomial distribution. We know that for the sample size n large the binomial can be approximated by the normal. Let $\hat{p}$ equal the proportion of + signs. From Section 7.5 we learned that for n large we can test

$$H_0: p = 0.5$$

versus a one-sided or two-sided alternative by using the value

$$z = \frac{(\hat{p} - 0.5) \pm 1/2n}{\sqrt{0.25/n}} \tag{12.2}$$

The test is now performed exactly as in Chapter 7.

Note that this z test for large sample sizes is applicable both to the one-sample and two-sample sign tests.

Example 12 ▪ 7   The daily production from two machines A and B is given in Table 12.3. We wish to test the hypothesis that machine A produces on the average the same number of items as B versus the alternative that it produces fewer items than

Table 12 ▪ 3 Daily Production from Two Machines

	Day														
	1	*2*	*3*	*4*	*5*	*6*	*7*	*8*	*9*	*10*	*11*	*12*	*13*	*14*	*15*
Machine A	6	5	1	6	3	8	7	1	3	2	8	6	3	3	4
Machine B	7	7	8	4	5	3	6	4	4	6	2	7	5	1	8
Sign $A - B$	−	−	−	+	−	+	+	−	−	−	+	−	−	+	−

	16	*17*	*18*	*19*	*20*	*21*	*22*	*23*	*24*	*25*	*26*	*27*	*28*	*29*	*30*
Machine A	6	6	1	7	5	4	1	8	6	1	4	5	3	6	0
Machine B	1	5	3	8	2	6	7	2	7	8	6	7	8	2	2
Sign $A - B$	+	+	−	−	+	−	−	+	−	−	−	−	−	+	−

machine B. Assuming that all the assumptions underlying the sign test are true, we let p denote the probability of a $+$ and test

$$H_0: p \geq 0.5$$

$$H_1: p < 0.5$$

Because $n = 30$ we may use the z of expression 12.2. Let us choose $\alpha = 0.05$. We note from Table 12.3 that $\hat{p} = \frac{10}{30}$ and

$$z = \frac{0.33 - 0.5 + 1/60}{\sqrt{0.25/30}} = \frac{-0.15}{0.09}$$

$$= -1.67$$

In this case $-z_{0.05} = -1.65$. Hence the z-value falls in the rejection region and we reject H_0. We conclude that the machines are significantly different.

12 ■ 3   RUNS TEST FOR RANDOMNESS

To use the sign test, we determine the proportion of $+$ signs. If this proportion is near 0.5, we accept the null hypothesis, H_0: proportion of $+$ signs $= 0.5$. However, it may happen that the proportion of plus signs may be near 0.5 but that the sequence of $+$ and $-$ signs may not appear randomly. For example, consider the following experience of an amateur statistician residing in the French town of Gex located near the Swiss border. In this town two rival wine tasters reside and are in constant animosity to each other. Each claims the other knows nothing of wines. The amateur statistician had each man taste 10 wines, 5 wines from Burgundy and 5 from the Rhine region. The wine tasters rated each wine on a scale from 0 to 7. The results are shown in Table 12.4.

There are an equal number of $+$ and $-$ signs. Using the sign test, the amateur statistician declared that the tasters have the same taste. But this result flies in

Table 12 ■ 4 Results of the Wine Tasting

	Wines from Burgundy					Wines from the Rhine				
	1	2	3	4	5	6	7	8	9	10
Taster A	1.1	2	1.1	0	1	4.3	7	5	6.1	5
Taster B	6.8	6.8	6	7	7	0.6	0	0.11	0.4	0.1
Sign of difference $A - B$	$-$	$-$	$-$	$-$	$-$	$+$	$+$	$+$	$+$	$+$

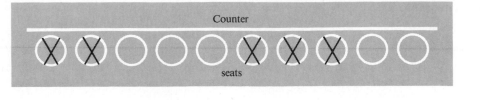

Figure 12 ▪ 1
Picture of seating at a
lunch counter (an ✕
indicates an occupied seat).

the face of the data! Clearly, taster A prefers Rhine wines; taster B prefers Burgundy wines. The amateur made a mistake because he failed to note that, in this determination of randomness, the proportion of + signs as well as the *order* of the + and − signs are important.

As another example, suppose we walk into a café that has a lunch counter. Figure 12.1 shows the counter and the seats indicated by circles. A cross on a seat indicates it is occupied. Are the people distributed randomly among the seats? Let + indicate an occupied seat and − an empty one. The sequence we observe is

$$+ + - - - + + + - -$$

Are the + signs scattered in a *random* manner among the − signs ? As a step in answering this question, we introduce the definition of a run.

> Run: Any consecutive set of signs of the same type in a row is called a run.

Example 12 ▪ 8   In the sequence taken from the preceding lunch counter example

$$+ + - - - + + + - -$$

there are four runs, the first is + +, the second is − − −, the third is + + +, and the fourth is − −.

The total number of runs is an indicator of randomness. For example, the sequence of signs + + + + + − − − − − indicates less randomness than the sequence + − + + − − + + − −. The former sequence has two runs, whereas the latter has six runs. The fact that there are very few runs indicates a lack of randomness; however, too many runs may indicate the same thing. For example, the sequence

$$+ - + - + - + - + -$$

has the maximum number of runs for five + signs and five − signs. If we observe such a distribution of five people among the 10 seats at a counter (a + indicates an occupied seat), we would conclude that the selection of seats by the five people among the 10 seats was not random but was dictated by the desire of the five to sit as far apart as possible from one another (perhaps because they are strangers to each other.)

Model for the Runs Test

The assumption underlying the runs test is that represented in the coin-tossing model, plus the assumption that a + and − sign are equally likely on each trial.

Runs Test: Small Samples

To test

$$H_0: \text{a sequence of} + \text{and} - \text{signs is random}$$

$$H_1: \text{the sequence is not random}$$

first the significance level α is chosen. Let n_1 be the number of + signs and n_2 the number of − signs. Let u equal the total number of runs among the $n_1 + n_2$ observations. Appendix A.10 gives the critical region for u for various values of α and n_1 and n_2.

Example 12 ■ 9 A supervisor wishes to know if there is a decreasing trend in production on one of the assembly lines. To answer this question, she directs that each day for 20 days the total number of items off a production line be counted. The median number for the distribution of the total production is known. If the number of items produced in a day is greater than the median, a + is assigned to the production for this day; if it is less, a − is assigned. If there is a decreasing trend in production, we should see mostly + signs at first and then − signs at the end of the sequences; hence there should be too few runs. The following sequence was obtained for a 1-month period of 20 working days:

$$+ - - + - - + + + - - - + + - + - - + +.$$

We test H_0: the sequence is random at $\alpha = 0.05$. There are 11 runs with $n_1 = n_2 = 10$; 11 runs correspond to the 58.6 percentile of u as shown in Appendix A.10. We do not reject H_0; and we cannot conclude that there is a nonrandom effect.

Runs Test: Large Samples

If n_1 and n_2 are both larger than 10, the observed value u comes from a distribution that is approximately normal with mean μ_U and σ_U^2, and where

$$\mu_U = \frac{2n_1n_2}{n_1 + n_2} + 1$$

$$\sigma_U^2 = \frac{2n_1n_2(2n_1n_2 - n_1 - n_2)}{(n_1 + n_2)^2(n_1 + n_2 - 1)}$$

To test H_0, we calculate

$$z = \frac{u - \mu_U \pm \frac{1}{2}}{\sigma_U}$$

where the continuity correction $\frac{1}{2}$ is subtracted when u is greater than μ_U and is added otherwise. We reject H_0 if $z < -z_{\alpha/2}$ or $z > z_{\alpha/2}$.

Example 12 ▪ 10   Thirty performers attend a screening audition for a part in a play. The purpose of the audition is to separate potentially gifted performers from those who are not gifted. It is of interest to determine if the order of appearance influences the decision. (Hence the standard phrase "that was a hard act to follow.") The following sequence was obtained (A means accepted, R indicates rejected):

AAAAAAAARRRAAAAAAAAARR

AARRRRRA

We wish to test for randomness with $\alpha = 0.01$.

The values μ_U and σ_U^2 are calculated by using $n_1 = 20$, $n_2 = 10$, and $u = 7$, yielding

$$\mu_U = 14.33$$

$$\sigma_U = 2.38$$

Hence

$$z = \frac{7 - 14.33 + 0.5}{2.38} = -2.87$$

This value is less than $-z_{0.005} = -2.58$. We reject H_0 and conclude that the order of appearance strongly influences the artistic decisions.

Other Uses of the Runs Test

We have shown how to test whether a sequence is random. The sequence could consist of *quantitative* information (such as the number of items produced off a production line) or *qualitative* information (such as accepted and rejected in example 12.10); the nature of the data does not matter as long as the assumptions underlying the runs test are valid.

The runs test has other uses. It can be used to test the hypothesis that two random samples come from populations having the same distribution. We arrange the observations of the two samples together according to increasing size. We then count the number of runs where a run is a sequence of consecutive observations all of which come from the same population. If the two populations really are the same, we expect a good interlacing of the two sets of observations. If there are too few runs under the null hypothesis H_0 that the two populations are the same, then we reject H_0.

Example 12 ▪ 11   Ten items off a production line of ball bearings are made by machine A and 10 are made by machine B. The diameters of the bearings from each of the machines are:

Diameters of the Bearings (mm)									
Machine A 26.3	28.6	25.4	29.2	27.6	25.6	26.4	27.7	28.2	29.0
Machine B 28.5	30.0	28.8	25.3	28.4	26.5	27.2	29.3	26.2	27.5

We wish to test

H_0: Bearings from A and B that come from the same distribution

The bearings are ordered by the size of the diameters, and under each we record the machine that produced that part.

25.3	25.4	25.6	26.2	26.3	26.4	26.5	27.2
B	A	A	B	A	A	B	B

27.5	27.6	27.7	28.2	28.4	28.5	28.6
B	A	A	A	B	B	A

28.8	29.0	29.2	29.3	30.0
B	A	A	B	B

There are 11 runs of the 10 symbols A and 10 symbols B. From Appendix

A.10 we see that this number of runs is not smaller than the critical value at $\alpha = 0.01$, so we fail to reject H_0. The distributions could be the same.

12 ■ 4   RANK TESTS

Information about the *order* of the elements of a sequence is not exploited in any way by the sign test. The runs test uses some of this information: it uses the *number* of runs, which incorporates only some of the *order* information. For example, the sequences $+ + + - - - +$ and $+ - - - - + + +$ both have three runs but the majority of the $+$ signs is to the left in the first situation, and to the right in the second situation. Descriptively, the sequence of signs is quite different, yet they have the same number of runs. As we see, the runs test uses some of the order information but also loses some of it.

We would like to devise a test that uses more information. A rough rule in statistics is that the more pertinent information from the sample that a statistical test uses, the better the test is in the sense of having a smaller probability of committing a type 2 error.

One way to incorporate more order information from a sample is to rank the observations in the sample. Let us take the data of example 12.11. Recall that we are testing the null hypotheses that the populations A and B have the same distribution. Pooling the data and ordering the values from the lowest to the highest, we find that the smallest of all observations comes from B, the next from A, and so on. The sequence

BAABAABBBAAABB

ABAABB

shows the order of the values from A and B with respect to relative magnitude. We assign the number 1 to the first member in the sequence, the number 2 to the second member of the sequence, and so forth. These values are called *ranks*. The utilization of the ranks permits us to incorporate more of the information about order into our tests than is possible by using either the sign or run tests. Tests that employ ranks are called *rank tests*. Rank tests that employ the *sum* of ranks are called *rank sum tests*. Some rank sum tests follow.

Rank Sum Test: Mann–Whitney Test

Continuing the preceding discussion, we wish to test H_0 that populations A and B have the same distribution. First, we arrange the entire set of numbers from both populations together in increasing order. If H_0 is true, and if again we have independent random sampling, then the average of the ranks assigned to the two samples should be approximately the same. For example, the ranks assigned to population B are $1, 4, 7, 8, 9, 13, 14, 16, 19, 20$, whereas the remaining ranks are assigned to A. The average of those assigned to B is

$$\frac{1 + 4 + 7 + 8 + 9 + 13 + 14 + 16 + 19 + 20}{10} = 11.1$$

whereas the average of the ranks assigned to A is 9.9. Is 9.9 sufficiently close to 11.1 to justify accepting H_0? This is the question the following test answers. This test is called the Mann–Whitney test.

An assumption underlying the Mann–Whitney test is that the two populations have continuous distributions. If this is not the case, the test can usually be used without serious consequences.

Ties may result even if the distributions are continuous due to round-off in recording the results. In the case of ties a rank is assigned to each of the tied observations that is equal to the average of the ranks for which the ties occur. For example, if the fifth and sixth values were tied, we would assign the rank $(5 + 6)/2 = 5.5$ to each. If the fifth, sixth, and seventh are all tied, we assign the rank 6 to each. This value is called the *midrank* value for ties. All ties are assigned the midrank value.

The sum of the ranks assigned to *one* of the populations, say, the first, is denoted R and is a random variable. Let n_1 and n_2 represent the number of observations for the first and second populations, respectively. For example, the number of observations from populations A and B are both 10; $n_1 = n_2 = 10$. The random variable used in the Mann–Whitney test is

Mann–Whitney rank test statistic:

$$U = n_1 \cdot n_2 + \frac{n_1(n_1 + 1)}{2} - R$$

where n_1 and n_2 are the sample sizes from population 1 and 2, respectively, and R is the sum of the ranks corresponding to the first population.

If H_0 is true, then it can be shown that the mean and variance for U are $E(U) = n_1 \cdot n_2/2$ and $V(U) = n_1 \cdot n_2(n_1 + n_2 + 1)/12$. If n_1 and n_2 are both greater than or equal to 10, then

$$Z = \frac{U - n_1 n_2/2}{\sqrt{n_1 n_2(n_1 + n_2 + 1)/12}}$$

has an approximately normal distribution with $\mu = 0$ and $\sigma = 1$. Let u represent the observed value of U. At a level of significance α, if the observed value

$$z = \frac{u - n_1 n_2/2}{\sqrt{n_1 n_2(n_1 + n_2 + 1)/12}}$$

is less than $-z_{\alpha/2}$ or greater than $z_{\alpha/2}$ (α is the significance level), then we reject the hypothesis H_0 that the two populations have the same distribution.

Example 12 ▪ 12 Let us continue our discussion of the application of the Mann–Whitney test to H_0 that populations A and B have the same distribution. We find the observed value of R (the sum of the ranks of A) to equal 99. Hence

$$u = 10 \cdot 10 + \frac{10 \cdot 11}{2} - 99 = 56$$

and

$$z = \frac{56 - 50}{\sqrt{175}} = 0.45$$

Let $\alpha = 0.01$ so that $z_{0.005} = 2.58$. Since $0.45 < 2.58$, we do not reject H_0. We cannot conclude that the two populations differ significantly.

Wilcoxon Signed Rank Test

This test is applicable to the situation where we wish to test that observations come from a population that has a *symmetrical* distribution with mean μ_0. The null hypothesis is

$$H_0: \mu = \mu_0$$

To perform this test, we first subtract μ_0 from each observation x_i. The resulting numbers $x_i - \mu_0$ are ordered according to their absolute values (the sign of each number is ignored). The ranks are then divided into two groups: those ranks associated with the values $x_i - \mu_0$ that are negative and those associated with value $x_i - \mu_0$ that are positive. The test statistic is

$$W = \text{sum of ranks associated with positive values of } x_i - \mu_0$$

The statistic W is used to test H_0, and we reject H_0 if its observed value w is greater than $w_{\alpha/2}$ or less than $w_{1-\alpha/2}$, where $w_{\alpha/2}$ and $w_{1-\alpha/2}$ are found in Appendix A.11.

The logic behind this test is that if the population is symmetrically distributed about μ_0, then any observed absolute value is equally likely to be from a positive or negative value of $x_i - \mu_0$. If the distribution is skewed, then w will be too large or too small.

Example 12 ▪ 13 A production process is supposed to produce items at the rate of one part per second. We wish to see if this is indeed the case. We know from the nature of the production process that the variation in the rate is small and the distribution of the rate, if H_0 is true, is symmetrical about 1. Since the rate is symmetrical about 1, we can use the signed rank test. We test

$$H_0: \mu = 1$$

$$H_1: \mu \neq 1$$

The observed values x_i are 2.1, 2.7, 1.8, 0.5, 1.2, 0.3, 0.81, 0.2, and 1.6. The numerical values of $(x_i - 1)$ are 1.1, 1.7, 0.8, -0.5, 0.2, -0.7, -0.19, -0.8, and 0.6. If we arrange these in increasing order by absolute value, and put the sign of the difference $(x_i - \mu_0)$ below each number, we have

$$0.19, 0.2, 0.5, 0.6, 0.7, 0.8, 0.8, 1.1, 1.7$$

$$-, \quad +, \quad -, \quad +, \quad -, \quad -, \quad +, \quad +, \quad +$$

The corresponding ranks are 1, 2, 3, 4, 5, 6.5, 6.5, 8, and 9, where we have used the midrank $(6 + 7)/2 = 6.5$ for the two values of 0.8 that were tied for sixth and seventh places. The sum of the values corresponding to the positive values is

$$w = 2 + 4 + 6.5 + 8 + 9 = 29.5$$

Let $\alpha = 0.05$. We use Appendix A.11 to test H_0. Since we are using a two-tailed test we use $w_{0.025}$ and $w_{1-0.025}$. For $n = 9$, Appendix A.11 yields $w_{0.027}$ and $w_{1-0.027}$ which we take as close enough to the values $w_{0.025}$ and $w_{1-0.025}$ that we want; $w_{0.027} = 6$ and $w_{1-0.027} = 39$. Since the observed value of 29.5 lies between these two values, we do not reject H_0. The process could indeed be producing items at one part per second.

The Spearman Rank Correlation

In Chapter 9 we could test for an expected *linear* relationship between two variables provided the variables under investigation came from a jointly normal population. The technique we used was to test the hypothesis $H_0 : \rho = 0$, where ρ was the population correlation coefficient. We used the sample correlation coefficient r in order to test for this linear relationship. If we rejected $H_0: \rho = 0$, we concluded that there was a significant linear relationship.

In the previous sections of this chapter we have seen that by transforming the original data to their ranks, we are able to eliminate the normal population assumption and still make inferences concerning the population or populations under investigation. We did this for comparing two populations (the Mann–Whitney U statistic) and also for investigating a single population distribution for departures from a hypothesized center μ_0 (the signed rank test). In this section, we examine testing the association between two variables in a manner analogous to the correlation coefficient of Chapter 9, but without the assumption of normality.

As always, when we go from the original data to their ranks, we lose some information. For example, in correlational analysis we lose the ability to detect a linear relationship between the underlying variables. If we replace all the variables by their ranks, then we can only detect a linear relationship between the *ranks* of the variables by calculating the correlation coefficient on the ranks. A linear relationship on the ranks is equivalent to a monotone relationship on the underlying original variables. This monotone relationship may be an increasing relationship (larger values of X tend to correspond to larger values of Y) or a monotone decreasing relationship (larger values of X tend to correspond to smaller values of Y).

The correlation coefficient computed on the ranks of the variables is called the Spearman rank correlation coefficient and is denoted by r_s. For sample sizes greater than 10 we can use the normal approximation described below. We summarize the technique as follows:

To test H_0: no monotone relationship exists between variables X and Y.
1. **Rank the observed X values among themselves, and rank the observed Y values among themselves.**
2. **Replace the observed numerical values (x_i, y_i) by their ranks calculated in (1), that is, (x_i, y_i) becomes $[R(x_i), R(y_i)]$, where $R(x_i)$ is the rank of x_i among the x's, and $R(y_i)$ is the rank of y_i among the y's.**
3. **Calculate the correlation coefficient r_s between the ranks $[R(x_i), R(y_i)]$ in the same manner used in Chapter 9. A computationally simpler formula for r_s is given below.**
4. **If $n \geq 10$, form the z-statistic $z = r_s\sqrt{n-1}$. This is the observed value of an approximately normal variable with mean zero and variance 1 if H_0 is true.**
5. **a. To test H_0: no monotone relationship versus H_1: some monotone relationship, at level α, reject H_0 if $z > z_{\alpha/2}$ or if $z < -z_{\alpha/2}$.**
   b. To test H_0: no monotone relationship versus H_1: a monotone increasing relationship, at level α, reject H_0 if $z > z_\alpha$.
   c. To test H_0: no monotone relationship versus H_1: a monotone decreasing relationship, at level α, reject H_0 if $z < -z_\alpha$.

Example 12 ▪ 14   We wish to determine if there is a significant monotone relationship between the scores on two language exams so we test H_0: no monotone relationship versus H_1: some monotone relationship exists. The data are (ranking from high to low):

Student	Grade French exam	Rank	Grade German exam	Rank
1	91	1	82	4
2	87	4	95	1
3	70	9	83	3
4	65	10	90	2
5	40	12	21	12
6	75	7	80	6
7	83	5	70	8
8	60	11	81	5
9	88	3	45	11
10	72	8	78	7
11	90	2	60	10
12	79	6	65	9

Calculating the ordinary correlation on the ranks $\{(1, 4), (4, 1), (9, 3), (10, 2), (12, 12), (7, 6), (5, 8), (11, 5), (3, 11), (8, 7), (2, 10),$ and $(6, 9)\}$, we obtain

$$r_s = -0.056$$

thus

$$z = -0.056\sqrt{11} = -0.186$$

Using $\alpha = 0.05$, we find that -0.186 lies between $z_{\alpha/2} = 1.96$ and $-z_{\alpha/2} = -1.96$. Hence we do not reject H_0. There is insufficient evidence to conclude there is a significant monotone relationship between the tests at $\alpha = 0.05$.

Our calculation of the sample correlation r_s can be simplified. It can be shown algebraically that the correlation on the ranks is always equivalent to the formula

$$r_s = 1 - \frac{6 \sum_{i=1}^{n} d_i^2}{n(n^2 - 1)}$$

where n is the number of pairs and d_i is the difference of the rank of x_i and the rank of y_i for the ith pair; that is, $d_i = R(x_i) - R(y_i)$.

When there are fewer than 10 observations of the populations, the normal approximation is not very good and we must use an alternative test statistic and rejection region. In this case, our tests for a monotone relationship now can be performed by using the value r_s and comparing this value with the critical values given in Appendix A.12. Summarizing, we have

> **For Small Sample Sizes**
> 1. To test H_0: no monotone relationship versus H_1: a monotone relationship, at level α, reject H_0 if $r_s > r(n, \alpha/2)$ or if $r_s < -r(n, \alpha/2)$, where $r(n, \alpha/2)$ is the upper $\alpha/2$ centile for r_s with sample size n given in Appendix A.12.
> 2. To test H_0: no monotone relationship versus H_1: monotone increasing relationship at level α, reject H_0 if $r_s > r(n, \alpha)$, the upper α centile from Appendix A.12.
> 3. To test H_0: no monotone relationship versus H_1: a monotone decreasing relationship, at level α, reject H_0 if $r_s < -r(n, \alpha)$, the negative of the upper α centile in Appendix A.12.

Example 12 ▪ 15 There are eight contestants in a contest. Two judges grade each contestant on a scale of 0 to 10. Note that a *quality* may be measured, as in a beauty contest or in some Olympic games. The grades may represent a qualitative, as opposed to a quantitative, statement about the relative standings of the contestants. The grading by each judge is shown in Table 12.5.

Using the computational formula,

$$r_s = 1 - \frac{6(17)}{8(63)} = 0.8$$

If we test H_0: no monotone relationship versus H_1 : a monotone relationship exists at $\alpha = 0.05$, we find that for the upper 0.025 centile, $r(8, 0.025) = 0.79$, and for the upper 0.975 centile, $r(8, 0.975) = -0.73$. Thus we would reject H_0 if $r_s > 0.79$ or if $r_s < -0.73$. Since we calculated $r_s = 0.8$, we reject H_0 and conclude that a significant monotonic relationship exists between the two judges. (It would be a poor contest indeed if this were not true.)

In this example and in example 12.5 we investigated whether the judges had "similar tastes." The preceding test is more refined than the sign test used in example 12.5. Using this test, we conclude that when one judge rates a contestant high (low), then the other judge rates the contestant high (low). The ranks

Table 12 ▪ 5 Grading by Two Judges in a Contest

	Contestant							
	A	B	C	D	E	F	G	H
Judge I (X)	5	8	3	8	5	7	2	8
Judge II (Y)	4	6	2	7	3	5	1	7
Difference of individual ranks (d_i)	+1	+2	+1	+1	+2	+2	+1	+1
d_i^2	1	4	1	1	4	4	1	1

$$d_1^2 + d_2^2 + \cdots + d_8^2 = \sum_{i=1}^{8} d_i^2 = 17$$

follow each other. In ordinary language, we conclude that the two judges have similar tastes. However, if we use the sign test on the previous data (Table 12.5), we find there are eight $+$ signs and no $-$ signs, and so conclude (at $\alpha < 0.05$) that there is a significant difference in the tastes of the judges. Why do the tests differ so in their conclusions? In the sign test we are assuming that if the judges have similar tastes, then their rankings will be about the same and the sign of the differences will vary randomly. The sign test does not take into account the possibility that the ranks of the two judges can follow each other but that one judge can consistently rank lower than the other. The sign test cannot take this possibility into account since it does not use the magnitude of the differences of the ranks. This is a weakness of the sign test that our preceding test does not share.

Rank Sum Test for Several Populations — The Kruskal – Wallis Nonparametric ANOVA

In the previous chapter we saw how to compare several normally distributed populations with the same variance to determine if they all had the same distribution. If there were k such populations, then it was sufficient to test $H_0: \mu_1 = \mu_2 = \cdots = \mu_k$. If the hypothesis that the data come from a normal distribution is in doubt, the Kruskal – Wallis test is usually the test used. For this reason it is often called the nonparametric analysis of variance test. We shall only consider the analog of the one-way ANOVA here.

If we have k populations and wish to test

$$H_0: \text{all } k \text{ populations have the same distribution}$$

we proceed as follows:

1. Considering all the data from all k of the populations as a single group, arrange them in increasing order from the smallest to largest (this is just like we did for the Mann – Whitney rank sum test).
2. Consider each sample separately now and sum the ranks found in (1) corresponding to that sample. Let R_i denote the sum of the ranks corresponding to the sample from the ith population.
3. Let n_i denote the number of observations from the ith population, and $n = n_1 + n_2 + \cdots + n_k$ represent the total sample size. We form the test statistic

$$H = \frac{12}{n(n+1)} \left(\frac{R_1^2}{n_1} + \frac{R_2^2}{n_2} + \cdots + \frac{R_k^2}{n_k} \right) - 3(n+1)$$

4. If H_0 is true and the populations really have the same distribution, then the test statistic H has approximately a χ^2-distribution with $k - 1$ degrees of freedom. We reject H_0 if the observed value of H is greater than χ_α^2 with df $= k - 1$.

5. The values of n_i should all be at least 5. Ties may be treated as we did in the Mann–Whitney test by substituting midranks for the tied rankings.

Example 12 ▪ 16   An individual is considering the purchase of life insurance and gets quotes from underwriters from different firms on the price per $1000 for each of three types of insurance. The results are the price quotes for

Ordinary term (OT)	Convertible term (CT)	Universal life (UL)
0.37	0.39	0.41
0.47	0.43	0.45
0.27	0.51	0.60
0.36	0.29	0.38
0.28	0.38	0.51
0.33	0.39	
0.43		

The individual wishes to test

H_0: the distribution of price is the same for the three types of insurance

The following calculation table is formed:

Rank position	Data point	Sample (OT, CT, UL)	Assigned rank	OT sample	CT sample	UL sample
1	0.27	OT	1	1		
2	0.28	OT	2	2		
3	0.29	CT	3		3	
4	0.33	OT	4	4		
5	0.36	OT	5	5		
6	0.37	OT	6	6		
7	0.38	CT	7.5		7.5	
8	0.38	UL	7.5			7.5
9	0.39	CT	9.5		9.5	
10	0.39	CT	9.5		9.5	
11	0.41	UL	11			11
12	0.43	CT	12.5		12.5	
13	0.43	OT	12.5	12.5		
14	0.45	UL	14			14
15	0.47	OT	15	15		
16	0.51	CT	16.5		16.5	
17	0.51	UL	16.5			16.5
18	0.60	UL	18			18

Sum of ranks: $R_1 = 45.5$; $R_2 = 58.5$; $R_3 = 67$
Sample sizes: $n = 18$; $n_1 = 7$; $n_2 = 6$; $n_3 = 5$

The observed value h of the test statistic H is

$$h = \frac{12}{18(19)} \left(\frac{(45.5)^2}{7} + \frac{(58.5)^2}{6} + \frac{(67)^2}{5} \right) - 3(19) = 4.89$$

If we are to test H_0 at $\alpha = 0.05$, then we compare h with $\chi^2_{0.05}$ with df $= 3 - 1 = 2$. Since $\chi^2_{0.05} = 5.99 > 4.89$, we fail to reject H_0 at $\alpha = 0.05$. We cannot reject the hypothesis of equal price distributions.

12 ▪ 5   NONPARAMETRIC CONFIDENCE INTERVALS

Confidence Intervals on the Cumulative Population Distribution

Recall that $F_X(x) = P(X \leq x)$ is the cumulative distribution of the random variable X. It is also called the cumulative population distribution. This distribution is estimated by the *sample cumulative polygon* which is found as follows:

1. Let $x_1, \ldots, x_n$ represent the sample consisting of n values arranged in increasing order.
2. The estimate $\hat{F}_X(x)$ of $F_X(x)$ is equal to n_x/n, where n_x is the number of values in the sequence $x_1, x_2, \ldots, x_n$ that are less than or equal to x; that is, $\hat{F}_X(x)$ is the proportion of the sample that is less than or equal to x.
3. Draw the cumulative polygon of $\hat{F}_X(x)$ by connecting the values $\hat{F}_X(x_i)$ by straight lines.

We would like to estimate $F_X(x)$ since if we know $F_X(x)$, then we can find the probability of any set. If we want to find $P[a < X \leq b]$, this is precisely $F_X(b) - F_X(a)$. Thus we can calculate the probability of anything related to X if we know F_X. If we can find confidence intervals for $F_X(x)$, then we can also estimate bounds on $P[a < X \leq b]$.

Example 12 ▪ 17   The following 10 values of blood pressure were obtained in a sample from a population of men between 45 and 55 years of age:

$$148, 120, 135, 140, 110, 125, 117, 109, 131, 160.$$

We find that $\hat{F}_X(109) = 0.1$, $\hat{F}_X(110) = 0.2$, $\hat{F}_X(117) = 0.3$, $\hat{F}_X(120) = 0.4$, $\hat{F}_X(125) = 0.5$, $\hat{F}_X(131) = 0.6$, $\hat{F}_X(135) = 0.7$, $\hat{F}_X(140) = 0.8$, $\hat{F}_X(148) = 0.9$, and $\hat{F}_X(160) = 1$. All other values of blood pressure below 109 have an estimated cumulative distribution of 0 and all above 160 have an estimated cumulative distribution of 1. All values between 109 and 160 take on one of the values between 0.1 and 1 and are determined by connecting the points (109, 0.1), (110, 0.2), (117, 0.3), (120, 0.4), (125, 0.5), (131, 0.6), (135, 0.7), (140, 0.8),

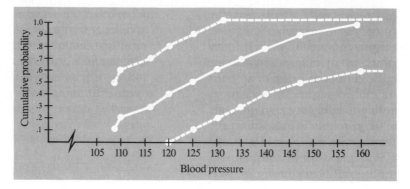

Figure 12 ▪ 2
Sample cumulative poly-
gon and the 95% confi-
dence band for cumulative
distribution function
(dotted lines).

(148, 0.9), and (160, 1) by straight lines. Figure 12.2 shows a graph of the
sample cumulative polygon.

   The question is, how close is the cumulative polygon to the actual cumulative
population distribution? One can use confidence intervals to answer this question. To find the confidence interval with confidence $1 - \alpha$, draw two parallel
lines above and below the cumulative polygon each at a distance $d_{1-\alpha}$ from the
polygon, where $d_{1-\alpha}$ is found from Appendix A.13. The actual cumulative
probability distribution lies with probability $1 - \alpha$ in the band dictated by these
upper and lower curves. Note that the band never goes above one or below zero
since we know $0 \le F_X(x) \le 1$.

Example 12 ▪ 18   Continuing example 12.17, we note that in Appendix A.13 we use $n = 10$. Let
$\alpha = 0.05$. Then $d = 0.41$. The upper and lower curves are shown as broken
lines in Figure 12.2. We are 95 percent confident that the cumulative population distribution is entirely inside this band.

Confidence Interval About the Median

Let X be a random variable with a continuous population distribution, and let
$x_1, x_2, \ldots, x_n$ represent n observations from this population arranged *in
order*, that is, $x_1 < x_2 < \cdots < x_n$. The sample median is the middle value in
the sequence and is an estimator of the population median. If we want a
confidence interval on the population median, it is natural to include more
observations on either side of the sample median. Thus if n is odd, then $x_{(n+1)/2}$
is the sample median. The interval from $x_{(n-1)/2}$ to $x_{(n+3)/2}$ has some chance of
containing the population median, $x_{(n-3)/2}$ to $x_{(n+5)/2}$ has a greater probability
of containing the population median, $x_{(n-5)/2} x_{(n+7)/2}$ has a still greater chance
of containing the population median, and so on. We can calculate each of these
probabilities explicitly, and we stop expanding the interval when the desired
level of confidence (e.g., $100(1 - \alpha)\%$, say, 0.95, or 0.99, and so on) is reached.
Actually, the level $1 - \alpha$ cannot be exactly reached, just as we have seen before

in the case of discrete variables. Accordingly, Appendix A.14 gives the values of k so that in a sample of size n, if you count down k-ordered values from the top and up k values from the bottom, then you have a chance of at least 0.95 of containing the population median. Also shown are the values of k to count up and down to get a chance of at least 0.99 of containing the population median. The confidence interval is from x_k to x_{n+1-k}.

Example 12 ■ 19

A survey of $n = 20$ new homes in a certain section of Austin, Texas, was made and the prices of the houses were recorded. The prices were: $102,300, $115,230, $116,580, $107,530, $107,240, $101,830, $98,630, $105,050, $104,480, $99,470, $108,390, $91,610, $98,650, $81,470, $87,240, $104,810, $98,680, $94,660, $104,030, and $103,700. To find an approximate 95% confidence interval for the median home price, we must first arrange the numbers in increasing order. Looking up $n = 20$ in Appendix A.14, we find $k = 6$ for a confidence level of $1 - 0.041 = 0.959$. Thus we must take the sixth and $n + 1 - 6 =$ fifteenth value when arranged in order. The bottom six values in ascending order are: 81470, 87240, 91610, 94660, 98630, and 98650; and the top six values arranged in descending order are: 116580; 115230; 108390; 107530; 107240; and 105050. Thus the approximate 95% confidence interval for the median home price is $98,650 to $105,050.

12 ■ 6 COMMENTS ON NONPARAMETRIC METHODS: EFFICIENCY

In general, nonparametric methods have two advantages over parametric methods: (1) They have wider applicability than do parametric methods. For example, for the *t*-test to be applicable to testing hypotheses about the mean, the data must be normally distributed, or at least approximately normal. (Occasionally some data manipulation, such as smoothing and or truncating extreme values, can be used to come closer to satisfying this condition.) The normality requirement does not apply, however, to nonparametric tests. (2) In general, nonparametric methods are easier to use.

A disadvantage of nonparametric methods is that they may be wasteful of information if indeed you are sure that the assumptions for a parametric test are valid. A general rule is that the test that employs the most information (and for which all underlying assumptions are satisfied) will yield the lowest probability of type 2 error for a *specified sample size*. Hence it appears that one should always use the test that uses the most information. But do not jump to conclusions. In some cases nonparametric methods of analysis permit an easier and cheaper collection of data. For example, no actual measures are needed if the sign test is to be used; we need only know whether each observation is above or below a given value. An electronic counter can be implemented in many cases to automatically record this type of information. The savings in time and money may permit a much larger sample size for the same fixed budget than would be practical if precise quantitative measurements were carefully made. The larger sample size may, in the end, produce a smaller probability of a type 2 error than that associated with a parametric test that uses a smaller number of samples. We return to this matter when we discuss "practical aspects" of statistics.

Efficiency. If a test employs the same information that another test uses as well as additional information, then this test should have a lower probability β of type 2 error than does the other test. For example, for a fixed sample size the β of the sign test should be higher than that of the t-test (if the assumptions underlying the t-test are valid). Usually one compares the power, which is $1 - \beta$, of each test. The ratio of the power for the sign test to the power of the t-test for a fixed sample size is called the Pitman efficiency. For large sample sizes, the efficiency of the sign test is approximately 64%. The efficiency of the Wilcoxon signed rank test is considerably higher, just as we would expect since it employs more information than does the sign test. (See the Bickel and Doksum reference listed at the end of Chapter 11 for details.)

Summary

The model underlying the sign and runs tests is the binomial model: namely, there are two possible outcomes $(+, -)$; we have independent trials, and the chance p of $+$ on each trial is the same as on every trial. In particular, for the sign tests we discussed statistics for testing H_0: $p = 0.5$ versus various alternatives.

We discussed for large and small samples the one-sample and two-sample (two populations) sign tests. If the original data (before they are transformed into signs) contain relevant magnitudes and/or order information, then the sign test loses this information. The runs test captures order information. The rank sum tests capture order information and some magnitude information as well.

The Spearman rank correlation coefficient r_s was introduced as one measure of association that permits a nonparametric test of a monotonic relationship between two variables.

Nonparametric confidence intervals were introduced through confidence intervals on the population cumulative distribution and on the median.

Key Terms

nonparametric analysis
continuous distribution
median
symmetric distribution
runs test

ranks
nonparametric confidence
    intervals
rank sum test
Mann–Whitney Test

signed rank test
midrank
sign test
one-sample sign test
two-sample sign test

Spearman's rank correlation coefficient
monotone relationship
ties
nonparametric ANOVA

References*

Iman, R. L., and W. J. Conover: The use of the rank transformation in regression. *Technometrics, 21*(4):499–510, 1979. This article shows how the rank techniques we have learned can be used for nonlinear regression.

Lehmann, E.: *Non-parametrics: Statistical Methods Based on Ranks.* San Francisco, Holden-Day, 1975. This is an excellent book at a more advanced level showing the power and usefulness of statistical methods using ranks. It discusses many more tests and confidence intervals than we cover here.

Noether, G.: *Introduction to Statistics: A Non-parametric Approach,* 2nd ed. New York, Houghton Mifflin, 1976. This book is at a lower level than Lehmann (1975). It covers most statistical problems from a nonparametric viewpoint. A good supplement for this text.

Exercises

Section 12.2

1. In many cultures it is believed that a full moon has an adverse effect on the mental stability of people. Indeed, the use of the word "lunatic" to describe mentally disturbed individuals is based on this supposed lunar influence. In a study by Blackman and Catalina, the number of admissions to the emergency room of a Staten Island, New York, mental clinic during the full moons from August 1971 to July 1972 was given as 5, 13, 14, 12, 6, 9, 13, 16, 25, 13, 14, and 20.

   During the remainder of the year the admission rate averaged 11.2 persons per night. Using the sign test, assuming the number of admissions per night has a symmetric distribution, test if there is a significantly higher admission rate during the full moon. (Use $\alpha = 0.1$.)

   (Blackman, S., and D. Catalina: The moon and the emergency room. *Perceptual and Motor Skills, 37*:624–626, 1973.)

2. A study presented by Forrester and Ury relates the tensile strength of tape-closed versus sutured wounds on rats. Each of 10 rats had a wound closed by stitches, and one held by tape. After 150 days the rats were sacrificed and the difference in tensile strength of the wounds was recorded. The following data are given in Colton:

---

* See also the sources cited in the problems.

Rat	Difference in tensile strength (tape-suture) (lb/sq in)
1	372
2	−107
3	0
4	564
5	−70
6	649
7	199
8	198
9	738
10	452

Subject	Determination 1	Determination 2
1	190	205
2	135	132
3	123	135
4	117	126
5	129	117
6	137	116
7	182	200
8	170	186
9	152	171
10	147	142
11	135	140
12	122	136

Test the hypothesis that the two methods yield the same median strength, using a two-tailed test and $\alpha = 0.1$.

(Forrester, J. C., and H. K. Ury: The signed-rank (Wilcoxon) test in the rapid analysis of biological data. *Lancet, 1*:239, 1969. Colton, T.: *Statistics in Medicine.* Boston, Little, Brown and Co., 1974.)

3. A paint manufacturer is comparing the glossiness of a new formula for varnish with that of the current product. To do this, a judge is presented with pairs of varnished objects made of various materials. One of the objects in the pair is varnished with the current product, whereas the other is varnished with the new formulation. The judge writes down which object is judged as glossier. If A represents the current and B represents the new formula, the following data were obtained:

ABBABBABBABBABB

Can we reject the hypothesis that the two formulas are equally glossy at $\alpha = 0.1$? How about at $\alpha = 0.4$?

4. Two different determinations of fasting levels of triglycerides were made for 12 male subjects. The following data were obtained.

Do these data show any significant difference in the median level of triglycerides as determined by the two methods? (Use $\alpha = 0.1$ and the two-sample sign test.)

5. Two different common stocks have the same risk level but possibly different returns. Naturally a security analyst working for a mutual fund company would like to determine if this is indeed the case (and invest accordingly). Since the stock returns are not normally distributed (they seem to have "fatter tails"), the analyst chooses to use a nonparametric test. The annualized returns for the two stocks for 12 weeks follow:

Week	Return stock A	Return stock B
1	0.10	0.09
2	0.11	0.12
3	0.10	0.11
4	0.09	0.10
5	0.08	0.09
6	0.10	0.09
7	0.11	0.10
8	0.12	0.11
9	0.13	0.12
10	0.12	0.12
11	0.10	0.11
12	0.09	0.10

Test the hypothesis that the two stocks give the same median return, using $\alpha = 0.1$ and the two-sample sign test. Do a two-tailed test.

6. An important parameter in cardiovascular analysis is the area below the cardiac output curve. Two methods for calculating this area are compared in Stedham and Blackwell.

Curve	Old method	New method
1	38.9	41.9
2	33.4	33.0
3	25.5	25.0
4	28.4	27.8
5	26.8	26.1
6	24.9	24.8
7	29.0	25.0
8	20.3	19.4
9	32.6	30.6
10	27.5	28.2

Test by using the two-sample sign test if the two methods yield the same median area. (Use $\alpha = 0.1$ and a two-tailed test.)

(Stedham, R. E., and L. H. Blackwell: A new method for determination of the area under a cardiac output curve. *IEEE Transactions on Bio-Medical Engineering BME-17*:4, 1970.

7. In a letter to the medical journal *Lancet,* West gave data on the length of remission between attacks of endogenous and neurotic depression. The following data are given in Colton:

Length of remission (days)

Twelve patients with endogenous depression	Twelve patients with neurotic depression
109	546
214	844
1818	602

(table continued)

(table continued)
Length of remission (days)

Twelve patients with endogenous depression	Twelve patients with neurotic depression
140	87
179	794
744	643
105	199
101	91
105	105
1547	479
529	1296
140	279

Using these data and random pairing [see problem 8(c)], test the hypothesis that the median length of remission is the same with endogenous and neurotic depression. (Use $\alpha = 0.1$ and a two-tailed test.)

(West, E. D.: The signed rank (Wilcoxon) test. *Lancet, 1*:526, 1969. Colton, T.: *Statistics in Medicine.* Boston, Little Brown and Co., 1974.)

8. The University Psychiatric Clinic of Geneva collected data on survival times for persons with various mental disorders. The data are described in Todorov, Go, Constantidinis, and Elston. For female patients aged 74 when they were diagnosed as having dementia, the following survival times in years were recorded*:

0.50	1.00	1.25	1.41	1.42
2.92	3.08	3.75	3.92	4.17
8.50	9.33	10.33	11.25	12.50
1.58	1.66	1.67	2.25	2.33
4.67	5.00	5.25	5.83	7.25

* This is just part of the study. Further data can be found in Elandt-Johnson, and Johnson: *Survival Models and Data Analysis.* New York, John Wiley & Sons, 1980, p. 132.

Test the hypothesis that the median patient survives at least 5 years, using $\alpha = 0.1$ and the one-sample sign test.

(Todorov, et al.: Specificity of the clinical diagnosis of dementia. *Journal of Neurological Science*, 98:26–81, 1975.)

9. An investigation was undertaken by Finn, Jones, Tweedie, Hall, Dinsdale, and Bourdillon to determine uric acid levels among 267 normal males. The object of the study concerns the genetic mechanism for gout, a disease characterized by high uric acid levels. The data they obtained are given in the following table:

Serum uric acid (mg/100 ml)	Number of men
3.0–3.4	2
3.5–3.9	15
4.0–4.4	33
4.5–4.9	40
5.0–5.4	54
5.5–5.9	47
6.0–6.4	38
6.5–6.9	16
7.0–7.4	15
7.5–7.9	3
8.0–8.4	1
8.5–8.9	3
	267

Under a simple genetic inheritance model, the distributional shape of the frequency distribution should be two humped (bimodal), whereas under a more complicated genetic model we expect unimodality.

Test the hypothesis that the previous data come from a distribution with a median of 5 mg/100 ml by using the one-sample sign test. (Use $\alpha = 0.05$.)

(Finn, et al.: Frequency curve of uric acid in the general population. *Lancet*, 2:185, 1966.)

10. Does attitude affect IQ scores? To see if discouragement affects performance on intelligence tests, Gordon and Durea gave Form L of the revised Stanford–Binet test to 40 subjects. Twenty of the subjects were discouraged by telling them they were doing poorly, and so on, and 20 (the control group) deliberately were not discouraged. They were then all given the test again 2 weeks later. The difference between their second and first scores was recorded as follows:

Controls	Discouraged	Controls	Discouraged
−1	7	14	−3
8	−5	1	−3
3	4	4	−3
13	−4	1	2
0	−5	−3	−4
1	−7	9	−3
6	−2	3	1
2	0	3	−9
16	−6	5	−4
3	6	2	0

(From Lehmann, E.: *Nonparametrics. Statistical Methods Based on Ranks*. San Francisco, Holden-Day, 1975.)

a. Test the hypothesis that the median of the control group is zero. Use $\alpha = 0.1$ for a one-tailed test with the alternative being greater than zero.

b. Test the hypothesis that the median of the discouraged group is zero. Use $\alpha = 0.1$, a one-tailed test, and the alternative hypothesis being less than zero.

c. Randomly pair* the controls with the dis-

* "Random pairing" is accomplished by pairing each member of one group randomly with a member of the other group. For example, consider the first member of the discouraged group who has score 7. We wish to pair this person randomly with a person in the control group. To do so, we go to the table of random numbers (Appendix A.3) and search a column until we come to a number between 1 and 20. Suppose the first number we encounter is 4. We then match up 7 with the score 13, which is the score of person 4 in the control group.

couraged group and use the sign test to test the hypothesis that the two groups have the same median improvement. (*Note:* You may want to use a normal approximation.)

(Gordon, and Durea: The effect of discouragement on the Revised Stanford-Binet scale. *Journal of Genetic Psychology, 73*:201–207, 1948.)

11. In a study concerning highway deaths in Trinidad in 1975 the following times between highway deaths were recorded:

Time in days between highway deaths	Frequency
0	50
1	48
2	24
3	14
4	15
5	5
6	6
7	4
8	0
9	2
10	0
11	0
12	1

Using the one-sample sign test, test the hypothesis that the median time between deaths is one day. *Note:* Since there are ties involved in the comparison of the data with the hypothesized median (e.g., in 48 cases the data are listed as 1), we obtain +'s, −'s, and 0's in the sign test. To implement the test, split the zero group in half, taking 24 of them as +'s and 24 as −'s, and then perform the sign test as usual. (Use $\alpha = 0.05$.)

(Richards, W.: The Poisson and exponential models. *MAYTC Journal,* Spring 1978, pp. 113–117.)

12. The following is a list of grades in a statistics class prior to the final examination and on the final exam itself. Using the sign difference between the grades, test if the median for the students prior to the exam is the same as that on the final exam. (Use $\alpha = 0.1$ and two tails.)

Student	Percentage prior to final	Final exam
1	59	35
2	94	91
3	87	93
4	80	76
5	89	89
6	93	65
7	96	92
8	99	92
9	98	93
10	75	80
11	93	96
12	96	91
13	87	83
14	69	44
15	88	97
16	86	88
17	73	55
18	88	82
19	85	75
20	72	51
21	93	86
22	71	54
23	87	86
24	96	87
25	83	86

Section 12.3

13. To test if admissions to hospitals on the nights of full moons has a temporal component (e.g., perhaps seasonal effects), compute a runs test for the data given in problem 1, using runs above and below the yearly average 11.2 admissions per night. Test for nonrandomness by using $\alpha = 0.1$.

14. The variables used in fitting a particular economic model with regression analysis were indexed by time. The residuals, $Y_t - (\alpha - \beta X_t) = \epsilon_t$ show how well the data fit the curve. For a particular economic model the following residuals were obtained. Using the runs test and $\alpha = 0.05$, check for any possible nonrandomness over time of the data.

Time	Residual	Time	Residual
1	0.464	11	−0.531
2	0.060	12	−0.634
3	1.486	13	−0.323
4	1.022	14	−0.194
5	1.394	15	2.455
6	0.137	16	1.279
7	−2.526	17	0.046
8	−0.354	18	0.697
9	−0.472	19	3.521
10	−0.555	20	0.321

Sex Ratios in 6115 Sibships of 12 in Saxony

Number of males	Observed frequency $6115\,\hat{f_i}$	Binomial fit $6115\,\hat{g_i}$	Sign of $\hat{f_i} - \hat{g_i}$
0	3	0.9	+
1	24	12.1	+
2	104	71.8	+
3	286	258.5	+
4	670	628.1	+
5	1033	1085.2	−
6	1343	1367.3	−
7	1112	1265.6	−
8	829	854.3	−
9	478	410.0	+
10	181	132.8	+
11	45	26.1	+
12	7	2.3	+
Total	6115	6115.0	

(From Shaked, M.: On mixtures from exponential families. *Journal of the Royal Statistical Society, B:*195, 1980.)

15. A production process in a styrene manufacturing plant produces plastic that is quality rated on a scale from 1 to 10. Any material with a rating of 7 or above is considered acceptable for shipment. Every half hour the production run is sampled and a quality rating is obtained. The following data come from the sample obtained during the 8-hour night shift:

6, 6, 7, 7, 8, 9, 8, 7, 7, 6, 5, 4, 5, 7, 8, 8

Do these data exhibit significant nonrandomness in the acceptability over time? (Use $\alpha = 0.1$ and the runs test.)

16. Sokal and Rohlf give the observed frequency of males in 6115 sibships of size 12 in Saxony from 1876 to 1885. The following table is extracted from Shaked:

The frequencies found in the third column are obtained using a binomial model for sex of offspring. Using the runs test on the signs in the fourth column, test the hypothesis that the binomial differs only randomly from the observed distribution. (Use $\alpha = 0.1$.)

(Sokal and Rohlf: *Introduction to Biostatistics.* San Francisco, Freeman and Co., 1973.)

Section 12.4

17. Return to the security analyst considering two stocks with returns as given in problem 5. Using the rank sum (Mann–Whitney) test, test the hypothesis that the two stocks have the same distribution of returns. (Use $\alpha = 0.05$.)

18. To test if discouragement affects scores on intelligence tests, use the Mann–Whitney test on the data given in problem 10 and test

if the distribution of scores for the control group and discouragement treated group have the same distribution. (Use $\alpha = 0.05$.)

19. Five samples each of two types of wine are scored as follows:

Wine A	75	78	81	80	74
Wine B	79	77	80	75	79

Are the two types of wine significantly different with respect to this scoring system?

20. Return to the data in problem 7 and test, using the Mann–Whitney test if the length of remission for endogenous depression is the same as that for patients with neurotic depression. (Use $\alpha = 0.1$.)

21. Using the signed rank test, test the hypothesis that the different determinations for triglyceride in problem 4 come from populations with the same mean. (*Hint:* Consider the differences and test if the differences have mean zero. (Use $\alpha = 0.1$.)

22. Test if the distribution of survival times in problem 8 is consistent with a model that says the distribution should be symmetric with a center at 5 yrs. Use the first 20 observations. (Use $\alpha = 0.1$.)

23. Using the stock return data given in problem 5, test the hypothesis that the two stocks have identical distributions. (Use the signed rank test at $\alpha = 0.05$.)

24. Using the signed rank test, test if the number of admissions to mental hospitals on full moons (problem 1) has a distribution that is symmetric about the mean admission rate of 11.2 persons per night that was found for the remainder of the year. (Use $\alpha = 0.1$.)

25. Use the signed rank test to ascertain if the discouragement significantly affects test performance on intelligence tests. (Use the data of problem 10 and $\alpha = 0.01$.)

26. In the investigation of sutures versus tape closing of wounds discussed in problem 2,

the null hypothesis is that the two methods yield the same distribution of tensile strengths. Using these data and the signed rank test at $\alpha = 0.05$, check if the null hypothesis is verified or rejected.

27. Using the pairing given in problem 7 for length of remission between attacks of endogenous and neurotic depression, use the signed rank test to investigate if there is any significant difference in remission times for the two forms of depression. (Use $\alpha = 0.1$.)

28. In problem 6 two methods of calculating the area below the cardiac output curve were listed. Using the signed rank test, check if the new and old methods for determining this area are significantly different. (Use $\alpha = 0.1$.)

29. Using the data of problem 4, find the rank correlation between the two different determinations of fasting levels of triglycerides for the 12 male subjects. Is this significantly different from zero at $\alpha = 0.05$. (Use the computational formula for r_s. Is the result consistent with that of problem 21?)

30. Common stock prices are often influenced by general economic conditions, and hence prices tend to move up and down together with the entire stock market. Using the data from problem 5, calculate the Spearman rank correlation coefficient for the two stocks in problem 5 and test if they have significant correlation at $\alpha = 0.05$.

31. Good students tend to be good throughout the entire course, and hence professors tend to believe that the grade prior to the final exam is very indicative of the final exam score. Using the data on statistics students' scores given in problem 12, determine if the rank correlation between scores prior to the final and on the final are significantly related at $\alpha = 0.1$.

32. In introducing a new method for calculating the area under the cardiac output curve, we must be sure that the new method is closely related to the old method. Using the data

from problem 6, test if there is a significant rank correlation between the new and old methods. (Use $\alpha = 0.1$.)

33. To determine if there is any difference between the perceived returns of three different common stocks A, B, and C, a questionnaire was sent out to 15 mutual fund managers. The numerical expected returns were then ordered according to their sizes and the following data were obtained on the relative rankings:

Stock A: $n_A = 5$; sum of ranks 49

Stock B: $n_B = 5$; sum of ranks 26

Stock C: $n_C = 5$; sum of ranks 45

Using the Kruskal–Wallis test, test the hypothesis that the three stocks have the same distribution of returns. (Use $\alpha = 0.1$.)

34. In comparing different diet groups, we must first see if the diets within each individual group are sufficiently close so as to be considered as essentially the same. For one diet A with four subdiets A_1, A_2, A_3, and A_4 the following growth patterns were found from Lehman.

Diet	Growth figures
A_1	257, 205, 206, 164, 190, 214, 228, 203
A_2	201, 231, 197, 185
A_3	248, 265, 187, 220, 212, 215, 281
A_4	202, 276, 207, 204, 230, 227

(Lehman, E.: *Non Parametrics: Statistical Methods Based on Ranks*. San Francisco, Holden-Day, 1975, p. 207. The original data are from Boer and Jansen: *Archives Neerlandaises Physiology*, 26:1, 1952.)

Use the Kruskal–Wallis test to determine if these subdiets differ significantly. (Use $\alpha = 0.1$.)

35. Are certain birth conditions associated with a lowering of IQ? To determine this, each of 24 girls had their IQ tested. The following data are taken from Lehman:

Birth condition		IQ Score				
Normal:	103	111	136	106	122	114
Anoxic:	119	100	97	89	112	86
Rh negative:	89	132	86	114	114	125
Premature:	92	114	86	119	131	94

(Lehman: *Non Parametrics: Statistical Methods Based on Ranks*. San Francisco, Holden-Day, 1975, p. 228. The original data come from Graham and are presented in Steel: A multiple comparison rank sum test: Treatment versus control. *Biometrics*, 15:560–572, 1959.)

Test the hypothesis that all the birth conditions have the same distribution of IQ scores. (Use the Kruskal–Wallis test and $\alpha = 0.1$.)

36. To determine if quantitative ability is significantly different among students already in quantitative areas, the investigators selected a random sample of students from four different majors. Each student was given a test and each score was recorded. The results were:

Major	Score
Mathematics	93, 106, 100, 112, 117, 101
Physics	90, 100, 113, 118, 111, 114, 103
Engineering	121, 102, 107, 95, 110, 118
Actuarial Science	119, 97, 98, 101, 120, 99, 123

Do these data show significant differences among the majors as far as scores on this quantitative ability test measure? (Use $\alpha = 0.1$.)

Section 12.5

37. Using the data given in problem 8 concerning survival times of female patients aged 74 when diagnosed as having dementia, find a

90% confidence interval for the distribution function for the survival time.

38. Using the data on traffic deaths in Trinidad given in problem 11, find a 95% confidence interval for the distribution of the time between highway deaths.

39. Two different determinations of triglyceride levels for male subjects are given in problem 4.

   a. Find a 90% confidence interval for the distribution function corresponding to determination method 1, and plot the curves for the boundaries.

   b. Plot the sample cumulative polygon corresponding to determination method 2 on top of the bounds found in part a.

   c. Does it appear that the distribution for method 1 differs from that of method 2 as exhibited in the picture given in part b?

40. Repeat steps a, b, and c of problem 39 by using the stock return data given in problem 5.

41. Repeat steps a, b, and c of problem 39 by using the IQ score differences for control and discouraged test takers (Problem 10).

Section 12.6

42. Under what circumstance would we prefer to find an estimate of the median instead of the mean? (See Section 2.4.)

43. Construct an example in hypothesis testing about the mean in which the sign test yields a different conclusion from that of the t-test.

CHAPTER · 13

Sampling Theory

It is a capital mistake to theorize before one has data.

Sherlock Holmes
Scandal in Bohemia

Throughout most of the text we have assumed that independent random sampling has been our mode of sampling. Even when admitting that most sampling in practice is done without replacement (and so the sampling is not independent random sampling), we have assumed that the population from which sampling is taken is so large that independent random sampling is a good *approximation* to sampling without replacement. Here for the first time we focus directly on what happens when using random sampling without replacement. Moreover, we discover in this chapter that there are different sampling procedures that enable us sometimes to sample with greater precision and often with greater convenience than we can do with independent random sampling or straightforward random sampling with replacement.

13 ■ 1   SURVEY SAMPLING

The general public over the last three decades has become well aware of the practice of survey sampling. Major political elections in the United States are preceded by polls or surveys in which a number of people are questioned about whom they will vote for on election day. The results of these surveys are used to predict who will win the election as well as the margin of victory.

The application of survey sampling is by no means restricted to political elections. Another obvious application is in the area of marketing research. A company develops a new product, say, a candy bar, and wishes to evaluate its acceptability by the consumer. Instead of producing the product in large quantities and distributing it throughout the market place, the company can instead select a relatively small number of people to taste the candy bar. On the basis of the reactions of this group of people, the company will either abandon the product or go ahead with full production.

As indicated, survey sampling finds direct application to affairs of politics and business. It is also important in scientific research, especially in sociology, medicine, economics, and psychology among many such areas.

In Chapter 14 we discuss broad problems associated with *survey design,* of which the *sampling design* constitutes one part. However, the design of the sample is usually a central issue in designing a survey and so we deal with this matter in some detail here.

13 ■ 2   TERMINOLOGY

In this section we present a summary of terms that were previously introduced in various portions of the text and also define a few new ones.

Population. The *population* is the set of all objects from which we select samples. Examples of populations are:

1. the human population of a nation
2. the dog population of a city
3. a species of laboratory animals
4. a file of patient hospital records

As the last example illustrates, a population may be a set of inanimate objects.

In this chapter and the next one we consider only populations that have a finite population size. We denote this finite population size by N.

Sample size. A *sample of size n* is a collection of n observations from a population.

Units and items. The members of the population to be sampled are called *units.* The units in a survey may be all men who suffer from heart disease; then in a study of heart disease we might measure each unit's blood pressure, cholesterol, and glucose level since it is conjectured that these factors are "risk" factors in the development of heart disease. Those properties of the units that are measured are called *items*. Thus in the preceding example, a unit's glucose level is an item, as is the unit's blood pressure and cholesterol level.

Item values. The numerical values obtained for any *specific* item in the N units that comprise the total population of such units are denoted by $y_1, y_2, \ldots, y_N$; that is, y_1 is the item value of the first unit in the population, y_2 is the item value of the next unit, and so on.

Simple random sampling. The theory of Chapters 6 to 12 has relied on the assumption of independent random sampling. This assumption does not apply in survey sampling where sampling is done without replacement. It is assumed here that simple random sampling (srs) is employed. By a simple random sample we mean one for which

1. sampling is without replacement
2. all possible samples of size n from a population of size N are equally likely; that is, the probability of any particular sample of size n is

$$1 \bigg/ \binom{N}{n}$$

Under the srs assumption, the individual units are also *equally likely* to be chosen in a sample; that is, the probability of obtaining a particular unit out of the population is n/N.

The *population mean μ* of item values is defined as before,

$$\mu = (y_1 + y_2 + \cdots + y_N)/N$$

The *population variance σ^2* of the item values is

$$\sigma^2 = \frac{(y_1 - \mu)^2 + (y_2 - \mu)^2 + \cdots + (y_N - \mu)^2}{N}$$

Inflation factor and sampling ratio. We let X_1 denote the random variable that represents the set of possible item values for the first sampled observation, X_2 represents item values for the second observation, and so on until X_n, which represents the possible item values of the last observation. (*Note:* $X_i = y_j$ if the jth unit is selected as the ith observation.) Let $\bar{X}$ again represent the sample mean: $\bar{X} = (1/n) \sum_{i=1}^{n} X_i$. The total sum for units in the population is also of interest in many situations. We define the population total τ to be the parameter

$$\tau = y_1 + y_2 + \cdots + y_N$$

The estimator for τ based on a sample of n units from the population is

$$T = N\bar{X} = \frac{N}{n}(X_1 + X_2 + \cdots + X_n)$$

and is called the inflated sample total. The factor N/n is called the *inflation factor* (or expansion factor) and its reciprocal n/N is called the *sampling ratio*.

Example 13 ■ 1 Health planners in the city of Cali, Colombia, were interested in establishing a family planning education clinic in a certain neighborhood. They wished to obtain an estimate of the number of women of reproductive age (15 to 49 years) living in the neighborhood in order to determine the number of personnel required to staff the clinic. To simplify the survey, they chose as the *unit* of the survey the "block." There are 120 blocks in the neighborhood. The blocks were assigned numbers and, using a table of random numbers, the planners chose 20 blocks. The *item value* of each unit (block) is the number of women of repro-

Table 13 ■ 1 Sampling of 20 Units (Blocks) for Estimating the Number of Women of Reproductive Age in a Cali, Colombia, Neighborhood

Sample number	Actual block number	Number of women of reproductive age on the block
1	76	5
2	12	9
3	32	18
4	78	33
5	27	31
6	6	47
7	89	11
8	28	1
9	17	1
10	92	5
11	113	29
12	7	30
13	85	5
14	51	1
15	48	5
16	102	61
17	65	15
18	39	20
19	20	1
20	42	29

ductive age in that unit. The parameter they wished to estimate was τ. The data obtained are shown in Table 13.1

For this sample $\bar{x} = 357/20 = 17.9$ women of reproductive age per block. The inflated sample total is $t = 120$ (number of blocks) times $17.9 = 2148$ women of reproductive age in the neighborhood. The value t is an *estimate* of τ. The value t is an observed value of the *estimator,* T.

The sampling ratio is $n/N = 0.167$; that is, the sample included 16.7% of the population.

Unbiased Estimates. Previously we derived the result that

$$E(\bar{X}) = \mu$$

for *any* random sample size n. An estimator that has the property that its expected value is equal to the population parameter it is supposed to estimate is called an *unbiased* estimator. We have at once that T is an unbiased estimator of the population total $y_1 + y_2 + \cdots + y_N$. To see this, note that $T = N\bar{X}$; hence

$$E(T) = NE(\bar{X}) = N\mu$$

But $\mu = (y_1 + y_2 + \cdots + y_N)/N$ so that

$$E(T) = N(y_1 + y_2 + \cdots + y_N)/N = y_1 + y_2 + \cdots + y_N = \tau$$

This property of unbiasedness is important for it assures us that on the average the statistic we are using to estimate a parameter is equal to that parameter and not to some other (unknown) quantity. Unbiased estimators are especially useful in situations where the same estimation technique is employed repeatedly as in continuing surveys. Using biased estimates in such cases tends to lead to an accumulation of errors.

13 ▪ 3   VARIANCE OF $\bar{X}$ FROM SIMPLE RANDOM SAMPLING

In Chapter 5 it was shown that when we use an independent random sampling technique, the variance of the sample mean $\bar{X}$ is given by σ^2/n, where σ^2 is the population variance and n is the sample size. On the other hand, if we use the *simple random sampling* (srs) technique, this formula does not hold. Essentially, this reflects the fact that when sampling without replacement from a population of size N, the samples are *not* independent.

When using srs, we find that the variance of the sample mean $\bar{X}$ turns out* to be

$$V(\bar{X}) = \sigma_{\bar{X}}^2 = \frac{\sigma^2}{n}\left(\frac{N-n}{N-1}\right) \tag{13.1}$$

* See Chapter 6, problem 37 for a derivation of this result for proportions.

To see that equation (13.1) is reasonable, note that when we sample only once ($n = 1$) there is no difference between sampling with replacement and sampling without replacement; both methods yield $\sigma_{\bar{X}}^2 = \sigma^2$. On the other hand, if we sample $n = N$ times without replacement, then we have sampled the entire population, yielding $\sigma_{\bar{X}}^2 = 0$ which is not equal to σ^2/N. If we sample the entire population, then $\bar{X}$ can only be equal to one number (namely, μ); hence there is no variation in $\bar{X}$ so that the result $\sigma_{\bar{X}}^2 = 0$ is what we should obtain. In general, the larger n is, the more of the population we have sampled and consequently the less room there is for error in $\bar{X}$. All these facts are incorporated in equation (13.1).

Finite population correction. The quantity $\sqrt{(N-n)/(N-1)}$ is called the *finite population correction* for the standard error of the mean. It is so named because this "correction factor" contributes to the standard error term only for relatively small finite population sizes N. For very large N we have that $\sqrt{(N-n)/(N-1)}$ is approximately 1, so we do not need to "correct" the standard error $\sigma/\sqrt{n}$ which was obtained in the independent random sampling case.

Note also that for the sample total we have

$$\sigma_T = N\sigma_{\bar{X}} = \frac{N\sigma}{\sqrt{n}}\sqrt{\frac{N-n}{N-1}}$$

Example 13 ■ 2 The finite population correction in a sample of size 20 taken from a population of size 120 is equal to $\sqrt{100/119} = 0.917$. Thus the standard error of $\bar{X}$ is $\sigma(0.917/\sqrt{20}) = 0.205\sigma$. The standard error of the estimator T of the total for the item values in a population is $N\sigma_{\bar{X}} = (120)(0.205\sigma) = 24.60\sigma$. Thus, referring back to example 13.1, we have a measure of the variability of the estimator of the number of women of reproductive age in the neighborhood in Cali, Colombia, once we have determined how to estimate σ.

13 ■ 4   AN UNBIASED ESTIMATE FOR A FINITE POPULATION σ^2

Recall that

$$S^2 = \frac{\sum_{i=1}^{n}(X_i - \bar{X})^2}{n-1}$$

is an estimator of the population variance σ^2 when we have independent random sampling; S^2 is an unbiased estimator of σ^2. In *simple random sampling* where we are sampling without replacement, the observations are *not* independent. What then is the expected value of S^2 in the case of simple random sampling? The answer is

$$E(S^2) = \frac{N}{N-1}\,\sigma^2$$

Hence $[(N-1)/N]S^2$ is an unbiased estimator of σ^2. Let us denote our unbiased estimator by $\hat{S}^2$;

$$\hat{S}^2 = \frac{N-1}{N}\,S^2$$

We notice that S^2 is unbiased in the independent random sampling case and $\hat{S}^2$ is unbiased in the simple random sampling case. This fact points out another difference between sampling with replacement and sampling without replacement.

Since $\hat{S}^2$ is an unbiased estimator of σ^2, we may use equation (13.1) to find that the unbiased estimators of the variance of $\overline{X}$ and T are, respectively,

$$\hat{S}^2_{\overline{X}} = \frac{\hat{S}^2}{n}\left[\frac{N-n}{N-1}\right] \quad \text{and} \quad \hat{S}^2_T = N^2\hat{S}^2_{\overline{X}} = \frac{N^2\hat{S}^2}{n}\left[\frac{N-n}{N-1}\right]$$

The estimated standard errors of X and T are $\hat{s}_{\overline{X}}$ and $\hat{s}_T$, respectively where

$$\hat{s}_{\overline{X}} = \frac{s}{\sqrt{n}}\sqrt{\frac{N-n}{N}}; \hat{s}_T = \frac{N}{\sqrt{n}}\,s\,\sqrt{\frac{N-n}{N}}.$$

and s is the familiar sample standard deviation.

Example 13 ▪ 3   Referring back to example 13.1, the value of s^2 is

$$s^2 = \frac{1}{19}[(5-17.9)^2 + (9-17.9)^2 + (18-17.9)^2 + (33-17.9)^2$$

$$+(31-17.9)^2 + (47-17.9)^2 + (11-17.9)^2 + (1-17.9)^2 + \cdots$$

$$+(15-17.9)^2 + (20-17.9)^2 + (1-17.9)^2 + (29-17.9)^2 = 286.56$$

so that the estimate of the standard error of $\overline{X}$ is $\hat{s}_{\overline{X}} = \sqrt{(286.56)(0.0417)} = 3.46$ women of reproductive age per block. The standard error of the total number of women in reproductive age in the neighborhood is $\hat{s}_T = (120)(3.46) = 415.2$ women. Recall that the estimate of the total number of these women in the neighborhood was 2148.

The formula for the finite population factor explains the seemingly surprising (to the layperson) fact that the precision of the estimator $\overline{X}$ depends mostly on

Table 13 ▪ 2 Finite Population Correction for Samples of Size 100 and Various Population Sizes

Sample size	Population size	Percent sampled	$\dfrac{N-n}{N-1}$	Finite population correction factor	$\sqrt{\dfrac{N-n}{N-1}}$
100	100	100	0		0
100	200	50	0.503		0.709
100	300	33.3	0.669		0.818
100	400	25	0.752		0.867
100	500	20	0.802		0.896
100	600	16.7	0.835		0.914
100	700	14.3	0.858		0.926
100	800	12.5	0.876		0.936
100	900	11.1	0.890		0.943
100	1000	10	0.900		0.949
100	10,000	1	0.990		0.995
100	100,000	0.1	0.999		0.9995
100	1,000,000	0.01	1.000		1.000
100	∞	0	1.000		1.000

the sample size and *not* on the percentage of the population that was sampled. For example, a sample of size 100 from a population of 10,000 yields approximately the same precision as a sample of size 100 from a population of size 1,000,000. Table 13.2 shows the value of the finite population correction for samples of size 100 and various population sizes. The increase in precision obtained by taking a larger sample occurs because the factor σ^2/n decreases, not because we have sampled a larger percentage of the population.

13 ▪ 5   CONFIDENCE INTERVALS

It turns out that if n and N are both large (but n is not too close to N), then the Central Limit Theorem still applies even with srs, and $\overline{X}$ and T are both at least approximately normally distributed. In this case confidence limits can be placed on both the theoretical mean μ of the population and the total of the item values which we denoted by $\tau = N\mu$. The $100(1 - \alpha)\%$ confidence interval on μ is given by the expression

$100(1 - \alpha)\%$ confidence interval for μ with srs from

$$\overline{x} - t_{\alpha/2}\,\hat{s}_{\overline{X}} \quad \text{to} \quad \overline{x} + t_{\alpha/2}\,\hat{s}_{\overline{X}} \tag{13.2}$$

The value $t_{\alpha/2}$ is the value taken from the t-table for the desired confidence probability using $(n - 1)$ degrees of freedom.

The corresponding confidence interval for the population total is

---

$100(1 - \alpha)\%$ confidence interval for the population total $N\mu = \tau$

$$t - t_{\alpha/2}\,\hat{s}_T \quad \text{to} \quad t + t_{\alpha/2}\,\hat{s}_T$$

Example 13 ■ 4   From example 13.1 to 13.3 we find that the estimated average number of women of reproductive age per neighborhood block is $\bar{x} = 17.9$, whereas the estimate of the total neighborhood population of these women is 2148. Are these good estimates? The 95% confidence interval on the true average number of these women per block is from

$$17.9 - t_{\alpha/2}\hat{s}_{\bar{X}} \quad \text{to} \quad 17.9 + t_{\alpha/2}\hat{s}_{\bar{X}}$$

or since we have calculated $\hat{s}_{\bar{X}} = 3.38$ in example 13.3, the 95% confidence interval is from

$$17.9 - t_{\alpha/2}\,3.46 \quad \text{to} \quad 17.9 + t_{\alpha/2}\,3.46$$

The value of $t_{\alpha/2} = t_{0.025}$ is 2.093 for 19 degrees of freedom. Thus the 95% confidence interval for the mean number per block is from

$$17.9 - (2.093)(3.46) \quad \text{to} \quad 17.9 + (2.093)(3.46)$$

or from

$$10.66 \quad \text{to} \quad 25.14$$

Of course, in this application, what we are really interested in is the *total* in the population. The 95% confidence interval on the population total is calculated by using $\hat{s}_T = 415.2$ from examples 13.3. It is from

$$2148 - (2.093)(415.2) \quad \text{to} \quad 2148 + (2.093)(415.2)$$

or upon doing the calculation, we have 95% confidence interval for our estimate of the number of women of reproductive age total from

$$1278 \quad \text{to} \quad 3017$$

13 ▪ 6 ESTIMATES OF PROPORTIONS

We can specialize the previous discussion about population means to study population proportions such as the proportion of married couples in which both partners work, or the proportion of parolees who commit an offense while on parole. Indeed, if we let $y_i = 1$ or 0 depending on whether ith unit does or does not possess the characteristic under study, then the *population mean* becomes

$$p = \frac{1}{N} \sum_{i=1}^{N} y_i = \frac{1}{N} \times \text{(number of units in the population possessing the characteristic)}$$

= population proportion of units with the characteristic.

The *sample mean* becomes

$$\hat{p} = (1/n) \sum_{i=1}^{n} x_i = \text{proportion in the sample with the characteristic}$$

where each x_i is also 0 or 1 since each x_i is one of the y_i's.

We summarize the previous results in Table 13.3. For certain types of surveys (analytic and descriptive surveys that we describe in the next chapter) these are the most important population parameters and sample estimators.

13 ▪ 7 SAMPLE DESIGNS

Introduction

We have discussed independent random sampling and simple random sampling. However, there are many other sampling schemes that are available to the scientist. One can choose the students in a first-year psychology class, or medical students, or the first n people who walk past the corner of Guadalupe and 24th Street in Austin, Texas. Such schemes are called *convenience sampling.* Their primary advantages are convenience, low cost, and speed. The primary disadvantages are that the survey results may not be repeatable, and since the sample obtained may not be representative, they may not be generalizable to the population under study, called the *target* population. Also, one has no way of assessing how good the estimates obtained are in terms of confidence intervals.

When the survey inflicts pain or inconvenience on the respondent it is often necessary to use volunteers. This type of sampling (often used in medicine and psychology) has disadvantages that are similar to convenience sampling. Another commonly used sampling procedure is *judgment sampling,* wherein a

Table 13 ▪ 3 Summary of Population and Sample Characteristics of Interest in Many Surveys

Population parameter	*Sample estimate*
Population total: $\tau = \sum\limits_{i=1}^{N} y_i$	Inflated sample total: $t = \dfrac{N}{n} \sum\limits_{j=1}^{n} x_j$
Population mean: $\mu = \dfrac{1}{N} \sum\limits_{i=1}^{N} y_i$	Sample mean $\bar{x} = \dfrac{1}{n} \sum\limits_{j=1}^{n} x_j$
Population proportion: $p = \dfrac{1}{N} \times$ (number in population with characteristic)	$\hat{p} = \dfrac{1}{n} \times$ (number in sample with characteristic)

supposedly knowledgeable person picks the *n* units to be sampled that he or she considers "representative" of the entire target population. As an example, each teacher might be asked to choose three "average" students from his or her class and the composite of these students would constitute a judgment sample from the school. Advantages of judgment sampling are that the expertise of a person is utilized and that completely nonrepresentative samples will never occur so the surveyor can feel that the result is in the right ball park at least. Disadvantages include the fact that no one can agree to exactly what it takes to be "representative" of the target population, and hence a repeat of the survey with another expert might yield a completely different result. Again, one cannot obtain measures of how good each estimate obtained is.

A combination of the preceding two procedures that is often useful is *quota sampling*. In this method a knowledgeable person lists several variables that he or she feels are strongly related to the issue under study. This person then instructs the interviewer how many individuals of each type to choose. For example, the person might say to pick 100 people, 50 of whom are women and 50 men, 10 blacks, 5 Mexican-Americans, 20 people in the $16,000 to $25,000 income bracket, and 20 people in the $5000 to $10,000 income bracket. Other than these requirements, the interviewer is free to choose the sample at his or her convenience. The benefits of this method are that by adjusting the significant variables for the sample to match those of the target population the knowledgeable person can (hopefully) ensure a somewhat representative sample and thus avoid some of the pitfalls of convenience sampling. Also, by allowing the interviewer freedom to select respondents within the bounds outlined by the quotas, one obtains many of the advantages of convenience sampling (speed and cost reduction). This is the method used by many of the national polling organizations for obtaining their samples. A disadvantage is that when the expert fails to allow for some significantly related variable, it is quite possible that there may be a systematic bias introduced into the sample. On the other side of the coin, if one becomes too specific in defining the quotas for each variable, this procedure may lose its cost reduction benefit since tremendous screening must take place to meet all the quotas simultaneously. It is also hard to assess the quality or accuracy of these types of samples.

The previous discussion indicates that there are many ways of selecting a sample. The preceding methods do not permit us to quantify how accurate we can expect the result to be. An advantage of independent random sampling (irs) and simple random sampling (srs) is that we are able to quantify (through the use of variances and confidence intervals) the accuracy of the result. We usually use random number tables or a random number computer program to determine which of the population members will be included in the sample. Once the selection is made, we can use the well-developed theory of probability to compute the probability distribution of various estimators and ascertain the probability of having randomly selected a sample with too large a bias. Various estimators can be compared objectively (by using variances) when using either irs or srs. A sampling procedure such as irs or srs that permits probability statements about the sample and estimators is called a *probability sampling design*. Various different sampling techniques, all of which employ probability sampling, can be compared with respect to the standard error (called the *precision*) of their estimators. For example, $\overline{X}$ is an estimator of the population mean μ. The standard error of $\overline{X}$ using irs is $\sigma/\sqrt{n}$ (where σ = population standard deviation, n = sample size), whereas for srs it is $\sigma\sqrt{(N-n)/(N-1)}/\sqrt{n}$. Hence srs yields a smaller standard error for the same sample size and so a smaller confidence interval. In this sense srs yields a better estimator of μ.

The irs and srs sampling designs often have the disadvantage that they may be expensive and time consuming. We concentrate now on some additional probability sampling designs that are often more cost efficient than irs and srs.

Systematic Sampling

In this type of sampling a starting place is chosen according to some procedure (which may be a random one) and then units are selected in a systematic way. For example, I wish to interview 30 heads of household in a given suburb. I choose a street at random and then pick every house on this street. If I need more houses, I go to the next consecutive street.

Example 13 ■ 5 I have a population of six numbers (1, 2, 3, 4, 5, 6). (Note that the population mean is $(1+2+3+4+5+6)/6 = 3.5$.) I do not know it, but these six numbers appear in order. I wish to select a sample of size 2. I select one at random (equally likely) from the first five members of the population and then choose the next consecutive one; for example, if I choose 2 I will then choose 3. My estimator is $\overline{X} = (X_1 + X_2)/2$. The possible values of $\overline{X}$ are:

$$\overline{X} = \begin{cases} \dfrac{1+2}{2} = 1.5 \\[2ex] \dfrac{2+3}{2} = 2.5 \\[2ex] \dfrac{3+4}{2} = 3.5 \\[2ex] \dfrac{4+5}{2} = 4.5 \\[2ex] \dfrac{5+6}{2} = 5.5 \end{cases}$$

Since all values of $\overline{X}$ are equally likely (each has the probability of $\frac{1}{5}$ of occurrence),

$$E(\overline{X}) = \frac{1.5 + 2.5 + 3.5 + 4.5 + 5.5}{5} = 3.5$$

Hence $E(\overline{X}) = 3.5 =$ the population mean and so $\overline{X}$ is unbiased. The standard error, $\sigma_{\overline{X}}$ of $\overline{X}$, is

$$\sigma_{\overline{X}} = \sqrt{\frac{(1.5 - 3.5)^2 + (2.5 - 3.5)^2 + (3.5 - 3.5)^2 + (4.5 - 3.5)^2 + (5.5 - 3.5)^2}{5}}$$
$$= 1.41$$

Note that for irs and srs

$$\text{irs } \sigma_{\overline{X}} = \frac{\sigma}{\sqrt{2}} = 1.2 \qquad (\sigma \text{ of the population} = 1.70)$$

$$\text{srs } \sigma_{\overline{X}} = \frac{\sigma}{\sqrt{2}} \sqrt{\frac{6-2}{6-1}} = 1.08$$

Hence the preceding systematic random sampling plan is not so *precise* (in the sense of standard error) as either irs or srs for a sample of the same size.

This is not always the case. The estimator when using systematic sampling can sometimes have greater precision [see problem 16(c)] than either irs or srs. A fault of this method, however, is that there is no sample based estimator of the precision; that is, we cannot get an estimate of the precision from the sample as we can from irs and srs. A saving grace of systematic random sampling is that it is less costly to implement than irs or srs, and hence for a fixed budget constraint it often permits us to have larger sample sizes.

The possible smaller precision of systematic sampling reflects a danger in using this procedure that is not present in irs or srs. If the population is ordered in some way, as was the population in example 13.5, then a result may be obtained that is far from the result we are seeking. Example 13.5 illustrates this point; because of systematic sampling we easily obtain samples [e.g., (1, 2) and (5, 6)] that yield extreme results for the estimate of the population mean. As another example, if we sample a population every 6 months, let us say in December and June, for prevalence of illness, we may in fact be sampling the 2 months in which major illnesses (say, pneumonia and diarrhea) are at their peaks. In general if there is a systematic arrangement for the population, then systematic random sampling may lead to extreme results or biased results [see problem 16(b)] or both.

Stratified Sampling

If we have some prior knowledge about the makeup of a population, we can use this knowledge to obtain greater precision of our estimates. One use of such prior knowledge might be to divide the population into *strata* (nonoverlapping populations) and then to sample from each by using srs. This type of sampling design is called *stratified random sampling*.

Example 13 ▪ 6   Suppose we know that we have a population that consists of six unknown numbers, and also that the three smaller ones reside in unit A (see Fig. 13.1), whereas the remainder reside in B. We take one observation each from A and B in an attempt to estimate the population mean μ. The actual values of the population that reside in A and B are shown in Figure 13.1. In actual practice we do not know these values for if we did there would be no need to sample in trying to estimate μ. We present these values to make the following point that the standard error of the mean when using stratified random sampling can be smaller than when using srs.

The possible values of $\overline{X}$ are (when taking one observation from each stratum):

$$\overline{X} = \begin{cases} \dfrac{1+4}{2} = 2.5 & \dfrac{2+4}{2} = 3 & \dfrac{3+4}{2} = 3.5 \\[2mm] \dfrac{1+5}{2} = 3 & \dfrac{2+5}{2} = 3.5 & \dfrac{3+5}{2} = 4 \\[2mm] \dfrac{1+6}{2} = 3.5 & \dfrac{2+6}{2} = 4 & \dfrac{3+6}{2} = 4.5 \end{cases}$$

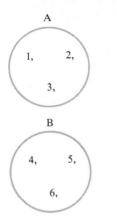

A

1,   2,

3,

B

4,   5,

6,

Figure 13 ▪ 1
The values of the population that reside in *A* and *B* are unknown. If they were known, we would not need to sample in order to estimate the population mean μ. We present the actual values in *A* and *B* only for illustrative purposes

Since all values of $\overline{X}$ are equally likely, we see that $E(\overline{X}) = 3.5$, the mean of the population. Thus $\overline{X}$ is unbiased. Let us find $\sigma_{\overline{X}}$:

$$\sigma_{\overline{X}} = \sqrt{\frac{(2.5 - 3.5)^2 + 2(3.0 - 3.5)^2 + 3(3.5 - 3.5)^2 + 2(4 - 3.5)^2 + (4.5 - 3.5)^2}{9}}$$

$$= 0.577 \qquad \text{(standard error of the mean)}$$

whereas

$$\text{irs } \sigma_{\overline{X}} = 1.2$$

$$\text{srs } \sigma_{\overline{X}} = 1.08$$

We see that stratified random sampling produces a marked decrease in the standard error of the mean (and so a marked increase in precision). The reason is that by correctly stratifying, and choosing a little bit from each stratum, we have eliminated the possibility of choosing a completely unrepresentative sample (one with all the units belonging to the same stratum). This possibility always exists with irs and srs.

Stratified Random Sampling with Proportional Allocation

Suppose we wish to estimate the mean income of the population of a city. We also know that the city is composed mainly of three ethnic groups that live in three distinct socioeconomic neighborhoods. Morever, the ethnic group *A* constitutes 50% of the population, ethnic group *B* constitutes 30% of the population, and *C* the remaining 20%. We have a budget that restricts us to a sample of size 100. It makes intuitive sense to divide the population into three strata *A*, *B*, and *C*, corresponding to the ethnic neighborhoods, and to choose a sample size from each group in proportion to its representation in the population; thus, we would select 50% of the observations from *A*, 30% from *B*, and 20% from *C*. This kind of selection is called *proportional allocation.*

Sample Size for Proportional Allocation:

n_i = sample size of *i*th stratum

n = total sample size

N_i = size of population in *i*th stratum

N = total population size

$n_i = (N_i/N)\, n$, the sample size n_i for the *i*th stratum is proportional to the population size N_i of the *i*th stratum

Problem 18 shows that for this sampling $\overline{X}$ is always unbiased.

Example 13 ▪ 7 Let us take our population of six numbers again and divide them into two unequal sized strata as shown here.

Stratum I	Stratum II
(1, 6)	(2, 3, 4, 5)

We wish to take a total sample of size $n = 3$. Using proportional allocation, we select $(2/6)(3) = 1$ observation from stratum I and $(4/6)(3) = 2$ from stratum II. Note

$$\overline{X} = \frac{X_{1,\mathrm{I}} + X_{1,\mathrm{II}} + X_{2,\mathrm{II}}}{3}$$

$X_{1,\mathrm{I}}$ = random variable representing observation from stratum I

$X_{i,\mathrm{II}}$ = random-variable representing ith observation from stratum II; $i = 1, 2$

Note that

$$E(X_{1,\mathrm{I}}) = \frac{1 + 6}{2} = 3.5$$

$$E(X_{i,\mathrm{II}}) = \frac{2 + 3 + 4 + 5}{4} = 3.5; \quad i = 1, 2$$

Hence

$$E(\overline{X}) = \left[\frac{X_{1,\mathrm{I}} + X_{1,\mathrm{II}} + X_{2,\mathrm{II}}}{3} \right]$$

$$= \frac{1}{3} [E(X_{1,\mathrm{I}}) + E(X_{1,\mathrm{II}}) + E(X_{2,\mathrm{II}})]$$

$$= 3.5$$

The mean μ of the six numbers is 3.5; hence $\overline{X}$ is unbiased. The standard error $\sigma_{\overline{X}}$ is found from the formula, problem 20, and is equal to 0.94.

For a sample of size 3

$$\mathrm{irs} \; \sigma_{\overline{X}} = \frac{\sigma}{\sqrt{3}} = 1$$

$$\mathrm{srs} \; \sigma_{\overline{X}} = \left(\frac{\sigma}{\sqrt{3}} \right) \sqrt{\frac{6 - 3}{6 - 1}} = 0.77$$

Thus proportional allocation can lead to a less precise estimate than srs. One advantage of proportional allocation is that it guarantees an unbiased estimator.

When to Stratify

Some of the reasons for using stratified sampling follow:

1. Stratified sampling can be used in some cases to eliminate the possibility of randomly selecting a totally unrepresentative sample. (This is always possible with simple random sampling).
2. Administrative considerations such as location of field offices of the agency conducting the survey can necessitate stratification. The training, supervision, and control of the survey might be under separate control in separate locations.
3. If data are desired for certain subgroups of the target population with each subgroup estimate being of a certain precision, then stratified sampling might be used to achieve this end.
4. The cost and practical problems associated with the survey can vary in different parts of the population.
5. The population characteristics might vary markedly across the population but vary much less within certain subgroups. For example, the accident rate of automobile drivers is quite variable, but stratifying on variables like age, marital status, and education yields subgroups that are much more homogeneous in response. As another example, the average age in Austin, Texas, is below 30 years, whereas in Miami, Florida, it is over 50 years. If age is suspected to be a contributing factor in responses, stratification should be used.
6. Partial lists of the target population when available can be used to draw part of the sample, and the remaining units are chosen in the field, the two methods of choosing respondents defining the strata.
7. In a survey without 100% completion rate, the nonresponders can be considered as a stratum, and on resampling the nonresponders, the surveyors can treat the survey as a stratified sample. (*Note:* with a very poor response rate, even poststratification according to response-nonresponse might not save the survey from obtaining biased results.)

What Can Go Wrong? Biased or Nonrepresentative Samples

In order to develop a legitimate stratified design, we must have accurate information on which to form the strata. If proportional allocation is used, we need accurate information on which to base the choice of sample size for each stratum. If the information is inaccurate, a nonrepresentative sample may result. Now the term *nonrepresentative* (or representative) sample is a vague term that is used with a variety of meanings. We use it here in the sense of producing a *biased estimator*. It is possible, for example, for a design to result in $\overline{X}$ such that

$$E(\overline{X}) \neq \mu$$

Such an estimator is said to be *biased*. The sample that comes from a sampling design that yields a biased estimator is called a *biased* sample.

Example 13 ▪ 8 Let us stratify the population of the six numbers {1, 2, 3, 4, 5, 6} as follows:

Stratum I	*Stratum II*
(1)	(2, 3, 4, 5, 6)

and let us take one observation from each.

$$\overline{X} = \frac{X_{\mathrm{I}} + X_{\mathrm{II}}}{2} \qquad X_i = \text{observation from stratum } i;\ i = \mathrm{I}, \mathrm{II}$$

$$E(\overline{X}) = \frac{1}{2}\{E(X_{\mathrm{I}}) + E(X_{\mathrm{II}})\}$$

$$= \frac{1}{2}(1 + 4) = 2.5$$

whereas $\mu = (1 + 2 + 3 + 4 + 5 + 6)/6 = 3.5$. Hence the sample design results in a *biased* estimator of μ. Notice that we did not use proportional allocation that would have assured us of an unbiased estimator. Note that the only possible samples are: (1, 2), (1, 3), (1, 4), (1, 5), and (1, 6); the sample is "biased" by the low value of 1 that appears in every possible sample.

Cluster Sampling

Random Cluster Sampling

There are times when suitable information is not available for use in designing a stratified sample. Such was the situation in which one of the authors found himself while trying to design the 1972 health survey of West Azerbaijan, the mountainous northern sector of Iran that is inserted along three borders: those of Iraq, Turkey, and the Soviet Union. Simple random sampling was impossible; villages were generally scattered over rugged mountainous terrain covering hundreds of miles. Even a very small sample size for this type of survey (say, 300 people) would have required traveling several days between interviews over difficult roads (impossible to traverse during the winter snows and spring rains) at great danger from bandits. Stratification was no help, for the strata would have covered many miles and srs from the strata would have again required long and dangerous travel. However, most of the population in this area was concentrated in small villages varying in size from 50 to 500 people. A sample of 200 villages was selected by using srs. Every family in each village was interviewed. This kind of sampling is called *cluster sampling.*

In cluster sampling the population is divided into subdivisions called clusters. These *clusters* are randomly selected (using srs) for inclusion into the sample. Every population member of the selected clusters, or a subpopulation, can be sampled by using srs or some other sampling design.

Example 13 ▪ 9   Let us divide our population of six numbers into three clusters. We choose two

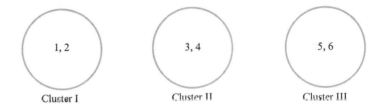

clusters by using srs, say, I and III, and select one member, by using srs from each cluster. Let us say 1 and 5; the estimate of the sample mean is $\bar{x} = (1 + 5)/2 = 3$.

Is the estimator $\bar{X}$ in this example unbiased? The possible values of $\bar{X}$ are:

$$\bar{X} = \begin{cases} \dfrac{(1 + 3)}{2} = 2 & \dfrac{(2 + 3)}{2} = 2.5 & \dfrac{(3 + 5)}{2} = 4 \\[2mm] \dfrac{(1 + 4)}{2} = 2.5 & \dfrac{(2 + 4)}{2} = 3 & \dfrac{(3 + 6)}{2} = 4.5 \\[2mm] \dfrac{(1 + 5)}{2} = 3 & \dfrac{(2 + 5)}{2} = 3.5 & \dfrac{(4 + 5)}{2} = 4.5 \\[2mm] \dfrac{(1 + 6)}{2} = 3.5 & \dfrac{(2 + 6)}{2} = 4 & \dfrac{(4 + 6)}{2} = 5 \end{cases}$$

All values of $\bar{X}$ are equally likely; hence

$$E(\bar{X}) = \frac{1}{12}(2 + 2.5 + \cdots + 5) = 3.5$$

Thus $\overline{X}$ is unbiased. What is $\sigma_{\overline{X}}$? Note, in this case

$$\sigma_{\overline{X}} = \sqrt{\frac{(2-3.5)^2 + \cdots + (5-3.5)^2}{12}} = 0.9$$

Note also that for samples of size 2 we have the standard deviations

$$\text{irs } \sigma_{\overline{X}} = 1.2$$

$$\text{srs } \sigma_{\overline{X}} = 1.08$$

whereas stratification (example 13.6) into two equal groups yields $\sigma_{\overline{X}} = 0.58$. In this case cluster sampling yields a better result than do irs and srs. This is not always the case. If clusters are poorly designed, biased samples may result. We leave the demonstration of this fact to the exercises (problem 23). One of the main advantages of cluster designs is the convenience they permit in sampling.

Examples and Reasons for Clustering

Perhaps one of the most compelling reasons for using cluster sampling is to save survey costs. Taking a simple random sample of residents of Texas and sending an interviewer out to personally interview each person is more expensive (travel costs and so on) than selecting certain cities and sending interviewers to each city, allowing them to obtain many interviews in practically the same location. Since the target population units are often found grouped naturally into geographical clusters, the researcher can take advantage of this by sampling these clusters. This allows larger sample sizes for the fixed cost. Thus in practice one selects the clusters for cost convenience, availability of information, and so on. Of course, if possible we would like the clusters to be heterogeneous; however, typically when sampling, for example, households, the geographical location is correlated with economic status, education, social status, age, and other variables so that the clusters actually are somewhat homogeneous. A list of target populations and possible clusters that could be used to study them is presented in Table 13.4.

In general, the choice of what to take as clusters depends on the following factors:

1. The clusters should be well defined and the number of units (or an estimate of it) must be available for each cluster.
2. The size of the clusters need not always be the same, but each should contain enough units to afford the interviewer multiple interviews, and the geographical spread of the cluster should be small enough to save some travel costs.
3. Clusters should be as heterogeneous as possible to reduce sampling variances.

Table 13 ▪ 4 Target Populations and Possible Clusters	
Studies	*Possible clusters*
Agricultural studies	Farms
	Acres
	Counties
Nationwide studies of adults	Cities
	Counties
	Standard metropolitan statistical areas (SMSAs)
	City blocks
	Neighborhoods
	Census tracts
	Households
Industrial studies	States
	Counties
	Cities
	Manufacturing plants
College students	Colleges

The homogeneity present in most geographical clusters has an adverse effect on the sample variance of $\overline{X}$ since as we have seen, we want the clusters to be as heterogeneous as possible. On the other hand, actual face-to-face interviewing costs do not constitute the entire cost of a completed interview, the rest of the cost comes from such factors as travel expenses, training, supervision costs, obtaining maps and directories, and so on. These costs remain the same no matter how many interviews are conducted at the final destination, and hence the use of cluster sampling by using geographical clusters can significantly reduce the *per unit* interviewing costs for a fixed budget by allowing more units to be sampled for the same cost. However, the homogeneity in the clusters reduces the amount of independent information available per completed interview, so these two forces (homogeneity and decreased costs) conflict. A resolution of this conflict is discussed in several of the references at the end of the chapter (e.g., Cochran).

Comments on Confidence Intervals for Alternative Sampling Designs

If the population is normally distributed and the population variance is unknown, the confidence interval on μ is given by expression (13.2), where $\sigma_{\overline{X}}$ is the standard error associated with the sample design; however, the formula for the estimate $\hat{s}_{\overline{X}}$ changes also with each sampling design. The appropriate expression for $\hat{s}_{\overline{X}}$ can be found for various designs in texts on sampling theory.

13 ▪ 8 SAMPLING AND OTHER ERRORS: METHODS FOR ACHIEVING A RANDOM SAMPLE

The difference $\overline{X} - \mu$ is a random variable: It is called the *sampling error, ϵ*; it arises because we have sampled only a portion of the population, not the entire population. If we sample the entire population, ϵ will be zero. If $\overline{X}$ is unbiased, then $E(\epsilon) = E(\overline{X} - \mu) = E(\overline{X}) - \mu = 0$, since $E(\overline{X}) = \mu$. However, $V(\epsilon) = V(\overline{X}) = \sigma_{\overline{X}}^2$, that is, the standard deviation of the sampling error ϵ is $\sigma_{\overline{X}}$. It depends on the sampling design used (e.g., cluster or stratified or srs or irs).

The sampling error is not the only type of error we encounter in practice. Suppose the items in the population are $y_1, y_2, \ldots, y_{100}$, where each item value i is a weight y_i of person i. Suppose when weighing each person in the sample we obtain the true weight plus 2 lb because the scale we are using is defective; it has a *bias:* It adds 2 lb to each person's weight. Hence

$$\overline{X}_b = \frac{(X_1 + 2) + (X_2 + 2) + \cdots + (X_n + 2)}{n} = \overline{X} + 2$$

If we use this scale, the estimator $\overline{X}_b$ has a bias of 2 lb. $E(\overline{X}_b) = E(\overline{X}) + 2 = \mu + 2$. This type of error is called a *systematic* measurement error because it is added on to each observation in the same way. The bias is called *bias due to a systematic error.*

Systematic errors can arise in many ways. In fact, systematic *sampling* can introduce systematic errors. This fact is left to the exercises to demonstrate (see problem 15).

There are many sources of error other than from sampling or systematic biases. There are *errors of coverage* resulting from members of the selected sample (respondents) who do not respond to questions. Errors arising from nonresponse are the bane of most surveys (there is an extensive literature on this subject). In addition, respondents can answer falsely either through misunderstanding what is being asked or on purpose. Errors also can be made in recording by enumerators and can arise in the analysis and even in the preparation of final tables.

Often, in the totality of all errors, the sampling error turns out to be small. Thus a design that is optimal in the sense of yielding the smallest sampling error, that is, the smallest standard error of the mean $\sigma_{\overline{X}}$, can, in practice, yield data contaminated by many larger errors because of the difficult circumstances under which the sampling must be conducted. On the other hand, a design that takes into account these circumstances can yield a larger $\sigma_{\overline{X}}$ but much more accurate data overall. Controlling the sampling error is like using a scalpel to cut away the controllable variation. All this expense is worthless if one has tried to cut away the nonsampling error with something as crude as a chain saw.

Methods for Achieving a Random Sample

Up to now we have assumed that the n units of the sample were selected randomly without replacement from among the population units. It is reasonable to ask by what mechanism this can be achieved. It is this mechanism ensuring randomness that differentiates probability sampling from haphazard judgment sampling.

One common method used to ensure randomness is to enumerate the population from 0 to $N - 1$ and then randomly choose n numbers between 0 and $N - 1$ and take the corresponding units into the sample. One way to choose n numbers randomly from among N is to use what is known as a random number table. This is a table of numbers in which the digits 0, 1, 2, . . . , 9 all appear with relative frequency about $\frac{1}{10}$ and in a random order. One way to construct such a table would be to roll a balanced 10-sided die and record the digit that appears. Of course, this is a rather time-consuming operation and is not done in practice. In applications we usually use either a reliable computer routine or a published list of random numbers such as "A Million Random Digits," printed by Rand Corporation. Two pages of random numbers are given in Appendix A.3. These lists have been checked for nonrandomness such as runs of digits, unequal distribution of digits, and so on.

To use the random number table, we first enumerate the population units. Suppose there are 250 members in the population labeled 000 to 249, each unit getting a three-digit label. We then start anywhere in the table and move in any direction in groups of three (since each unit has a three-digit label). It is usually convenient to move downward in the table from the starting point and read off labels as you go. If we come to a label such as 372 that does not correspond to a population unit, or if we encounter a duplication, these values are ignored and we just continue on until the sample of size n is selected. If there are 6000 units in the population, labels of four digits are used, whereas if there are 100,000 units, labels of six digits are used.

Example 13 ■ 10   Suppose you wish to take a simple random sample of size 20 from the student directory of the University of Texas at Austin, 1982 to 1983 edition. One way to do this is to label the listed students from 0 to 49,000 and use the random number table to select 20 students, each student having a five-digit label. If the directory was typeset by a computer (as many directories are), and had the same number of entries per page, then the labeling of each student is easily accomplished without actual enumeration. Suppose, for example, there are 100 entries per page and the random number selected is 31,252. You would then select the fifty-second student on page 312 since that is the 31,252 entry in the book.

Even if the directory is not computer typeset, and the number of students per page varies, you can select a simple random sample without actually engaging in the long drawn-out process of enumerating the entire student body. You proceed as follows. Each page has some maximum number of students (say, no more than 150 students per page), and there are a certain number of pages (say, 400) in the directory. Each student will get a six-digit label, the first three digits representing the page number and the last three digits representing the line number on the page. Thus the number 120,024 represents the twenty-fourth name on page 120. Numbers like 701,050 and 100,182 are ineligible and are thrown out. You proceed by picking six-digit numbers until 20 eligible numbers are determined.

Another method used to obtain samples is systematic random sampling. If asked to pick 30 students from an elementary school of 300 students, many people would say to "select every tenth pupil." This technique is the basis of systematic sampling; one selects every $i = N/n$ units from the list. You calculate $i = N/n$ (called the sampling interval) and then uses the random number table to select a number between 1 and i as a starting point. Suppose, for example, you wish to select a sample of size 10 from a population of 500 units. The sampling interval is $i = 500/10 = 50$ so you select every fiftieth unit. Using the random number table, you pick a number at random between 1 and 50 and obtain 37. Thus the units to be included in the sample are numbers 37, 87, 137, 187, 237, 287, 337, 387, 437, and 487. If the sampling interval $i = N/n$ turns out not to be an integer, you simply take the integer part of $[N/n]$. This will result in more than n units selected and one can then use a random number table to delete randomly the extra units to make the sample size n.

Some cautions should be made about systematic sampling. It is not really random, and if the list of units contains a periodicity, there is a danger of obtaining a nonrepresentative sample. For example, if a sample of 500 voters is needed from a voter's registration list of 1500 people, the sampling interval is $i = 3$. The voter's registration list, however, is according to households so that husbands and wives are listed in order. Hence using systematic sampling would mean that we would seldom get two members from the same family in the sample. Usually if there is a periodicity present in the list this shows up by examining the list or the sample before any harm is done. Nevertheless, the problems of periodicity should be kept in mind when using systematic sampling. The cost and convenience of systematic sampling versus simple random sampling make it an often used technique for sampling when a list of units from the target population is available. In most cases it yields samples that are random.

If you are sampling from a list and come to an ineligible entry, the correct thing to do is to throw out that choice. Simply proceeding to the next eligible unit can produce a bias. For example, if you are sampling from a telephone book to obtain a sample of n people and your sampling technique (random or systematic) leads you to "John Doe's Appliance Repair Shop," the next eligible entry is more than likely John Doe, the shop owner. Proceeding to the first eligible entry will tend to oversample the doctors, lawyers, dentists, and small businesspeople. Thus ineligible entries are simply omitted from the start; they are not included in the list at all. Going through the list and deleting the ineligible units can be a time-consuming and expensive job, hence if you are using systematic sampling, the following procedure is used. Take a random sample of, say, 8 to 12 pages and estimate the proportion p of eligible entries on the list. Then use the sampling interval $i = Np/n$ and expect to get approximately n eligible units for the sample.

13 ■ 9   REMARKS

The study of sampling theory is extensive. We have only discussed the most rudimentary elements of this beautiful theory in the hope of providing an introduction that will permit the reader to approach specialized books armed with some of the terminology and theoretical background to make subsequent readings easier. In the next chapter we discuss some of the broader aspects of survey design in order to place the area of sampling theory in its proper context as part of the larger picture of conducting surveys. This discussion will also emphasize the difficulty of implementing a sample design in a practical world that has little sympathy for the beauty of any theory that is not subjected to practical application.

Summary

We have now faced the issue of what happens to the standard error $\sigma_{\bar{X}}$ of $\bar{X}$, when we do not sample with independent random samples but use simple random sampling, that is, random sampling without replacement. We find that in this case

$$\sigma_{\bar{X}} = \frac{\sigma}{\sqrt{n}}\sqrt{(N-n)/(N-1)}$$

where N is the population size and n is the sample size. We now understand why, when N is much larger than n, independent random sampling approximates simple random sampling. In such a case the standard error of $\bar{X}$ for simple random sampling (srs) is approximately equal to the standard error of $\bar{X}$ for independent random sampling.

We found an unbiased estimate when using irs for the population variance σ^2. Using this estimate, we were able to find confidence limits on the population mean μ when σ^2 is unknown and when using srs. We were also able to place confidence limits on the total of the item values.

We next discussed various sampling designs: systematic sampling, convenience sampling, stratified sampling, cluster sampling, and other designs. For some of the more frequently used designs we compared their standard errors of

$\overline{X}$ under various conditions. In particular, we discussed proportional allocation in stratified sampling and showed that for this design $\overline{X}$ is an unbiased estimate of μ.

We ended by noting that the best sampling design is not necessarily the one with the smallest $\sigma_{\overline{X}}$; it might be the one that permits greatest accuracy at the least cost.

Key Terms

simple random sampling	convenience sampling
item	quota sampling
item values	judgment sampling
unit	stratified sampling
finite population correction	proportional allocation
inflation factor	cluster sampling
sampling ratio	systematic error
unbiased estimate	sampling errors
sample design	nonsampling errors
systematic sampling	random numbers

References

Casley, D. J., and D. A. Jury: *Data Collection in Developing Countries.* Oxford, Clarendon Press, 1981. This is an excellent and readable book that covers in detail the matters developed in this chapter and many that were not.

Cochran, W. G.: *Sampling Techniques,* 3rd ed. New York, John Wiley & Sons, 1977. A very good (but somewhat more advanced) book covering most aspects of sampling theory. Formulas for the variance of $\overline{X}$ using each of the sampling designs introduced in this chapter are presented in this book.

Hansen, M. H., and W. G. Madow: Some important events in the historical development of sample surveys. *In* Owen, D. R. (ed.): *On the History of Statistics and Probability.* Ann Arbor, MI, M. Dekker, Inc., 1976, p. 87.

Hess, I., D. C. Reidel, and T. B. Fitzpatrick: *Probability Sampling of Hospitals and Patients.* Ann Arbor, MI, Health Administration Press, 1975. This book presents long and interesting (but technically detailed) case histories involving discussions of various sampling designs.

Kish, L.: *Survey Sampling.* New York, John Wiley & Sons, 1965. A classic text, one that goes into detail (but remains interesting) in deriving $\sigma_{\overline{X}}$ for many useful sampling designs.

Moser, S. C., and G. Kaltin: *Survey Methods in Social Investigation.* New York, Basic Books, Inc., 1972.

Partner, M.: *Surveys, Polls, and Samples.* New York, Harper & Row, 1950.

Stephan, F. F., and P. J. McCarthy: *Sampling Opinions.* New York, John Wiley & Sons, 1958. A readable text on the methodology underlying opinion polls.

Stuart, A.: *Basic Ideas of Scientific Sampling,* 2nd ed. Griffin's Statistical Monographs and Courses, No. 4. Charles Griffin & Co., 1976.

Sudman, S.: *Applied Sampling*. New York, Academic Press, 1976. An applied introduction, giving good rules of thumb and formulas. It is not at an advanced mathematical level.

Exercises

Sections 13.1 and 13.2

1. In each of the following descriptions, specify the population, the units, items, the item values, and the sample size.

   a. A *waybill* is a document issued with every shipment of freight sent by train. It gives details about the goods shipped, the route, and the charges. If the freight is shipped over a route by several different railroad companies (as is often the case in shipments over long distances), then the amount of money due each company is recorded on the waybill.

      From June 26, 1981 to December 30, 1981, in a district of Southern Illinois serviced by two railroads, called *A* and *B* here, there were 32,000 waybills. It is an extremely time-consuming and expensive process to go through every waybill to determine the amount of money due each railroad. The two railroads decided to sample 10% of the waybills to determine how much money was due each of them on these waybills; from this sample, they could estimate what money was due each of them *per* waybill. By multiplying the estimates obtained from this sample by 32,000 (the number of waybills), they could get an estimate of the total money due each railroad.

   b. It might not be possible to determine all the bad side effects of a drug before it is put on the market. One reason is that the side effects may appear rarely and only after the drug has been used for a very long time, like 10 years. Usually drug companies do not test a drug for such long periods of time; in addition, the tests may take place on a relatively few number of people, several hundred, and so if the side effects are rare, they will not be detected.

      A certain drug has been used consistently by 100,000 people for more than 10 years. A sample of 10,000 is to be taken to determine if the drug has severe side effects such as producing kidney failure, blindness, or other disabling effects.

   c. A firm has hired 3000 people in a given job category over the last 20 years. Seventy-five percent of the qualified applicants for the job were women. A sample of 300 people is to be selected from the 3000 hired to obtain an estimate of the proportion of women hired in this job category over the last 20 years.

2. Using simple random sampling (srs), how many samples of size 2 can be drawn from a population of size
   a. $N = 3$? b. $N = 5$? c. $N = 10$? d. $N = 100$?

3. Using srs, how many different samples of size 5 can be drawn from a population of size
   a. $N = 5$? b. $N = 10$? c. $N = 100$?

4. There are 100 people living in my neighborhood. The city is going to interview 10 of these people concerning the effectiveness of public services. The city will sample by using srs. What is the probability that I will be interviewed?

5. For srs to be applied, we must be sure that all

samples of the population are equally likely. Explain in each of the following cases why the sampling is not srs:

a. To predict a municipal election I ask the opinions of 10 of my friends and ask each of them in turn to obtain the opinions of each of their friends to get a total sample of size 110.

b. A university wishes to obtain an estimate of the income of its 1960 graduating class. It looks through the 1979 issue of *Who's Who in Science* and finds 10 of its 1960 graduates in this issue. It then sends queries to these 10 graduates concerning their incomes and averages the responses.

c. Interviewers are sent into the population to determine opinions about a proposed new real estate tax. They are only sent into low crime areas, in order to ensure the safety of the interviewers.

d. An announcement is made over a cable television that all viewers should send in their opinions to the station as to the best candidate in a local election.

6. The following table shows the monthly incomes of a hypothetical population of 10 individuals:

Individual	Income (dollars/month)
1	1,300
2	5,500
3	2,400
4	1,400
5	2,000
6	2,500
7	900
8	4,400
9	1,000
10	1,200
total income per month	$22,600
mean income per month	$ 2,260

a. Find the population variance of the

monthly incomes. Using simple random sampling (srs), we have $\binom{10}{2}$ or 45 samples of size 2. One possible sample of size 2 is person 8 and person 3. The average monthly income of these two people is ($4400 + $2400)/2 = $3400. Another possible sample is person 6 and person 9. These two have an average monthly income of $1750. Make a table of all possible samples of size 2, by using srs and, with each, the average monthly income of the pair.

b. Find the average and variance of all the sample averages in the table constructed in (a). Is the estimator $\overline{X}$, which is the set of all averages of monthly earnings in samples of size 2, an unbiased estimator of the population average of monthly incomes?

c. Selection of a simple random sample: Use the Table of Random Numbers to select a simple random sample of size 2 from the population of the incomes of the 10 individuals shown in the table. Is the estimate $\overline{x}$ you obtain close to the population mean of $2260 per month? Is $10\overline{x}$ a close estimate of the total income per month of $22,600 per month? What is the value of the sampling ratio in this problem? The finite population correction?

Section 13.3

7. A population consists of the numbers {1, 2, 3, 4, 5, 6}
   a. Show that the mean of this population is $\mu = 3.5$ and that its standard deviation is $\sigma = 1.71$.
   b. Using srs, choose a sample of size 2 from the population. List the 15 possible samples that can be so drawn from this population and calculate the sample mean for each.
   c. Note that a sample mean equal to 3.5 occurs three times in the list of sample

means in (b), the mean 2.5, 3, 4, and 4.5 each occur twice, whereas 1.5, 2, 5, and 5.5 each occur once. What is the mean of the sample mean? Show that the standard deviation of the sample mean is equal to 1.08 and verify that this is the result that can be obtained by using formula (13.1).

Section 13.4

8. A study is to be launched of the frequency of heart attacks in men between the ages of 45 and 55 years of age in a given 200 block area of a city. In order to determine the resources needed to conduct the study (the number of investigators, the time period of the investigation, and so on), we must first obtain an estimate of the number of such men in the 200 block area. Ten blocks are sampled. The results are

Block	Numbers of men between 45 and 55
1	8
2	17
3	3
4	22
5	14
6	35
7	11
8	9
9	49
10	60

Find a 95% confidence interval on the number of men in the specific age group in the 200 block area.

9. There are 23,522 men in a given area between the ages 45 and 55. A sample of 1200 such men is taken by using srs. The sample average $\bar{x}$ of diastolic blood pressure is 132 and s^2 is 14. Find 95% confidence limits on the value of the true mean diastolic pressure of men in this age group.

10. There are 40 black jack tables in a certain casino. Two gamblers watch 15 tables be-

tween them during an evening and note the total winnings of the house on each table. The results are:

Table	1	2	3	4	5
Winnings (thousands of dollars)	20	35	6	80	30

Table	6	7	8	9	10
Winnings (thousands of dollars)	22	29	18	31	41

Table	11	12	13	14	15
Winnings (thousands of dollars)	26	32	22	18	35

Find 95% and 99% confidence levels on the average house winnings per table in an evening.

11. Two hundred bags of coffee are being readied for shipment. A sample (srs) of 50 bags is found to have mean weight of 30 lb with standard deviation s of 8 lb. If we take the 30 lb as an estimate of the average weight of all the bags, what can we assert with probability 0.95 about the possible size of the error of this estimate?

12. As described in problem 1(a), a waybill is a document issued with every railroad freight shipment that shows how much is to be paid to each company that has had a share in the shipment. Suppose there are 32,000 waybills and 3200 are sampled by using srs. The total amount of money t due Railroad A is estimated from t = $32,000 \bar{x}$, where $\bar{x}$ equals the sample average due per waybill; $\bar{x}$ was found to be $35.20 with the sample standard deviation s equal to $5.30. Hence t = $1,126,400 with standard deviation $32,000(\$5.30) = \$169,600$.

    a. Find 99% confidence intervals on τ the actual total amount due Railroad A.

b. Suppose it costs 0.25 per waybill to process all the waybills to obtain the *exact* amount due Railroad *A*. It would then cost $32,000 \times 0.25 = \$8000$ to process all the waybills. Is Railroad *A* likely to lose money by using the sampling procedure to estimate the money due it?

13. In problem 6 we assumed a population of 10 individuals. Let us assume instead a population of 10,000,000 individuals; 1,000,000 receive each of the incomes shown in the table in problem 6. With this size of population it would be an impossible task to list all possible samples of even moderate sample size. The utility of the formula for the standard deviation of the sample mean becomes apparent in the case. For a sample size 6, using srs, find the standard deviation of the sample mean where the population is that as previously described.

Section 13.6

14. An interviewer wishes to obtain an estimate of the proportion of the population in favor of reducing social security benefits to the elderly. The interviewer decides to obtain a "convenience" sample by going to an area in which large numbers of people congregate and are not too busy to answer questions. He knows that every Sunday morning a large number of joggers run a race in the local park and afterward congregate about the finish line to socialize. He intends to choose his sample from among this group. What are the dangers in doing this?

15. The following lists the total number of cases of pneumonia and diarrhea in children between 1 and 5 years of age in a certain city by month:

A sample of size 4 is to be taken of the months and the number of cases in each month recorded.

a. (i) Calculate the sample average of the systematic sample obtained by starting with Dec. (no. 1) and then going to the fourth month in the sequence (March) to the seventh month and then to the tenth month.

   (ii) Calculate the sample average using the same procedure but starting with January; February. If each of the first 3 months is equally likely as the starting point, what is the expected value of the sample mean? Is the sample mean unbiased?

b. Find the average of the sample mean obtained, using srs, samples of size 4.

16. There are 10 movie houses in a given city: in order of their distance from the population center of the city, they are: Bijous, Central, Flick Fun, Grand Theatre, Orpheum, Stillwell, King & Highway, Nite Out, Lost Pitcher Show, and Sun. Each theatre had, respectively, the following attendance last Monday night: 300, 60, 75, 140, 55, 40, 25, 60, 10, 400.

a. List all possible systematic samples of size 4 by starting with one of the first three theatres and then taking every other theatre on the list.

b. List all possible values of the sample mean obtained from sampling as in (a). Is the sample mean here unbiased?

c. What is the standard deviation of $\overline{X}$?

d. If srs is used, what is $\sigma_{\overline{X}}$?

17. This problem illustrates that a biased estimate of the population mean can result from

Dec.	Jan.	Feb.	March	April	May	June	July	Aug.	Sept.	Oct.	Nov.
220	240	210	180	122	110	125	208	170	200	110	212

a stratified sample if proportional allocation is not used.

Population stratum A consists of the three numbers $\{1, 2, 3\}$ and stratum B consists of the three numbers $\{4, 5, 6\}$. A sample of size 2 (using srs) is selected from A, whereas a sample of size 1 is chosen from B.

a. Write down all possible samples and the sample average of each.

b. Show that the mean of the sample averages in (a) is 3, whereas that of the population of the six numbers is 3.5 so that the sample average in this case is biased.

18. We now show through this problem that an unbiased estimator of the population mean is obtained if proportional sampling from strata is employed. Let A represent the population of the first stratum consisting of the N_1 numbers, $A_1, A_2, \ldots, A_{N_1}$; let B represent the population of the second stratum consisting of numbers $B_1, B_2, \ldots, B_{N_2}$, and so forth until S consisting of $S_1, S_2, \ldots, S_{N_s}$. The total population size is $N = N_1 + N_2 + \cdots + N_s$.

a. The population means of stratum 1, 2, . . . , s, respectively, are

$$\mu_1 = \frac{A_1 + A_2 + \cdots + A_{N_1}}{N_1}$$

$$\mu_2 = \frac{B_1 + B_2 + \cdots + B_{N_2}}{N_2}$$

$$.$$
$$.$$
$$.$$

$$\mu_s = \frac{S_1 + S_2 + \cdots S_{N_s}}{N_s}$$

Show that the mean of entire population is

$$\mu = \frac{(N_1\mu_1 + N_2\mu_2 + \cdots + N_s\mu_s)}{N}$$

b. Let sampling be done by proportional allocation and let $\overline{X}_1$ be the sample mean of the sample taken from the first stratum, $\overline{X}_2$ be the sample mean of the sample taken from the second stratum, and so forth. Show that an unbiased estimator of the population mean is

$$\overline{X} = \frac{(N_1\overline{X}_1 + N_2\overline{X}_2 + \cdots + N_s\overline{X}_s)}{N}$$

19. Construct simple examples to illustrate each of the following points:

a. With proportional stratified sampling, as with srs, each unit in the population has an equal chance of being included in the sample.

b. Certain samples possible under srs are impossible with stratified sampling; moreover, the strata can be chosen so that the more extreme samples (those containing extreme or unusual values in the population) that are possible under srs are impossible under the stratified sampling. It is these more extreme samples that contribute more heavily to the sampling variance; that is, to $\sigma_{\overline{X}}^2$;

c. The variance is smaller when we are able to stratify the units into groups so that the differences within each group are relatively small, whereas at the same time differences between groups (measured by the differences between their averages) are large.

d. Stratification can be particularly effective when there are extreme values in the population that can be segregated into separate strata.

20. Derive the result that the variance of the sample mean with proportional stratified sampling is

$$\sigma_{\overline{X}}^2 = \frac{(1-f)\, S^2}{n} \qquad (n = \text{sample size})$$

where

$$S^2 = \frac{N_1 S_1^2 + N_2 S_2^2 + \cdots + N_k S_k^2}{N}$$

k = number of strata

N_i = number of population members in strata i; $i = 1, 2, \ldots, k$

$N = N_1 + N_2 + \cdots + N_k$

$S_i^2 = \dfrac{N_i}{N_i - 1} \sigma_i^2$; σ_i^2 = variance of population in ith stratum

$f = \dfrac{n}{N}$; (f is called the sampling ratio)

21. A stratified sample of size $n = 120$ is to be taken from a population of size $N = 4000$ that consists of five strata for which $N_1 = 500$, $N_2 = 2200$, $N_3 = 600$, $N_4 = 200$, and $N_5 = 500$. If proportional allocation is to be employed, how large a sample must be taken from each stratum?

22. Explain the difference between cluster sampling and stratified sampling.

23. The sampling plan is as follows: I have three (3) clusters of numbers as shown:

The clusters are equally likely to be selected. I select one at random. I choose two observations from the selected cluster without replacement and calculate $\bar{x}$.

a. Find the expected value of the sample mean.

b. Is the sample mean biased or unbiased?

24. Consider the following two ways of drawing a sample of households throughout the United States:

1. Using srs, a sample of 2000 families is drawn from a list of every family in the United States.

2. A sample of 50 counties in the United States is drawn and a sample using srs of 2000 families is taken from within the sample of counties.

The first way will result in a sample with smaller sampling error (smaller $\sigma_{\bar{x}}^2$). Explain why the second alternative might be preferred, especially if money and time for the implementation of the sampling phase are in short supply.

CHAPTER · 14

Practical Aspects of Statistics; Survey Design

See the land, what it is, and the people that dwelleth therein, whether they are strong or weak, whether they are few or many. . . .

Numbers XIII:18–20

In previous chapters the reader has been introduced only to the skeleton of statistical knowledge—fundamental statistical concepts and techniques. A few hints have been given concerning the excitement, difficulties, and sometimes dangers one encounters in practice. This chapter attempts to put a bit of flesh on the skeleton by communicating some of the practical realities through a discussion of an important area, survey design. Thus we hope to avoid the impression that mastery of statistical analyses (z-test, t-test, and so on) is the most important matter, whereas practical matters such as collecting and reducing data are child's play.

14 ■ 1   INTRODUCTION

It has become an accepted practice in a wide variety of fields to collect the data necessary for decision making by taking a sample from the group under study. A manufacturer will randomly sample a number of parts from a batch presented by his supplier and will accept or reject the batch based on this sample. An epidemiologist samples the population of a certain city to ascertain the prevalence of a particular disease. Auditors sample entries from the books of their clients to estimate the dollars in error in their accounts. There is virtually no area of study in which sampling has not been applied with success in both accuracy and cost effectiveness. Indeed the generalizability of studies such as the Salk polio vaccine test, the National Crime survey, and, of course, social science studies depends to a large extent on the sampling techniques used.

In an age in which our lives are touched daily by samples such as the Gallup and Roper polls, the network news polls, and various political polls, it is important to note that sampling is a relatively new and rapidly growing field of science. It has become such an important tool for researchers in social science fields that it is hard to imagine that as recently as 1930 the *Encyclopedia of the Social Sciences* recommended a complete enumeration of the population during studies since, they claimed, the difficulties of sampling human populations were too great. Today these difficulties have largely been overcome.

14 ■ 2   CENSUS OR SAMPLE?

There are many reasons to conduct a sample survey instead of a complete enumeration or census. Perhaps the most important is that of money. It is often possible to achieve the desired accuracy in the survey at a much lower cost when a sample is taken. In fact, a sample can sometimes yield more accurate results than can a complete census or enumeration!

The data collected from a complete census require a large well-trained and widely distributed staff. If you are not able to pay well, or hire an existing survey organization, the data collected may be full of errors. Moreover, the enormous mass of data collected require enumerators, key punch operators, interviewers, and supervisors. At any step, errors can be introduced. Some of these errors are systematic and can introduce biases into the results. In contrast, the survey sample is often small enough in size so that it is possible to hire only top-notch interviewers, supervisors, and data analysts. Also, in a sample we can go back and check the quality of the work and reduce the systematic error appreciably. The main error in a census is likely to be a systematic error, the size of which cannot really be estimated exactly. By judicious choice of personnel and by checking and editing the data received in a sample, we can reduce the systematic error made in a sample survey so that the major portion of the error in a random sample occurs due to sampling variability (sampling error). Then by using probabilistic sampling models, we can estimate the size of the error.

In summary, sampling is advantageous when budgetary resources are limited. A complete census requires much more financial support than does a sample. A limited budget may require that resources be primarily spent on obtaining the data, and very little may be left for checking and editing. In a sample survey, however, we can adjust our sample size so that the data collection takes, say, only half our budget, with the other half going for checking, editing, and analysis.

Another advantage of sampling is *greater speed.* Clearly, if less data need to be collected, edited, and reported than is required in a complete enumeration, then the time period from start to finish in a survey will be less than in a census. This saving of time may be crucial to the timely presentation of data. For example, to obtain data on "cost of living," one does not wish to spend more than 1 month or so collecting data and developing a report of the results. Otherwise the results may be out of date by the time they are released. As another example, the 1964 Colombian census has never been reported! Only a 4% sample of the census, which was reported within months of the completion of the census, has ever been made public. The Colombian government could have saved considerable time and money by originally taking a 4% sample instead of conducting a census.

The sample survey and the complete census are not always in competition. There are times when the goal of the study necessitates a complete census, such as when information (say, about "quality of life") is desired for every village or town. Since local information is wanted, only a complete census will fill the bill. It is also desirable to have a census of the population every so often (every 10 years in the United States) since the information obtained on the number of persons, housing cost, population distribution, and so on can be used as a measuring stick for future studies and for measurements of change in time. To gather more intimate or detailed information, one can combine a sample survey with the census by having the interviewers ask additional questions of randomly selected persons. These data can then be matched up with the census results to give quite accurate estimates of the desired information for the population. Since the 1970 U.S. census, almost the entire census is now performed on a sampling basis (some information is, of course, required of each individual for legal reasons).

Even with the census, sampling is used to speed up preliminary publication of the results by as much as 1.5 years over the final enumeration of the results.

14 ▪ 3   HISTORICAL SUMMARY

The Central Limit Theorem and the Law of Large Numbers form the theoretical foundations on which the modern theory of sampling is based. The latter theorem states that repeated observations under the same conditions when averaged together will yield a value close to the population mean. The Central

Limit Theorem states that, provided the samples are reasonably large, the distribution of the averages so obtained can be approximated by the normal distribution. These theorems date back to Gauss and Laplace, about 1800. It was almost one hundred years before these results were used to provide the basis for a theory of sample surveys. A. N. Klaer in the 1895 meeting of the International Statistical Institute described work he carried out in Norway in which he used the proportional allocation method in stratification. His work was generally reviled; the census method was regarded as "sacred." No doubt much of this resistance was due to entrenched interests; those who made their livelihood and reputation through the administration of census studies might have consciously or unconsciously felt threatened by the introduction of new concepts and techniques that they would be required to master in order to maintain leadership in their field. Although shattering to our concept of how science should work, entrenched interests many times block scientific advance. (Here we find the major utility of the university — for it is able to present new ideas to young people who are not afraid of them.)

Bowley in about 1900 showed that a standard deviation of the average calculated from the sample information provides a measure of precision. This was a decisive step, since it basically introduced the idea of the probability sample. However, it was 25 years before the International Statistical Institute recognized previous advances by appointing a committee to study and report on them. Not until 35 years after Bowley's result, with a paper by Neyman (1934), was the importance of the probability sample recognized and institutionalized.

While this was going on, Gossett (who developed the *t*-test) and Fisher were laying the foundations of modern statistical theory in the context of experimental work in agriculture and medicine (circa 1920). They showed how even small sample sizes could be used in estimation and testing hypotheses.

Progress was slow in the 30's. In the United States "sampling was trusted neither by the public nor by members of Congress. . . . A common view of that time was that others could take risks with loose methods, and thereby undermine the basis for confidence in their data, but not the census bureaus."*

In 1947 the United Nations set up a Sub-Commission on Statistical Sampling Methods involving Fisher and other notable statisticians. Progress increased rapidly from that point; sampling methods became widely used and were authoritatively supported.

14 ▪ 4   STAGES IN A SAMPLE SURVEY

In general, one encounters various degrees of complexity in surveys, ranging from the relatively easy task of sampling from a hospital's files, which are

---

* Hansen, M. H., and W. G. Madow: Some important events in the historical development of sample surveys. *In* Owen, D. R. (ed.): *On the History of Statistics and Probability.* Ann Arbor, MI, M. Dekker, 1976.

numbered and kept in a single file case or microfilm record, to the difficult task of assessing the living habits of Indians living deep in the Amazon jungle where no maps are available and where suspicion of overly inquisitive strangers can be deadly. Nevertheless, there are certain principal steps that we will now discuss that are more or less common in implementing all surveys.

Preliminary Planning and Organization

Purpose of the Survey

The beginning stage is when the exact motivation behind the study is discussed.

> **Basic questions in the development of any survey are: Why is the survey being conducted? What is the nature of the problem being examined? Hence, what are the objectives of the survey?**

Application: A Case History

One of the authors spent a year in Iran (1978) designing and implementing a National Survey of Mental Health. The survey was encouraged and supported by the Shah's sister, Princess Farah, who felt that discontent was being created by the speed of modernization in Iran and that this modernization might be associated with an apparent increase (as reported by Iranian psychiatrists) in mental illness, especially clinical depression and anxiety. The main objectives of the survey (among several others) were to estimate the prevalence of these and other illnesses in various socioeconomic and religious groups and to find if these increases were associated with "modernity" and the subsequent alienation people might have felt resulting from the changing social and economic system.

We cannot, of course, report even superficially all the planning that went into this survey. The statement of the survey's objectives alone constituted a report of 60 pages. We present in several sections of this chapter snippets of memos and reports written in the course of the study design to try to give the reader some picture of the problems faced in a survey.

Memo written in the earliest days of the study:

> To: Deputy Minister of Health for Mental Health
> From: Chief Statistician
> Subject: Development of Survey Strategy
> (i) The first order of business is to develop a set of clearly stated objectives . . . and a detailed protocol of the survey. . . .

(ii) The ultimate use of the survey, as I understand it, is in developing a five-year plan for improving the mental well-being through mental and other health services. This issue should be spelled out as clearly as possible; namely, how the survey results will be used as input to the development of such a plan.

(iii) In light of (ii) above, an important issue is the *coverage* of the survey:

(a) One strategy is to design a survey from which results can be extrapolated to over 90% of the population. Past experience has shown that such a survey from start to finish will take about five years; much too long a period for practical purposes. Internal migration in Iran is so great and changes in the structure of the society so pronounced that by the time the results are reported, disseminated, and digested, they will no longer be pertinent. . . .

(b) Another strategy is to restrict the survey to two years and to design it so that results can be extrapolated to the most important sectors of the nation, about 75% of the population. . . .

(There follows a discussion of several other strategies and the strengths and weaknesses of each. Finally, it is stated that any final choice must wait upon the statement of objectives.)

(iv) Sampling Plan. The sampling plan must be based upon the objectives and strategy. However, it is already clear that the development of any plan requires estimates of the division of the population into urban and rural groups. Estimates should also be obtained of the proportion of major religious groups other than Muslim in the urban areas (Zorastrians, Jews, . . .) and ethnic groups (Arabs, Armenians, Kurdish tribesmen, Nomadic groups, etc.). . . .

(ix) Miscellaneous. . . . Will we be able to obtain data near the Iraq border? We will need personnel who can communicate with the Kurds. One specialist must be trained in the technique of getting into walled villages. . . .

After a great many memos and meetings, the objectives were finally developed in great detail. The following is the opening paragraph of a 60-page report on objectives:

The main objective of the study is to develop factual data concerning the prevalence of mental illness, drug abuse, alcoholism, and mental retardation in Iran, as a foundation for developing a national plan for mental health care and evaluating the effectiveness of those programs. The data base for planning must sample from a universe which permits the development of relevant information for program impact 5–10 years from now. Thus, it is not sufficient merely to count numbers of persons who are ill by geographic region. It is also necessary to examine the effects of urbanization, industrialization, modernity, and cross-cultural issues towards the goal of future planning.

Library Search

An initial step in the planning process concerns identifying any other studies that have been undertaken on the same subject in the same geographic area. Insights into the nature of the population and possible problems in implementing the survey may come to light.

Case History Continued

The chief statistician in charge of the Iranian survey previously discussed wrote himself an "aide memoire" which contained the following insight:

> In reviewing the literature on the 1972 epidemiological survey conducted by WHO* for improving health services in West Azerbaijan, and in speaking to colleagues who took part in this survey and the later health survey conducted by the Public Health School of the University of Teheran, it appears that there is an attitude perpetrated by unknown groups for reasons I cannot understand that any survey is to be mistrusted because it is used by the SAVAK† to obtain information on the population. I fear that should this attitude be widespread it may affect the validity of the results.

Precision

The survey sponsor should decide on the precision required to meet the goals. Every survey has a certain amount of error present due to sampling variability and nonsampling errors. By taking larger samples, we can reduce the sampling error, and by using more expensive measuring techniques and highly trained personnel, we can also reduce the nonsampling error. Of course, all of this requires money so that at this preliminary stage we should discuss the budget as well as how much error is tolerable for the client in order to still make valid decisions based on the results of the survey. If it appears impossible to perform the survey with the desired accuracy within the budget or time allotted, the planners should discuss this problem at this stage and either get the sponsors to relax their accuracy requirements, come up with more money, or look for alternate techniques for collecting the decision-making information.

Target Population

The preliminary planning stage is also the appropriate time to define the population that is to be studied. Are we concerned with institutions, households, cities, individuals, or accounts? Are we to restrict our attention to certain age groups or geographic areas or perhaps to other demographic variables such as

* World Health Organization
† The secret police of the Shah.

sex or education? The determination of the population to be studied may not be so clear as it seems at first glance. Suppose for example that the population of interest is adult household members. How do we define "members"? Are 16-year-olds adults? Are students who live at school considered as members of their parents' household? How do we handle hotel and dormitory residents? These types of questions should be considered to ensure that the population actually sampled is the same population as the one about which we want to obtain the information. The population of interest is called the *target population,* and the generalizability of the results from the sample to the target population depends on correctly defining the population to be sampled. Often the population to be sampled is chosen by convenience (e.g., psychologists often run experiments on first-year psychology students, and medical students are a favorite of certain medical studies). In these cases the generalizability from the sampled group to the target population may require a bit of argument.

Choice of a Sampling Plan

Once the precision and target population are determined, a decision is made of which type of sampling plan should be used. Different plans are discussed and the cost for achieving the desired accuracy for each plan is computed. The most accurate plan for the money is generally chosen. In surveys with multiple goals sometimes a compromise is necessary among the various goals. At this stage also one defines the *sampling units* or parts into which the target population is divided, and the *frame* or list of sampling units from which the sample is to be taken. In an employment questionnaire the sampling units may be the employers, the employees, the household personnel, or even the block of a neighborhood. On the other hand, sometimes the choice of a sampling unit is obvious, as when sampling transistors to determine the proportion of defectives (each transistor is a unit).

Protocol

The protocol is a document that details the implementation of the survey. It includes the sampling plan, questionnaire, pre-testing procedures, interviewer training, logistics of field work, data analysis, editing, and reporting.

Questionnaire Design: Coding

In this stage of the survey the objectives of the study are transformed into the actual questions that will be used to obtain the information. Often a trial-and-error process is involved with presampling and elimination or rewording of nonproductive questions. The method of administration (e.g., mailed out, telephone interviews, personal interviews, and so on) affects the actual question design, and psychology plays a part in the design. The length of the questionnaire, the type of questions asked (e.g., multiple choice or essay), and the sequence of presentation of the questions is decided.

The *coding* of the data goes together with the design, especially if large samples are taken requiring many observations of each member of the sample. For example, Figure 14.1 shows a questionnaire designed to obtain information on all members of a household. Note that the possible responses are assigned a number. If the first individual is male, a 1 is placed in box 16. If that person is not at home but is visiting someone, an 03 is placed in boxes 23 and 24. If there is a response that is not listed on the form (e.g., suppose the person is in the hospital), the instructions on the form read "or write (see code)"; this means look up the code for that response in the code book provided to the interviewer.

Coding of information enables the quick and easy transfer of data to com-

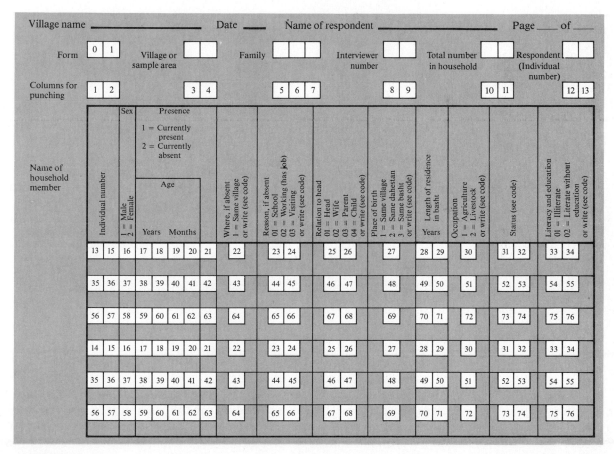

Figure 14 ■ 1
Case history survey form.

Organizational Chart

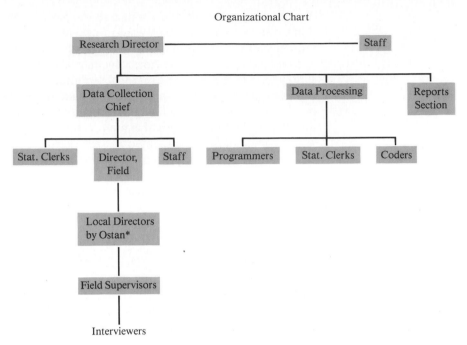

Figure 14 ■ 2
Case history continued.

puter storage and so makes the editing and analyses of large data sets possible in a reasonable time.

Codes need not be filled in directly on the questionnaire as is done on the previous one. Nevertheless, codes must be provided for responses, which should be done before the survey begins, if possible.

Organization

Figure 14.2 displays the outline of a typical organizational chart for a large-scale research survey. (This framework was used for the Iranian survey previously discussed.)

Pretesting

It is usually productive to pretest the questionnaire on a small scale prior to the full-scale study. Ambiguous questions and fieldwork problems (interviewer problems, organizational gaps, and so on) can be detected early and corrected before they cost time, money, and accuracy in the actual survey. Serious problems are sometimes uncovered by pretesting which would destroy the survey

* An Ostan in Iran is akin to a "state" in the United States.

itself (e.g., the interviewer cost may be many times larger than projected, or the nonresponse rate may be too high.)

Fieldwork

This stage of the survey concerns the actual questionnaire administration and data collection. The interviewers find (hopefully) the respondents chosen for the sample. Occasional checks on the interviews and the progress of the survey are made and corrections are implemented along the way as necessary. Remailing is made for mailed out questionnaires, and call backs may be necessary for telephone interviews.

Editing

During the fieldwork of a lengthy survey, or at the end of a shorter one, the data are edited by computer for errors. Logical errors and coding errors (e.g., a 3-month pregnant male, or an unmarried man whose spouse is at home) are detected by comparing questions. These errors also may be corrected (if possible) by either referring back to the original interviewer's notes or reinterviewing if necessary. Editing can also be used to signal highly unlikely responses for future checking. For example, suppose one question is "How often do you read the morning paper? (a) every day, (b) three to six times per week, (c) one to three times per week, or (d) never." Another question is "How many years of schooling have you completed? (1) 0 to 2 years, (2) 2 to 6 years, (3) 7 to 12 years, or (4) more than 12 years." If a large proportion of respondents answer (a) and (1), respectively, then one should be suspicious, for although this is not a logical contradiction in answers, it is an unlikely pair. Editing can be utilized to discover such illogical or unlikely responses and the reasons searched out and referred back to the fieldwork process. This stage can be a check on the fieldwork as well as a preliminary step to the actual analysis.

Analysis of the Data

There are two "analysis" stages. The first is the planning stage where the exact analysis to be used should be stated, including the reasons for using the analysis and the objectives of the analysis. The second stage is the actual analysis.

This second stage may consist only of presentation of a tabulation of the results of the survey, or could involve complex statistical analysis based on the data. For example, regression analysis, analysis of variance, discriminant analysis using information gain (DIG) (discussed in the next chapter), and so on can be performed on the data. Of course, the exact method of analysis depends on the purposes of the study.

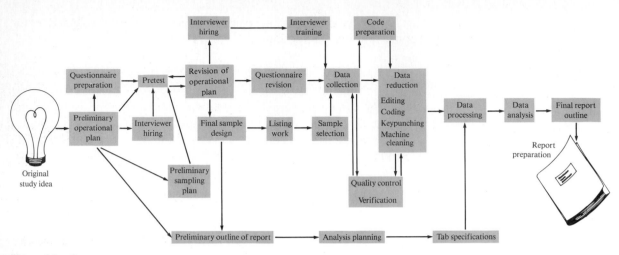

Figure 14 ▪ 3
Stages of a survey. (Taken from "What Is a Survey?" by Ferber, Sheatsley, Turner, and Waksberg, American Statistical Association, Washington, D.C.)

Reporting the Results

Again, there are two stages; the first is the planning of the final report even before the data are collected. What kind of tables will be used, how many, in how much detail? The second stage is the actual writing of the report.

The final report on the survey is a summary of the findings. This information may be of interest in its own right and is also useful for any future studies planned on the target population. Any information gained in a survey is potentially useful in improving the next survey.*

We summarize the preceding stages of a survey in Figure 14.3.

14 ▪ 5 TYPES OF SURVEYS

We generally divide surveys into five types depending on the purposes and information desired. The first type is a *descriptive* questionnaire. The purpose of this type of survey is to describe the target population. Questions such as the average expenditure per household for food, the average number of acres of farmland under production, the total unemployment, the total dollar error in an accounting balance, the proportion of the population earning between

---

* A useful reference is "Recommendations for the preparation of sample survey reports." Department of Economic and Social Affairs Statistical Office of the United Nations, Statistical Papers, Series C, No. 1, Rev. 2, United Nations, New York, 1964, reprinted in *Readings and Applied Statistics* by W. S. Peters, ed., Englewood Cliffs, NJ, Prentice-Hall, 1969.

$15,000 to $25,000 per year, and the proportion of school children who are adequately inoculated against polio are addressed by a descriptive survey.

The *analytical survey* is used to compare measurements on a single population unit. Typical analytical surveys might compute the ratio of the dollars spent on entertainment versus housing per family, or the ratio of the cost of safety features versus nonfunctional chrome accessories per car produced by the Detroit automobile manufacturers.

Repeated surveys are those for which there will be a repeated administration at a later date. Examples of this are the presidential popularity poll, the consumer price index, the unemployment rate, and the current population survey. An important type of repeated survey is the "longitudinal study." This is a study of a fixed group, called a *cohort,* in time. For example, in studying long-term side effects of drugs, a group of people who had taken a drug may be given a physical examination every year for a period of 10 years. Another group similar in age, sex, and so on but that did not take the drug may also be so examined. This latter group is the *control* group.

Two other important types of surveys are the *classification* and *discrimination* surveys. In the *classification questionnaire* one takes a sample from the target population and divides (or classifies) the sample into groups, each of which is as nearly homogeneous as possible with respect to response pattern. The responses should be as different as possible between groups. Examples of classification questionnaires might include studies on "adaptation" of rural migrants into urban centers. The questionnaire is given to the migrants and one attempts to split the respondents into "good adaptors," "moderate adaptors," and "poor adaptors." The divisions are not given to us so we must make them based on the response patterns found in the data. Another example might be a multiple-choice test (questionnaire). The administrator must divide the respondents into groups (grades) based on their response patterns.

The *discrimination questionnaire* on the other hand takes samples from known groups and gives them the questionnaire. If a new respondent is given the questionnaire, we wish to place him in the correct group based on his questionnaire response as compared to the responses from the *known* groups. An example of a discrimination questionnaire would be a psychiatric screening exam. One administers the questionnaire to a group of known mentally ill patients and to a group of known "normal" individuals. Using the responses from these two groups, we can decide to allot a new individual to the ill or normal group based on his or her questionnaire score as it compares to the scores of the ill and normal groups. This can result in considerable savings since the questionnaire may be administered by a trained staff member, whereas a mental exam may require a psychiatrist.

A major difference between the classification and discrimination questionnaire is that on the discrimination questionnaire the respondent groups are known ahead of time and are given to us. In the classification questionnaire the groups themselves must be determined from the data.

14 ▪ 6   CHECKING INTERVIEWERS: WHO'S LYING, WHO'S UNPOPULAR?

We have presented an overview of "survey planning." We now treat a specific area—methods for checking on interviewers. We chose this area to demonstrate both an important source of problems in survey design and the usefulness of the methodological tools we have previously studied.

The following is taken from a memo written (1978) to the Deputy Minister of Health for Mental Health in Iran regarding the National Survey of Mental Health discussed previously:

> . . . interviewer supervisors should be shifted for a couple of days so that we don't develop a whole section that could "fake" interviews. The gravest mistake made by neophytes in field surveys is to develop too much trust in the interviewers. Every field survey has some interviewers who do not conduct their interviews or fill in answers to questions that were never asked. This creates an unacceptable source of non-sampling error. The only solution to this is to check, recheck, and recheck again . . .

One method of checking interviewers is for supervisors to reinterview some respondents to determine if previous answers were recorded accurately. This kind of checking is akin to "quality control" on production lines where a sample is taken to determine if the line is operating as it should. This type of checking must be done but can be supplemented by analyses of the data that can be done in both the field and the data collection center.

Suppose each of six interviewers have the task of measuring the weight of the population members of a village. They have a quota of at least 90 people per week. The following table shows the number of people each weighed in a 3-day period in two weight intervals:

		Weight Interval (lb)		
		50–150	150–200	Total
	1	10	35	45
	2	6	42	48
	3	8	37	45
Interviewer	4	30	20	50
	5	3	32	35
	6	9	63	72

This is a 6×2 contingency table and we can use a χ^2-analysis to determine if the proportions in each weight group are the same for each interviewer. If not, then we are suspicious about the results. We next examine the data more closely to determine which interviewers *may* have been faking data.

We find that χ^2 is approximately 52 which exceeds $\chi^2_{0.05}$ with 5 degrees of freedom. On examining the data, we become suspicious about the work of interviewers 4 and 6 — 4 because of the many people with their weight between 50 to 150 lb; 6 because of the large number of people he weighed. On investigating, we find that an elementary school was included in 4's interviewing schedule. However, we find that 6 did not weigh the 72 people on his list (he wanted to finish his quota early in the week so that he could go home for the weekend), and so 6 was fired.

Another problem faced in a survey are "refusals," those people who refuse to be interviewed. Refusals are a source of error. Some interviewers, because of their personality, are refused more often than others. It is important to detect early an interviewer who is refused too much. The following table shows a 2×6 contingency table of acceptability by interviewers:

		Interviewer					
		1	2	3	4	5	6
Acceptability	No. of refusals	8	6	7	2	14	0
	No. of nonrefusals	32	29	38	30	20	35

A large value of χ^2 leads us to suspect that acceptability is associated with the interviewer. The value of χ^2 here is about 24 which is significant at the 0.05 level. Further investigation shows that interviewer 5 has low acceptability and 6 has high acceptability. Interviewer 6 is congratulated, whereas 5 is sent back for retraining.

The problems at the end of the chapter present examples in which other types of analyses are used.

14 ▪ 7 APPLICATION: THE BREAST-FEEDING CONTROVERSY (OPTIONAL)

A worldwide controversy simmered between commercial baby food companies and some nutritionists and pediatricians for many years. It finally broke out into the world press in 1980 when the World Health Assembly of the World Health Organization (WHO) voted to oppose the replacement of breast milk for infants by baby foods that were advertised as breast milk substitutes. WHO believes that breast milk is necessary for the health of infants. Without going into the arguments put forth by the antagonists in this controversy, we bring this matter up because it motivated WHO in 1981 to begin studies that were aimed among other things at determining the prevalence of breast-feeding in many countries. This prevalence figure serves as a baseline figure for each country; subsequent surveys would determine if the prevalence were increasing or decreasing. If decreasing, and the Ministry of Health in the country decided that breast milk is necessary for infant health, then the country would need to expend resources in time, personnel, and money in a program to convince mothers to breast-feed their children.

Jamaica was one of the countries chosen. As you recall from your reading, a statement of the survey objectives and the general survey plan should be written first. This statement by the Jamaicans is now given.

1. Note where in the report each of the following points is explained:
   a. purpose or objective of the survey
   b. background literature
   c. desired precision and accuracy
   d. universe* and target population
   e. choice of sampling plan
   f. funding
   g. survey organization
2. Suppose you were designing such a survey in the region in which you live. Which characteristics of the society described in the report would be much different in your society?

Determination of Breast-Feeding Patterns in Jamaica

by H. C. Fox and D. Ashley†
Statement of *General Plan*

Description of Country

Jamaica is an island with a total land area of 4243 sq mi. It is very mountainous with almost 50% of the land being more than 1000 ft above sea level. Approximately 1.68 million acres or 62% of the land is suitable for some agriculture, but increasing and unplanned urbanization has been continually encroaching on good agricultural land. The country has relatively high population density and urban/rural ratios. The ratio of population to gross land space was 1.3 persons per acre in 1978. Of the total population, 45% live in the urban areas in settlements of relatively low population densities.

The total population is 2.2 million, with an age breakdown as follows:

---

* The universe consists of selected places where the target population resides.
† Maternal and Child Health Care Division of the Ministry of Health, Jamaica, 1982. This report has been edited by us. Any errors are our responsibility.

0 – 11 months	58,600
1 – 4 years	193,400
5 – 14 years	609,100
15 – 19 years	253,700
20 – 29 years	331,700
30 – 39 years	184,600
40 – 49 years	161,900
50 – 59 years	142,600
60 – years	228,900

The ethnic composition is predominantly persons of African descent with small proportions of East Indians, Chinese, Europeans, and those of mixed descent. Although of several ethnic origins, the population does not have social divisions based purely on these lines.

Purpose of Survey

Since the 1950's the pattern of malnutrition in infants between 0 and 12 months has changed to reflect a higher incidence of the disease, and a marked decrease in the age at which malnutrition was most commonly found (Reddy, 1971; Fox, 1963; Fox, 1970). This shift in the pattern of malnutrition has been associated with the decline in breast-feeding that seems to have occurred as women increasingly became a part of the work force, as industrialization progressed, and economic constraints forced them to seek wage-earning employment outside the home. This together with commercialization of infant formula, which appeared to enable mothers to work, freeing them from the ties of breast-feeding, had deleterious effects on breast-feeding patterns (Marchione, 1980). More recently, the Jamaican government, recognizing the harmful effects of non-breast-feeding on the health of infants, introduced several policies to counteract the trend and to stimulate greater breast-feeding in all segments of the population. Although studies of breast-feeding patterns have shown that a high proportion of the population is at least partially breast-fed for variable but usually short durations, little or no data exist on the breast-feeding patterns among different socioeconomic groups of the population. Given the heterogeneity of the population in terms of socioeconomic status inter alia, it is important that planners have some knowledge of the variations between groups in order to plan effectively their strategy for increasing the prevalence of breastfeeding. With the implementation of a national nutrition education program that has as one of its objectives the promotion of breast-feeding, especially through community education activities, a study of breast-feeding patterns among different groups is apropos. Health workers who have a greater understanding of the knowledge, attitudes, and practices of their community, and the determinants of its behavior, will be better equipped to address the problems of the community.

The purposes of the survey are:

1. to determine the prevalence and duration of breast-feeding both at the national level and within different socioeconomic groups in Jamaica
2. to identify the main factors associated with breast-feeding patterns in the different groups

The survey is the responsibility of the Ministry of Health, with all staffing, that is, interviewers, supervisors, coordinators, coders, and editors, being persons within that ministry.

Age Groups of Interest

Traditionally, Jamaican women breastfed for prolonged periods. However, the pattern of breast-feeding has changed considerably, with much shorter durations now being the norm and few mothers breast-feeding beyond 1 year. Given the desirability of having breast-feeding for 1 year and that the Ministry of Health's nutrition education program encourages breast-feeding for that duration, exclusively for the first 4 months, the age group of interest in this study is 0 to 11 months, with the breakdown as follows: under 1, 2, 3, 4, 5, 6 to 8, 9 to 11.

Selection of Universe

The prevalence and duration of breast-feeding appear to vary according to socioeconomic conditions of the population in Jamaica. Although the population is quite small, there is considerable diversity in the way of life, health, and behavior of the people. Thus the breast-feeding patterns can be expected to vary according to the lifestyle of the different population groups. For this reason, as well as for the need for interregional comparison, more than one universe was selected and the sample design was planned accordingly. The universes selected were:

1. urban, middle, and upper income households
2. urban, low income households
3. rural households
4. rural households in bauxite industry
5. households in tourism

These universes do not exhaust all the population groups existing in Jamaica but are representative of the majority. Given the limited resources available to the Ministry of Health (MOH), it was felt that selection of these universes would allow for workable implementation of the survey while providing useful information both about important groups and at a national level. Urban areas have been selected because of the general trend of declining breast-feeding with increasing urbanization. The urban area has been broken down into two income groups since income levels and related characteristics appear to be associated with patterns of breast-feeding. Rural agricultural households have also been selected because they are an 'at risk' group nutritionally and it can be expected that urban influences may have affected breast-feeding patterns. The mining area is a relatively newly emergent subgroup that has a peculiar feature that may influence their breast-feeding pattern. In comparison with the rest of the rural population, they earn much higher wages and their aspirations and acquisitions resemble those of the urban middle-class population. To what extent this has influenced breast-feeding patterns is of interest to health planners, especially since increasing industrialization of the rural areas is expected with economic development. Lastly, the tourism area has been chosen because of its special characteristics, that is, high but seasonal employment, high level of foreign influence, and greater access to consumer goods, in particular, commercial infant preparations.

Sample Size

Bearing in mind the objectives of the survey and the age group of interest, the sample size had to be large enough to permit statistically valid individual rates for each age group within each universe. The sample should include a total of 2500 children divided among five universes with 500 children in each universe in order to obtain a reasonably small confidence interval (approximately plus or minus 5%) of confidence level 90%. Although recognizing the general rule that a universe sample size of approximately 900 is appropriate, MOH realized that its sample size could be reduced as fewer age groups of interest had been selected.

Sample Design

With the exception of the urban universes, each universe was comprised of randomly selected subdivisions located in more than two parishes (political divisions) in the country. The urban universe was also composed of randomly selected subdivisions, but all were located within a single parish. Although other urban areas existed, it was felt that representative data could be obtained from a single urban area.

The 1970 census, although conducted several years ago, provided reasonable divisions that could be used to select the clusters for the survey. These *enumeration divisions* (ED) are the smallest political divisions, redefined during the 1970 census for equitable distribution of work among field-workers. Population estimates of these EDs were obtained by projecting their 1970 census figures. For each subdivision eight enumeration divisions were selected. All the EDs in a universe were listed with their cumulative populations. The sampling units were determined by multiplying the number of subdivisions in each universe by 4. The sampling interval was calculated by dividing the total population by the population of the sampling unit. A random number between 1 and the sampling interval was selected by using a table of random numbers. The first cluster was the ED containing the cumulative population of the random number. Next, the sampling interval was added to the random number and the total used to determine the second cluster that was the ED whose cumulative population was less than or equal to that total. The sampling interval was added sequentially to the previous figure until all the sampling units were identified. The location of each sampling unit was in the ED whose cumulative population is less than or equal to the number resulting from adding the sampling interval.

The number of children in each cluster varied depending on the total number of subdivisions in the universe. For example, the universe of rural agricultural households had 10 subdivisions and approximately 4 enumeration divisions in each, for a total of 40 enumeration divisions. Therefore the number of children in each cluster was the required sample size per universe (500), divided by the number of enumeration divisions, that is, approximately 12 children per cluster. Each cluster consisted of contiguous households obtained by proceeding in a predetermined direction from the starting point until the required number of children under 12 months was obtained.

The method of selection of the starting point was the same regardless of the universe. The census publication included detailed maps of all enumeration divisions and indicated by roads, tracks, river, and so on the boundary of each ED. The starting point for each unit was the beginning of the first boundary described in the census book.

Questionnaire Development

The questionnaire* at the end of this report was used to obtain the prevalence plus information in the socioeconomic situation, attitudes, and practices regarding breast-feeding. In addition, background demographic and ecological data of the sample area were included.

Selection and Training of Survey Personnel

All survey personnel are selected from the Ministry of Health (MOH) staff. The field team will consist of 25, two-person interviewing teams and 4 supervisors who are responsible for logistical arrangements and technical aspects. Interviewers will be Community Health Aides who are front-line workers. All field personnel will be given intensive training by the project coordinator on (1) the purpose of the survey, (2) their role as interviewers or supervisors and (3) implementation of the survey including, (a) selection of the sample and (b) interviewing and completion of the questionnaire. The trained field staff will perform a pretest on 10% of the survey sample, using enumeration divisions that are not included in the sample. The pretest is made to (1) evaluate the

---

* These questionnaires are not presented here.

questionnaire, (2) test sampling methodology, and (3) test the interviewers. Field staff will then be retrained immediately (if necessary) before the actual survey. At the second training session, problems encountered in the pretest will be ironed out, knowledge acquired at the first training session reinforced, and the revised questionnaire reviewed.

Data Analysis

The cross-tabulations to be performed include:

1. prevalence of breast-feeding by age of infant
2. duration of exclusive breast-feeding by infant age
3. duration of breast-feeding with or without supplementation
4. duration of breast-feeding without artificial milk supplementation
5. the introduction of various supplements to the diet by age of introduction
6. the proportion of mothers in each maternal age group
7. socioeconomic data for each universe
   a. average household size, range ± SD
   b. average per capita food expenditure
   c. proportion of food expenditure: total expenditure
   d. employment status of (i) main provider and (ii) mother
   e. social index — a composite of presence or absence of electricity, toilet, water supply, specific household amenities, and crowding index
8. duration of diarrhea by age
9. episode of diarrhea by (i) breast-feeding status and (ii) type of water supply
10. knowledge and attitudes regarding (i) breast-feeding and (ii) diarrhea
11. contraceptive use by mother's age — type of contraceptive used by mother's age

Where it is expected that too little data will be obtained to allow for the age group breakdown specified, age groups will be collapsed into larger intervals to provide large enough numbers for statistically valid conclusions.

Use of Information

It is expected that the information will be used by MOH planners to determine their strategy in accordance with national health policies. The information should be especially useful to those involved in health education.

Summary

We have discussed the advantages and disadvantages of sample surveys over complete enumerations of a population. The advantages include greater speed, less cost, and sometimes, greater accuracy.

The stages, generally, in implementing a survey were described. These include: (1) a statement of purposes and objectives, (2) the library search, (3) specification of the universe, (4) statement of precision and accuracy, (5) sampling plan, (6) development of a protocol, (7) design of questionnaires and their coding, (8) description of the survey organization, (9) plans for pretesting, (10) description of fieldwork to be done, (11) reduction, editing, and analysis of the data and (12) reporting.

A brief description of different types of surveys was given. Finally, techniques for checking the reliability of interviewers were discussed.

Key Terms

sample survey	analytical survey
census	repeated survey
target population	classification survey
universe	classification questionnaire
protocol	discrimination question-
coding	naire
pretesting	cohort
editing	

References

Fox, H. C., et al.: Jamaica. *Inf. Bull. Sci. Res. Council, 8*:33, 1968.

Fox, H. C., et al.: A rapid survey to assess nutritional status of Jamaican infants and young children. *TRS of Trop. Med., 66*(4):653–662, 1970.

Marchione, T.: A history of breastfeeding practices in the English-speaking Caribbean in the twentieth century. *Food and Nutrition Bulletin, 2*(2):9–18, 1980.

Reddy, S. T.: Artificial feeding in Jamaica and Barbados. *W.I. Med. Journal, 20*:198–207, 1971.

Exercises

Section 14.1

1. In performing a sample survey in a small town in Colombia, the investigators found that the survey angered the population strongly enough to influence the results. Those not selected to be interviewed were angry because they believed that they were being excluded from a good thing, and that those selected were being paid for their time. Thus they tried to influence those selected into not cooperating with the survey. Those selected had their anger nurtured by a suspicion as to why they were picked to be interviewed in the first place. The investigators tried to solve the problem by distributing information to the population about the goals of the survey and the logic behind the sampling. It did not work and the investigators

finally had to make a complete census of the area. What might the investigators have done to detect the problem and solve it before it affected the survey?

2. (i) The oldest recorded total population census is found in the Old Testament. This census, and many other types of census counts, are described in *Numbers,* the fourth book of the Pentateuch. This book begins with the words, "And the Lord spoke unto Moses in the wilderness of Sinai . . . 'Take ye the sum of all the congregation of the children of Israel, by their families, by their fathers' houses, according to the number of names, every male, by their polls' "* (*Numbers,* Chap. 1, V. 1,2.)

Notice that Moses was invoked to record the census from a number of different points of view.

a. Families and fathers' houses were to be counted. Why both?

b. Why count families and the number of names? (Hint: Not all individuals lived in their father's house; servants who lived in a house were not members of the family.)

c. How does one obtain the number of males and females from the data Moses was asked to collect?

Notice that the age distribution of the population cannot be ascertained from the type of census information previously described. Clearly, age distribution by sex is an important characteristic of any population. This information was supplied by special census counts that were smaller and more frequent than the census already discussed. One such special census was that of warriors (males 20 years upward). The ingenious way in which warriors were counted is described in *Exodus* XXX, V. 11–15.

(ii) Moses also performed surveys. In *Numbers* XIII, verses 17–20, he sends 12 specially picked men to do a surreptitious survey of Canaan. He gives them the following instructions:

Get you up here into the South, and go up into the mountains; and see the land, what it is; and the people that dwelleth therein, whether they are strong or weak, whether they are few or many; and what the land is they dwelleth in, whether it is good or bad; and what cities they are that they dwell in, whether in camps, or in strongholds; and what the land is, whether it is fat or lean, whether there is wood therein or not, And be ye of good courage, and bring of the fruit of the land.

a. What are the potential dangers of this type of survey? (The children of Israel suffered dire consequences when some of the potential dangers turned into realities.)

b. Do you believe that Moses, by sending his men to look over the land from some mountain vantage point, wished them to map out some plan of sampling before they went to look over the land in detail?

c. What sample did Moses ask his men to bring back? Did Moses want a representative sample? Why did he want the sample?

Section 14.4

3. A small university in a southern city in the United States has only limited funds; the administration must decide whether to continue losing one million dollars per year on its football team or to give up football and spend the money it saves on expanding library facilities. One faction in the administration that is favorable to keeping the football team argues that many students are attracted to the university because of its football team and that admissions applications will decrease if football is abolished. The administration wants to survey the students to determine the validity of this argument.

* Literally "skulls"; individual persons as in the English *poll* tax or *head* money.

The university has a school of Arts & Sciences (1000 freshmen, 700 sophomores, and 1100 upper-class students of which 600 are juniors, 500 seniors). Ten percent of upper-class students are English majors, 15% are history and philosophy majors, 11% are mathematics majors, 20% are economics majors, 30% are psychology majors, 20% are biology majors, and the rest are distributed among other fields; it has an engineering school (600 students), a school of architecture (200 students), a medical school (150 students), and a graduate school of 2000 students.

The administration has only $5000 to spend on the survey.

a. What additional information is needed in order to develop a practical survey plan?

b. Suppose you have obtained this information (make up some hypothetical information), state clearly the purpose of the survey, the target population, and the sampling plan.

c. With respect to questionnaire design, would you ask

"Did you come to — University because of its football team?" or

"Which of the following attracted you to — University?

(Rank in order of importance, (1) most important, (2) next in importance, and so on.)

    —— the faculty
    —— the campus
    —— the football team
    —— the library
    —— the social life

or would you ask different questions? (If so, what question or questions?)

d. What background information would you request of each respondent; (e.g., sex, age, class, school)?

e. Suppose it costs $10 to interview a student (this includes the cost of the interviewer, the cost of printing the questionnaires, the editing and reduction of data, and so on). What is the maximum possible sample size?

f. What sampling design would you use? Explain.

g. Would you pretest the questionnaire? If so, what sample size would you use in the pretest?

h. What tables would you prepare for your report to show the characteristics of the respondents? For example,

Students interviewed

No. of males	No. of females

or

Responses to question no. 5 of Arts & Science students

	Yes	No	Do not know	Refusal
Lower-class students				
Upper-class students				

i. What "checking tables" would you prepare to pick up mistakes made in the data collection and reduction?

4. Do problems 12 and 14 of Chapter 1. You now have had some instruction in sampling and survey design. If you had answered these problems while studying Chapter 1, how do your answers then differ from your answers now?

Problems 5 and 6 deal with "Reporting of Results."

5. Do problem 9 of Chapter 1. What additional questions might you have about the survey?

6. An important part in reporting results is the proper preparation of tables. It is assumed that any literate person can read a table of information intelligently and even prepare one if necessary. As any unscrupulous advertiser or propoganda expert can tell you, tables and charts are wonderful instruments for spreading false information; or, at least, false impressions. The island of Fictionia published the following table:

Crimes per 100,000 Population in *Fictionia* vs. *USA* (1994)		
	Fictionia	*USA*
Violent crimes		
Murder	0	3
Rape	1	20
Assault	0	60
Nonviolent Crimes		
Robbery	1	40
Fraud	0	150
Corruption	0	200

Fictionia is indeed a wonderland according to this table. The table fails to transmit some vital information. Is "crime" defined as "reported incidents," "arrests," or "convictions"? Find other faults with this table.

7. Problem 6 has been contrived to display dangers in accepting a table of results without critical scrutiny. Many tables either convey incorrect information or no information because they are incompetently prepared. Consider the table shown here that was introduced by the Texas Department of Public Safety, March 1974, in a report entitled "A Study of Police Height Requirement" in an attempt to relate the effectiveness of police officers to their height. Why is the table confusing and misleading?

Height Compared to Accidents			
Height (in.)	*I* *Percent of all assaults*	*II* *Percent of all officers*	*III* *Difference (I–II)*
68	9.9	5.0	+4.9
69	13.7	13.6	+0.1
70	17.6	24.8	−7.2
71	23.5	18.6	+4.9
72	7.8	14.0	−6.2
73	17.6	15.8	+1.8
74	7.8	4.5	+3.3
75	0.02	2.2	−2.18
76	0.0	1.1	−1.1

Height (in.)	*IV* $(I-II)^2$	$\dfrac{(I-II)^2}{II}$
68	24.91	4.98
69	0.01	0.01
70	51.84	2.09
71	24.01	1.29
72	38.44	2.74
73	3.24	0.20
74	10.89	2.42
75	4.75	2.15
76	1.21	1.10

$$X^2 = 16.98$$

$$P = 0.05$$

(significant)

8. A type of table used throughout this text is the *special purpose table*. This is a table intended to emphasize certain special, numerical features of collected data, either special relationships or characteristics of those data. This is the kind of table used in published papers, in the preparation of reports, and for the tabulations that should accompany whatever graphs are incorporated in a report.

The parts of a model "special purpose table." This type of table has (i) a *title*. The title may be a simple statement of content, consisting of a title and subtitle. It should be *brief* and *precise*.

The table has (ii) a boxhead. By a boxhead is understood the series of column captions running crossways of the table. What are the features of a good boxhead? The labels should include as few words as possible and yet be sufficiently clear to explain exactly what data are included in the several columns.

A common mistake is the failure to show the units of measurement for items of a column. For instance, if you put down only the word *age,* there is a possibility for confusion, as to whether it is age in months or in years. The proper label is "age in years." Be as specific as you can.

The (iii) stub is the technical term for the captions along the left vertical margin of the table. It commonly contains a series of row captions or side headings. Whenever possible, it is desirable for easy reading to arrange the material in the stub in groups, for example, by a certain series of years, instead of having one long whole column.

Go through the *Wall Street Journal* or other newspaper and magazines and find some special purpose tables that do and do not obey the rules outlined in (i) to (iii).

9. The reporting of statistics is a delicate matter that must be done with professional care. The following is reprinted from the *International Herald Tribune* (January 12, 1984, page 2) under a banner headline reading "Saudis Credit Islamic Law With Keeping Crime Rate Down."

"According to [Saudi] government statistics, there were only 14,220 major and minor crimes committed in 1982 in Saudi Arabia, a country of seven million people. In comparison, Los Angeles County, also with a population of seven million, recorded 159,662 arrests for felony crimes in 1982. Misdemeanor arrests totalled 339,837, and there were 1,415 murders committed."

Why is this reporting misleading?

CHAPTER · 15

Questionnaires

Professional pollsters will tell you that in any survey of public opinion, two factors are paramount. The first is the selection of a sample; the second is the wording of the question. Of the two, the question probably counts for more.

James Kilpatrick
Austin American Statesman

I hate quotations. Tell me what you know.

Ralph Waldo Emerson
Journal

In most applications of statistics in the field, the statistician does not work alone but, instead, works closely with those who have expertise in other areas. One important area where statisticians work in an interdisciplinary environment is that of questionnaire design. They often work with sociologists, psychologists and psychiatrists, and other professionals who use questionnaires as instruments in gathering information.

Two main issues that confront the statistician in questionnaire design are: (1) How can the questionnaire be made a more efficient instrument for collecting information? (2) What are the best methods for analyzing questionnaire information? In this chapter we develop some of the background that is necessary to explore these issues.

15 ▪ 1 INTRODUCTION

We have discussed the planning and analysis of surveys. One important tool for collecting survey information is the *questionnaire*. The construction of a questionnaire may appear to be easy—after all, is not one simply asking questions to obtain desired information? True, but experience often shows the matter is not simple—questions whose meaning seems perfectly obvious to the interviewer are not clearly understood by others, questions that seem entirely objective in content appear as highly biased to the listener, innocuous sounding questions may be so offensive that respondents may refuse to answer, and, worst of all, questions do not, on analysis of the results, provide answers to the problem under investigation.

Usually before the construction of a questionnaire can begin, many weeks of planning, reading, and exploratory sampling must be done. We must identify the population under investigation called the *target population*. For example, will we be dealing with children, adults, company directors, women of reproductive age, and so on? In addition, we must ascertain how large the sample will be, if respondents will be reinterviewed, and if our investigation is influenced by the season as, for example, in inquiring about the prevalence of pneumonia.

Many of these matters have been discussed in Chapter 14 on survey planning and analysis. Assuming that this aspect of questionnaire construction has been completed, we go on to discuss fundamental principles of questionnaire design and analysis. Our objective here is to clarify these principles as well as to explain the basic terminology and methodology used in the study of questionnaires so that the student may advance to more detailed texts and be able to employ this instrument in fields such as psychology, business, government, and so on.

15 ▪ 2 THE STRUCTURE OF A QUESTIONNAIRE

In general, the word *questionnaire* as used here refers to a list of questions on a form that a respondent fills in himself or herself or that is filled in by an interviewer in a face-to-face situation. A questionnaire may be (1) mailed, (2) handed out to a captured audience as is sometimes done by some airlines that, during the flight, give their passengers questionnaires to fill in about the service, or (3) filled in through an interview. There are any number of ways questionnaire data can be distributed and collected. The *method* of data collection (mail questionnaires, direct interviews, telephone interviews, and so on) should be decided before the first question is written, since certain types of questions should be filled out by an impartial interviewer, certain types of questions are too involved for a mailed-out questionnaire, and so on.

The next decision that must be made concerns the *structure* of the questionnaire. The structure includes the buildup of question sequences and the order of the questions within each question sequence. For example, one must decide whether or not to begin with attitudinal questions (e.g., "Do you believe politicians are honest?") as opposed to factual questions ("How old are you?"); or to begin with *funneling questions* we now describe.

Example 15 ▪ 1 "Funneling" questions are those questions that are designed to divide respondents into several groups. For example, the following set of questions divides respondents into groups having certain tastes in recreational activity:

1. With respect to sports, do you prefer to be a participant or a spectator?

                         participant   observer

Check one   [   ]   [   ]

If the answer to 1 is "participant" proceed to Section A; otherwise go to Section B.

Section A	*Section B*
2A Do you prefer (a) winter (b) summer sports (a) [ ] (b) [ ]	2B Do you prefer to (a) watch live sports events (b) watch them on television (a) [ ] (b) [ ]
3A If the answer to 2A is (a), do you prefer skiing or ice skating . . . ?	3C If the answer to 2B is (a), which of the following sports do you prefer . . . ?
3B If your answer to 2A is (b), do you . . . ?	3D If your answer to 2B is (b) do you . . . ?

As another example, consider the tax form that is a questionnaire containing many funneling questions. The following are questions 9, 10, and 11a that appeared on the 1968 1040 Federal Tax Return form:

    9. Total income. . . . . [   ]
  10. If you do not itemize deductions and line 9 is under $5000, find tax in tables . . . Omit lines 11a, b, c, or d. Enter tax on line 12a.
11a. If you itemize deductions, enter total from page 2, Part IV, line 17. If you do not itemize deductions, and line 9 is $5000 or more, enter the larger of (1) 10% of line 9 or (2) $200 ($100 if married and filing separate returns plus $100 for each exemption claimed in line 4 above).

(For those who believe it is easy to construct a lucid questionnaire, please study the tax form.)

In constructing a questionnaire, we should note that the buildup of questions is important. For example, in a psychiatric questionnaire meant to determine if a person is psychotic, innocuous questions are asked at first such as "Do you have a poor appetite?" and "Do you have heartburn?" This gets the respondent involved. He or she is more likely to continue through the questionnaire if he or she has already invested some time answering the first part of the questionnaire. At the end of the questionnaire, questions having more emotional implications can be asked, such as "Are you afraid of dying?" "Have you ever tried to put an end to your life?" and so on.

The structure of the questionnaire also refers to the *question structure.* This means that some ways of asking a question leave only a few alternative ways of answering it, whereas others allow a wide variety of responses. A question that permits only specified responses is called a *categorical* question since it places the respondent in one of several categories. For example, "Are you at present: Single ___? Married ___? Divorced ___? Separated ___? Widowed ___?" is a categorical question. The question: "Under what circumstances do you go to the movies?" is called an *"open-ended"* question because it permits many different answers. We may also have *continuous response questions* such as "What is your weight?" or "List your blood pressure."

There is a great deal of thinking and research that goes into selecting questions and ordering them within the questionnaire. Some of the references in the bibliography at the end of this chapter discuss these matters in detail. However, the ultimate criterion by which a question will be judged as pertinent and useful is whether or not it contributes to resolving those problems that motivated the questionnaire survey. Consequently, the choice of questions must be related to the kind of analyses that will be performed on the resulting data. It follows that methods of analyses of the questionnaire should be specified before the survey is made. (This is not to imply that other analyses that are not considered prior to the data collection cannot also be used; we are simply saying here that to ensure that the questions *are* pertinent to the type of data analyses, these analyses should be specified ahead of time.) In short, the choice of data analysis should influence the questionnaire design. A different mathematical analysis is needed for categorical questionnaires than for open-ended questionnaires or funneling questionnaires.

15 ▪ 3   NONSAMPLING ERRORS

Most of the types of errors discussed thus far have been due to sampling fluctuations and hence could be avoided if we were able to observe and perfectly measure every item in the target population. However, in many large surveys it is not possible to use highly trained interview personnel and expensive measuring devices. Further, there may not be a uniform attitude of the interviewed and interviewers to the questions, and if the data obtained from the survey must be coded for the computer, there is always the possibility of coding errors. Thus a complete enumeration (or census) may be *less* accurate than a sample. In practice, a quality check is usually made after large-scale surveys to discover any errors. These checks are small surveys taken on the target population. This methodology has been used to check the U.S. census; for example, these checks revealed an underenumeration of 6 million people in the 1960 U.S. census. These types of errors are not in the target population data themselves, but show up in our sampled data. Such errors are called *nonsampling* errors and should be eliminated as much as possible. We now consider various sources of such errors.

Nonsampling error can arise when a large portion of the target population simply refuses to respond to the questionnaire. A high *nonresponse* rate means that the target population may not be randomly sampled, and hence one should be certain that the reason for nonresponse has no effect on the conclusions of the survey. Otherwise the conclusions drawn about the *entire population* may prove to be fallacious.

Example 15 ■ 2   A marketing survey company mailed 8000 questionnaires to a randomly selected group in the suburbs of a large city concerning their TV viewing habits. The questionnaire had 40 questions and the company did not include a self-addressed stamped return envelope for the return of the questionnaire. Only 320 questionnaires were returned. The average age of those returning the questionnaires was 55 and they averaged 5 viewing hours of TV per day. Does this mean that the average suburbanite watches television 5 hours per day? This figure would imply that suburbanites are certainly an indolent lot. But consider those who chucked the questionnaire into the nearest wastebasket. It is fair to guess that they were simply too busy to answer 40 questions or to take the time (or money) to mail it. One could suppose then that the nonrespondents have different viewing habits from those who had the leisure time to answer and mail this long questionnaire. Since the nonrespondents formed such a large percentage of the sample, one cannot conclude anything from the survey about viewing habits.

Another source of error is *interviewer bias.* As an example, suppose an interviewer is standing on a street corner and is supposed to select respondents randomly. The interviewer may unconsciously select those who are well dressed as opposed to passersby who are shabbily dressed. Such a selection does not yield a random sample. In fact, there may well be a systematic selection of those from a higher economic class (those who are better dressed), thereby invalidating any conclusions of the survey that are supposed to be applicable to the population as a whole.

Yet another possible source of error could arise from "volunteerism." For example, those people who volunteer to respond to a questionnaire often have a vested interest in, or are quite partial to, the outcome of the survey. (Something had to motivate the person to volunteer. The statistician should make certain it has nothing to do with the purpose of the questionnaire.)

Example 15 ■ 3   The classical example of volunteerism bias is the *Literary Digest*'s survey for the 1936 presidential race between Landon and Roosevelt. On the basis of its survey, the magazine predicted that Landon would soundly beat Roosevelt by 370 electoral votes to 161 votes. The results were the reverse of what was predicted. What went wrong? The commonly held explanation is that the *Digest* for some reason neglected to sample the poor or lower strata voters. This explanation, however, does not hold water. The lower economic strata voters *were* reached. According to the 1936 *Digest,* they polled *every third registered voter* in Chicago, *every other* registered voter in Scranton, PA, and in smaller cities, *every* registered voter was polled. All in all, 10 million sample ballots were mailed to registered voters, and these voters were picked from the lists of registered voters as well as from telephone books, *Literary Digest* subscribers, and motor vehicle registration lists. *The fault lies in the fact that only 2.3 million ballots were returned.* Those respondents who *voluntarily* returned the ballots represented that subset of the target population that had a relatively intense interest in the outcome of the election. One can imagine that the minority of anti-Roosevelt voters felt more strongly about the election than did the majority of voters who were pro-Roosevelt (Bryson, 1976).

Since mailing out questionnaires is a common method of data collection for social scientists, it is important that the possibility of volunteerism bias be recognized. One should resample a portion of nonresponders to ascertain the extent of the bias introduced by voluntary responses. The reader interested in the method of analysis that is then called for may consult the references at the end of the chapter.

As a final example of what not to do, we can observe the survey conducted by the mayor of a city in Ohio. As part of his campaign for reelection in 1977, he launched an "antipornography" campaign. To show that the majority of voters agreed with his views, he had the city garbage collectors give each household a questionnaire. He prefaced the questionnaire with an attack on decaying moral values and the corrupting effect of pornography. The questions were phrased like "Do you agree with me that . . . ?" Additionally, the questionnaire was to be returned by mail (a stamp was not included). This survey has virtually every flaw we have discussed and enough to make any data collected worthless. It was not only a waste of the taxpayer's money, it was a waste of the garbage collector's time. (Incidentally, the mayor was not reelected.)

There are other sources of error as well. One such source is misunderstanding or reticence on the part of the respondent. For example, a questionnaire on sexual habits given to teenagers by their parents would tend to inhibit certain responses, whereas the same questionnaire administered by a peer would yield completely different results. Another error can arise because of inaccurate recording of responses. Our space is too limited here to give these matters the attention they deserve. Here we look at two general areas to which the beginning students should be introduced before they become involved in questionnaire design. These are the *reliability* and *validity* problems raised by questionnaire techniques.

15 ▪ 4 RELIABILITY AND VALIDITY

First of all, we should distinguish between reliability and validity. Reliability refers to consistency, to obtaining the same results under the same conditions. Validity tells us whether the question really measures what it is supposed to measure. For example, a ruler is supposed to measure true distance in terms of meters. If the scale were incorrectly made we would say that the distance measured by the ruler was *invalid.* If this incorrect scale were easy to read, everyone who used it to measure a specific length would get the same answer. Thus the scale yields consistent answers and is said to be *reliable* (even though the answers are not *valid*).

Now suppose we have a valid scale. If the ruler is divided into very small parts that are difficult to read, two different users would easily come up with two different measures of the same distance. We would say the ruler is *unreliable* even though the scale is *valid.* Of course, what we want is both a valid and reliable measuring device.

Reliability

Three methods of measuring reliability are in general use. These are the test–retest method, the multiple-form method, and the split-half method.

Test–Retest (matching pairs)

In the "test–retest" method a questionnaire is applied twice to the same population and the results are compared. A high level of association or correlation between results is demanded before reliability can be assumed. This method of measuring reliability has an inherent flaw since the first application of the questionnaire may affect the responses in the second application. This happens, for example, in learning situations such as students learning math problems by doing them. As illustrated in example 15.4, one way to overcome this difficulty is to divide the sample into subsamples of matched pairs. Then the overall association between the pairs would be measured. If the association is high, then the questionnaire is reliable.

Example 15 ▪ 4 The following is a portion of a questionnaire designed to study attitudes toward the church (Thurstone, 1929).

Check (✓) every statement below that expresses your sentiment toward the church. Interpret the statement in accordance with your own experience with churches.

1. I think the teaching of the church is altogether too superficial to have much social significance.
2. I feel the church services give me inspiration and help me to live up to my best during the following week.
3. I think the church keeps business and politics up to a higher standing.
.
.
.
9. I am careless about religion and church relationships but I would not like to see my attitude become general.
10. I regard the church as a static, crystallized institution, and as such it is unwholesome and detrimental to society and the individual.
.
.
.
28. I respect any church member's beliefs but I think it is all "bunk."
29. I enjoy my church because there is a spirit of friendliness there.
.
.
.
31. I believe the church is the greatest institution in America today.

The following 10 people responded to the questionnaire (M = male, F = female):

Respondent	Sex	Age	Occupation
R. L.	M	45	Doctor
L. T.	M	52	Farmer
M. V.	M	26	Divinity student
R. Q.	M	33	Construction worker
A. L.	M	38	Professor of mathematics
D. R.	M	30	Taxi driver
L. B.	F	24	Theology student
M. C.	F	33	Teacher
G. P.	F	40	Lawyer
C. A.	F	30	Machine operator

We wish to pair the respondents as closely as possible with respect to those characteristics we feel influence attitude toward the church. If we feel profession and age are dominating factors we might pair as shown in Table 15.1.

Observe that R. L., a 45-year-old doctor, is paired with G. P., a 40-year-old lawyer; L. T., a 52-year-old farmer is paired with a 33-year-old construction worker, and so forth.

Table 15.1 also shows the score of each respondent. (How the scores were obtained is irrelevant to our analysis; hence we do not discuss this matter here.)

	Table 15.1	
Pair	Score of first member of pair x_i	Score of second member of pair y_i
R. L., G. P.	30	28
L. T., R. Q.	35	41
M. V., L. B.	60	58
A. L., M. C.	22	45
D. R., C. A.	28	29

A measure of reliability can be obtained by correlating the first column of scores with the second. If the correlation is greater than 0.5, we consider the questionnaire to be reliable. Using the data in Table 15.1, we may calculate the correlation coefficient r as in Chapter 9 to obtain $r = 0.72$. We conclude that the questionnaire is reliable.

The data raise questions about the value of the pairing criterion. The scores of A. L. and M. C. are markedly different even though both are academics. Perhaps profession is not so dominating a factor as supposed. This example is intended to emphasize that great care must be taken in selecting the pairs to be used in measuring reliability and to ensure that the two groups selected are really well matched.

Multiple Forms

This method of measuring reliability consists of administering questionnaire A and then administering questionnaire B which is supposed to measure the same thing as A (such as attitude toward the church). A correlation between the two sets of scores produces a "reliability coefficient" called the *"coefficient of equivalence."* This method does not completely solve the problem of the first test affecting the retest. The two forms, since they are measuring the same thing, may be sufficiently close in structure and content that their connection may well be obvious to the respondent so that his or her answers to B will be influenced by his or her answers to A. This effect may be partially eliminated by randomly selecting which questionnaire will be given first.

Example 15 ▪ 5   The following questions are taken from a questionnaire designed to determine the attitude of respondents to religion:

	Yes	No
1. Do you believe in life after death?	☐	☐
2. Do you believe that spiritual values take precedence over materialistic values?	☐	☐
3. Is there a true religion?	☐	☐

Twelve people responded to this questionnaire (religion) and to the previously described questionnaire on attitude toward the church. Six of the respondents took the religion questionnaire first, and six took the church questionnaire first. The scores of the 12 to the two questionnaires are recorded here.

Respondent	1	2	3	4	5	6	7	8	9	10	11	12
Score on religion	26	9	15	29	20	22	34	23	17	12	2	23
Score on church	28	6	12	36	8	7	30	25	20	5	7	18

The sample correlation coefficient between the two scores is 0.8. If one assumes that the two questionnaires are measuring the same thing, then 0.8 represents what is known as the *"coefficient of equivalence"* of the questionnaires (i.e., the correlation between the questionnaires) and is taken as a measure of their reliability.

Split Half

The questionnaire is divided randomly into two halves. Each of the two is treated as a separate questionnaire and scored accordingly. The association of the two sets of scores is then taken as a measure of reliability. Again, if the association is high, the questionnaire is said to be reliable.

There is, for each of the methods previously described, a measure of reliability given by the correlation coefficient or some other measure of association. Another measure of reliability that is commonly used for a variety of questionnaires is called the *coefficient of reliability.*

The Coefficient of Reliability

To emphasize the meaning of reliability, let us consider a patently ridiculous situation where one fills out a questionnaire but does not recall the answers on filling it out a second time, a third time, and so forth. Suppose Jones *repeatedly* obtains a 50% score, Smith a 90% score, Harris a 20% score, and so on. If such is the case, we say that the questionnaire is perfectly reliable since it gives the same score, X_i, to person i each time he takes it. We call each score X_i the *true* score of person i. The collection of true scores, $X_1, X_2, \ldots$, has a variance that we denote as σ_t^2.

Now suppose that Jones, Smith, and the others do not obtain the same score each time, so the questionnaire is not perfectly reliable. For each person i we obtain the true score X_i *plus* an error E_i. The total score is the sum of all the individual scores, $X_i + E_i$. We assume that variance of the total score is equal to $\sigma_t^2 + \sigma_e^2$, where σ_e^2 is the variance of the sum of the errors. Let σ^2 represent the variance of the total score:

$$\sigma^2 = \sigma_t^2 + \sigma_e^2$$

Divide both sides of this equation by σ^2 to obtain

$$1 = \frac{\sigma_t^2}{\sigma^2} + \frac{\sigma_e^2}{\sigma^2}$$

so that

$$\frac{\sigma_t^2}{\sigma^2} = 1 - \frac{\sigma_e^2}{\sigma^2}$$

The reliability R is defined as that part of the score variance that is the true variance, namely,

$$R = \frac{\sigma_t^2}{\sigma^2}$$

Since $\sigma_t^2/\sigma^2 = 1 - (\sigma_e^2/\sigma^2)$ we estimate the reliability by 1 minus the ratio of the *estimate* s_e^2 of σ_e^2 to the estimate s^2 of σ^2, namely, by

$$1 - \frac{s_e^2}{s^2}$$

This quantity is called the coefficient of reliability and is represented by the symbol r_{tt}. Thus

$$r_{tt} = 1 - \frac{s_e^2}{s^2}$$

where

$$s^2 = \sum_{i=1}^{n} \frac{(x_i - \bar{x})^2}{n-1}$$

x_i = observed overall score of the ith person on the entire questionnaire

$\quad = \sum_j x_{ij}$, where x_{ij} is the score of the ith person on the jth question

$\bar{x} = \left(\sum_{i=1}^{n} \frac{x_i}{n} \right)$, the average overall score for all n respondents.

It remains to describe how to calculate s_e^2. Let us suppose there are c questions or items in the questionnaire. Each question is answered by each respondent, so we have the following table:

	Question					Respondent's overall score
	1	2	3		c	
1	x_{11}	x_{12}	x_{13}	$\cdots$	x_{1c}	$x_1 = \sum_j x_{1j}$
2	x_{21}	x_{22}	x_{23}	$\cdots$	x_{2c}	$x_2 = \sum_j x_{2j}$
.	.	.	.	.	.	.
.	.	.	.	.	.	.
n	x_{n1}	x_{n2}	x_{n3}	$\cdots$	x_{nc}	$x_n = \sum_j x_{nj}$
Question totals	$\sum_i x_{i1}$	$\sum_i x_{i2}$	$\sum_i x_{i3}$		$\sum_i x_{ic}$	

(Respondent)

(This is just the type of table we encountered in studying two-way ANOVA with one observation per cell.) If we assume all the questions are approximately equivalent, then the internal variation of scores within each question is a measure of the variance of the error, and hence so is the average of these internal variations:*

$$s_e^2 = \frac{1}{c} \sum_{j=1}^{c} \text{Var(question } j \text{ scores)} = \frac{1}{c} \sum_{j=1}^{c} \left[\frac{n \sum_i x_{ij}^2 - \left(\sum_i x_{ij} \right)^2}{n(n-1)} \right]$$

Thus we arrive at the formula

$$r_{tt} = 1 - \frac{\left(\dfrac{\text{sum of individual question variances}}{\text{number of questions}} \right)}{\text{variance of respondent's overall scores}}$$

Referring back to the ANOVA table, we calculate the row totals $x_1, \ldots, x_n$,

---

* The mathematical explanation of this result is: *Assume* that the expected value of the error E_i is zero. Let $X_{ij} + E_i$ represent the random variable with range the score of the ith person on the jth question. Clearly, $E(\sum_i X_{ij} + E_i) = cE(\bar{X}_j)$. Hence $E[(\sum_i X_{ij} + E_i) - c\bar{X}_j]^2 = E[E_i^2]$ from which it follows that s_e^2 is an estimate of the variance of the error.

and then the variance for these n numbers and call it s^2. For each column we calculate the variance, call it s_j^2 for the answers to question j. We then have

$$s_e^2 = \frac{1}{c} \sum_{j=1}^{c} s_j^2$$

and

$$r_{tt} = 1 - \frac{s_e^2}{s^2}$$

For questionnaires where each question score is either zero or one; that is, $x_{ij} = 0$, if the answer is wrong and $x_{ij} = 1$ if the answer is right, we know the variance of the score in question j is approximately $\hat{p}_j(1 - \hat{p}_j)$, where $\hat{p}_j$ is the proportion of individuals who answer question j correctly (c.f. the binomial variance with $n = 1$ trial). Then we have

$$s_e^2 = \frac{1}{c} \sum_{j=1}^{c} \hat{p}_j(1 - \hat{p}_j)$$

and

$$r_{tt} = 1 - \sum_{j=1}^{c} \frac{\hat{p}_j(1 - \hat{p}_j)}{cs^2}$$

To aid in calculating r_{tt}, we could set up a summary of the results of the questionnaire as in the following table:

Question No.	$\hat{p}$	$(1 - \hat{p})$	$\hat{p}(1 - \hat{p})$
1	0.6	0.4	0.24
2	0.1	0.9	0.09
3	0.5	0.5	0.25
.	0.2	0.8	0.16
.	.	.	.
.	.	.	.
.	.	.	.
100	.	.	.
			Total = $\Sigma\,\hat{p}(1 - \hat{p}) = 1.2$

Suppose the test scores of 15 people were as follows: 98%, 92%, 92%, 90%, 89%, 85%, 80%, 79%, 75%, 75%, 68%, 58%, 55%, 50%, 50%, and $\Sigma\,\hat{p}(1 - \hat{p})/100 = 0.012$. Then using the 15 overall scores, we have $s^2 = 0.026$

and hence

$$r_{tt} = 1 - \frac{0.012}{0.026} = 0.54$$

that is, 54% of the variation in test scores is due to "true variation." This is considered good reliability for this type of test.

Validity

Recall that a questionnaire possesses *validity* when it actually measures what it claims to measure. For example, we might inquire "do IQ tests really measure intelligence?" How can we judge whether an IQ test measures intelligence in the absence of any *external* measure of intelligence? There is much research being done in this area of determining questionnaire validity. Thus far, three approaches have generally been taken.

Jury Opinion

This is a method whereby a group of experts judge the validity of the questionnaire. For example, if a questionnaire were aimed at determining "teaching ability," a panel consisting of educators, sociologists, psychologists, students, teachers in various fields, and so on might constitute a jury to determine validity. Experts, of course, can err so this approach might determine that a questionnaire has validity yet it may, in fact, have no relation to the matter it is supposed to be measuring. Moreover, expert opinion can change through the years, for example, IQ scores are now under attack as culture-biased. For this reason more objective techniques are usually demanded. One of these is now discussed.

Known Groups

Suppose a questionnaire is designed to determine the attitude of respondents to a political party, say, the Democratic party. One might test the questionnaire on members of the Democratic party and opposing Republican party and members of other opposing parties. The answers would then be compared by using statistical tests discussed in other chapters. If the answers of Democratic party members did not significantly differ from those belonging to opposing parties, we would conclude that the questionnaire was not valid in ascertaining a person's attitude to the Democratic party.

Example 15 ▪ 6   A questionnaire to determine attitude toward the Democratic party is given to 50 members of the Democratic party and 50 members of the Republican party.

The data obtained are as follows:

$$\bar{x}_D = \text{average score of Democrats} = 90$$

$$\bar{x}_R = \text{average score of Republicans} = 50$$

$$s_D^2 = \text{sample variance of Democrats' scores} = 64$$

$$s_R^2 = \text{sample variance of Republicans' scores} = 100.$$

We wish to test the hypothesis that the questionnaire does not distinguish between the attitudes of Republicans and Democrats; that is, we test the hypotheses that the true mean score (μ_D) of Democrats is equal to the true mean score (μ_R) of Republicans.

$$H_0: \mu_D = \mu_R$$

From a pretest of this questionnaire we can make the assumption that the variance of Democrats' scores is equal to the variance of Republicans' scores. We also assume that the sample averages are approximately normally distributed. To test the hypotheses $H_0: \mu_D = \mu_R$, we use the statistic

$$t = \frac{\bar{x}_D - \bar{x}_R}{s_p \sqrt{\dfrac{1}{50} + \dfrac{1}{50}}}$$

where s_p is the usual pooled estimate of the standard deviation,

$$s_p = \sqrt{\frac{(50-1)s_D^2 + (50-1)s_R^2}{50 + 50 - 2}}$$

$$= \sqrt{\frac{(49)64 + (49)100}{98}} = 9.06$$

so that

$$t = \frac{90 - 50}{9.06 \sqrt{\dfrac{2}{50}}} = 22$$

Using the t-table, we observe that we can certainly reject the hypothesis at the 0.01 level.

This method always carries with it the danger that there might be other differences between the groups besides belonging to different political parties which might account for the differences in their responses to the questions. For example, Democrats may largely be younger, or members of a lower economic class than Republicans. The questionnaire might inadvertently be measuring this economic class difference rather than the attitude toward the Democratic party.

Independent Criteria

This is the ideal technique but it is not always practical. The questionnaire discussed here is one in which independent criteria are available.

A questionnaire* was designed to detect psychotic members of Iranian society. The questions are of a categorical nature in that the only responses permitted are "almost always," "frequently," "occasionally," and "never." The reason for using categorical questionnaires in this context is that a shortage of trained psychiatrists exists in Iran so that interviewers who may only have a college or high school level education must be used to administer the questionnaire. They are not permitted to interpret the answers nor to write down full answers from the respondents for fear that their reports might be inaccurate. Thus it is necessary to permit fixed, categorical responses.

The following are examples of questions found in the questionnaire:

1. Is your appetite poor?
2. Is your mouth dry, or does it taste bitter?
3. Do you sleep too much?
4. Do you feel inclined to tears?

The questionnaire once devised was pretested to determine its reliability and validity. Note that in this case validity could be tested against independent criteria since the psychotic or other symptoms of each member of the sample were known. The test of validity was obtained by measuring the scores of each respondent and comparing the result to his or her known mental condition. The proportion of correct classifications of subjects using their questionnaire scores was taken as a measure of validity (see Cronbach, 1970, for further information on reliability and validity).

15 ■ 5 GENERAL PURPOSES OF QUESTIONNAIRES; ANALYSIS OF CATEGORICAL QUESTIONNAIRES

General Purposes of Questionnaires

There are many different types of questionnaires with regard to their purposes. However, there are three general types of questionnaires; descriptive, classifica-

---

* This section was written while one of the authors was living in Iran in 1978, shortly before the revolution.

tion, and discrimination questionnaires. The purpose of a descriptive questionnaire is to determine certain characteristics of the population as a whole. For example, the proportion of males between 21 and 35 years of age who favor gun control, or the proportion of prospective customers who favor a package of type *A* versus type *B* for a new product, and so on. For the analysis of questionnaires of a descriptive nature we may utilize the techniques of Chapters 6 and 7 to obtain confidence intervals and perform tests of hypothesis about the proportion p under consideration.

A classification questionnaire is used for classifying respondents into groups, each of which is as much alike as possible. For this purpose we generally assign a numerical value to each question response and then give an overall score on the questionnaire. We utilize this score for making our classification.

A discrimination questionnaire is used for the expressed purpose of setting up a rule that will enable us to allot a new individual to some known group. We are given the responses of samples from each of the known groups and must compare a new individual's questionnaire answers with those of the known group to ascertain into which group he or she best fits.

Next, we discuss a class of questionnaires called categorical questionnaires and concentrate in Section 15.6 on an important matter related to these questionnaires, namely, which questions should be eliminated from the questionnaire. We consider this for categorical questionnaires in general but give special consideration (in Section 15.7) to this subject as it relates to discrimination questionnaires.

Categorical Questionnaires; Types of Data

As previously stated, a categorical questionnaire is one that contains questions that only permit specified responses, and the permitted responses (married, single, and so on) to each question are qualities or categories. Each respondent on answering a question falls into one of the categories. An important problem in questionnaire analysis is how to assign *values* to each response. For example, if a person answers "single" we may assign the value "0" to the answer, the value "1" to the answer "married," and so forth. (See Brockett, 1981 for further results on how to assign numerical values to categories to facilitate statistical analysis.)

Types of Data

Nominal data: A nominal level variable is one whose values indicate categories and *not* magnitudes. A number is sometimes assigned to each category. The student may very well ask "What is the virtue of assigning numerical values to the categorical qualities?" One immediate advantage is that the assignment of quantitative values permits an overall score to be calculated for each respondent such as a numerical score which is calculated for a student on an examination. The score serves to summarize the performance of each respondent on the questionnaire and thereby permits respondents to be easily and quickly compared.

Although the development of a score is important, there are more fundamental reasons for transforming qualitative information into quantitative information. All sciences move in the direction of greater precision in measurement. This takes many forms, but one fundamental form is measuring gradations. As an example, suppose we have given a number of respondents a questionnaire that is supposed to measure their attitude toward the church. On analyzing the results, we may divide the respondents into three groups, "approve of," "indifferent to," and "disapprove of" the church. However, the more we probe, the more "in betweens" we find. The preceding classification becomes too crude as well as inconvenient for descriptive purposes; we feel intuitively that we should be able to divide the respondents into more and more categories and finally, place them along some quantitative scale. This is the motivation for assigning numbers to *qualitative* data.

Ordinal Data: Ranking means arranging in order with regard to some common aspect. In a science fair contest, we rank the contestants 1 = best project, 2 = next best project, and so forth. Question responses may also be ranked. As an example, consider the question "Do you have trouble sleeping?" with the permitted responses "never," "occasionally," "frequently," and "almost always." This question comes from a classification questionnaire that is supposed to measure "mental stress." The responses are ranked from "low stress" to "high stress." The response never is given rank 0 (lowest stress), occasionally, rank 1; frequently, rank 2; and almost always, rank 3 (highest stress). The score of a respondent is the sum of the ranks of all responses to all the questions. By using ranked responses, we see that people with high scores will have a high level of stress and those with low scores have low stress. The high and low scorers constitute two groups.

Note that "ranking" values indicates *order;* that is, 0 indicates less mental stress than 1, 1 is less than 2, and so forth. This type of data is called *ordinal* data since it allows an ordering of the data.

Ranking Data

In many situations the rankings for the responses are obvious to the questionnaire designer. For some questions, however, the designer must utilize an external method for ranking the answers. One way of determining the ranks for answers is to use the method of *"paired comparisons"* between responses. To understand this method, consider the following question that was part of a questionnaire designed to measure the ability of nurses to deal effectively with patients having clinical sexual problems.

Mrs. Brown, age 40, is informed that she should have a hysterectomy because

Table 15 ▪ 2 Response Pattern to a Paired Comparison of Nurses' Responses to Patients' Problems

	Preferred response			
	1	*2*	*3*	*4*
1	. . .			
2	0.9	. . .		
3	0.95	0.89	. . .	
4	0.98	0.93	0.8	. . .

her uterus is prolapsed beyond repair. She tells the nurse that she just could not do that. Her husband would not approve. As the nurse you would:

1 ___ . Find out why Mrs. Brown feels her husband would not approve.
2 ___ . Suggest to Mrs. Brown that she ask her husband to call or come in to discuss the hysterectomy with her (the nurse).
3 ___ . Suggest the advantages that the patient would have, for example, no fear of becoming pregnant. . . .
4 ___ . Reassure Mrs. Brown that the uterus is only good for babies. . . .

   In order to rank the responses from a good to bad reaction on the part of the nurse, a representative sample of nurses was told to compare the responses in pairs and to note which of the two responses was preferable. Table 15.2 illustrates the results of the sample.

   The values in the table indicate the proportion of the sample that preferred the response listed in the heading to that listed in the stub of the table. For example, the value 0.9 in the table means that 90% of the sample preferred response 1 to response 2, 95% preferred 1 to 3, 98% preferred 1 to 4, 89% preferred 2 to 3, and so on.

   Note from the table that the pattern of responses is consistent in the sense that if response *a* is preferred to *b*, and *b* preferred to *c*, then *a* is preferred to *c*. The pattern need not have been consistent. For example, suppose results shown in Table 15.3 had been obtained.

Table 15 ▪ 3 Inconsistent Response Pattern

	Preferred response			
	1	*2*	*3*	*4*
1	—	—	—	—
2	0.9	—	—	—
3	0.95	0.89	—	—
4	0.98	0.43	0.9	—

In this case 2 is preferred to 3, and 3 is preferred to 4, but 2 is not preferred to 4. If the pattern is inconsistent, the question is usually omitted from the questionnaire. If the pattern is consistent, the responses are assigned values according to their ranks. In the case of Table 15.2, question 1 would be given value 4, since it is most preferred, 2 would be given value 3, and so forth. (For a more detailed discussion of analysis of classification questionnaires with ranked responses and a different approach, see Luber, 1974.)

15 ▪ 6 INTERNAL CONSISTENCY—EVALUATING EACH QUESTION FOR WORTH IN CATEGORICAL QUESTIONNAIRES

A question in most of the literature on questionnaires is called an "item." The analysis described subsequently is called "item analysis." Each item or question is analyzed to determine whether or not it should be included in the questionnaire. Because of the time and money involved in administering a questionnaire, we wish to throw out any question that does not help resolve the problem that prompted the questionnaire. The first method of item analysis discussed here is called "internal consistency." The idea was introduced by L. L. Thurston and E. J. Chave in 1929. They suggested the general proposition that ideally the values assigned to responses should be constructed from the responses of those to whom it was administered. This led to the concept of internal consistency. To illustrate this method, consider one of the questions in the psychiatric screening questionnaire discussed previously:

	never	sometimes	often	always
Do you suffer heartburn?	☐	☐	☐	☐

Note again that all responses in the question are ordered from "low stress" to "high stress." Thus a person with a high total score overall would be a person exhibiting high stress. The preceding question is said to be "consistent" if those with overall low scores responded with a low score (i.e., 0 or 1) to the question and those with overall high scores responded with a high score (i.e., 2 or 3). On the other hand, if all high scorers answered with a 0 or 1, and all low scorers answered with a 2 or 3, we would say that the question was inconsistent.

To put this another way, if the observed responses to a particular question have high positive correlation with the overall score, we say that this question is consistent. Otherwise it is inconsistent.

To illustrate this method, let us suppose the questionnaire was given to 10 respondents and we wish to assess the worth of question 1. Table 15.4 lists the respondents, their total score on the questionnaire, their score on the first question, and their total score minus their score on the first question. Table 15.5

	Table 15 ▪ 4		
Respondent	Total score	Score on question 1	Total score minus score on question 1
A	120	3	117
B	115	3	112
C	100	3	97
D	75	2	73
E	63	2	61
F	58	2	56
G	40	1	39
H	45	3	42
I	19	0	19
J	10	0	10

(called a scattergram) illustrates the relationship between the responses to question 1 and the total score minus question 1 score. The body of the table lists the number of persons with a given overall score and the question 1 score.

The relationship between total score and the score on the first question appears strong. In fact, the correlation between the "total score minus question 1 score" and the question 1 score is $r = 0.835$. In practice, this analysis is performed for each question. The questions are ranked in order from high to low in accordance with the value of the correlation coefficients. Those with low coefficients are discarded and the selection made from the remainder.

The method previously described was developed by Likert and the resulting assignment of responses is called the "Likert scale."

Discriminative power. Another measure of how well a question discriminates between high and low scores is called the discriminative power (DP) measure.

In calculating DP, we first divide the sample into those with scores in the highest (Q_H) and lowest (Q_L) quartiles, that is, in highest quarter and lowest quarter. Table 15.6 is then constructed for each item. Here there were 40 respondents. Of the top 10 overall scorers (Q_H) there were 3 with scores of 5, 4

	Table 15 ▪ 5 Total Score Minus Question 1 Score					
Question 1 score	0 to 19	20 to 39	40 to 59	60 to 79	80 to 99	100 to 119
0	2	—	—	—	—	—
1	—	1	—	—	—	—
2	—	1	1	2	—	—
3	—	—	—	—	1	2

Table 15 ▪ 6

		Number of respondents with question rank scores					*x̄ average question score*	$\bar{x}_H - \bar{x}_L = DP$
	Question score	*1*	*2*	*3*	*4*	*5*		
Number of respondents	In Q_H 10	0	1	2	4	3	3.9	$3.9 - 2 = 1.9$
	In Q_L 10	3	5	1	1	0	2	

with scores of 4, 2 with scores of 3, and 1 with a score of 2. The average score for Q_H is $\bar{x}_H = 3.9$. For Q_L there were 10 members with an average $\bar{x}_L = 2$. The *DP*-value is $3.9 - 2 = 1.9$. Once the *DP*-value is calculated for each question, these values are then arranged in order as were the correlation coefficients in the use of the method described previously in the internal consistency method. Questions with a *DP*-value below 0.5 are usually eliminated from the questionnaire since, for these questions, the high and low scorers do approximately the same.

Divergence Weighting. In discrimination questionnaires with two groups of respondents, we have a sample of size n_1 from group 1 and of size n_2 from group 2. Using these samples, we can assess how well a particular question discriminates between these two groups. Namely, for a question with k answers, we let $(\hat{p}_1, \hat{p}_2, \ldots, \hat{p}_k)$ and $(\hat{q}_1, \hat{q}_2, \ldots, \hat{q}_k)$ denote the empirical estimates of the group 1 and group 2 probabilities for the question; that is,

$\hat{p}_i =$ (number of respondents from group 1 who answer i)/n_1 and

$\hat{q}_i =$ (number of respondents from group 2 who answer i)/n_2.

The measure of the value of the question is known as divergence weighting. The odds ratio in favor of group one membership given the response j to the question is $(\hat{p}_j/\hat{q}_j)$. Thus the expected log odds ratio in favor of group one membership is $\Sigma \, \hat{p}_j \, \ell n \, p_j/q_j$, whereas for the expected log odds ratio in favor of group two membership is $\Sigma \, \hat{q}_j \, \ell n \, q_j/p_j = -\Sigma \, \hat{q}_j \, \ell n \, p_j/q_j$. Thus the log odds ratio for distinguishing between the two groups is formed by adding these two values. We define the *D*-value to be

$$D = \frac{n_1 n_2}{n_1 + n_2} \sum_{i=1}^{k} (\hat{p}_i - \hat{q}_i) \ell n(\hat{p}_i/\hat{q}_i) \qquad \hat{p}_i, \hat{q}_i \neq 0$$

D-value for determining the discriminatory power of a question.

Here ℓn stands for the natural logarithm function. For example, if $\hat{p}_1 = 0.8$ and $\hat{q}_1 = 0.2$, then $\ell n(\hat{p}_1/\hat{q}_1) = \ell n4 = 1.386$ so that $(\hat{p}_1 - \hat{q}_1)\ell n(\hat{p}_1/\hat{q}_1) = (0.6)(1.386) = 0.832$. If $n_1 = 10$ and $n_2 = 20$, then $D = (10 \cdot 20)/(10 + 20)(0.832) = 5.55$. It can be shown that the random variable with range the values of D has (for n_1 and n_2 large) a χ^2 distribution with $k - 1$ degrees of freedom. Only those questions are included in the questionnaire for which there is significant evidence of discriminatory ability, that is, those with $D > \chi_\alpha^2$ with $k - 1$ degrees of freedom. An added advantage of this procedure is that it automatically ensures the *validity* of the resulting questionnaire for distinguishing between the two known groups. Alternatively, the questions may be arranged in decreasing order of D-values and those with low D-values thrown out.*

15 ▪ 7   DISCRIMINATION CATEGORICAL QUESTIONNAIRES (OPTIONAL)

After a discrimination questionnaire has been designed, it is administered to samples from each of the two known groups. Using these results, one can determine the reliability for each group. Using the divergence weighting, we can find which questions should be included in the questionnaire and which should be thrown out. Once this is done, it remains to score the questionnaire and to give a rule for classifying new respondents into one of the two known groups.

Suppose a new respondent takes the questionnaire and answers i for the tth question. What numerical score should we give him for this question? Raw scoring (0, 1, 2, and so on) is always a possibility; however, there is a better method possible. If he answers i to the tth question, we give him the score $s_t = \ell n(\hat{p}_i/\hat{q}_i)$ where $\hat{p}_i$ and $\hat{q}_i$ are the relative frequencies of group 1 and group 2 responses to answer i of question t. His overall score on the questionnaire is obtained by summing his scores on all individual questions: $s = \Sigma_t s_t$.

Now that we have his score, how do we determine whether he belongs in group 1 or group 2? One method is as follows: Let π_1 denote the proportion of respondents that belong to group 1 and $\pi_2 = 1 - \pi_1$ the proportion of respondents that belong to group 2. If the new respondent's score satisfies $s \geq \ell n(\pi_2/\pi_1)$, we classify the respondent as coming from group 1, and if $s < \ell n(\pi_2/\pi_1)$, we classify the respondent into group 2. It may be shown that if the questions are assumed to be statistically independent, then this classification procedure minimizes the probability of assigning the respondent to the wrong group.

---

* For additional discussion of this procedure, see Brockett, Levine, and Haaland (1981) and Haaland, Brockett, and Levine (1979). Note, the symbol D represents the observed value in accordance with the use of this symbol in the questionnaire literature.

Table 15 ▪ 7 Preliminary Analysis of Responses of Successful (Group 1) and Unsuccessful (Group 2) Applicants to Engineering Screening Questionnaire.*

Question no.	Answer 1 Group 1	Answer 1 Group 2	Answer 2 Group 1	Answer 2 Group 2	Answer 3 Group 1	Answer 3 Group 2	Answer 4 Group 1	Answer 4 Group 2	D-value for question	Critical $\chi^2_{0.1}$ df	Critical $\chi^2_{0.1}$ Value
1	0.2	0.5	0.2	0.3	0.4	0.1	0.2	0.1	8.006	3	6.25
2	0.1	0.2	0.3	0.4	0.1	0.3	0.5	0.1	9.62	3	6.25
3	0.3	0.4	0.7	0.6	—	—	—	—	0.442	1	2.71
4	0.2	0.2	0.4	0.6	0.4	0.2	—	—	2.20	2	4.61
5	0.5	0.6	0.3	0.1	0.2	0.3	—	—	2.79	2	4.61
6	0.3	0.3	0.2	0.3	0.2	0.2	0.3	0.2	0.8	3	6.25
7	0.4	0.5	0.1	0.1	0.3	0.2	0.2	0.2	0.63	3	6.25
8	0.6	0.5	0.1	0.4	0.3	0.1	—	—	6.54	2	4.61
9	0.1	0.1	0.2	0.3	0.1	0.5	0.6	0.1	15.8	3	6.25
10	0.3	0.2	0.1	0.1	0.4	0.5	0.2	0.2	0.628	3	6.25
11	0.2	0.5	0.5	0.1	0.3	0.4	—	—	9.47	2	4.61
12	0.1	0.2	0.7	0.8	—	—	—	—	0.827	1	2.71
13	0.2	0.1	0.4	0.5	0.4	0.4	—	—	0.916	2	4.61
14	0.2	0.3	0.2	0.1	0.3	0.4	0.3	0.2	1.79	3	6.25
15	0.1	0.2	0.3	0.7	0.6	0.1	—	—	13.04	2	4.61
16	0.3	0.5	0.1	0.25	0.6	0.25	—	—	5.46	2	4.61
17	0.4	0.6	0.6	0.4	—	—	—	—	1.62	1	2.71
18	0.5	0.5	0.3	0.1	0.2	0.4	—	—	3.58	2	4.61
19	0.1	0.3	0.5	0.5	0.3	0.1	0.1	0.1	4.39	3	6.25
20	0.2	0.1	0.3	0.3	0.3	0.2	0.2	0.4	2.48	3	6.25
21	0.3	0.2	0.1	0.7	0.6	0.1	—	—	21.04	2	4.61
22	0.4	0.5	0.5	0.1	0.1	0.4	—	—	10.82	2	4.61
23	0.2	0.1	0.8	0.9	—	—	—	—	0.811	1	2.71
24	0.3	0.2	0.5	0.3	0.2	0.5	—	—	4.17	2	4.61
25	0.1	0.1	0.3	0.3	0.4	0.2	0.2	0.4	2.77	3	6.25

* Table lists the proportion in each group who give the indicated answer.

Example 15 ▪ 7   A questionnaire consisting of 50 questions was designed to screen applicants for an engineering program This questionnaire was then given to a group of 20 successful engineers and to a group of 20 persons who were unsuccessful in their attempts to become engineers. The results are listed in Table 15.7.

The calculations of the *D*-value for each question are made as shown here. For question 1

$$D = \frac{20 \cdot 20}{20 + 20} \left[(0.2 - 0.5)\ell n \left(\frac{0.2}{0.5} \right) + (0.2 - 0.3)\ell n \left(\frac{0.2}{0.3} \right) \right.$$

$$+(0.4-0.1)\ell n\left(\frac{0.4}{0.1}\right)+(0.2-0.1)\ell n\left(\frac{0.2}{0.1}\right)\Big]$$

$$=10[(-0.3)(-0.916)+(-0.1)(-0.405)+(0.3)(1.386)$$

$$+(0.1)(0.693)]$$

$$=10[0.275+0.0405+0.4158+0.0693]=10[0.8006]=8.006$$

For question 7 the calculation of D-value is

$$D=\frac{20\cdot 20}{20+20}\left[(0.4-0.5)\ell n\left(\frac{0.4}{0.5}\right)+(0.1-0.1)\ell n\left(\frac{0.1}{0.1}\right)\right.$$

$$\left.+(0.3-0.2)\ell n\left(\frac{0.3}{0.2}\right)+(0.2-0.2)\ell n\left(\frac{0.2}{0.2}\right)\right]$$

$$=10[(-0.1)(-0.223)+0+(0.1)(0.405)+0]=0.63$$

and so on for all the questions.

Observing the last two columns of Table 15.7 we see that only nine questions have a D-value that is significantly different from zero at $\alpha=0.1$, namely, questions numbered 1, 2, 8, 9, 11, 15, 16, 21, and 22. Eliminating all other questions from the final questionnaire reduces its size to nine questions that (when renumbered) have the group 1 and group 2 response probabilities given in Table 15.8.

This new reduced questionnaire can now be applied to classify a new respondent. If it is known that 30% of the applicants become successful engineers

Table 15 ■ 8 Response Probabilities of Successful (Group 1) and Unsuccessful (Group 2) Applicants for the Reduced Final Screening Questionnaire

Question no.	Answer 1		Answer 2		Answer 3		Answer 4	
	Group 1	Group 2	Group 1	Group 2	Group 1	Group 2	Group 1	Group 2
1	0.2	0.5	0.2	0.3	0.4	0.1	0.2	0.1
2	0.1	0.2	0.3	0.4	0.1	0.3	0.5	0.1
3 (old no. 8)	0.6	0.5	0.1	0.4	0.3	0.1	—	—
4 (old no. 9)	0.1	0.1	0.2	0.3	0.1	0.5	0.6	0.1
5 (old no. 11)	0.2	0.5	0.5	0.1	0.3	0.4	—	—
6 (old no. 15)	0.1	0.2	0.3	0.7	0.6	0.1	—	—
7 (old no. 16)	0.3	0.5	0.1	0.25	0.6	0.25	—	—
8 (old no. 21)	0.3	0.2	0.1	0.7	0.6	0.1	—	—
9 (old no. 22)	0.4	0.5	0.5	0.1	0.1	0.4	—	—

(belong to group 1) and 70% do not become successful engineers (belong to group 2), then the procedure is to classify a respondent to group 1 if and only if his score satisfies $s \geq \ell n \frac{\pi_2}{\pi_1} = \ell n(0.70/0.30) = 0.85$. (If the group 1 and group 2 proportions π_1 and π_2 are unknown, then take $\pi_1 = \pi_2 = 0.5$ and classify into group 1 if $s \geq 0 = \ell n \ 1$.)

Suppose now an applicant wanders in and is given the screening exam. He gives the following answers to the questions:

Question	Answer given	Question	Answer given
1	1	6	2
2	1	7	3
3	1	8	2
4	4	9	3
5	1		

To ascertain whether or not this applicant will become a successful engineer, we calculate his score as in Table 15.9 by using the probabilities from Table 15.8. Since $s = -3.855 < 0.85$, we classify this respondent into group 2 (i.e., chances are he will not become a successful engineer).

It should be remarked that although the calculations exhibited here seem somewhat tedious, they can easily be performed on a hand calculator. Moreover, for larger scale questionnaire applications it is easy to program this procedure onto a computer, essentially freeing the user from doing any calculations at all.

The field of discrimination research in statistics is currently under rapid growth. (For other methods that are applicable to continuous response questionnaires, and questionnaires with a mixture of continuous and categorical response questions see Goldstein and Dillon, 1978.)

15 ▪ 8   A FINAL WORD

In questionnaire analysis (as in most statistical problems) it is wise to utilize several methods of analysis. The answer to the problem is taken as the consensus of the statistical procedures. For example, when determining whether or not a question should be eliminated from a questionnaire, the *DP*-value, the *D*-value, and the internal consistency of the question should all be calculated (if time and money allow). If all these measures agree that the question is worthless, then indeed it should be eliminated. In borderline cases further investigation is indicated. Do not be afraid to withhold judgment if it is possible to do so.

Question	Answer	Question scores
1	1	$s_1 = \ell n\left(\dfrac{0.2}{0.5}\right) = -0.916$
2	1	$s_2 = \ell n\left(\dfrac{0.1}{0.2}\right) = -0.693$
3	1	$s_3 = \ell n\left(\dfrac{0.6}{0.5}\right) = 0.182$
4	4	$s_4 = \ell n\left(\dfrac{0.6}{0.1}\right) = 1.792$
5	1	$s_5 = \ell n\left(\dfrac{0.2}{0.5}\right) = -0.916$
6	2	$s_6 = \ell n\left(\dfrac{0.3}{0.7}\right) = -0.847$
7	3	$s_7 = \ell n\left(\dfrac{0.6}{0.25}\right) = 0.875$
8	2	$s_8 = \ell n\left(\dfrac{0.1}{0.7}\right) = -1.946$
9	3	$s_9 = \ell n\left(\dfrac{0.1}{0.4}\right) = -1.386$

$$s = -3.855$$

Table 15 ▪ 9 Calculating a Respondent's Score

We have not discussed the exact method of analysis for many types of questionnaires. For classification questionnaires see Picard (1972) and for further information on branching type questionnaires see Picard (1972) and Terrenoire (1976).

SUMMARY

In this chapter we have presented methods for analysis of questionnaire data. Various types of errors that can arise in practice and how to identify them have been discussed. Several different methods were presented for ascertaining when a particular question is fulfilling its purpose and when it should be thrown out. Different types of questions and questionnaires were defined and examples given. For a categorical discrimination questionnaire explicit scoring and classification procedures were outlined.

Key Terms

questionnaire	paired comparison method
target population	discriminative power
branching question	interviewer bias
funneling questions	volunteerism bias
categorical question	validity
open-ended question	split-halves method
nonsampling errors	coefficient of equivalence
nonresponse bias	jury opinion
reliability	descriptive questionnaire
test–retest method	discrimination question-
coefficient of reliability	naire
known groups method	ordinal data
classification questionnaire	internal consistency
nominal data	divergence weighting

References

Brockett, P. L.: A note on the numerical assignment of scores to ranked categorical data. *Journal of Mathematical Sociology, 8:*91–101, 1981.

Brockett, P. L., A. Levine, and P. Haaland: Information theoretic analysis of questionnaire data. *IEEE Transactions on Information Theory, I-27*(4):438–446, 1981.

Bryson, M. C.: The *Literary Digest* poll: Making of a statistical myth. *The American Statistician, 30*(4):1976.

Cronbach, L. J.: *Essentials of Psychological Testing,* 3rd ed. New York, Harper & Row, 1970. See p. 161 for a development of measures of reliability and validity.

Goldstein, M., and W. R. Dillon: *Discrete Discriminant Analysis.* New York, John Wiley & Sons, 1978.

Greenberg, B. G., J. R. Abernathy, and D. G. Howitz: A method for estimating the incidence of abortion in an open population. Paper read at the 37th session of the International Statistical Institute, London, 1969.

Haaland, P., P. L. Brockett, and A. Levine: A characterization of divergence with applications to questionnaire information. *Information and Control, 41*(1):1–8, 1979.

Kalton, G., M. Collins, and L. Brook: Experiments in wording opinion questions. *Applied Statistics, 27*(2):149–161, 1978.

Levine, A., et al.: A mathematical method for analysing questionnaires. *Bulletin of the World Health Organization, 47*(1):87–91, 1972.

Luber, T.: Sexuality of nurses: Correlations of knowledge, attitudes and behavior. Dissertation Doctor Philosophy, School of Public Health, Topical Medicine, Tulane University, New Orleans.

Niel, B. D., and B. S. Thomas: *The Logic of Questions and Answers.* New Haven, Yale University Press, 1976.

Oppenheim, A. N.: *Questionnaire Design and Attitude Measurement.* New York, Basic Books, 1966.

Payne, S. L.: *The Art of Asking Questions.* Princeton, N.J., Princeton University Press, 1951.

Picard, P. F.: *Graphes et Questionnaires.* Vols. 1, 2. Paris, Gauthier-Villars, 1972.

Sellitz, C., et al.: *Research Methods in Social Relations.* New York, Holt, Rinehart and Winston, 1959.

Terrenoire, M., and D. Tounissoux: Pseudoquestionnaires and information. *In* Le Dombal and Gremy (eds.): *Decision Making and Medical Care — Can Information Science Help?* Amsterdam, North-Holland, 1976.

Thurstone, L. L., and E. J. Chave: *The Measurement of Attitude.* Chicago, University of Chicago Press, 1929.

Warner, S. L.: Randomized response: A survey technique for eliminating evasive answer bias. *Journal of the American Statistical Association,* 60:63–69, 1965.

Young, D. W.: Assessment of questions and questionnaires. *Methods of Information in Medicine* (Heidelberg, Germany), *10*(4):222–228, 1971.

Exercises

Section 15.2

1. Consider the questionnaire of example 15.1. We may represent this question as "branching" in the following way:

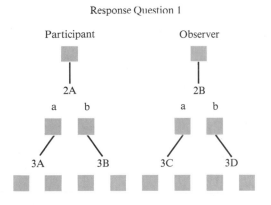

Response Question 1

Notice, for the first question there are $2^1 = 2$ possible answers or categories a respondent may fall into. If there are two responses for the second question a person must answer, then there are $2^2 = 4$ possible categories. At the third stage there are $2^3 = 8$ possible categories.

a. Suppose there are 10 stages, how many possible categories are there?

b. If there are 10 stages, what is the minimum sample size you would need to have some chance of having at least one respondent in each of the final categories.

2. Consider again the funneling questionnaire described in problem 1. We wish to estimate p_j, the probability of falling in the *final* category j. Even for a funneling questionnaire with a small number of questions, say, 10, with only two responses per question, unless the sample size is greater than 2^{10} many of the final categories will contain no respondents; they will be empty.

(i) How many categories will certainly be empty if the sample size is 800?

(ii) For every empty category j the estimate $\hat{p}_j$ of p_j will be zero. Thus our estimate cannot distinguish the values of p_j in these empty categories. In order to attempt to obtain a nonzero estimate of p_j, we must increase our sample size. What sample size do we need if the smallest p_j we must estimate is 2^{-20}. Discuss why funneling questionnaires with a large number of questions are impractical if we wish to obtain estimates of p_j.

Section 15.3

3. *Embarrassing questions:* Respondents are often embarrassed to answer questions truthfully that relate to personal habits (How often do you bathe?), to socially unacceptable behavior (Do you steal?), to their standing with the community or their colleagues (Do you feel your job as a cage-keeper at the zoo earns the respect of your neighbors?), to personal beliefs and preferences. A technique, known as the randomized response technique was designed to treat such questions. For example, suppose we wish to estimate the proportion of females who were pregnant in the past 12 months and had terminated the pregnancy with an abortion. We would want to estimate the proportion of females who would answer yes to the question (Greenberg, 1969).

(1) "I was pregnant at some time during the past 12 months and had an abortion that terminated the pregnancy."

If we asked this question directly, we might very well not get a truthful answer. However, let us pose another question.

(2) "I was born in the month of April."

We provide the respondent with a box containing 35 red and 15 blue balls. The respondent selects a ball at random. The color of the ball is unknown to the interviewer. If the ball is red, she answers question 1, if blue she answers 2. She *knows* which question she is answering, and she knows that the interviewer *does not* know.

a. What is the chance, p, that the respondent answers question 1?

b. Let $P(\text{yes}|i) =$ probability of answering yes to question i, given i is chosen. Show

that $\lambda = P(\text{yes})$ is given by

$$\lambda = p\, P(\text{yes}|1) + (1 - p)\, P(\text{yes}|2)$$

c. We know p and can estimate λ with $\hat{\lambda} = (\text{number who say yes}/n)$, where n is the sample size. Greenberg, et al. (1969) took a survey of females aged 18 to 44 in North Carolina in 1969. They estimated $P(\text{yes}|2)$ from vital statistics on live births in April and not-April in North Carolina between 1924 and 1950; they found the estimate to be $P(\text{yes}|2) = 0.0826$. Find an estimate for $P(\text{yes}|1)$.

d. Show that the estimate of $P(\text{yes}|1)$ has a larger sampling error than would be the case if the women were to answer 1 directly and truthfully?

e. Why might the actual error in estimating $P(\text{yes}|1)$ be smaller than if women were asked to answer directly?

4. Give examples of questions where nonsampling error can arise because

a. The answer requires recall of distant or indistinct events. (*Hint:* Use a question such as "When did you receive your last inoculation?").

b. The question is ambiguous.

c. The question presumes knowledge the respondent need not have.

d. The question is leading.

e. The question is misleading.

Section 15.4

5. Determine whether each of the following instruments is reliable and/or valid:

a. This watch has been checked against a standard time piece and has been found to be accurate; however, its face is so poorly designed, it is easy to make mistakes in reading it.

b. The thermometer I bought yesterday was

miscalibrated; in addition, there were not enough calibrations to easily determine a fraction of a degree.

c. This voltmeter has a scale that only goes up to 10 sec.

d. The following questions were submitted to a magazine publisher to determine the total number of different readers of its three magazines (the number of different readers is important to advertisers):

1. How many subscribers do you have to *Mechanics Monthly*?

2. How many subscribers do you have to *Mechanics Digest*?

3. How many subscribers do you have to *Woman's Point of View*?

(To answer the question, determine what is being measured; then ask yourself, "Do people necessarily subscribe to only one magazine?" The instrument is the questionnaire consisting of these questions.)

6. Even though the ANOVA table appears in the discussion of reliability, the model underlying the ANOVA model is different from that underlying r_{tt}. Discuss the differences in the models.

7. A questionnaire was given to migrants from rural areas to the city of Dakar, Senegal, which was meant to measure their adaptation to urban life.*

Some typical questions from the questionnaire are:

a. Do you belong to social clubs?

b. Are your friends mainly of the village or mainly from the town?

c. Do you own a radio?

d. Do you speak French or Woloff (the language of the village)?

Discuss how you might design such a ques-

* This questionnaire and the results of the survey can be found in Benyousseff, et al., Health Effects of Rural-Urban Migration in Developing Countries. *Social Science and Medicine,* 8:243–254, 1974.

tionnaire and the procedures you might use so that the reliability and validity of the questionnaire could be measured?

Section 15.5

8. a. Which of the questionnaires described in Section 15.4 are categorical questionnaires?

b. Develop a questionnaire to measure the "interest" of the class in statistics. Try to develop questions that reveal the "interest" or "disinterest" of the class. Start the development of the questionnaire by dividing your inquiry into the "dimensions" of interest such as "motivation" (I study statistics because it is part of my professional training, I study simply to obtain a good grade), "sources of inspiration" (teachers, books, and magazine or newspaper articles, personal experiences), and so on.

9. We have defined nominal and ordinal data in this chapter. We have also encountered two other types of data in this text: interval and ratio data. An example of interval data is temperature readings from a centigrade scale. The zero position on the temperature scale is arbitrary. Hence we obtain only difference or interval information from such a scale. On the other hand, height data have a nonarbitrary zero; this type of data is called ratio data. In short, a continuous scale with an arbitrary zero yields interval data; a continuous scale with a nonarbitrary zero yields ratio data.

a. Why is 120° centigrade not twice as hot as 60° centigrade? (*Hint:* change the centigrade readings to Fahrenheit and see if the first temperature is still twice the second.)

b. Why can I assert that a 6-ft tall person is twice the height of a 3-ft tall person? (*Note:* This example motivates the name "ratio data" for data such as height since the ratio of two heights has physical meaning. This is not the case with interval data.)

c. State in each of the following cases whether the data are nominal, ordinal, interval, or ratio data.
   i. the number of passengers carried by the airlines in one year.
   ii. the responses to the mental health questionnaire in Section 15.4.
   iii. the responses to the questionnaire described in Section 15.5.
   iv. license plate numbers
   v. military ranks
   vi. weight
   vii. IQ scores

Section 15.7

10. As an exercise for class participation, we present here a questionnaire meant to discriminate between those who come from large cities (populations over one million) and those who do not (don't know responses should be grouped with no responses).

a. Do you have easy access to performances of a symphony orchestra?
   ○ yes   ○ no   ○ sometimes
   ○ don't know

b. Do you have more than four large department stores to which you are easily accessible?
   ○ yes   ○ no   ○ sometimes
   ○ don't know

c. Is it easy to travel from your home to city hall?
   ○ yes   ○ no   ○ sometimes
   ○ don't know

d. Are there several large hospitals that you are easily accessible to?
   ○ yes   ○ no   ○ sometimes
   ○ don't know

e. To obtain fresh eggs and produce, must you shop in a chain supermarket?
   ○ yes   ○ no   ○ sometimes
   ○ don't know

If you do not have enough respondents in each group, develop two other groups and develop a questionnaire to distinguish between them; then use the method of Section 15.7 to evaluate the questions.

APPENDIX · A

Appendix A.1 Normal Curve Areas and Common Upper Centiles

z	$F(z)$ Area — Probability or area to left of z	z	$F(z)$ A — Probability or area to left of z	z	$F(z)$ A — Probability or area to left of z
−3.00	0.0013	−0.75	0.2266	1.55	0.9394
−2.95	0.0016	−0.70	0.2420	1.60	0.9452
−2.90	0.0019	−0.65	0.2578	1.65	0.9505
−2.85	0.0022	−0.60	0.2743	1.70	0.9554
−2.80	0.0026	−0.55	0.2912	1.75	0.9599
−2.75	0.0030	−0.50	0.3085	1.80	0.9641
−2.70	0.0035	−0.45	0.3264	1.85	0.9678
−2.65	0.0040	−0.40	0.3446	1.90	0.9713
−2.60	0.0047	−0.35	0.3632	1.95	0.9744
−2.55	0.0054	−0.30	0.3821	2.00	0.9772
−2.50	0.0062	−0.25	0.4013	2.05	0.9798
−2.45	0.0071	−0.20	0.4207	2.10	0.9821
−2.40	0.0082	−0.15	0.4404	2.15	0.9842
−2.35	0.0094	−0.10	0.4602	2.20	0.9861
−2.30	0.0107	−0.05	0.4801	2.25	0.9878
−2.25	0.0122	0.00	0.5000	2.30	0.9893
−2.20	0.0139	0.05	0.5199	2.35	0.9906
−2.15	0.0158	0.10	0.5398	2.40	0.9918
−2.10	0.0179	0.15	0.5596	2.45	0.9929
−2.05	0.0202	0.20	0.5793	2.50	0.9938
−2.00	0.0228	0.25	0.5987	2.55	0.9946
−1.95	0.0256	0.30	0.6179	2.60	0.9953
−1.90	0.0287	0.35	0.6368	2.65	0.9960
−1.85	0.0322	0.40	0.6554	2.70	0.9965
−1.80	0.0359	0.45	0.6736	2.75	0.9970
−1.75	0.0401	0.50	0.6915	2.80	0.9974
−1.70	0.0446	0.55	0.7088	2.85	0.9978
−1.65	0.0495	0.60	0.7257	2.90	0.9981
−1.60	0.0548	0.65	0.7422	2.95	0.9984
−1.55	0.0606	0.70	0.7580	3.00	0.9987
−1.50	0.0668	0.75	0.7734		
−1.45	0.0735	0.80	0.7881		
−1.40	0.0808	0.85	0.8023		
−1.35	0.0885	0.90	0.8159		
−1.30	0.0968	0.95	0.8289		
−1.25	0.1056	1.00	0.8413		
−1.20	0.1151	1.05	0.8531		
−1.15	0.1251	1.10	0.8643		
−1.10	0.1357	1.15	0.8749		
−1.05	0.1469	1.20	0.8849		
−1.00	0.1587	1.25	0.8944		
−0.95	0.1711	1.30	0.9032		
−0.90	0.1841	1.35	0.9115		
−0.85	0.1977	1.40	0.9192		
−0.80	0.2119	1.45	0.9265		
		1.50	0.9332		

Common upper centiles z_α

z_α	α
0.000	0.50
0.842	0.20
1.282	0.10 — 90%
1.645	0.05 → 90%
1.960	0.025 → 95
2.054	0.02
2.326	0.01
2.576	0.005 → 99%
3.090	0.001
3.719	0.0001
4.265	0.00001

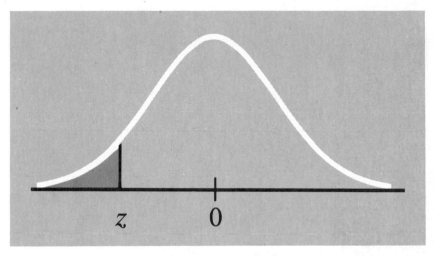

Standard Normal Distribution

Appendix A.2 Combination Values

$$\text{Binomial Coefficients } \frac{n!}{k!(n-k)!} = \binom{n}{k}$$

n \ k	2	3	4	5	6	7	8	9	10
2	1								
3	3	1							
4	6	4	1						
5	10	10	5	1					
6	15	20	15	6	1				
7	21	35	35	21	7	1			
8	28	56	70	56	28	8	1		
9	36	84	126	126	84	36	9	1	
10	45	120	210	252	210	120	45	10	1
11	55	165	330	462	462	330	165	55	11
12	66	220	495	792	924	792	495	220	66
13	78	286	715	1,287	1,716	1,716	1,287	715	286
14	91	364	1,001	2,002	3,003	3,432	3,003	2,002	1,001
15	105	455	1,365	3,003	5,005	6,435	6,435	5,005	3,003
16	120	560	1,820	4,368	8,008	11,440	12,870	11,440	8,008
17	136	680	2,380	6,188	12,376	19,448	24,310	24,310	19,448
18	153	816	3,060	8,568	18,564	31,824	43,758	48,620	43,758
19	171	969	3,876	11,628	27,132	50,388	75,582	92,378	92,378
20	190	1,140	4,845	15,504	38,760	77,520	125,970	167,960	184,756

Appendix A.3 Table of Random Numbers

03991	10461	93716	16894	66083	24653	84609	58232	88618	19161
38555	95554	32886	59780	08355	60860	29735	47762	71299	23853
17546	73704	92052	46215	55121	29281	59076	07936	27954	58909
32643	52861	95819	06831	00911	98936	76355	93779	80863	00514
69572	68777	39510	35905	14060	40619	29549	69616	33564	60780
24122	66591	27699	06494	14845	46672	61958	77100	90899	75754
61196	30231	92962	61773	41839	55382	17267	70943	78038	70267
30532	21704	10274	12202	39685	23309	10061	68829	55986	66485
03788	97599	75867	20717	74416	53166	35208	33374	87539	08823
48228	63379	85783	47619	53152	67433	35663	52972	16818	60311
60365	94653	35075	33949	42614	29297	01918	28316	98953	73231
83799	42402	56623	34442	34994	41374	70071	14736	09958	18065
32960	07405	36409	83232	99385	41600	11133	07586	15917	06253
19322	53845	57620	52606	66497	68646	78138	66559	19640	99413
11220	94747	07399	37408	48509	23929	27482	45476	85244	35159
31751	57260	68980	05339	15470	48355	88651	22596	03152	19121
88492	99382	14454	04504	20094	98977	74843	93413	22109	78508
30934	47744	07481	83828	73788	06533	28597	20405	94205	20380
22888	48893	27499	98748	60530	45128	74022	84617	82037	10268
78212	16993	35902	91386	44372	15486	65741	14014	87481	37220
41849	84547	46850	52326	34677	58300	74910	64345	19325	81549
46352	33049	69248	93460	45305	07521	61318	31855	14413	70951
11087	96294	14013	31792	59747	67277	76503	34513	39663	77544
52701	08337	56303	87315	16520	69676	11654	99893	02181	68161
57275	36898	81304	48585	68652	27376	92852	55866	88448	03584
20857	73156	70284	24326	79375	95220	01159	63267	10622	48391
15633	84924	90415	93614	33521	26665	55823	47641	86225	31704
92694	48297	39904	02115	59589	49067	66821	41575	49767	04037
77613	19019	88152	00080	20554	91409	96277	48257	50816	97616
38688	32486	45134	63545	59404	72059	43947	51680	43852	59693
25163	01889	70014	15021	41290	67312	71857	15957	68971	11403
65251	07629	37239	33295	05870	01119	92784	26340	18477	65622
36815	43625	18637	37509	82444	99005	04921	73701	14707	93997
64397	11692	05327	82162	20247	81759	45197	25332	83745	22567
04515	25624	95096	67946	48460	85558	15191	18782	16930	33361

83761	60873	43253	84145	60833	25983	01291	41349	20368	07126
14387	06345	80854	09279	43529	06318	38384	74761	41196	37480
51321	92246	80088	77074	88722	56736	66164	49431	66919	31678
72472	00008	80890	18002	94813	31900	54155	83436	35352	54131
05466	55306	93128	18464	74457	90561	72848	11834	79982	68416
39528	72484	82474	25593	48545	35247	18619	13674	18611	19241
81616	18711	53342	44276	75122	11724	74627	73707	58319	15997
07586	16120	82641	22820	92904	13141	32392	19763	61199	67940
90767	04235	13574	17200	69902	63742	78464	22501	18627	90872
40188	28193	29593	88627	94972	11598	62095	36787	00441	58997
34414	82157	86887	55087	19152	00023	12302	80783	32624	68691
63439	75363	44989	16822	36024	00867	76378	41605	65961	73488
67049	09070	93399	45547	94458	74284	05041	49807	20288	34060
79495	04146	52162	90286	54158	34243	46978	35482	59362	95938
91704	30552	04737	21031	75051	93029	47665	64382	99782	93478
94015	46874	32444	48277	59820	96163	64654	25843	41145	42820
74108	88222	88570	74015	25704	91035	01755	14750	48968	38603
62880	87873	95160	59221	22304	90314	72877	17334	39283	04149
11748	12102	80580	41867	17710	59621	06554	07850	73950	79552
17944	05600	60478	03343	25852	58905	57216	39618	49856	99326
66067	42792	95043	56280	46780	56487	09971	59481	37006	22186
54244	91030	45547	70818	59849	96169	61459	21647	87417	17198
30945	57589	31732	57260	47670	07654	46376	25366	94746	49580
69170	37403	86995	90307	94304	71803	26825	05511	12459	91314
08345	88975	35841	85771	08105	59987	87112	21476	14713	71181
27767	43584	85301	88977	29490	69714	73035	41207	74699	09310
13025	14338	54066	15243	47724	66733	47431	43905	31048	56699
80217	36292	98525	24335	24432	24896	43277	58874	11466	16082
10875	62004	90391	61105	57411	06368	53856	30743	08670	84741
54127	57326	26629	19087	24472	88779	30540	27886	61732	75454
60311	42824	37301	42678	45990	43242	17374	52003	70707	70214
49739	71484	92003	98086	76668	73209	59202	11973	02902	33250
78626	51594	16453	94614	39014	97066	83012	09832	25571	77628
66692	13986	99837	00582	81232	44987	09504	96412	90193	79568
44071	28091	07362	97703	76447	42537	98524	97831	65704	09514

Appendix A.4 Tabulated Values of e^{-x} (e is approximately equal to 2.71828)

x	e^{-x}	x	e^{-x}	x	e^{-x}	x	e^{-x}
0.0	1.000	2.5	0.082	5.0	0.0067	7.5	0.00055
0.1	0.905	2.6	0.074	5.1	0.0061	7.6	0.00050
0.2	0.819	2.7	0.067	5.2	0.0055	7.7	0.00045
0.3	0.741	2.8	0.061	5.3	0.0050	7.8	0.00041
0.4	0.670	2.9	0.055	5.4	0.0045	7.9	0.00037
0.5	0.607	3.0	0.050	5.5	0.0041	8.0	0.00034
0.6	0.549	3.1	0.045	5.6	0.0037	8.1	0.00030
0.7	0.497	3.2	0.041	5.7	0.0033	8.2	0.00028
0.8	0.449	3.3	0.037	5.8	0.0030	8.3	0.00025
0.9	0.407	3.4	0.033	5.9	0.0027	8.4	0.00023
1.0	0.368	3.5	0.030	6.0	0.0025	8.5	0.00020
1.1	0.333	3.6	0.027	6.1	0.0022	8.6	0.00018
1.2	0.301	3.7	0.025	6.2	0.0020	8.7	0.00017
1.3	0.273	3.8	0.022	6.3	0.0018	8.8	0.00015
1.4	0.247	3.9	0.020	6.4	0.0017	8.9	0.00014
1.5	0.223	4.0	0.018	6.5	0.0015	9.0	0.00012
1.6	0.202	4.1	0.017	6.6	0.0014	9.1	0.00011
1.7	0.183	4.2	0.015	6.7	0.0012	9.2	0.00010
1.8	0.165	4.3	0.014	6.8	0.0011	9.3	0.00009
1.9	0.150	4.4	0.012	6.9	0.0010	9.4	0.00008
2.0	0.135	4.5	0.011	7.0	0.0009	9.5	0.00008
2.1	0.122	4.6	0.010	7.1	0.0008	9.6	0.00007
2.2	0.111	4.7	0.009	7.2	0.0007	9.7	0.00006
2.3	0.100	4.8	0.008	7.3	0.0007	9.8	0.00006
2.4	0.091	4.9	0.007	7.4	0.0006	9.9	0.00005

Appendix A.5 *t*-Distribution Upper Centiles, t_α

Degrees of freedom df	Upper-centile t_α				
	$t_{0.100}$	$t_{0.050}$	$t_{0.025}$	$t_{0.010}$	$t_{0.005}$
1	3.078	6.314	12.706	31.821	63.657
2	1.886	2.920	4.303	6.965	9.925
3	1.638	2.353	3.182	4.541	5.841
4	1.533	2.132	2.776	3.747	4.604
5	1.476	2.015	2.571	3.365	4.032
6	1.440	1.943	2.447	3.143	3.707
7	1.415	1.895	2.365	2.998	3.499
8	1.397	1.860	2.306	2.896	3.355
9	1.383	1.833	2.262	2.821	3.250
10	1.372	1.812	2.228	2.764	3.169
11	1.363	1.796	2.201	2.718	3.106
12	1.356	1.782	2.179	2.681	3.055
13	1.350	1.771	2.160	2.650	3.012
14	1.345	1.761	2.145	2.624	2.977
15	1.341	1.753	2.131	2.602	2.947
16	1.337	1.746	2.120	2.583	2.921
17	1.333	1.740	2.110	2.567	2.898
18	1.330	1.734	2.101	2.552	2.878
19	1.328	1.729	2.093	2.539	2.861
20	1.325	1.725	2.086	2.528	2.845

$t_{\alpha/2}$

Appendix A.5 continued

Degrees of freedom	Upper-centile t_α				
	$t_{0.100}$	$t_{0.050}$	$t_{0.025}$	$t_{0.010}$	$t_{0.005}$
	1.323	1.721	2.080	2.518	2.831
	1.321	1.717	2.074	2.508	2.819
	1.319	1.714	2.069	2.500	2.807
	1.318	1.711	2.064	2.492	2.797
	1.316	1.708	2.060	2.485	2.787
5	1.315	1.706	2.056	2.479	2.779
7	1.314	1.703	2.052	2.473	2.771
8	1.313	1.701	2.048	2.467	2.763
9	1.311	1.699	2.045	2.462	2.756
0	1.310	1.697	2.042	2.457	2.750
31	1.310	1.696	2.040	2.453	2.744
32	1.309	1.694	2.037	2.449	2.739
33	1.308	1.692	2.035	2.445	2.733
34	1.307	1.691	2.032	2.441	2.728
35	1.306	1.690	2.031	2.438	2.724
60	1.296	1.671	2.000	2.390	2.660
120	1.289	1.658	1.980	2.358	2.617

For values of the degrees of freedom greater than 35 the upper centiles z_α can be used from Appendix A.1. Compare the values given in Appendix A.1 with those given in this table for df = 60 and df = 120 to convince yourself that the use of Appendix A.1 introduces a very small error in the critical value for df greater than 35.

Appendix A.6 Upper Centile Values χ^2_α

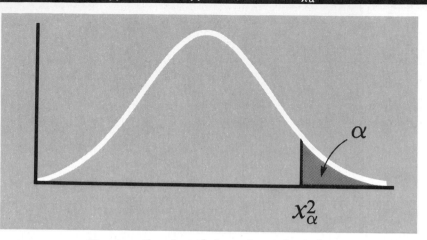

Upper centile values χ^2_α for various values of α

df	0.995	0.99	0.975	0.95	0.9	0.1	0.05	0.025	0.01	0.005
1	0.0000	0.0002	0.001	0.0039	0.0158	2.71	3.84	5.02	6.63	7.88
2	0.0100	0.0201	0.0506	0.1026	0.2107	4.61	5.99	7.38	9.21	10.60
3	0.0717	0.115	0.216	0.352	0.584	6.25	7.81	9.35	11.34	12.84
4	0.207	0.297	0.484	0.711	1.064	7.78	9.49	11.14	13.28	14.86
5	0.412	0.554	0.831	1.15	1.61	9.24	11.07	12.83	15.09	16.75
6	0.676	0.872	1.24	1.64	2.20	10.64	12.59	14.45	16.81	18.55
7	0.989	1.24	1.69	2.17	2.83	12.02	14.07	16.01	18.48	20.28
8	1.34	1.65	2.18	2.73	3.49	13.36	15.51	17.53	20.09	21.96
9	1.73	2.09	2.70	3.33	4.17	14.68	16.92	19.02	21.67	23.59
10	2.16	2.56	3.25	3.94	4.87	15.99	18.31	20.48	23.21	25.19
11	2.60	3.05	3.82	4.57	5.58	17.28	19.68	21.92	24.73	26.76
12	3.07	3.57	4.40	5.23	6.30	18.55	21.03	23.34	26.22	28.30
13	3.57	4.11	5.01	5.89	7.04	19.81	22.36	24.74	27.69	29.82
14	4.07	4.66	5.63	6.57	7.79	21.06	23.68	26.12	29.14	31.32
15	4.60	5.23	6.26	7.26	8.55	22.31	25.00	27.49	30.58	32.80
16	5.14	5.81	6.91	7.96	9.31	23.54	26.30	28.85	32.00	34.27
18	6.26	7.01	8.23	9.39	10.86	25.99	28.87	31.53	34.81	37.16
20	7.43	8.26	9.59	10.85	12.44	28.41	31.41	34.17	37.57	40.00
24	9.89	10.86	12.40	13.85	15.66	33.20	36.42	39.36	42.98	45.56
30	13.79	14.95	16.79	18.49	20.60	40.26	43.77	46.98	50.89	53.67

Appendix A.6 continued

Upper centile values χ^2_α for various values of α

df	0.995	0.99	0.975	0.95	0.9	0.1	0.05	0.025	0.01	0.005
31	14.46	15.66	17.54	19.28	21.43	41.42	44.99	48.23	52.19	55.00
32	15.13	16.36	18.29	20.07	22.27	42.59	46.19	49.48	53.49	56.33
33	15.82	17.07	19.05	20.87	23.11	43.75	47.40	50.73	54.78	57.65
34	16.50	17.79	19.81	21.66	23.95	44.90	48.60	51.97	56.06	58.96
35	17.19	18.51	20.57	22.47	24.80	46.06	49.80	53.20	57.34	60.28
36	17.89	19.23	21.34	23.27	25.64	47.21	51.00	54.44	58.62	61.58
37	18.59	19.96	22.11	24.08	26.49	48.36	52.19	55.67	59.89	62.88
38	19.29	20.69	22.88	24.88	27.34	49.51	53.38	56.90	61.16	64.18
39	20.00	21.43	23.65	25.70	28.20	50.66	54.57	58.12	62.43	65.48
40	20.71	22.16	24.43	26.51	29.05	51.81	55.76	59.34	63.69	66.77

Appendix A.7 Upper Centiles $F_\alpha(df(N), df(D))$

$df(N)$, degrees of freedom for numerator

$df(D)$	α	1	2	3	4	5	6	10	12	15	20	30	50	Large sample size
1	0.995													
	0.99	—	0.010	0.029	0.047	0.062	0.073	0.100	0.107	0.115	0.124	0.132	0.139	0.151
	0.975	—	0.026	0.057	0.082	0.100	0.113	0.144	0.153	0.161	0.170	0.180	0.187	0.199
	0.95	—	0.054	0.099	0.130	0.151	0.167	0.201	0.211	0.220	0.230	0.240	0.248	0.261
	0.90	0.025	0.117	0.181	0.220	0.246	0.265	0.304	0.315	0.325	0.336	0.347	0.350	0.370
	0.10	39.9	49.5	53.6	55.8	57.2	58.2	60.2	60.7	61.2	61.7	62.3	62.7	63.3
	0.05	161.	200.	216.	225.	230.	234.	242.	244.	246.	248.	250.	252.	254.
	0.025	648.	800.	864.	900.	922.	937.	969.	977.	985.	993.	1000	1010	1020
	0.01	4050	5000	5400	5620	5760	5860	6060	6110	6160	6210	6260	6300	6370
	0.005													
3	0.995	—	0.005	0.021	0.041	0.060	0.077	0.124	0.138	0.154	0.172	0.191	0.211	0.234
	0.99	—	0.010	0.034	0.060	0.083	0.102	0.153	0.168	0.185	0.203	0.222	0.238	0.264
	0.975	0.001	0.026	0.065	0.100	0.129	0.152	0.207	0.224	0.241	0.259	0.289	0.295	0.321
	0.95	0.005	0.052	0.108	0.152	0.185	0.210	0.270	0.287	0.304	0.323	0.342	0.358	0.384
	0.90	0.019	0.109	0.185	0.239	0.276	0.304	0.367	0.384	0.402	0.420	0.439	0.455	0.480
	0.10	5.54	5.46	5.39	5.34	5.31	5.28	5.23	5.22	5.20	5.18	5.17	5.15	5.13
	0.05	10.1	9.55	9.28	9.12	9.01	8.94	8.79	8.74	8.70	8.66	8.62	8.58	8.53
	0.025	17.4	16.0	15.4	15.1	14.9	14.7	14.4	14.3	14.3	14.2	14.1	14.0	13.9
	0.01	34.1	30.8	29.5	28.7	28.2	27.9	27.2	27.1	26.9	26.7	26.5	26.4	26.1
	0.005	55.6	49.8	47.5	46.2	45.4	44.8	43.7	43.4	43.1	42.8	42.5	42.2	41.8
5	0.995	—	0.005	0.022	0.045	0.067	0.087	0.146	0.165	0.186	0.210	0.237	0.260	0.299
	0.99	—	0.010	0.035	0.064	0.091	0.114	0.177	0.197	0.219	0.244	0.270	0.293	0.331
	0.975	0.001	0.025	0.067	0.107	0.140	0.167	0.236	0.257	0.280	0.304	0.330	0.353	0.390
	0.95	0.004	0.052	0.111	0.160	0.198	0.228	0.301	0.322	0.345	0.369	0.395	0.417	0.452
	0.90	0.017	0.108	0.188	0.247	0.290	0.322	0.397	0.418	0.440	0.463	0.501	0.514	0.541
	0.10	4.06	3.78	3.62	3.52	3.45	3.40	3.30	3.27	3.24	3.21	3.17	3.15	3.10
	0.05	6.61	5.79	5.41	5.19	5.05	4.95	4.74	4.68	4.62	4.56	4.50	4.44	4.36
	0.025	10.0	8.43	7.76	7.39	7.15	6.98	6.62	6.52	6.43	6.33	6.23	6.14	6.02
	0.01	16.3	13.3	12.1	11.4	11.0	10.7	10.1	9.89	9.72	9.55	9.38	9.24	9.02
	0.005	22.8	18.3	16.5	15.6	14.9	14.5	13.6	13.4	13.1	12.9	12.7	12.5	12.1

$df(D)$, degrees of freedom for denominator

df(D), degrees of freedom for denominator

df = 10

α													
0.995	—	0.005	0.023	0.048	0.073	0.098	0.171	0.197	0.226	0.260	0.299	0.334	0.397
0.99	—	0.010	0.037	0.069	0.100	0.127	0.206	0.233	0.263	0.297	0.336	0.370	0.431
0.975	0.001	0.025	0.069	0.113	0.151	0.183	0.269	0.296	0.327	0.360	0.398	0.431	0.488
0.95	0.004	0.052	0.114	0.168	0.211	0.246	0.336	0.363	0.393	0.426	0.462	0.493	0.546
0.90	0.017	0.106	0.191	0.255	0.303	0.340	0.430	0.457	0.486	0.516	0.549	0.578	0.625
0.10	3.28	2.92	2.73	2.61	2.52	2.46	2.32	2.28	2.24	2.20	2.16	2.12	2.06
0.05	4.96	4.10	3.71	3.48	3.33	3.22	2.98	2.91	2.85	2.77	2.70	2.64	2.54
0.025	6.94	5.46	4.83	4.47	4.24	4.07	3.72	3.62	3.52	3.42	3.31	3.22	3.08
0.01	10.0	7.56	6.55	5.99	5.64	5.39	4.85	4.71	4.56	4.41	4.25	4.12	3.91
0.005	12.8	9.43	8.08	7.34	6.87	6.54	5.85	5.66	5.47	5.27	5.07	4.90	4.64

df = 20

α													
0.995	—	0.005	0.023	0.050	0.077	0.104	0.190	0.221	0.258	0.301	0.354	0.405	0.500
0.99	—	0.010	0.037	0.071	0.105	0.135	0.227	0.259	0.297	0.340	0.392	0.441	0.532
0.975	0.001	0.025	0.071	0.117	0.158	0.193	0.292	0.325	0.363	0.406	0.456	0.503	0.585
0.95	0.004	0.051	0.115	0.172	0.219	0.258	0.360	0.393	0.430	0.471	0.518	0.562	0.637
0.90	0.016	0.106	0.193	0.260	0.312	0.353	0.454	0.485	0.520	0.557	0.600	0.637	0.704
0.10	2.97	2.59	2.38	2.25	2.16	2.09	1.94	1.89	1.84	1.79	1.74	1.69	1.61
0.05	4.35	3.49	3.10	2.87	2.71	2.60	2.35	2.28	2.20	2.12	2.04	1.97	1.84
0.025	5.87	4.46	3.86	3.51	3.29	3.13	2.77	2.68	2.57	2.46	2.35	2.25	2.09
0.01	8.10	5.85	4.94	4.43	4.10	3.87	3.37	3.23	3.09	2.94	2.78	2.64	2.42
0.005	9.94	6.99	5.82	5.17	4.76	4.47	3.85	3.68	3.50	3.32	3.12	2.96	2.69

df = 40

α													
0.995	—	0.005	0.024	0.051	0.080	0.108	0.237	0.201	0.279	0.331	0.396	0.463	0.599
0.99	—	0.010	0.038	0.073	0.108	0.140	0.276	0.240	0.319	0.371	0.435	0.498	0.628
0.975	0.001	0.025	0.071	0.119	0.162	0.199	0.344	0.307	0.387	0.437	0.498	0.556	0.674
0.95	0.004	0.051	0.116	0.175	0.224	0.265	0.412	0.376	0.454	0.502	0.558	0.613	0.717
0.90	0.016	0.106	0.194	0.263	0.317	0.360	0.504	0.469	0.542	0.585	0.636	0.683	0.772
0.10	2.84	2.44	2.23	2.09	2.00	1.93	1.76	1.71	1.66	1.61	1.54	1.48	1.38
0.05	4.08	3.23	2.84	2.61	2.45	2.34	2.08	2.00	1.92	1.84	1.74	1.66	1.51
0.025	5.42	4.05	3.46	3.13	2.90	2.74	2.39	2.29	2.18	2.07	1.94	1.83	1.64
0.01	7.31	5.18	4.31	3.83	3.51	3.29	2.80	2.66	2.52	2.37	2.20	2.06	1.80
0.005	8.83	6.07	4.98	4.37	3.99	3.71	3.12	2.95	2.78	2.60	2.40	2.23	1.93

Large sample size

α													
0.995	—	0.005	0.024	0.052	0.082	0.113	0.216	0.256	0.307	0.372	0.460	0.559	1.00
0.99	—	0.010	0.038	0.074	0.111	0.145	0.256	0.298	0.349	0.413	0.499	0.595	1.00
0.975	0.001	0.025	0.072	0.121	0.166	0.206	0.325	0.367	0.418	0.480	0.560	0.645	1.00
0.95	0.004	0.051	0.117	0.178	0.229	0.273	0.394	0.436	0.484	0.543	0.617	0.694	1.00
0.90	0.016	0.105	0.195	0.266	0.322	0.367	0.487	0.525	0.570	0.622	0.687	0.752	1.00
0.10	2.71	2.30	2.08	1.94	1.85	1.77	1.60	1.55	1.49	1.42	1.34	1.26	1.00
0.05	3.84	3.00	2.60	2.37	2.21	2.10	1.83	1.75	1.67	1.57	1.46	1.35	1.00
0.025	5.02	3.69	3.12	2.79	2.57	2.41	2.05	1.94	1.83	1.71	1.57	1.43	1.00
0.01	6.63	4.61	3.78	3.32	3.02	2.80	2.32	2.18	2.04	1.88	1.70	1.52	1.00
0.005	7.88	5.30	4.28	3.72	3.35	3.09	2.52	2.36	2.19	2.00	1.79	1.59	1.00

Interpolation should be used to obtain values of F_α not tabulated here. The identity $F_\alpha(v_1, v_2) = 1/F_{1-\alpha}(v_2, v_1)$ can also be used to obtain nontabulated upper centile values.

Appendix A.8 Binomial Probabilities*

n	x	0.1	0.2	0.3	0.4	0.5		n	x	0.1	0.2	0.3	0.4	0.5
7	0	0.478	0.210	0.082	0.028	0.008		12	0	0.282	0.069	0.014	0.002	
	1	0.372	0.367	0.247	0.131	0.055			1	0.377	0.206	0.071	0.017	0.003
	2	0.124	0.275	0.318	0.261	0.164			2	0.230	0.283	0.168	0.064	0.016
	3	0.023	0.115	0.227	0.290	0.273			3	0.085	0.236	0.240	0.142	0.054
	4	0.003	0.029	0.097	0.194	0.273			4	0.021	0.133	0.231	0.213	0.121
	5		0.004	0.025	0.077	0.164			5	0.004	0.053	0.158	0.227	0.193
	6			0.004	0.017	0.055			6		0.016	0.079	0.177	0.226
	7				0.002	0.008			7		0.003	0.029	0.101	0.193
									8		0.001	0.008	0.042	0.121
8	0	0.430	0.168	0.058	0.017	0.004			9			0.001	0.012	0.054
	1	0.383	0.336	0.198	0.090	0.031			10				0.002	0.016
	2	0.149	0.294	0.296	0.209	0.109			11					0.003
	3	0.033	0.147	0.254	0.279	0.219								
	4	0.005	0.046	0.136	0.232	0.273		13	0	0.254	0.055	0.010	0.001	
	5		0.009	0.047	0.124	0.219			1	0.367	0.179	0.054	0.011	0.002
	6		0.001	0.010	0.041	0.109			2	0.245	0.268	0.139	0.045	0.010
	7			0.001	0.008	0.031			3	0.100	0.246	0.218	0.111	0.035
	8				0.001	0.004			4	0.028	0.154	0.234	0.184	0.087
									5	0.006	0.069	0.180	0.221	0.157
9	0	0.387	0.134	0.040	0.010	0.002			6	0.001	0.023	0.103	0.197	0.209
	1	0.387	0.302	0.156	0.060	0.018			7		0.006	0.044	0.131	0.209
	2	0.172	0.302	0.267	0.161	0.070			8		0.001	0.014	0.066	0.157
	3	0.045	0.176	0.267	0.251	0.164			9			0.003	0.024	0.087
	4	0.007	0.066	0.172	0.251	0.246			10			0.001	0.006	0.035
	5	0.001	0.017	0.074	0.167	0.246			11				0.001	0.010
	6		0.003	0.021	0.074	0.164			12					0.002
	7			0.004	0.021	0.070								
	8				0.004	0.018		14	0	0.229	0.044	0.007	0.001	
	9					0.002			1	0.356	0.154	0.041	0.007	0.001
									2	0.257	0.250	0.113	0.032	0.006
10	0	0.349	0.107	0.028	0.006	0.001			3	0.114	0.250	0.194	0.085	0.022
	1	0.387	0.268	0.121	0.040	0.010			4	0.035	0.172	0.229	0.155	0.061
	2	0.194	0.302	0.233	0.121	0.044			5	0.008	0.086	0.196	0.207	0.122
	3	0.057	0.201	0.267	0.215	0.117			6	0.001	0.032	0.126	0.207	0.183
	4	0.011	0.088	0.200	0.251	0.205			7		0.009	0.062	0.157	0.209
	5	0.001	0.026	0.103	0.201	0.246			8		0.002	0.023	0.092	0.183
	6		0.006	0.037	0.111	0.205			9			0.007	0.041	0.122
	7		0.001	0.009	0.042	0.117			10			0.001	0.014	0.061
	8			0.001	0.011	0.044			11				0.003	0.022
	9				0.002	0.010			12				0.001	0.006
	10					0.001			13					0.001

* If $p \geq .5$, the desired probability for x may be obtained by looking up $n - x$ under column $1 - p$.

Appendix A.8 continued

n	x	p 0.1	0.2	0.3	0.4	0.5		n	x	p 0.1	0.2	0.3	0.4	0.5
11	0	0.314	0.086	0.020	0.004									
	1	0.384	0.236	0.093	0.027	0.005		15	0	0.206	0.035	0.005		
	2	0.213	0.295	0.200	0.089	0.027			1	0.343	0.132	0.031	0.005	
	3	0.071	0.221	0.257	0.177	0.081			2	0.267	0.231	0.092	0.022	0.003
	4	0.016	0.111	0.220	0.236	0.161			3	0.129	0.250	0.170	0.063	0.014
	5	0.002	0.039	0.132	0.221	0.226			4	0.043	0.188	0.219	0.127	0.042
	6		0.010	0.057	0.147	0.226			5	0.010	0.103	0.206	0.186	0.092
	7		0.002	0.017	0.070	0.161			6	0.002	0.043	0.147	0.207	0.153
	8			0.004	0.023	0.081			7		0.014	0.081	0.177	0.196
	9			0.001	0.005	0.027			8		0.003	0.035	0.118	0.196
	10				0.001	0.005			9		0.001	0.012	0.061	0.153
									10			0.003	0.024	0.092
									11			0.001	0.007	0.042
									12				0.002	0.014
									13					0.003

Values of 0.0005 or less are omitted from this table.

Appendix A.9 Values of $F(v) = \frac{1}{2}ln(1 + v)/(1 - v)$

For each value v, the entry in the table is $F(v)$.

v	0.00	0.01	0.02	0.03	0.04	0.05	0.06	0.07	0.08	0.09
0.0	0.000	0.010	0.020	0.030	0.040	0.050	0.060	0.070	0.080	0.090
0.1	0.100	0.110	0.121	0.131	0.141	0.151	0.161	0.172	0.182	0.192
0.2	0.203	0.213	0.224	0.234	0.245	0.255	0.266	0.277	0.288	0.299
0.3	0.310	0.321	0.332	0.343	0.354	0.365	0.377	0.388	0.400	0.412
0.4	0.424	0.436	0.448	0.460	0.472	0.485	0.497	0.510	0.523	0.536
0.5	0.549	0.563	0.576	0.590	0.604	0.618	0.633	0.648	0.662	0.678
0.6	0.693	0.709	0.725	0.741	0.758	0.775	0.793	0.811	0.829	0.848
0.7	0.867	0.887	0.908	0.929	0.950	0.973	0.996	1.020	1.045	1.071
0.8	1.099	1.127	1.157	1.188	1.221	1.256	1.293	1.333	1.376	1.422
0.9	1.472	1.528	1.589	1.658	1.738	1.832	1.946	2.092	2.298	2.647

For negative values of v put a minus sign in front of the tabled numbers.

Appendix A.10 Distribution of Total Number of Runs u in Samples of Size (n_1, n_2)

n_1, n_2 \ u	2	3	4	5	6	7	8	9	10	11	12	13	14	15	16	17	18	19	20
(2,3)	0.200	0.500	0.900	1.000															
(2,4)	0.133	0.400	0.800	1.000															
(2,5)	0.095	0.333	0.714	1.000															
(2,6)	0.071	0.286	0.643	1.000															
(2,7)	0.056	0.250	0.583	1.000															
(2,8)	0.044	0.222	0.533	1.000															
(2,9)	0.036	0.200	0.491	1.000															
(2,10)	0.030	0.182	0.455	1.000															
(3,3)	0.100	0.300	0.700	0.900	1.000														
(3,4)	0.057	0.200	0.543	0.800	0.971	1.000													
(3,5)	0.036	0.143	0.429	0.714	0.929	1.000													
(3,6)	0.024	0.107	0.345	0.643	0.881	1.000													
(3,7)	0.017	0.083	0.283	0.583	0.833	1.000													
(3,8)	0.012	0.067	0.236	0.533	0.788	1.000													
(3,9)	0.009	0.055	0.200	0.491	0.745	1.000													
(3,10)	0.007	0.045	0.171	0.455	0.706	1.000													
(4,4)	0.029	0.114	0.371	0.629	0.886	0.971	1.000												
(4,5)	0.016	0.071	0.262	0.500	0.786	0.929	0.992	1.000											
(4,6)	0.010	0.048	0.190	0.405	0.690	0.881	0.976	1.000											
(4,7)	0.006	0.033	0.142	0.333	0.606	0.833	0.954	1.000											
(4,8)	0.004	0.024	0.109	0.279	0.533	0.788	0.929	1.000											
(4,9)	0.003	0.018	0.085	0.236	0.471	0.745	0.902	1.000											
(4,10)	0.002	0.014	0.068	0.203	0.419	0.706	0.874	1.000											

	c1	c2	c3	c4	c5	c6	c7	c8	c9	c10	c11	c12	c13	c14	c15	c16	c17
(5,5)	0.008	0.040	0.167	0.357	0.643	0.833	0.960	0.992	1.000								
(5,6)	0.004	0.024	0.110	0.262	0.522	0.738	0.911	0.976	0.998	1.000							
(5,7)	0.003	0.015	0.076	0.197	0.424	0.652	0.854	0.955	0.992	1.000							
(5,8)	0.002	0.010	0.054	0.152	0.347	0.576	0.793	0.929	0.984	1.000							
(5,9)	0.001	0.007	0.039	0.119	0.287	0.510	0.734	0.902	0.972	1.000							
(5,10)	0.001	0.005	0.029	0.095	0.239	0.455	0.678	0.874	0.958	1.000							
(6,6)	0.002	0.013	0.067	0.175	0.392	0.608	0.825	0.933	0.987	0.998	1.000						
(6,7)	0.001	0.008	0.043	0.121	0.296	0.500	0.733	0.879	0.966	0.992	0.999						
(6,8)	0.001	0.005	0.028	0.086	0.226	0.413	0.646	0.821	0.937	0.984	0.998						
(6,9)	0.000	0.003	0.019	0.063	0.175	0.343	0.566	0.762	0.902	0.972	0.994						
(6,10)	0.000	0.002	0.013	0.047	0.137	0.288	0.497	0.706	0.864	0.958	0.990						
(7,7)	0.001	0.004	0.025	0.078	0.209	0.383	0.617	0.791	0.922	0.975	0.996	0.999	1.000				
(7,8)	0.000	0.002	0.015	0.051	0.149	0.296	0.514	0.704	0.867	0.949	0.988	0.998	1.000				
(7,9)	0.000	0.001	0.010	0.035	0.108	0.231	0.427	0.622	0.806	0.916	0.975	0.994	0.999				
(7,10)	0.000	0.001	0.006	0.024	0.080	0.182	0.355	0.549	0.743	0.879	0.957	0.990	0.998				
(8,8)	0.000	0.001	0.009	0.032	0.100	0.214	0.405	0.595	0.786	0.900	0.968	0.991	0.999	1.000			
(8,9)	0.000	0.001	0.005	0.020	0.069	0.157	0.319	0.500	0.702	0.843	0.939	0.980	0.996	1.000			
(8,10)	0.000	0.000	0.003	0.013	0.048	0.117	0.251	0.419	0.621	0.782	0.903	0.964	0.990	0.998			
(9,9)	0.000	0.000	0.003	0.012	0.044	0.109	0.238	0.399	0.601	0.762	0.891	0.956	0.988	0.997	1.000		
(9,10)	0.000	0.000	0.002	0.008	0.029	0.077	0.179	0.319	0.510	0.681	0.834	0.923	0.974	0.992	0.999	1.000	
(10,10)	0.000	0.000	0.001	0.004	0.019	0.051	0.128	0.242	0.414	0.586	0.758	0.872	0.949	0.981	0.996	0.999	1.000

By permission from C. Eisenhart and F. Swed, "Tables for testing randomness of grouping in a sequence of alternatives," *Annals of Mathematical Statistics*, vol. 14 (1943), p. 66.

Appendix A.11 Distribution of the Signed-Rank Statistic W

For various sample sizes n. Values W_α are such that the probability is α that the signed-rank statistic is less than or equal to W_α. The values $W_{1-\alpha}$ are such that the probability is α that W is greater than or equal to $W_{1-\alpha}$.

W_α	$W_{1-\alpha}$	α	W_α	$W_{1-\alpha}$	α	W_α	$W_{1-\alpha}$	α	W_α	$W_{1-\alpha}$	α	W_α	$W_{1-\alpha}$	α
	$n=5$			$n=8$ (Cont.)			$n=10$ (Cont.)			$n=13$ (Cont.)			$n=15$ (Cont.)	
0	15	0.031	1	35	0.008	14	41	0.097	10	81	0.005	16	104	0.005
1	14	0.062	2	34	0.012	15	40	0.116	12	79	0.009	19	101	0.009
2	13	0.094	4	32	0.027		$n=11$		13	78	0.011	20	100	0.011
3	12	0.156	6	30	0.055	5	61	0.005	17	74	0.024	21	99	0.013
	$n=6$		8	28	0.098	7	59	0.009	21	70	0.047	25	95	0.024
0	21	0.016	9	27	0.125	8	58	0.012	22	69	0.055	31	89	0.053
1	20	0.031		$n=9$		9	57	0.016	23	68	0.064	36	84	0.094
2	19	0.047	1	44	0.004	11	55	0.027	26	65	0.095	37	83	0.104
3	18	0.078	2	43	0.006	14	52	0.051	27	64	0.108		$n=20$	
4	17	0.109	3	42	0.010	15	51	0.062		$n=14$		37	173	0.005
5	16	0.156	4	41	0.014	18	48	0.103	12	93	0.004	38	172	0.005
	$n=7$		5	40	0.020		$n=12$		13	92	0.005	42	168	0.009
0	28	0.008	8	37	0.049	7	71	0.005	16	89	0.010	43	167	0.010
1	27	0.016	9	36	0.064	10	68	0.010	21	84	0.025	44	166	0.011
2	26	0.023	11	34	0.102	14	64	0.026	25	80	0.045	52	158	0.024
3	25	0.039		$n=10$		17	61	0.046	26	79	0.052	53	157	0.027
4	24	0.055	3	52	0.005	18	60	0.055	27	78	0.059	60	150	0.049
5	23	0.078	5	50	0.010	19	59	0.065	31	74	0.097	61	149	0.053
6	22	0.109	6	49	0.014	22	56	0.102	32	73	0.108	69	141	0.095
	$n=8$		8	47	0.024		$n=13$			$n=15$		70	140	0.101
0	36	0.004	11	44	0.053	9	82	0.004	15	105	0.004			

Appendix A.12 Percentile of r when $\rho = 0$

Upper-centile values $r(n, \alpha)$ for the Spearman Correlation for various sample sizes n

Sample Size (n)	α 0.999	0.975	0.95	0.05	0.025	0.001
5	—	−0.85	−0.77	0.87	0.95	—
6	−0.94	−0.79	−0.71	0.77	0.79	1
7	−0.93	−0.75	−0.66	0.69	0.79	0.96
8	−0.9	−0.73	−0.62	0.64	0.74	0.95
9	−0.88	−0.67	−0.58	0.59	0.68	0.90
10	−0.84	−0.64	−0.54	0.56	0.65	0.89

Appendix A.13 Percentiles of the Maximum Deviation Statistics

Values of $d_{1-\alpha}$ for finding $1 - \alpha$ confidence bands about an empirical distribution function

Sample size n	α 0.1	0.05	0.01
10	0.37	0.41	0.49
20	0.26	0.29	0.35
30	0.22	0.24	0.29
40	0.19	0.21	0.25
50	0.17	0.19	0.23
for larger sample sizes	$1.22/\sqrt{n}$	$1.36/\sqrt{n}$	$1.63/\sqrt{n}$

Appendix A.14 Confidence Intervals for the Median

If the observations are arranged in order of size $X_1 < X_2 < X_3 < \cdots < X_n$, then we are at least $100(1 - \alpha)$ percent confident that the population median will be between X_k and X_{n-k+1}, where k and α are given in the table headings. The exact value of α is given in the table.

Sample size n	Largest k	Near $\alpha \le$ 0.05	Largest k	Near $\alpha \le$ 0.01	N	Largest k	Near $\alpha \le$ 0.05	Largest k	Near $\alpha \le$ 0.01
6	1	0.031			36	12	0.029	10	0.004
7	1	0.016			37	13	0.047	11	0.008
8	1	0.008	1	0.008	38	13	0.034	11	0.005
9	2	0.039	1	0.004	39	13	0.024	12	0.009
10	2	0.021	1	0.002	40	14	0.038	12	0.006
11	2	0.012	1	0.001	41	14	0.028	12	0.004
12	3	0.039	2	0.006	42	15	0.044	13	0.008
13	3	0.022	2	0.003	43	15	0.032	13	0.005
14	3	0.013	2	0.002	44	16	0.049	14	0.010
15	4	0.035	3	0.007	45	16	0.036	14	0.007
16	4	0.021	3	0.004	46	16	0.026	14	0.005
17	5	0.049	3	0.002	47	17	0.040	15	0.008
18	5	0.031	4	0.008	48	17	0.029	15	0.006
19	5	0.019	4	0.004	49	18	0.044	16	0.009
20	6	0.041	4	0.003	50	18	0.033	16	0.007
21	6	0.027	5	0.007	51	19	0.049	16	0.005
22	6	0.017	5	0.004	52	19	0.036	17	0.008
23	7	0.035	5	0.003	53	19	0.027	17	0.005
24	7	0.023	6	0.007	54	20	0.040	18	0.009
25	8	0.043	6	0.004	55	20	0.030	18	0.006
26	8	0.029	7	0.009	56	21	0.044	18	0.005
27	8	0.019	7	0.006	57	21	0.033	19	0.008
28	9	0.036	7	0.004	58	22	0.048	19	0.005
29	9	0.024	8	0.008	59	22	0.036	20	0.009
30	10	0.043	8	0.005	60	22	0.027	20	0.006
31	10	0.029	8	0.003	61	23	0.040	21	0.010
32	10	0.020	9	0.007	62	23	0.030	21	0.007
33	11	0.035	9	0.005	63	24	0.043	21	0.005
34	11	0.024	10	0.009	64	24	0.033	22	0.008
35	12	0.041	10	0.006	65	25	0.046	22	0.006

By permission of S. K. Banerjee from K. R. Nair, "Table of confidence interval for the median in samples from any continuous population," *Sankhya*, vol. 4 (1940), pp. 551–558.

APPENDIX · B

Derivation of the Binomial Distribution

We wish to develop a formula for the probability of exactly k successes out of n independent trials. Let 1 denote a success and 0 denote a failure. Using this notation, we wish to find

$$P(k \text{ 1's in } n \text{ trials})$$

Let us be specific; suppose $k = 3$ and $n = 5$. One way to obtain three successes in 5 trials is as follows:

$$11100$$

that is, the first three trials produce successes and the last two failures. The probability of this event, denoted $P(11100)$ is, by the independence of the trials,

$$P(11100) = P(1)P(1)P(1)P(0)P(0)$$

where $P(1)$ = probability of success on a trial; $P(0)$ = probability of failure on a trial. To simplify the notation, let $p = P(1)$, $q = P(0)$. Then

$$P(11100) = pppqq = p^3q^2 = p^3q^{5-3}$$

Now consider another way one can obtain three successes out of five trials, namely, 10011. The probability of this event is, by the independence of trials,

$$P(10011) = pqqpp = p^3q^2 = p^3q^{5-3}$$

Notice that $P(11100) = P(10011)$. If we consider any other event of the same type, say, 11010, it too will have the same probability. Now

$$P(3 \text{ 1's in 5 trials}) = P(11100 \text{ or } 10011 \text{ or } 11010 \text{ or } \cdots)$$

where the three dots indicate all the other ways of getting three successes in 5 trials. All these events are mutually exclusive. Hence

$$P(3 \text{ 1's in 5 trials}) = P(11100) + P(10011) + P(11010) + \cdots$$
$$= p^3 q^{5-3} + p^3 q^{5-3} + p^3 q^{5-3} + \cdots$$
$$= r\, p^3 q^{5-3}$$

where r is the number of ways of getting 3 successes in 5 trials. To find r, consider the 5 trials and the placement of the three 1's among the 5 trials. There are 5 positions to choose for the placement of the first object 1, $5 - 1$ for the second and $5 - 2 = 5 - 3 + 1$ for the last. Hence there are $5(5 - 1)(5 - 3 + 1) = 5!/2!$ different positional arrangements for these objects. Now for every fixed position of the three 1's, the 1's may be permuted among themselves. There are 3! such permutations. All such permutations are indistinguishable since the objects are identical. Hence there are $5!/2!3!$ distinguishable arrangements among the places. Hence

$$r = \frac{5!}{2!3!} = \binom{5}{3}$$

and

$$P(3 \text{ successes in 5 trials})$$
$$= \binom{5}{3} p^3 q^{5-3}$$

Replace 5 by n and 3 by k in the preceding arguments and you will obtain

BINOMIAL DISTRIBUTION

$$P(k \text{ successes in } n \text{ trials}) = \binom{n}{k} p^k q^{n-k}; \qquad k = 0, 1, \ldots, n$$

where $\qquad q = 1 - p$

APPENDIX · C

Confidence Belts for Proportions

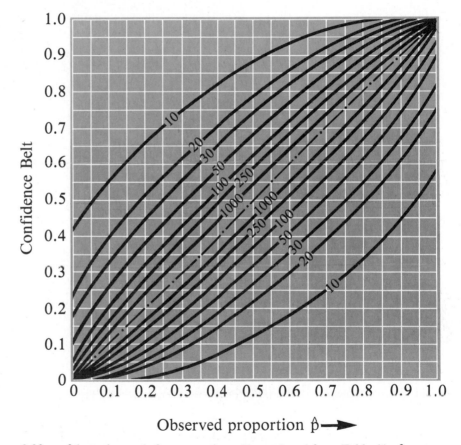

0.99 confidence intervals for proportions. (Reproduced from Table 41 of *Biometrika Tables for Statisticians,* Volume I. Cambridge: Cambridge University Press, 1954. By permission of the *Biometrika* trustees.)

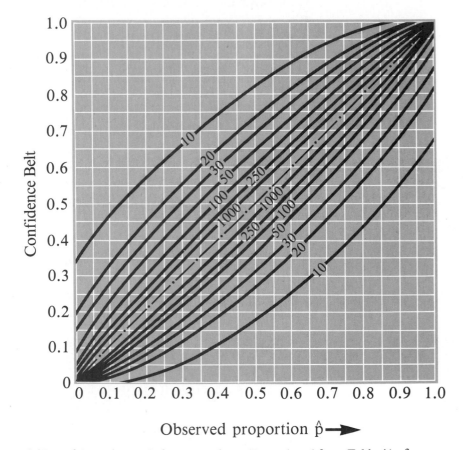

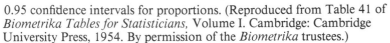

0.95 confidence intervals for proportions. (Reproduced from Table 41 of *Biometrika Tables for Statisticians,* Volume I. Cambridge: Cambridge University Press, 1954. By permission of the *Biometrika* trustees.)

Answers to Selected Odd-numbered Problems

CHAPTER 1

3. a. Suppose the employer hired 1 woman and 99 men for a job for which 100 qualified women and 100 qualified men applied. Hence 1% of those hired were women although 50% of those qualified were women. These data suggest that hiring may not have been fair to women. Now suppose also that before the hiring in 1966 the employer had 1 woman working in the firm. Then hiring 1 woman does increase the number of women in the firm by 100%. The employer's claim is true but it is misleading in the sense that it tends to lead us away from the actual issue: Does the proportion of women hired correspond to the proportion of qualified women applicants?

   b. To avoid being misled, we need enough information to enable us to compare the proportion of females hired in 1966 with the proportion of qualified female applicants. We could find the proportion of females hired by knowing the number of women employed at the start of 1966 and the total number of people hired in 1966 and then using the information that the increase in women was 100% in 1966.

   c. No. One needs to know the proportion of qualified females available to work for the company when positions become available.

5. The following list contains only a few of the many summary measures that could be developed. Let us assign numbers to the "responses" in order to simplify our discussion: let
   0 = absolutely must not, 1 = preferably should not, 2 = may or may not, 3 = preferably should, 4 = absolutely must.

   a. Remembering that each response is now a "number," for each respondent we can add all the responses to obtain a total score and then divide by the total number of questions to obtain an average score for each person.

   b. An overall average can be obtained by adding up all the individual respondent averages and dividing by the number (106) of such averages.

   c. For each question we can list the proportion of each response given to the question by the respondents as shown below.

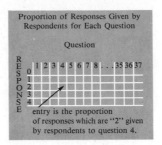

Proportion of Responses Given by Respondents for Each Question

    d. We may develop a table such as the one here for each level of educational attainment; that is, less than high school, high school diploma only, vocational school, 1 or 2 years of college, 3 or 4 years of college, masters degree, Ph.D., and so on. By comparing such tables the differences in "decisions" by "education" may immediately become apparent. The reason we might want to construct such a summary is to reach the objective of the questionnaire, which is to determine what motivates people to make certain types of decisions. Eductional background may be one important factor. The lesson to be learned here is that the types of summaries we use often are dictated by the objectives of the study.

7.   Inference (b) is a statistical inference. The others are not.

9.   **a.** (iii); **b.** No; **c.** Unknown; **d.** A cursory reading of the quote above might give the impression that all brands were defective. The given information does not convince us that this is so, however. Information on the level of radiation is not given; in fact, no indication is given that the levels are harmful. In addition, the proportion of defective sets is not given. The reporter may have discovered that each manufacturer has 1 set of 1000 giving off harmless radiation, in which case we would have a case of "much ado about nothing."

13.   Price of food, appliances, gasoline.

CHAPTER 2

3. **a.**

Births per day	Number of days
0–4.9	3
5.0–9.9	21
10.0–14.9	37
15.0–19.9	17
20.0–24.9	7
25.0–29.9	5
30.0–34.9	1

b.

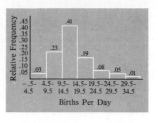

c. About 12 or 13.

d. No.

5. a.

Miles per gallon	Frequency
27.0 – 30.9	6
31.0 – 34.9	6
35.0 – 38.9	7
39.0 – 42.9	5
43.0 – 46.9	1

b.

c. Most test runs produced between 35 and 40 miles per gallon; the variability in the runs ran between 27 and 47 miles per gallon.

7. a.

Age Group 21 to 44

Pressure	Frequency	Relative frequency
71.0 – 90.9	0	0
91.0 – 110.9	13	.21
111.0 – 130.9	36	.57
131.0 – 150.9	13	.21
151.0 – 170.9	1	.02
171.0 – 190.9	0	0

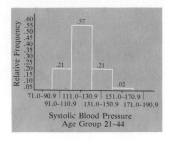

Age Group 45 to 59

Pressure	Frequency	Relative frequency
71.0–90.9	1	.03
91.0–110.9	7	.19
111.0–130.9	11	.30
131.0–150.9	8	.22
151.0–170.9	8	.22
171.0–190.9	2	.05

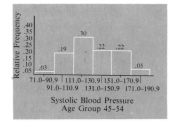

b. Age Group 21 to 44
c. Age Group 45 to 59
d. The younger age group (21 to 44) has blood pressure closer to the 120 "normal." The blood pressure seems to be age dependent.

9. a.

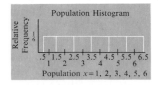

b.

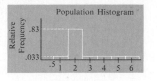

13. a.

b. .5, which is the size of the shaded area in part a.

15. a. $\bar{x} = \dfrac{1208}{91} = 13.27$; **b.** 13; **c.** 14;

d. $s^2 = \dfrac{\sum\limits_{i=1}^{n}(x_i - \bar{x})^2}{n-1} = 34.6$

$s = \sqrt{34.63} = 5.9$

e. 13.3 or about 13; **f.** There are about 13.27 ± 5.9 babies born per day. The answer is no.

17. a. $\bar{x}_g = 551.7$; **b.** The median is in the interval 550 to 600; **c.** We cannot tell which value(s) constitute the mode(s) since the data are grouped. The largest number of observed values, however, is in the interval 550 to 600. We take as the mode of the *histogram* for this particular grouping of the data the midpoint of this interval.

19. $\bar{x} = \dfrac{10(0) + 5(1)}{15} = .33$

I would estimate about $\frac{1}{3}$ of the ball bearings produced to be defective.

23. a. Each number occurs 16.7% of the tags.
$1(.167) + 2(.167) + 3(.167) + 4(.167) + 5(.167) + 6(.167) = 3.507$

b. 2 appears on 83.33% of the tags; each of the others appears on 3.33% of the tags.
$1(0.333) + 2(.8333) + 3(.0333) + 4(.0333) + 5(.0333) + 6(.0333) = 2.299$

c. 1 and 6 each appear on 10% of the tags; 2, 3, 4, 5 each appear on 20% of the tags.
$1(.1) + 2(.2) + 3(.2) + 4(.2) + 5(.2) + 6(.1) = 3.5$

    d. 2 and 5 each appear on 30% of the tags; 1, 3, 4, 6 each appear on 10% of the tags.

$$1(.10) + 2(.30) + 3(.10) + 4(.10) + 5(.30) + 6(.10) = 3.5$$

25. Yes; Yes.

27. No.

29. (3) 34.6; 5.9 (4) 4854.6; 69.7 (5) 22.4; 4.73

31. $\bar{x} = 1.87$; $s = 1.28$

33.

    a.

	Frequency
1–5	7
6–10	13
11–15	10
16–20	5

    b.

Apparent interval limits	Real interval limits	Midpoint	Frequency	Relative frequency
1–5	0.5–5.5	3.0	7	.20
6–10	5.5–10.5	8.0	13	.37
11–15	10.5–15.5	13.0	10	.29
16–20	15.5–20.5	18.0	5	.14

    c. $\bar{x}_g = \dfrac{(3.0 \cdot 7) + (8 \cdot 13) + (13 \cdot 10) + (18 \cdot 5)}{35} = \dfrac{345}{35} = 9.86$

$$s_g^2 = \frac{35(4205) - (345)^2}{35(34)} = \frac{28150}{1190} = 23.66$$

    d. $\bar{x} = \dfrac{349}{35} = 9.97$, $s^2 = \dfrac{35(4313.46) - (349)^2}{35(34)} = 24.51$

35. a. 3.5 **c.** .71 **d.** The population of problem 9a.

37. a. $\mu = .46$, $\sigma = .5$ **b.** $\mu = .46$, $\sigma = .5$

39. b. Midpoints*

(Class marks)	Relative frequency
2.45	.03
7.45	.23
12.45	.41
17.45	.19
22.45	.08
27.45	.05
32.45	.01

* The midpoint of each real interval is sometimes called the class mark of that interval.

c.

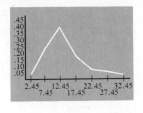

43. a.

Score	Frequency	Relative frequency
20–25	3	.06
25–30	4	.08
30–35	18	.36
35–40	16	.32
40–45	6	.12
45–50	1	.02
50–55	2	.04

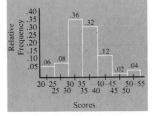

43. b.

Midpoints (Class marks)	Frequency	Relative frequency
22.5	3	.06
27.5	4	.08
32.5	18	.36
37.5	16	.32
42.5	6	.12
47.5	1	.02
52.5	2	.04

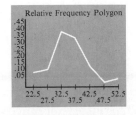

   c. $\bar{x}_g = 35.4$, $s_g^2 = 40.9$

45. a. Skewed to the right; symmetric; symmetric
    b. Median or mode; mean, median, or mode; mean or median
    c. Histogram a has the largest; b has the smallest.

CHAPTER 3

1. Statements a and c do not since the "events" do not represent outcomes of repeated trials. Statement b does since one can have a long sequence of repetitions of the experiment of drawing 1 card from a deck of 52. Statement d does also since one conceivably (although, in reality, only approximately) can encounter, at different times, weather maps that are the same with respect to the factors that cause rain in a given locality. In the long run, predictions of rain based on these maps will be correct $\frac{1}{20}$th of the time.

3. a. .1

5. a. $(\frac{1}{2})^2 = \frac{1}{4}$ b. $(\frac{1}{2})^3 = \frac{1}{8}$ c. $(\frac{1}{2})^{10}$

7. a. 6 b. 18 c. 6 d. 18 e. 0 f. 12 g. 18 h. 6 i. 24

9. a. 10 b. no; there are only 80 people in the set of "smokers and/or drinkers." If every spouse beater were in this set, it would mean that there is a total of 80 persons, which we know is not the case.

11. b.

13. $\frac{12}{35}$

15. a. $\frac{1}{2}$ b. $\frac{9}{10}$ c. $\frac{9}{10}$ d. $\frac{8}{10}$ e. $\frac{8}{10}$ f. $\frac{2}{10}$

17. a. $\frac{1}{4}$ b. $\frac{1}{2}$ c. $\frac{1}{4}$

19. a. $\frac{1}{36}$ b. $\frac{1}{36}$ c. $\frac{3}{36}$ d. $\frac{6}{36}$ e. $\frac{6}{36}$ f. $\frac{30}{36}$ g. The number of outcomes of four tosses of a die is 6^4. The probability of one 6 in these four tosses is (no. of ways of obtaining one 6 in four tosses)/6^4. The number of ways in the numerator is found by noting that the 6 can occur on any of the four tosses so there are 4 ways the 6 can occur. Given that it occurs on a specific toss, there are five different possibilities (excluding the 6) on each of the three remaining tosses; hence, there are 5^3 possibilities given a 6 on a specific toss. The total possibilities for obtaining one 6 is $4(5^3)$. Hence the probability we seek is $4 \cdot 5^3/6^4 = .386$.

   On each toss of a pair of dice there are 36 outcomes. For 24 tosses, there are $(36)^{24}$ total outcomes. The number of outcomes that occur in which there is exactly one double 6 is (using the reasoning outlined before) $24(35)^{23}$. Hence, the probability we seek is $24(35)^{23}/(36)^{24} = .349$.

23. .012, .32

25. $\frac{5}{16}$

27. $\frac{12}{36} + \frac{18}{36} - \frac{6}{36} = \frac{24}{36}$

29. $P(A$ and $B) = -.9 + .5 + .8 = .4$; $P(A'$ or $B') = 1 - P(A$ and $B) = .6$

31. $P(D) = \frac{12}{36}$; $P(D \mid E) = (\frac{6}{36})/(\frac{18}{36}) = \frac{6}{18} = P(D)$

33. P(male | colorblind) $= (\frac{5}{100})/.0525 = .95$

35. a. $(\frac{10}{16})(\frac{12}{18}) = .417$

b. $P(\text{red ball on second trial} \mid \text{red on first trial}) = \frac{12}{18} = .667$
$P(\text{red ball on second trial}) = P(\text{red ball on first and second trial}) + P(\text{black ball on first and red on second trial}) = (\frac{10}{16})(\frac{12}{18}) + (\frac{6}{16})(\frac{10}{18}) = .625 \neq P(\text{red on second} \mid \text{red on first trial}) = .667.$

37. $(.9)(.95)(.99) = .85$

39. a. .026; .061 **b.** .435

43. a. $\frac{8}{28}$ **b.** $\frac{1}{35}$ **c.** $\dbinom{48}{1} \Big/ \left[\dbinom{52}{5} - \dbinom{48}{1}\right]$ **d.** $1/(2^{10} - 1)$

45. $P(A \text{ bluffing} \mid A \text{ bets big}) = \dfrac{P(A \text{ bluffing and bets big})}{P(A \text{ bets big})}$

$P(A \text{ bets big}) = P(A \text{ bets big and has a good hand}) + P(A \text{ bets big and doesn't have a good hand}) = P(A \text{ has a good hand})P(A \text{ bets big} \mid \text{good hand}) + P(A \text{ does not have a good hand})P(A \text{ bets big} \mid \text{no good hand}) = (\frac{1}{6}) \cdot 1 + (\frac{5}{6})(\frac{1}{5}) = \frac{2}{6}.$ Now $P(A \text{ bluffing and bets big}) = P(A \text{ bluffing}) \times P(A \text{ bets big} \mid A \text{ bluffing}) = P(A \text{ bluffing}) = P(A \text{ bets big and does not have a good hand}) = (\frac{5}{6})(\frac{1}{5}).$

The answer is $(\frac{5}{6})(\frac{1}{5})(\frac{2}{6}) = \frac{1}{2}$, which means that half the time when A bets big he is bluffing. Is this not a surprising result and doesn't A's strategy have the effect of keeping you off balance if you're in a six-person game with him?

47. a. $P(d \mid +') = P(d) \cdot P(+' \mid d)/P(+')$. From the example $P(d) = .001$, $P(+') = 1 - P(+) = 1 - .01097$. The value $P(+' \mid d)$ is given in the statement of the problem as .02. Hence, $P(d \mid +') = (.001)(.02)/.989 = .00002$. **b.** $Sn = 98\%$, $Sp = 99\%$. **c.** Incidence $= .1\%$ **d.** From Table 3.1, $1 - FP = a/(c + a)$ and $FN = b/(b + d)$ and $(1 - FP)/FN = a/(a + c) \div b/(b + d)$.

49. a. 20!/10! **b.** 20!/10! 10!

51. 2^{10}

53. $\dbinom{6}{2}\dbinom{6}{3} \Big/ \dbinom{12}{5} = .379$

55. $\dbinom{2i}{p}\dbinom{2N - 2i}{N - p} \Big/ \dbinom{2N}{N}$

57. a. 100^2 **b.** $100 \cdot 99$

59. $2^{2 \times 10} = 2^{20}$

61. $\frac{1}{4}$; $\frac{1}{4}$; $(1 - \frac{1}{4})^2 = .563$

63. 150

65. a. $\dbinom{25}{6}$ **b.** $\dbinom{10}{3}\dbinom{15}{3}$ **c.** $\dbinom{10}{3}\dbinom{15}{3} \Big/ \dbinom{25}{6}$

67. a. 39 **b.** $\frac{9}{39}$ **c.** $\dbinom{4}{3} \Big/ \dbinom{5}{3} = .4$

69. $\frac{1}{6}$

71. $2 \cdot 26^3$; 26^3

73. a. $\frac{2}{2000}$ b. $\frac{513}{2000}$ c. $\frac{1487}{2000}$

75. a. $\begin{pmatrix} 10 \\ 5 \end{pmatrix}$ b. $6 \cdot \begin{pmatrix} 3 \\ 2 \end{pmatrix} = 18$ c. 2^5

77. 9×10^6; 10^4

CHAPTER 4

1. a. $X = \begin{cases} 1 \text{ if the outcome of the toss is "heads"} \\ 0 \text{ if the outcome of the toss is "tails."} \end{cases}$

$$P(X=0) = P(X=1) = .5$$

b. $X = i$ if the number of defectives is i; $i = 0, 1, \ldots, 100$. $P(X = i)$ is some value p_i, $0 \le p_i \le 1$, $\Sigma p_i = 1$

c. $X = i$ if the outcome of the toss is i; $i = 0, 1, 2, \ldots, 6$. $P(X = i) = \frac{1}{6}$ each i.

d. $X = \begin{cases} 1 \text{ if the item is defective} \\ 0 \text{ if the item is nondefective.} \end{cases}$

$$P(X = i) = p, \; P(X = 0) = 1 - p, \; 0 \le p \le 1$$

3.

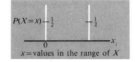

5. a, c, f, j are discrete while the others are continuous.

9. a. $\frac{1}{4}$ b. $\frac{1}{4}$ c. $\frac{3}{4}$

d.

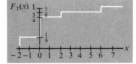

11. a. $\frac{1}{2}$ b. 3

13. $X = \begin{cases} -3 \text{ if "7 or 11"} & P(X = -3) = \frac{8}{36} \\ 6 \text{ if "2 same faces"} & P(X = 6) = \frac{6}{36} \\ 0 \text{ otherwise} & P(X = 0) = \frac{22}{36} \end{cases}$

Expected winnings $= -3(\frac{8}{36}) + 6(\frac{6}{36}) + 0(\frac{22}{36}) = \$\frac{12}{36} = \$.33\frac{1}{3}$; $\$.33\frac{1}{3}$.

15. $-\$1$.

17.

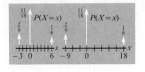

19. a. 4.5 **b.** 4 **c.** −4.6

23. b, d illustrate independent random variables.

25. $\binom{10}{3}(.2)^3(.8)^7 = .201$

27. Density Histogram of Binomial; $n = 8$, $p = 0.1$

Density Histogram of Binomial; $n = 10$, $p = 0.4$

29. 0.69

31. $1; \frac{1}{19}$

33. .285; .002

35. a. .25 **b.** .27

37. $10; 700 square dollars.

39. a. $\dfrac{1}{n}\left(1 - \dfrac{1}{n}\right)^{r-1}$ **b.** $1/n; 0$

41. $\binom{50}{10} \Big/ \binom{100}{10} = .0006$

43. a. .005 **b.** .36 **c.** .13

45. .63

47. Approximately .00004

49. Using Tchebychev's Approximation, the probability of getting as large a deviation as 20 from $\mu = 30$ is less than 9%. The traffic system certainly deserves some looking into considering that lives are at stake.

CHAPTER 5

1. **a.** (i) .4505; (ii) .4744; (iii) .9010; (iv) .9488; (v) .0239 **b.** (i) .9505; (ii) .9744; (iii) .0495; (iv) .0265 **c.** (i) .0495; (ii) .0265; (iii) .9505; (iv) .9744
3. **a.** -2.326 **b.** -1.96. You can find these results in Appendix A.1 using the small table denoted "common upper centiles" and the fact that $z_{1-\alpha} = -z_\alpha$.
5. **a.** 1.645 **b.** $-.842$ **c.** -1.960 **d.** 1.282 **e.** -1.645 **f.** -2.326 **g.** -2.576 **h.** 2.576
7. **a.** .06 **b.** .15 **c.** .21 **d.** 21,000. From part c the chance is .21. Considering the large number of births per year, 21% of the total is also a large number. If a majority of these fathers suffered suspicions of their wives, there would be a considerable strain on the fabric of our society.
9. **a.** .0016 **b.** .004
11. **a.** .383 **b.** 9987; $(.9987)^{128} = .85$
13. **a.** .995 **b.** 11.7
15. **a.** about 1 **b.** about 0 in both cases **c.** 0; 0; 0
17. **a.** about .99 **b.** about .98 **c.** about .34 **d.** 12
19. .426
25. .85
27. **a.** 2005.8; 38.8 **b.** .13 **c.** .1 **d.** 100

CHAPTER 6

1. **b.** Probability of $-\$2$ = Probability of $\$2$ = .25; Probability of $\$0$ = .5
3. **a.** $E(X_i) = 1(\frac{1}{2}) - 1(\frac{1}{2}) = 0, i = 1, 2$
   b. $E(\bar{X}) = E[(X_1 + X_2)/2] = \frac{1}{2}\{E(X_1) + E(X_2)\} = 0$
   c. $V(X_i) = (1-0)^2(\frac{1}{2}) + (-1-0)^2(\frac{1}{2}) = 1, i = 1, 2$
   d. $V(\bar{X}) = V[(X_1 + X_2)/2]$. Since X_1 and X_2 are independent,
   $$V\left(\frac{X_1}{2} + \frac{X_2}{2}\right) = V\left(\frac{X_1}{2}\right) + V\left(\frac{X_2}{2}\right) = \frac{1}{4}[V(X_1) + V(X_2)]$$
   $$= \frac{1}{4}(1 + 1) = .5$$
   e. Yes.
5. $\$25.19$; 2.46 square dollars; $\$1.57$
7. **a.** No **b.** approximately 2.5 and .25 respectively **c.** approximately 2.5 **d.** approximately .25
9. **a.** 11.34 to 12.66 **b.** 29.63 to 32.37 **c.** 81.91 to 118.1 **d.** 210.19 to 213.81
11. 111.94 to 119.62 months
13. 18.85 to 23.55 miles per gallon
15. $n - 2$
17. .01
19. 110.5 to 121.0 months

21. We assume the population of values of "breaking strength" is normally distributed: 38.11 to 42.57 lbs.
23. The population of values of "absentee rates" is normally distributed: 5.72 to 9.28 days/last 6 months.
25. We assume that the lengths of the produced parts are normally distributed: 1.27 to 3.69 mm^2.
27. 105263.2 to 272241.12 square dollars
29. Using Appendix C, about .45 to .74; using the normal approximation, about .46 to .74.
31. We assume that the "coin tossing" or binomial model describes the phenomenon under study: the estimate is .65 and the confidence interval goes from .56 to .74.
33. We assume that the population of grades is normally distributed: n about equal to 107.
35. n about 6147.

CHAPTER 7

1. **a.** He will purchase an unsafe rope; he will not purchase a safe rope.
   b. Type 1.
3. If you do not reject H_0, then you simply do not have enough evidence to be definitive as to whether it is not true.
5. H_0: drug is not "at least 90%" effective. **a.** result of making a type 1 error: marketing an ineffective drug and being subject to numerous lawsuits; result of making a type 2 error: not marketing an effective drug (but not so effective that experiments easily reveal its efficaciousness). **b.** Type 2
7. Do not reject H_0; reject H_0; do not reject H_0; reject H_0.
9. Do not reject H_0: $\mu = 5$. (Company should not switch.)
11. Yes.
13. Reject H_0. Average nonconformity rating has changed.
15. No.
17. No.
19. Yes.
21. Reject H_0 at both significance levels.
23. Reject H_0 at $\alpha = .05$ but not at $\alpha = .01$.
25. Reject H_0 at both significance levels.
27. Do not reject H_0.
29. Reject H_0.
31. The law significantly reduces the length of stay.
33. Replace with new desserts. There does not seem to be much of an improvement in the sales of the new dessert over the old, but McWendy has established a .01 significance level that tells us that they feel even a small improvement in sales is commercially desirable.

35. Reject H_0.
37. Reject H_0.
39. Reject H_0.
41. Reject H_0.
43. Reject H_0.
45. **a.** p_1 = proportion of nightmare sufferers (those who responded "often") among men; p_2 = proportion of nightmare sufferers among women. $H_0: p_1 = p_2$; $H_1: p_1 \neq p_2$ ($p_2 = .306$).
   b. Do not reject H_0. Proportion of nightmare sufferers is not sex dependent ($\alpha = 0.01$).
47. Do not reject H_0: $p \leq .9$. They should not advertise .9 effectiveness ($\alpha = 0.01$).
49. Do not reject H_0. Sex is not related to the response.
51. Do not reject H_0 at $\alpha = 0.05$.
53. Reject H_0: $p \geq .5$ if the continuity correction is not used. If it is, we do not reject H_0.
55. He can conclude that the majority of the town is in favor of the issue.

CHAPTER 8

1. **a.** -3.96 to $-.04$; -6.84 to $.86$; -3.38 to 8.38; -5.14 to -1.86; -8.17 to $-.33$ **b.** Reject H_0; Do not reject H_0; Reject H_0; Reject H_0; Reject H_0. **c.** Reject H_0.
3. No; -1.22 to 0.07.
5. No.
7. **a.** Do not reject H_0 **b.** Reject H_0 **c.** Reject H_0 **d.** Reject H_0 **e.** Do not reject H_0.
9. We cannot conclude that Rhine wines are more consistent than Moselle wines.
11. Stock A is significantly riskier than B.
13. Do not reject H_0.
15. **a.** .022; **b.** .0012; **c.** 1.21; **d.** 3.39 **e.** 3.35; **f.** 10991.6
17. **a.** .255 **b.** Cod caught close to shore have a significantly higher tissue concentration of the metal.
19. Moselle wines do not have significantly less ash.
21. The pupae raised in the higher humidity have significantly larger weights.
23. $H_0: \mu_D = 0$ where μ_D is the average difference of the before and after scores: $H_1: \mu_D \neq 0$. Reject H_0 at $\alpha = .1$ but do not reject at $\alpha = .01$.
25. No.
27. Yes; The 95% confidence interval on proportion "insolvent bank" $-$ proportion "successful bank" is from .66% to 3.28%.
29. Yes; The 80% confidence limit on μexpensive $- \mu$ is from 5.7% to 18.3%.

CHAPTER 9

1. It certainly appears from the graph that we do gain some indication of the final grade from the average grade prior to the final. It appears also that one can fit a straight line to these data. However, one can reasonably disagree with both these previous assertions. We have no way at this point of determining how to fit a line to these data or of even determining whether it is reasonable to assert there is a linear relationship between the prior average and final grade.

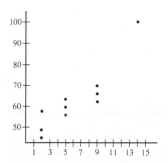

3.

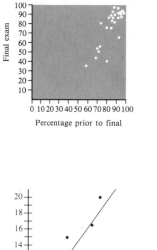

$\bar{x} = 5.14$
$\bar{y} = 12.71$

5.

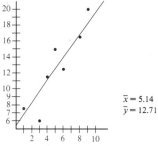

7. a.

x_i	y_i	$(x_i - \bar{x})$	$(x_i - \bar{x})^2$	$(y_i - \bar{y})$	$(x_i - \bar{x})(y_i - \bar{y})$
-1	1	-1.8	3.24	-1.2	2.16
-1	2	-1.8	3.24	$-.2$	.36
-1	1.5	-1.8	3.24	$-.7$	1.26
-1	2.3	-1.8	3.24	.1	$-.18$
1	1.9	.2	.04	$-.3$	$-.06$
1	2.5	.2	.04	.3	.06
2	2	1.2	1.44	$-.2$	$-.24$
2	2.8	1.2	1.44	.6	.72
3	3	2.2	4.84	.8	1.76
3	3.1	2.2	4.84	.9	1.98
8	22.1		25.6		7.82

$\bar{x} = .8 \qquad \bar{y} = 2.2$

$b = \dfrac{7.82}{25.6} = .3 \qquad a = 2.2 - (.3)(.8) = 2.2 - .24 = 1.96$

$\hat{y}_x = 1.96 + .3x$

$\hat{y}_0 = 1.96 + .3(0) = 1.96$

b.

x_i	y_i	$(x_i - \bar{x})$	$(x_i - \bar{x})^2$	$(y_i - \bar{y})$	$(x_i - \bar{x})(y_i - \bar{y})$
1	3	-2	4	2	-4
2	2.5	-1	1	1.5	-1.5
3	1	0	0	0	0
4	0	1	1	-1	-1
5	-1.5	2	4	-2.5	-5
15	5		10		11.5

$\bar{x} = 3 \qquad \bar{y} = 1$

$b = \dfrac{11.5}{10} = 1.15 \qquad \begin{aligned} a &= 1 - (1.15)(3) \\ &= 1 - 3.45 \\ &= -2.45 \end{aligned}$

$\hat{y}_x = -2.45 + (1.15)x$

$\hat{y}_0 = -2.45 + 0 = -2.45$

9. $b = 1.19$, $a = -26.5$; 68.7 millions of gallons of ice cream

11. $\hat{y}_x = -.029x + .378$; $\hat{y}_2 = .32$

13. $\hat{y}_{2.5} = 33.8$ sec.

15. b. $\hat{y}_x = 5.11x - 9940.9$; **c.** \$90 billion.

17. .61; Reject the null hypothesis that the two are independent.

19. .95; Reject H_0.

21. a. $\hat{y}_x = .615x + .248$; $\hat{y}_{18} = 11.32$ sec. **b.** $r = .97$ **c.** $s_X^2 = 106.74$; $s_Y^2 = 42.69$; $s_e^2 = 2.51$; $s_A^2 = 1.32$; $s_B^2 = .0018$ **d.** for γ, from -1.8 to 2.3; for β_1, from .54 to .69. **e.** 10.2 to 12.5 sec. **f.** 7.7 to 14.9 sec.

23. 63.2 to 74.2 gal.

25. 303 to 388 g.

27. a. $\hat{y}_x = -.0013x + .25$ **b.** Reject the null hypothesis, $H_0: \beta_1 = 0$. **c.** from .211 to .222 mm **d.** from .196 to .237 mm

29. a. $\hat{y}_x = -1.41x + 6.62$ **b.** from 1.1 to 1.6 days.

CHAPTER 10

1. a. $H_0: P(4) = .1, \ P(3) = .36, \ P(2) = .41, \ P(1) = .1, \ P(0) = .03$ where $P(i) = $ probability of i eggs hatching after the chemical spill. Reject H_0. **b.** $x^2 = 17.92$ while $\chi^2_{.05} = 9.49$ (df $= 4$) **c.** In rejecting H_0 we conclude that the chemical spill significantly changed the probability distribution of egg hatchings. From an examination of the data, it appears that the spill has lowered the total expected number of hatches. We expected a total of 480 hatches but there were only 427 observed.

3. No.

5. No.

7. Reject H_0 that the distribution is uniform.

9. The results are not consistent with the equal likelihood of "true" "false" answers.

11. The data are not consistent with the normal distribution.

13. Survival is not independent of treatment.

15. No.

17. Yes.

19. Yes.

21. $l_A = 0.2$.

23. $l_A = 0.28$.

25. $l_A = .187$.

CHAPTER 11

1. a.

	sum of squares	degrees of freedom	mean square	F ratio
between	22.35	2	11.18	2.80
within	87.87	22	3.99	

b. Do not reject H_0. Note that $F_{.05}(2, 22)$ is not in Appendix A.7 but we can reach a conclusion from $F_{.05}(2, 20)$ and $F_{.05}(2, 40)$, which are in this table.

3. a.

	sum of squares	degrees of freedom	mean square	F ratio
between	80	2	40	1.64
within	660	27	24.4	

b. Do not reject H_0.

5. a.

	sum of squares	degrees of freedom	mean square	F ratio
between	53.7	4	13.43	1.10
within	183.5	15	12.2	

b. Do not reject H_0.

7. a. Tests are not all the same. (Tests are significantly different.) **b.** Students are not all the same.

9. Reject the hypothesis that there are no differences among the row means (average family size seems to vary by city size) but do not reject the hypothesis that there is no difference among the column means; $(\alpha = .05)$.

11. Do not reject H_0, no interaction effect; reject the hypotheses "no tire effect" and "no road effect."

CHAPTER 12

1. Reject H_0.

3. Do not reject H_0 at $\alpha = .1$.

5. Do not reject H_0.

9. Reject H_0.

13. Do not reject at $\alpha = .1$.

15. These data exhibit nonrandomness at $\alpha = .1$.

17. The two stocks do not have significantly different distributions of return.

19. No.

21. Do not reject H_0.

23. Do not reject H_0.

25. Do not reject H_0.

27. Do not reject H_0.

29. $r_s = .76$; Yes; Yes.

31. The scores are significantly related.

33. Do not reject H_0.

35. Do not reject H_0.

43. Assume independent random sampling from a normal distribution. $H_0: \mu = 0$. The data are $(.01, -.1, .04, .03, .01, .02, -.1, .02, .03, .04)$. In this case, if we use the sign test we reject H_0 at $\alpha = .1$ whereas we do not reject H_0 if we employ the t-test.

CHAPTER 13

1. a. Population: 32,000 waybills issued between 26 June and 30 Dec., 1981. Unit: waybill. Item: money due. Item value: dollar value of money due on each waybill. Sample size: 3200.

b. Population: people who have used the drug consistently for more than 10 years. Unit: person from the population. Item: eyesight, kidney failure, and so forth. Item value: with respect to eyesight, one measure is the diopter that is the refractive power of the lens; there are measures of cloudiness of the cornea, and so forth. There are many measures of kidney malfunction including the pH of the urine, the number of red blood cells in the urine. Sample size: 10,000.

   c. Population: those hired by the firm in a given job category over the last 20 years. Unit: a person so hired. Item: sex. Item value: male or female. Sample size: 300.

3. a. 1 b. 252 c. 75,287,520.

5. In each case, not all samples are equally likely since a sample of a special population is selected. The special population is (a) my friends and friends of my friends, (b) "successful" graduates, (c) the population living in low crime areas (d) cable television viewers.

7. a. $\mu = (1 + 2 + 3 + 4 + 5 + 6)/6 = 3.5$
$\sigma = \sqrt{[(1 - 3.5)^2 + (2 - 3.5)^2 + \cdots + (6 - 3.5)^2]/6}$

   b. without regard to order, the samples are (1, 2), (1, 3), (1, 4), (1, 5), (1, 6), (2, 3), (2, 4), (2, 5), (2, 6), (3, 4), (3, 5), (3, 6), (4, 5), (4, 6), (5, 6).

   c. 3.5

9. From 131.8 to 132.2

11. With probability .95 the estimate of 30 lb. is within 1.96 lb. of the true mean.

13. $597.83

15. a. (i) 181.25 (ii) Yes b. 175.6 c. Both pneumonia and diarrhea are seasonal diseases. If a systematic sample is taken so that the period between observations corresponds to the seasonal fluctuations of these diseases, it will produce a biased sample mean. One must take care in systematically sampling diseases like pneumonia, which is periodic in nature. It should be noted that if the starting point is randomly selected among the 3 months December, January, and February, we obtain an unbiased estimator. (Note that these 3 months are the only possible starting months in which we can obtain a sample of size 4 and still keep all the observations within the same year.) However, if we choose December or January as equally likely starting points and select every fifth month to obtain a sample of size 3, then the sample mean is a biased estimator. On the other hand, if a sample of size 3 is taken using s.r.s. the estimator is not biased. One must take care in using systematic sampling procedures.

17. a. (1, 2, 4), (1, 2, 5), (1, 2, 6), (1, 3, 4), (1, 3, 5), (1, 3, 6), (2, 3, 4), (2, 3, 5), (2, 3, 6). The sample averages are respectively 2.33, 2.67, 3, 2.67, 3, 3.33, 3, 3.33, 3.67.

   b. The expected value μ of the six numbers is $(1 + 2 + 3 + 4 + 5 + 6)/6 = 3.5$. The expected value of the sample mean is $(7 + 2 \cdot 8 + 3 \cdot 9 + 2 \cdot 10 + 11)/27 - 3$.

21. $n_1 = 15$, $n_2 = 66$, $n_3 = 18$, $n_4 = 6$, $n_5 = 15$

23. 5; unbiased.

CHAPTER 14

1. **a.** The investigators should have discussed the survey with political, religious, and labor leaders and others active in the community (including teachers) to try to determine what the reaction of the populace might be to the type of survey they were going to conduct and how to handle possible adverse reactions.

5. The article does not make clear the purpose of the survey, although we might surmise that its objective was to estimate the probability of purchasing a color television set that emitted harmful radiation. On the other hand, the objective might have been to determine the proportion of television viewers who were subject to harmful radiation. The article might have given us some insight into the validity of the results by describing the sampling design. Some pertinent questions that the article never treated are: Was there proportional allocation? Were the number of sets tested for each manufacturer chosen in proportion to the numbers sold?

7. **a.** "Accidents" is in the title of the table, but the heading of column I is "percent of all assaults."

   b. The sample size is not given. There might have been only one officer with the height of 76 in., which would have made this height category virtually useless from a statistical standpoint.

   c. The period of the study is not given. Were all tall officers sampled in one period (perhaps in a season when crime was generally low), whereas short officers were investigated in another period?

   d. A good table should be self-contained and self-explanatory. This table is not self-explanatory. What is the purpose of columns IV and V? (It is not clear that the Pearson χ^2 statistic discussed in Chapter 10 is applicable here.) Column V is not numbered, whereas all the others are. Is there a reason for this omission, or is it due to poor proofreading? We cannot tell.

9. Notice that the number of *arrests* in Los Angeles is compared to the number of *crimes* in Saudi Arabia. Anyone familiar with criminal statistics in the United States is well aware of the fact that arrests far exceed convictions.

   The crime statistics are simply not comparable between the two societies. Here are some of the reasons (find others).

   a. Crimes in one society are not necessarily crimes in the other society. The very fact that the population of Los Angeles is mobile compared to that of Saudi Arabia creates a situation that produces different laws on the books in Los Angeles than in Saudi Arabia. It is a crime not to stop your car when a pedestrian walks onto the road in Los Angeles. Such a law does not exist in Saudi Arabia.

   b. In the article, Los Angeles and Saudi Arabia are said to have the same population. Yet women in Los Angeles are not restricted in their activities as are women in Saudi Arabia. Los Angeles, from the standpoint of producing crime, has about twice the population of Saudi Arabia.

c. Sociologists, and the layman, recognize that population density and pressure create greater opportunities for crime. Los Angeles is a highly dense urban area. Saudi Arabia is a low-density nation with few and relatively small urban centers.

d. Some crimes may not exist in a society simply because of the state and organization of the society. Check forgery is a major crime in Los Angeles. The very fact that the population of Los Angeles is more literate than that of Saudi Arabia creates the means for such a crime in Los Angeles. Moreover, almost every neighborhood in Los Angeles contains a bank. The organization of the society thereby permits the perpetration of this crime, almost unknown in Saudi Arabia.

The reader can come up with many other reasons for not making the comparisons made in the article. Can you think of some way that one can develop valid comparisons?

CHAPTER 15

1. a. $2^{10} = 1024$ b. 1024

3. a. $\frac{35}{50} = .7$ c. The estimate is $(\hat{\lambda} - .0248)/.7$. d. If they answered directly, the estimate of the women who had abortions would be $\hat{\lambda}_1 =$ (number who say yes/n). The estimator $(\hat{\lambda} - .0248)/.7$ has variance $V(\hat{\lambda})/(.7)^2$. [Recall the formula $V(aX + b) = a^2 V(X)$ where X is a random variable and a, b are constants: V represents "variance of." We are assuming that the value given for $P(\text{Yes} \mid 2)$ is exact. The reader can show that if we regard the value given to $P(\text{Yes} \mid 2)$ as an estimate, then the variance of our estimator of $P(\text{Yes} \mid 1)$ is larger than the value shown below, and our contention is still true.] Note also that $\hat{\lambda}$ is an unbiased estimator of $P(\text{Yes}) = .7P + .0248$ where $P = P(\text{Yes} \mid 1)$. Hence, $V(\hat{\lambda}) = P(\text{Yes}) [1 - P(\text{Yes})]/n = [(.7P + .0248][1 - .7P - .0248]/n = (.6653P - .49P^2 + .0242)/n$ and so $V(\hat{\lambda})/(.7)^2 = [1.358P - P^2 + .0494]/n = P(1 - P)/n + (.358P + .0494)/n$. Note that $V(\hat{\lambda}_1) = P(1 - P)/n$ so that $V(\hat{\lambda}) > V(\hat{\lambda}_1)$, which was to be shown.

The conclusion is true in general, that is without respect to particular values of p and $P(\text{Yes} \mid 2)$. The reader is urged to prove this fact. e. A large number of them would lie; hence, the estimate would be biased.

5. a. valid but unreliable. b. both not valid and not reliable. c. not valid. d. The instrument (questionnaire) is not a valid measure of total readership.

9. a. The zero is arbitrary. b. The zero is not arbitrary on this scale. c. (i) ratio, (ii) ordinal, (iii) ordinal, (iv) nominal, (v) ordinal, (vi) ratio, (vii) interval.

Index